Lecture Notes in Computer Science 1879

Edited by G. Goos, J. Hartmanis and J. van Leeuwen

T0216410

Springer
Berlin
Heidelberg
New York
Barcelona
Hong Kong
London
Milan
Paris
Singapore
Tokyo

Mike Paterson (Ed.)

Algorithms –
ESA 2000

8th Annual European Symposium
Saarbrücken, Germany, September 5-8, 2000
Proceedings

 Springer

Series Editors

Gerhard Goos, Karlsruhe University, Germany
Juris Hartmanis, Cornell University, NY, USA
Jan van Leeuwen, Utrecht University, The Netherlands

Volume Editor

Mike S. Paterson
University of Warwick
Department of Computer Science
Coventry CV4 7AL, United Kingdom
E-mail: msp@dcs.warwick.ac.uk

Cataloging-in-Publication data applied for

Die Deutsche Bibliothek - CIP-Einheitsaufnahme

Algorithms : 8th annual European symposium ; proceedings / ESA 2000,
Saarbrücken, Germany, September 5 - 8, 2000. Mike Paterson (ed.). -
Berlin ; Heidelberg ; New York ; Barcelona ; Hong Kong ; London ;
Milan ; Paris ; Singapore ; Tokyo : Springer, 2000
  (Lecture notes in computer science ; Vol. 1879)
  ISBN 3-540-41004-X

CR Subject Classification (1998): F.2, G.1-2, E.1, F.1.3, I.3.5

ISSN 0302-9743
ISBN 3-540-41004-X Springer-Verlag Berlin Heidelberg New York

Springer-Verlag Berlin Heidelberg New York
a member of BertelsmannSpringer Science+Business Media GmbH
© Springer-Verlag Berlin Heidelberg 2000
Printed in Germany

Typesetting: Camera-ready by author
Printed on acid-free paper   SPIN 10722400   06/3142   5 4 3 2 1 0

Preface

This volume contains the 39 contributed papers and two invited papers presented at the 8th Annual European Symposium on Algorithms, held in Saarbrücken, Germany, 5–8 September 2000. ESA 2000 continued the sequence:

- 1993 Bad Honnef (Germany)
- 1994 Utrecht (The Netherlands)
- 1995 Corfu (Greece)
- 1996 Barcelona (Spain)
- 1997 Graz (Austria)
- 1998 Venice (Italy) and
- 1999 Prague (Czech Republic).

The proceedings of these previous meetings were published as Springer LNCS volumes 726, 855, 979, 1136, 1284, 1461, and 1643.

Papers were solicited in all areas of algorithmic research, including approximation algorithms, combinatorial optimization, computational biology, computational geometry, databases and information retrieval, external-memory algorithms, graph and network algorithms, machine learning, on-line algorithms, parallel and distributed computing, pattern matching and data compression, randomized algorithms, and symbolic computation. Algorithms could be sequential, distributed, or parallel, and should be analyzed either mathematically or by rigorous computational experiments. Experimental and applied research were especially encouraged.

Each extended abstract submitted was read by at least three referees, and evaluated on its quality, originality and relevance to the Symposium. The entire Programme Committee met at the University of Warwick on 5–6 May 2000 and selected 39 papers for presentation. These, together with the two invited papers by Monika Henzinger and Thomas Lengauer, are included in this volume.

The Programme Committee was:

Lars Arge	(Duke)	Mike Paterson	(Chair; Warwick)
Yossi Azar	(Tel Aviv)	Marco Pellegrini	(CNR, Pisa)
Leslie Goldberg	(Warwick)	R. Ravi	(Carnegie-Mellon)
Mike Jünger	(Köln)	Jan van Leeuwen	(Utrecht)
Danny Krizanc	(Wesleyan)	Emo Welzl	(ETH, Zürich)
Alessandro Panconesi	(Bologna)		

As an experiment, ESA 2000 was held as a combined conference (CONF 2000) together with the Workshops on Algorithmic Engineering (WAE 2000) and Approximation Algorithms for Combinatorial Optimization Problems (APPROX 2000), in Saarbrücken. The Organizing Committee consisted of Uwe Brahm, Jop Sibeyn (Chair), Christoph Storb, and Roxane Wetzel.

ESA 2000 was sponsored by the European Association for Theoretical Computer Science (EATCS), and was supported by the Max-Planck-Institut für Informatik.

We hope that this volume offers the reader a representative selection of some of the best current research on algorithms.

Warwick University   Mike Paterson
July 2000   Programme Chair
                                                  ESA 2000

Referees

Pankaj K. Agarwal
Susanne Albers
Helmut Alt
Christoph Ambühl
Gunnar Andersson
Mark de Berg
Claudson Bornstein
Gerth S. Brodal
Christoph Buchheim
Alberto Caprara
Joseph Cheriyan
Paolo Cignoni
Mary Cryan
Andrzej Czygrinow
Elias Dalhaus
Bhaskar Dasgupta
Devdatt Dubhashi
Alberto Del Lungo
Matthias Elf
Leah Epstein
Jeff Erickson
Thomas Erlebach
Amos Fiat
Alan Frieze
Ashim Garg
Bernd Gärtner
Cyril Gavoille

Michel Goemans
Paul Goldberg
Roberto Grossi
Bjarni Halldorsson
Dan Halperin
Rafael Hassin
Michael Hoffmann
Thomas Hofmeister
David Hutchinson
Mark Jerrum
Adam Kalai
Viggo Kann
Gyula Karolyi
Philip Klein
Jochen Konemann
Goran Konjevod
Joachim Kupke
Giuseppe Lancia
Sebastian Leipert
Stefano Leonardi
Frauke Liers
Rune Bang Lyngsoe
Madhav Marathe
Jiri Matoušek
Alexander May
Seffi Naor
Gabriele Neyer

Ojas Parekh
Boaz Patt-Shamir
Sasa Pekec
Antonio Piccolboni
Bert Randerath
Theis Rauhe
S. S. Ravi
Oded Regev
Süleyman Cenk Şahinalp
Fatma Sibel Salman
Amitabh Sinha
Steven Skiena
Michiel Smid
Jack Snoeyink
Joel Spencer
Aravind Srinivasan
Christian Storm
Maxim Sviridenko
Laura Toma
Falk Tschirschnitz
Uli Wagner
Rajiv Wickremesinghe
Gerhard Woeginger
Martin Wolff
Afra Zomorodia

Table of Contents

Invited Papers

Web Information Retrieval - an Algorithmic Perspective 1
    Monika Henzinger

Computational Biology – Algorithms and More 9
    Thomas Lengauer

Contributed Papers (ordered alphabetically by first author)

Polygon Decomposition for Efficient Construction of Minkowski Sums 20
    Pankaj K. Agarwal, Eyal Flato, Dan Halperin

An Approximation Algorithm for Hypergraph Max k-Cut with Given Sizes
of Parts. 32
    Alexander A. Ageev, Maxim I. Sviridenko

Offline List Update Is NP-Hard 42
    Christoph Ambühl

Computing Largest Common Point Sets under Approximate Congruence . . 52
    Christoph Ambühl, Samarjit Chakraborty, Bernd Gärtner

Online Algorithms for Caching Multimedia Streams 64
    Matthew Andrews, Kamesh Munagala

On Recognizing Cayley Graphs 76
    Lali Barrière, Pierre Fraigniaud, Cyril Gavoille, Bernard Mans,
    John M. Robson

Fast Algorithms for Even/Odd Minimum Cuts and Generalizations. 88
    András A. Benczúr, Ottilia Fülöp

Efficient Algorithms for Centers and Medians in Interval and Circular-Arc
Graphs 100
    S. Bespamyatnikh, B. Bhattacharya, J. Mark Keil, D. Kirkpatrick,
    M. Segal

Exact Point Pattern Matching and the Number of Congruent Triangles in
a Three-Dimensional Pointset 112
    Peter Brass

Range Searching Over Tree Cross Products. 120
    Adam L. Buchsbaum, Michael T. Goodrich, Jeffery R. Westbrook

A $2\frac{1}{10}$-Approximation Algorithm for a Generalization of the Weighted
Edge-Dominating Set Problem 132
    Robert Carr, Toshihiro Fujito, Goran Konjevod, Ojas Parekh

The Minimum Range Assignment Problem on Linear Radio Networks 143
    A.E.F. Clementi, A. Ferreira, P. Penna, S. Perennes, R. Silvestri

Property Testing in Computational Geometry 155
    Artur Czumaj, Christian Sohler, Martin Ziegler

On R-trees with Low Stabbing Number 167
    Mark de Berg, Joachim Gudmundsson, Mikael Hammar,
    Mark Overmars

K-D Trees Are Better when Cut on the Longest Side 179
    Matthew Dickerson, Christian A. Duncan, Michael T. Goodrich

On Multicriteria Online Problems 191
    Michele Flammini, Gaia Nicosia

Online Scheduling Revisited 202
    Rudolf Fleischer, Michaela Wahl

Constant Ratio Approximation Algorithms for the Rectangle Stabbing
Problem and the Rectilinear Partitioning Problem 211
    Daya Ram Gaur, Toshihide Ibaraki, Ramesh Krishnamurti

I/O-Efficient Well-Separated Pair Decomposition and Its Applications 220
    Sathish Govindarajan, Tamás Lukovszki, Anil Maheshwari,
    Norbert Zeh

Higher Order Delaunay Triangulations 232
    Joachim Gudmundsson, Mikael Hammar, Marc van Kreveld

On Representations of Algebraic-Geometric Codes for List Decoding 244
    Venkatesan Guruswami, Madhu Sudan

Minimizing a Convex Cost Closure Set 256
    Dorit S. Hochbaum, Maurice Queyranne

Preemptive Scheduling with Rejection 268
    Han Hoogeveen, Martin Skutella, Gerhard J. Woeginger

Simpler and Faster Vertex-Connectivity Augmentation Algorithms 278
    Tsan-sheng Hsu

Scheduling Broadcasts in Wireless Networks 290
    Bala Kalyanasundaram, Kirk Pruhs, Mahe Velauthapillai

Jitter Regulation in an Internet Router with Delay Consideration 302
  Hisashi Koga

Approximation of Curvature-Constrained Shortest Paths through a
Sequence of Points 314
  *Jae-Ha Lee, Otfried Cheong, Woo-Cheol Kwon, Sung Yong Shin,
  Kyung-Yong Chwa*

Resource Constrained Shortest Paths 326
  Kurt Mehlhorn, Mark Ziegelmann

On the Competitiveness of Linear Search 338
  J. Ian Munro

Maintaining a Minimum Spanning Tree Under Transient Node Failures ... 346
  Enrico Nardelli, Guido Proietti, Peter Widmayer

Minimum Depth Graph Embedding 356
  Maurizio Pizzonia, Roberto Tamassia

New Algorithms for Two-Label Point Labeling 368
  Zhongping Qin, Alexander Wolff, Yinfeng Xu, Binhai Zhu

Analysing the Cache Behaviour of Non-uniform Distribution Sorting
Algorithms 380
  Naila Rahman, Rajeev Raman

How Helpers Hasten h-Relations 392
  Peter Sanders, Roberto Solis-Oba

Computing Optimal Linear Layouts of Trees in Linear Time 403
  Konstantin Skodinis

Coloring Sparse Random Graphs in Polynomial Average Time 415
  C.R. Subramanian

Restarts Can Help in the On-Line Minimization of the Maximum Delivery
Time on a Single Machine 427
  Marjan van den Akker, Han Hoogeveen, Nodari Vakhania

Collision Detection Using Bounding Boxes: Convexity Helps 437
  Yunhong Zhou, Subhash Suri

Author Index 449

Web Information Retrieval - an Algorithmic Perspective

Monika Henzinger[1]

Google, Inc., Mountain View, CA 94043 USA,
monika@google.com,
WWW home page: http://www.henzinger.com/monika/index.html

Abstract. In this paper we survey algorithmic aspects of Web information retrieval. As an example, we discuss ranking of search engine results
using connectivity analysis.

1   Introduction

In December 1999 the World Wide Web was estimated to consist of at least one
billion pages, up from at least 800 million in February 1999 [28]. Not surprisingly
finding information in this large set of pages is difficult and many Web users turn
to Web search engines for help. An estimated 150 million queries are currently
asked to Web search engines per day – with mixed success.

In this paper we discuss algorithmic aspects of Web information retrieval
and present an algorithm due to Brin and Page [6] that is very successful in
distinguishing high-quality from low-quality Web pages, thereby improving the
quality of query results significantly. It is currently used by the search engine
Google[1].

This paper is loosely based on part of a tutorial talk that we presented with
Andrei Broder at the 39th Annual Symposium on Foundations of Computer
Science (FOCS 98) in Palo Alto, California. See [8] for write-up of the remaining
parts of this tutorial.

We will use the terms *page* and *document* interchangeably.

2   Algorithmic Aspects of Web Information Retrieval

The goal of general purpose search engines is to index a sizeable portion of the
Web, independently of topic and domain. Each such engine consists of three
major components:

- A *crawler* (also called *spider* or *robot*) collects documents by recursively fetch-
  ing links from a set of starting pages. Each crawler has different policies with
  respect to which links are followed, how deeply various sites are explored, etc.
  Thus, the pages indexed by various search engines differ considerably [27,2].

---

[1] http://www.google.com/

M. Paterson (Ed.): ESA 2000, LNCS 1879, pp. 1–8, 2000.

- The *indexer* processes the pages collected by the crawler. First it decides which of them to index. For example, it might discard duplicate documents. Then it builds various data structures representing the pages. Most search engines build some variant of an inverted index data structure (see below). However, the details of the representation differ among the major search engines. For example, they have different policies with respect to which words are indexed, capitalization stemming, whether locations within documents are stored, etc. The indexer might also build additional data structures, like a repository to store the original pages, a Web graph representation to store the hyperlinks, a related-pages finder to store related pages, etc. As a result, the query capabilities and features of the result pages of various engines vary considerably.
- The *query processor* processes user queries and returns matching answers, in an order determined by a *ranking* algorithm. It transforms the input into a standard format (e.g. to lower-case terms), uses the index to find the matching documents, and orders (*ranks*) them.

Algorithmic issues arise in each part. We discuss some of them in the following.

2.1   Crawler

The crawler needs to decide which pages to crawl. One possible implementation is to assign to each page a priority score indicating its crawling importance. and to maintain all pages in a priority queue ordered by priority score. This priority score can be the PageRank score (defined in the next section) of the page [12] if the goal is to maximize the quality of the pages in the index. Alternatively, the score can be a criterion whose goal is to maximize the freshness of the pages in the index [14,11].

The crawler also has to consider various load-balancing issues. It should not overload any of the servers that it crawls and it is also limited by its own bandwidth and internal processing capabilities. An interesting research topic is to design a crawling strategy that maximizes both quality and freshness and respects the above load-balancing issues.

2.2   Indexer

The indexer builds all the data structures needed at query time. These include the *inverted index*, a *URL-database*, and potentially a *graph representation*, a *document repository*, a *related-pages finder*, and further data structures.

The inverted index contains for each word a list of all documents containing the word, potentially together with the position of the word in the document. This list is sorted lexicographically according to the (document-id, position in the document)-pair. (See [1] for a detailed description of this data structure.)

To save space, documents are represented by document-ids in the index and the other data structures. When results are displayed, these document-ids need to be converted back to the URL. This is done using the URL-database.

The graph representation keeps for each document all the documents pointing to it and all the documents it points to. (See [4] for a potential implementation.)

The document repository stores for each document-id the original document.

The related-pages finder stores for each document-id all document-ids of pages that are related to the document. A related page is a page that addresses the same topic as the original page, but is not necessarily semantically identical. For example, given www.nytimes.com other newspapers and news organizations on the Web would be related pages. See [15] for algorithms that find related pages.

Of course all or some of the latter four data structures can be combined.

Before building the data structures the indexer needs to determine which pages to index. For this, it can assign a numerical score to each page and then index a certain number of top-ranked pages. For example this score can be 0 for all but one of a set of duplicate pages. The score might also try to measure the query-independent quality of a page. This can for example be done by the PageRank measure (see below).

An interesting algorithmic question for each of these data structures is how to compress the space as much as possible without affecting the average look-up time. Furthermore, the data structure should remain space-efficient when insertions and deletions of documents are allowed.

2.3   Query Processor

The main challenge of the query processor is to rank the documents matching the query by decreasing value for the user. For this purpose again a numerical score is assigned to each document and the documents are output in decreasing order of the score.

This score is usually a combination of query-independent and query-dependent criteria. A *query-independent* criterion judges the document regardless of the actual query. Typical examples are the length, the vocabulary, publication data (like the site to which it belongs, the date of the last change, etc.), and various connectivity-based techniques like the number of hyperlinks pointing to a page or the PageRank score. A *query-dependent* criterion is a score which is determined only with respect to a particular query. Typical examples are the cosine measure of similarity used in the vector space model [34], query-dependent connectivity-based techniques [25], and statistics on which answers previous users selected for the same query.

The algorithmically most interesting of these techniques are query-independent and query-dependent connectivity-based techniques. We will describe the query-independent approach below. The assumption behind the connectivity-based ranking techniques is that a link from page A to page B means that the author of page A recommends page B. Of course this assumption does not always hold. For example, a hyperlink cannot be considered a recommendation if page A and B have the same author or if the hyperlink was generated for example by a Web-authoring tool.

The idea of studying "referrals" is not new. There is a subfield of classical information retrieval, called bibliometrics, where citations were analyzed. See, e.g., [23,17,37,18]. The field of sociometry developed algorithms [24,30] very similar to the connectivity-based ranking techniques described in [6,25]. Furthermore, a large amount of Web-related research exploits the hyperlink structure of the Web.

A Graph Representation for the Web The Web can be represented as a graph in many different ways. Connectivity-based ranking techniques usually assume the most straightforward representation: The graph contains a node for each page u and there exists a directed edge (u, v) if and only if page u contains a hyperlink to page v.

Query-Independent Connectivity-Based Ranking The assumption of connectivity based techniques immediately leads to a simple query-independent criterion: The larger the number of hyperlinks pointing to a page the better the page [9]. The main drawback of this approach is that each link is equally important. It cannot distinguish between the quality of a page pointed to by a number of low-quality pages and the quality of a page that gets pointed to by the same number of high-quality pages. Obviously it is therefore easy to make a page appear to be high-quality – just create many other pages that point to it.

To remedy this problem, Brin and Page [6,31] invented the PageRank measure. The PageRank of a page is computed by weighting each hyperlink proportionally to the quality of the page containing the hyperlink. To determine the quality of a referring page, they use its PageRank recursively. This leads to the following definition of the PageRank $R(p)$ of a page p:

$$R(p) = \epsilon/n + (1 - \epsilon) \cdot \sum_{(q,p) \text{ exists}} R(q)/outdegree(q),$$

where

- ϵ is a dampening factor usually set between 0.1 and 0.2;
- n is the number of pages on the Web; and
- $outdegree(q)$ is the number of hyperlinks on page q.

Alternatively, the PageRank can be defined to be the stationary distribution of the following infinite random walk $p_1, p_2, p_3, \ldots$, where each p_i is a node in the graph: Each node is equally likely to be node p_1. To determine node p_{i+1} a biased coin is flipped: With probability ϵ node p_{i+1} is chosen uniformly at random from all nodes in the graph, with probability $1 - \epsilon$ it is chosen uniformly at random from all nodes q such that edge (p_i, q) exists in the graph.

The PageRank measure works very well in distinguishing high-quality Web pages from low-quality Web pages and is used in the Google search engine[2].

---

[2] http://www.google.com/

Recent work refined the PageRank criterion and sped up its computation, see e.g. [33,19]. In [20,21] PageRank-like random walks were performed to sample Web page almost according to the PageRank distribution and the uniformly distribution, respectively. The goal was to compute various statistics on the Web pages and to compare the quality, respectively the number, of the pages in the indices of various commercial search engines.

2.4   Other Algorithmic Work Related to Web Information Retrieval

There are many other algorithmic challenges in Web information retrieval. We list several below, but this list is by no means complete.

Near-duplicates: To save space in the index search engines try to determine near-duplicate Web pages and completely or partially near-duplicate Web hosts (called *mirrors*). See [5,7,35] for algorithms for the near-duplicates problem and see [3,13] for algorithms for the mirrored host problem.

Clustering: Clustering query results has lead to an interesting application of suffix trees [38].

Web page categorization: There are various hierarchical directories of the Web, for example the open directory hierarchy[3]. They are usually constructed by hand. Automatically categorizing a web, i.e. placing it at a node(s) in the hierarchy to which it belongs, is a challenging problem. There has been a lot of work on text-only methods. See [10] for a first step towards a text and link-based approach.

Dynamically generated Web content: Trying to "learn" to crawl dynamically generated Web pages is an interesting topic of future research. A first step in this direction is taken by [16].

Web graph characterization: Characterizing the Web graph has led to a sequence of studies that are algorithmically challenging because of the pure magnitude of the data to be analyzed. See [26] for a survey.

Web user behavior: Mining user query logs shows that web users exhibit a different query behavior than users of classical information retrieval systems. An analysis of different query logs is given in [22,36].

Modeling: Different users asking the same query can widely disagree on the relevance of the answers. Thus, it is not possible to prove that certain ranking algorithms return relevant answers at the top. However, there has been some recent work on trying to find appropriate models for clustering problems in information retrieval and to use them to explain why certain algorithms work well in practise. See, e.g. [32]. This is certainly an interesting area of future research.

3   Conclusions

There are many interesting algorithmic questions related to Web information retrieval. One challenge is to find algorithms with good performance. The per-

---

[3] http://www.dmoz.org/

formance of these algorithms is usually validated by experimentation. The other challenge is to theoretically model information retrieval problems in order to explain why certain algorithms perform well.

References

1. R. Baeza-Yates and B. Ribeiro-Neto. *Modern Information Retrieval*. Addison-Wesley, 1999. 2
2. K. Bharat and A. Z. Broder. A technique for measuring the relative size and overlap of public Web search engines. In *Proceedings of the Seventh International World Wide Web Conference* 1998, pages 379–388. 1
3. K. Bharat, A. Z. Broder, J. Dean, and M. Henzinger. A comparison of Techniques to Find Mirrored Hosts on the World Wide Web. To appear in the *Journal of the American Society for Information Science*. 5
4. K. Bharat, A. Z. Broder, M. Henzinger, P. Kumar, and S. Venkatasubramanian. The connectivity server: Fast access to linkage information on the Web. In *Proceedings of the Seventh International World Wide Web Conference* 1998, pages 469–477. 3
5. S. Brin, J. Davis, and H. García-Molina. Copy detection mechanisms for digital documents. In M. J. Carey and D. A. Schneider, editors, *Proceedings of the 1995 ACM SIGMOD International Conference on Management of Data*, pages 398–409, San Jose, California, May 1995. 5
6. S. Brin and L. Page. The anatomy of a large-scale hypertextual Web search engine. In *Proceedings of the Seventh International World Wide Web Conference* 1998, pages 107–117. 1, 4
7. A. Z. Broder, S. C. Glassman, M. S. Manasse, and G. Zweig. Syntactic clustering of the Web. In *Proceedings of the Sixth International World Wide Web Conference* 1997, pages 391–404. 5
8. A. Z. Broder and M. R. Henzinger. Algorithmic Aspects of Information Retrieval on the Web. In *Handbook of Massive Data Sets*. J. Abello, P.M. Pardalos, M.G.C. Resende (eds.), Kluwer Academic Publishers, Boston, forthcoming. 1
9. J. Carriere and R. Kazman. Webquery: Searching and visualizing the web through connectivity. In *Proceedings of the Sixth International World Wide Web Conference* 1997, pages 701–711. 4
10. S. Chakrabarti, B. Dom, and P. Indyk. Enhanced hypertext categorization using hyperlinks. in *Proceedings of the ACM SIGMOD International Conference on Management of Data*, 1998, pages 307–318. 5
11. J. Cho and H. García-Molina. The Evolution of the Web and Implications for an incremental Crawler. *Proceedings of the 26th International Conference on Very Large Databases* (VLDB), 2000. 2
12. J. Cho, H. García-Molina, and L. Page. Efficient crawling through URL ordering. In *Proceedings of the Seventh International World Wide Web Conference* 1998, pages 161–172. 2
13. J. Cho, N. Shivakumar, and H. García-Molina. Finding replicated Web collections. *Proceedings of the 2000 ACM International Conference on Management of Data* (SIGMOD), 2000. 5
14. E. G. Coffman, Z. Liu, and R. R. Weber. Optimal robot scheduling for Web search engines. Technical Report 3317, INRIA, Dec. 1997. 2
15. J. Dean and M. R. Henzinger. Finding Related Web Pages in the World Wide Web. In *Proceedings of the 8th International World Wide Web Conference* 1998, pages 389–401. 3

16. R. B. Doorenbos, O. Etzioni, and D. S. Weld. A scalable comparison-shopping agent for the World-Wide Web. In W. L. Johnson and B. Hayes-Roth, editors, *Proceedings of the 1st International Conference on Autonomous Agents*, pages 39–48, New York, Feb. 1997. ACM Press. 5

17. E. Garfield. Citation analysis as a tool in journal evaluation. *Science*, 178, 1972. 4

18. E. Garfield. *Citation Indexing*. ISI Press, 1979. 4

19. T. Haveliwala. Efficient Computation of PageRank. Technical Report 1999-31, Stanford University, 1999. 5

20. M. R. Henzinger, A. Heydon, M. Mitzenmacher, and M. Najork. Measuring Search Engine Quality using Random Walks on the Web. In *Proceedings of the 8th International World Wide Web Conference* 1999, pages 213–225. 5

21. M. R. Henzinger, A. Heydon, M. Mitzenmacher, and M. Najork. On near-uniform URL sampling. In *Proceedings of the Ninth International World Wide Web Conference* 2000, pages 295–308. 5

22. B. J. Jansen, A. Spin, J. Bateman, and T. Saracevic. Real Life Information Retrieval: A Study of User Queries on the Web. *SIGIR FORUM,* 32 (1):5–17, 1998. 5

23. M. M. Kessler. Bibliographic coupling between scientific papers. *American Documentation*, 14, 1963. 4

24. L. Katz. A new status index derived from sociometric analysis. *Psychometrika*, 18(1):39-43, March 1953. 4

25. J. Kleinberg. Authoritative sources in a hyperlinked environment. In *Proceedings of the 9th Annual ACM-SIAM Symposium on Discrete Algorithms*, pages 668–677, January 1998. 3, 4

26. J. Kleinberg, S.R. Kumar, P. Raghavan, S. Rajagopalan, and A. Tomkins. The Web as a graph: Measurements, models and methods. Invited survey at the *International Conference on Combinatorics and Computing*, 1999. 5

27. S. Lawrence and C. L. Giles. Searching the World Wide Web. *Science*, 280(5360):98, 1998. 1

28. S. Lawrence and C. L. Giles. Accessibility of Information on the Web. *Nature*, 400(6740):107–109, 1999. 1

29. Dharmendra S. Modha and W. Scott Spangler. Clustering Hypertext with Applications to Web Searching. *Proceedings of the ACM Hypertext 2000 Conference, San Antonio, TX*, 2000. Also appears as IBM Research Report RJ 10160 (95035), October 1999.

30. M. S. Mizruchi, P. Mariolis, M. Schwartz, and B. Mintz. Techniques for disaggregating centrality scores in social networks. In N. B. Tuma, editor, *Sociological Methodology*, pages 26-48. Jossey-Bass, San Francisco, 1986. 4

31. L. Page, S. Brin, R. Motwani, and T. Winograd. The PageRank citation ranking: Bringing order to the Web. *Stanford Digital Library Technologies*, Working Paper 1999-0120, 1998. 4

32. C. Papadimitriou, P. Raghavan, H. Tamaki, and S. Vempala. Latent Semantic Indexing: A Probabilistic Analysis. In *Proceedings of the 17th ACM Symposium on the Principles of Database Systems*, 1998. 5

33. D. Rafiei, and A. Mendelzon. What is this page known for? Computing Web page reputations. In *Proceedings of the Ninth International World Wide Web Conference* 2000, pages 823–836. 5

34. G. Salton. *The SMART System – Experiments in Automatic Document Processing*. Prentice Hall. 3

35. N. Shivakumar and H. García-Molina. Finding near-replicas of documents on the Web. In *Proceedings of Workshop on Web Databases (WebDB'98)*, March 1998. 5

36. C. Silverstein, M. Henzinger, H. Marais, and M. Moricz. Analysis of a Very Large AltaVista Query Log. Technical Note 1998-014, Compaq Systems Research Center, 1998. To appear in *SIGIR FORUM*. 5

37. H. Small. Co-citation in the scientific literature: A new measure of the relationship between two documents. *J. Amer. Soc. Info. Sci.*, 24, 1973. 4

38. O. Zamir and O. Etzioni. Web document clustering: A feasibility demonstration. In *Proceedings of the 21st International ACM SIGIR Conference on Research and Development in Information Retrieval (SIGIR'98)*, pages 46–54. 5

Computational Biology – Algorithms and More

Thomas Lengauer

Institute for Algorithms and Scientific Computing (SCAI)
GMD – German National Research Center for Information Technology GmbH
Sankt Augustin, Germany
and
Department of Computer Science
University of Bonn, Germany
lengauer@gmd.de

1   Introduction

Computational biology is an area in applied computer science that has gained much attention recently. The reason is that new experimental methods in molecular biology and biochemistry have afforded entirely novel ways of inspecting the molecular basis of life's processes. Experimental breakthroughs have occurred in quick succession, with the first completely sequenced bacterial genome being published in 1995 (genome length 1.83 Mio bp, 1700 genes) [8], the first eukaryote yeast following in 1996 (genome length 13 Mio bp, 6275 genes) [9], the first multicellular organism C. elegans being sequenced in late 1998 (97 Mio bp, 19000 genes) [5], the fruitfly coming along this February (120 Mio bp, 13600 genes) [1] and man being pre-announced in April 2000. Several dozen completely sequenced microbial genomes are available today. This puts biology on a completely new footing since, for the first time, it can be ascertained not only which components are necessary to administer life but also which ones suffice.

But the experimental scene has moved beyond sequencing whole genomes. Today, it is possible to measure the protein populations inside a cell and to do so specific to certain cell types (tissues) and cell states (normal, stressed, diseased etc.) In so-called differential expression displays, the relative count of different proteins is monitored comparing two different cell states [6]. The experimental data are still quite noisy but, for the first time in history, it is now possible to query a cell as to the role of the proteins that it manufactures. This has interested many researchers and pharmaceutical companies in exploiting these data to search for fundamentally new approaches to the diagnosis and therapy of diseases. It is this not only scientifically but also economically quite important background that motivates us to discuss computational biology within a pharmaceutical context for the purpose of this article. Rather than giving a detailed description of a particular problem in computational biology, we aim at putting into context the set of problems to be solved and the solution methods applied.

M. Paterson (Ed.): ESA 2000, LNCS 1879, pp. 9–19, 2000.

2   Two Stages in Drug Development

The development of a new drug is performed in two basic steps. The first is the identification of a key molecule, usually a protein, the so-called *target protein*, whose biochemical function is causative of the disease. The second step is the search for or development of a drug that moderates – often blocks – the function of the target protein.

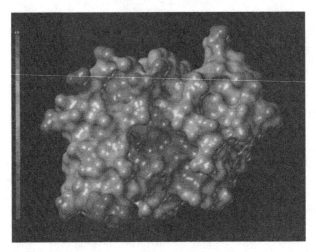

Fig. 1. 3D structure of the protein dihydrofolate-reductase (DHFR)

Figure 1 shows the three-dimensional shape of the protein *dihydrofolate-reductase (DHFR)* which catalyzes a reaction that is important in the cell division cycle. DHFR has a prominent binding pocket in its center that is specifically designed to bind to the substrate molecule *dihydrofolate* and induce a small modification of this molecule. This activity of DHFR can be blocked by administering the drug molecule *methotrexate (MTX)* (Figure 2). MTX binds tightly to DHFR and prevents the protein from exercising its catalytic function. MTX is a commonly administered drug in cancer treatment, where our goal is to break the (uncontrolled) cell division cycle.

This example shows both the benefits and the problems of current drug design. Using MTX, we can in fact break the cell division cycle. However, DHFR is actually the wrong target molecule. It is expressed in all dividing cells, thus a treatment with MTX not only affects the tumor but all dividing cells in the body. This leads to severe side effects such as losing one's hair and intestinal lining. What we need is a more appropriate target protein – one that is specifically expressed inside the tumor and whose inhibition does not cause side effects in other tissues.

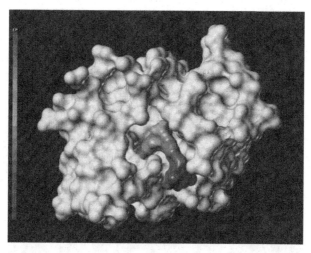

Fig. 2. The inhibitor methotrexate (MTX) bound to the binding pocket of DHFR.

3 Search for Target Proteins

Until a few years ago target proteins could not be systematically searched for. Now new experimental methods in molecular biology have rendered a cell-wide search for target proteins feasible. Today, we can search for a suitable target protein among a few ten thousand protein candidates. The selection of promising candidates from such a large set of molecules can only be done with the help of the computer.

The basis for such screens are so-called *microarrays* or *DNA-chips* (Figure 3). The actual image which can be downloaded from the website mentioned in the figure caption is an additive overlay of two pictures in the colors green (cell state 1: yeast in the presence of glucose) and red (cell state 2: yeast deprived of glucose), respectively. Bright dots correspond to proteins that are expressed highly (in large numbers) in the respective cell state.

In principle, microarrays afford a comprehensive picture of the protein state of the cell. (Actually what is measured here are mRNA levels, not the proteins. We comment further on this biological detail in the Summary of this article.) The problem is that we learn very little about each protein: its expression level and its amino-acid sequence (or even only part of the sequence) are the only data we are given. In order to select a suitable target protein, we need to know much more. Where does the protein occur in the cell (cellular localization)? It would help to know the 3D structure. What is the protein function? What are its binding partners? What biochemical pathway is it involved in? These are all very difficult questions whose exact answer would entail extensive laboratory procedures over a long time period with a lot of human resources – something we

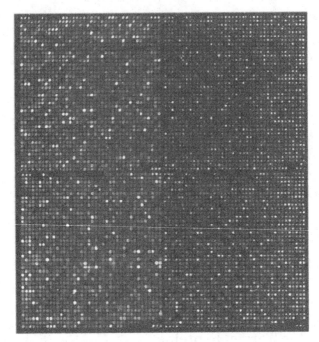

Fig. 3. A DNA chip containing all yeast genes (taken from http://cmgm.stanford.edu/pbrown/explore/index.html).

can afford to do with one or a few proteins but not with hundreds and thousands of interesting protein candidates coming from a DNA-chip experiment.

Therefore, these questions have to be answered by computer. Usually, a first step is to analyze the protein sequences in order to elucidate evolutionary relationships. This is done by a version of sequence alignment. If a high level similarity of the protein sequence under study to some sequence in a database is established then we can hope to infer the function of the sequence under study from biological knowledge of the sequence hit in the database. The process of inferring this kind of biological information is not automated, however.

Often, proteins are assembled from building blocks called *domains*. Thus, identifying the domains of a protein is a first problem to be solved in sequence analysis [2].

As the level of sequence similarity decreases, inferring function from sequence alone become increasingly hazardous. Thus, it helps to predict the 3D structure of the protein. At GMD, we have developed different methods for doing this. One program (123D) is based on an efficient dynamic programming scheme and can make a protein structure prediction within minutes [15]. Another program (RDP) is based on a more detailed NP-hard formulation of the protein folding problem and runs within hours [24]. From the protein structure, we can attempt to reason about the function, again in a process that has not been automated, so far.

Besides the protein structure, other sources of information can help to infer function such as the use of phylogenetic analyses [14] and the analysis of which protein domains are fused or separated in different organisms [12]. The localization of the protein inside the cell can be predicted by a special kind of sequence analysis [13].

Finally, we have developed a method to expand hypotheses on the localization of proteins in biochemical pathways using differential expression data. This method involves a statistical correlation analysis between the differential expression levels of the protein under study and those of the proteins that we have already assigned to the biochemical pathway [26].

4   Demands on Computational Biology

Of course, nature is much too complex to be modeled to any sufficiently accurate detail. And we have little time to spend on each molecular candidate. Thus we mostly do not even attempt to model things in great physical detail, but we use techniques from statistics and machine learning to infer "signals" in the data and separate them from "noise". Just as people interpret facial expressions of their dialog partners not by a thorough physiological analysis that reasons backwards from the shape of the muscles to the neurological state of the brain but learn on (usually vast amounts of) data how to tell whether somebody is happy or sad, attentive or bored, so do computational biology models query hopefully large sets of data to infer the signals. Here *signal* is a very general notion that can mean just about anything of biological interest, from a sequence alignment exhibiting the evolutionary relationship of the two proteins involved over a predicted 2D or 3D structure of a protein to the structure of a complex of two molecules binding to each other. On the sequence level, the splice sites in complex eukaryotic genes, the location and makeup of promoter sequences or the design of signal peptides giving away the final location of the protein in the cells are examples of interesting signals.

Statistical methods that are used to learn from biological data have classically included neural nets and genetic algorithms. Hidden-Markov models [11,7] are a very popular method of generating models for biological signals of all kinds. Recently support vector machines have been applied very successfully to solving classification problems in computational biology [10,25].

As the methods of analysis are inexact so are the results. The analyses yield predictions that cannot be trusted, in general. This is quite different from the usual situation in theoretical computer science, where you are either required to compute the optimum solution or, at least, *optimum* means something and so does the distance of the computed solution to the optimum, in case that you do not hit the optimum. Not so here. Cost functions in computational biology usually miss the goal. Notions such as *evolutionary distance* or *free energy* are much too complex to be reflected adequately by easy-to-compute cost functions.

Thus, computational biology is dominated by the search for suitable cost functions. Those cost function can be trained, just as the models in toto. We have used a training procedure based on linear programming to improve the predictive power of our protein structure prediction methods [27]. Another possibility is to leave the mystical parameters in the cost function variable and study the effect of changing them on the outcome. A method for doing this in the area of sequence alignment is presented in [28].

Whether a method or a cost function is good or bad cannot be proved but has to be validated against biologically interpreted data that are taken as a gold standard for purposes of the validation. Several respective data sets have evolved in different bioinformatics domains. Examples are the SCOP (http://scop.mrc-lmb.cam.ac.uk/scop/) and CATH (http://www.biochem.ucl.ac.uk/bsm/cath/) structural protein classifications for validating methods for protein structure prediction and analysis. These sets are not only taken to validate the different methods but also to compare them community-wide.

For more details on algorithmic aspects of computational biology, see also [20].

Validating methods and cost functions on known biological data has a serious drawback. One is not prepared to answer the question whether one has used the biological knowledge in the method, either on purpose or inadvertently. Therefore, the ultimate test of any computational biology methods is a "blind prediction", one that convincingly makes a prediction without previous knowledge of the outcome. To stage a blind prediction experiment involves a certification authority that vouches for the fact that the knowledge to be predicted was not known to the predictor. The biannual CASP (Critical Assessment of Structure Prediction Methods [3,4] experiment series that was started in 1994, performs this task for protein structure prediction methods. The CASP team provides a world-wide clearing house for protein sequences whose structures are in the process of being resolved, e.g. by crystallographers. The group that resolves the structure communicates the protein sequence to the CASP team that puts it on the web up for prediction. Sufficiently before the crystallographers resolve the structure, the prediction contest closes on that sequence. After the structure is resolved it is compared with the predictions. CASP has been a tremendous help in gaining acknowledgement for the scientific discipline of protein structure prediction.

5   Searching For New Drugs

The search for drugs that bind to a given target protein also has been systematized greatly with the advent of very efficient methods for synthesizing new compounds (combinatorial chemistry) and testing their binding properties to the protein target (high-throughput screening). Combinatorial libraries provide a carefully selected set of molecular building blocks – usually dozens or hundreds – together with a small set of chemical reactions that link the modules. In this way, a combinatorial library can theoretically provide a diversity of up to billions of molecules from a small set of reactants. Up to millions of these molecules can be synthesized daily in a robotized process and submitted to chemical test in a high-throughput screening procedure. In our context, the objective of the test is to find out which compounds bind tightly to a given target protein.

Here we have a similar situation as in the search for target proteins. We have to inspect compounds among a very large set of molecular candidates, in order to select those that we want to inspect further. Again, computer help is necessary for preselection of molecular candidates and interpretation of experimental data.

In the computer, finding out whether a drug binds tightly to the target protein can best be done on the basis of knowledge of the protein structure. If the spatial shape of the site of the protein to which the drug is supposed to bind is known, then we can apply docking methods to select suitable lead compounds which have the potential of being refined to drugs. The speed of a docking method determines whether the method can be employed for screening compound databases in the search for drug leads. At GMD, we developed the docking method FlexX that takes a minute per instance and can be used to screen up to thousands of compounds on a PC or hundreds of thousands of drugs on a suitable parallel computer. Docking methods that take the better part of an hour cannot suitably be employed for such large scale screening purposes. In order to screen really large drug databases with several hundred thousand compounds or more we need docking methods that can handle single protein/drug pairs within seconds. The high conformational flexibility of small molecules as well as the subtle structural changes in the protein binding pocket upon docking (induced fit) are major complications in docking. Furthermore, docking necessitates careful analysis of the binding energy. The energy model is cast into the so-called scoring function that rates the protein-ligand complex energetically. Challenges in the energy model include the handling of entropic contributions, solvation effects, and the computation of long-range forces in fast docking methods.

Here is a summary of the state of the art in docking. With our program FlexX, handling the structural flexibility of the drug molecule can be done within the regime up to about a minute per molecular complex on a PC [22]. (FlexX has also been adapted to screening combinatorial drug libraries [23].) A suitable analysis of the structural changes in the protein still necessitates more computing time. Today, tools that are able to dock a molecule to a protein within seconds are still based on rigid-body docking (both the protein and ligand conformational flexibility is omitted) [18].

Even if the structure of the protein binding site is not known, we can use computer-based methods to select promising lead compounds. Our program FlexS compares the structure of a molecule with that of a ligand that is known to bind to the protein, for instance, its natural substrate [19]. The runtime is again between one and two minutes on a PC.

Finally, the program Ftrees [21] can speed up this comparison to 50 ms per molecular pair, on the average. Molecular structure is still analyzed, on some level, which is an exception in this time-regime. Drug molecules are abstracted to tree-structures, for this purpose. Heuristic tree matching algorithms provide the basis for the molecular comparison.

The accuracy of docking predictions lies within 50% to 80% "correct" predictions depending on the evaluation measure and the method [17]. That means, that docking methods are far from perfectly accurate. Nevertheless, they are very useful in pharmaceutical practice. The major benefit of docking is that a large drug library can be ranked with respect to the potential that its molecules have for being a useful lead compound for the target protein in question. The quality of a method in this context can be measured by an enrichment factor. Roughly, this is the ratio between the number of active compounds (drugs that bind tightly to the protein) in a top fraction (say the top 1%) of the ranked drug database divided by the same figure in the randomly arranged drug database. State-of-the-art docking methods in the middle regime (minutes per molecular pair), e.g. FlexX, achieve enrichment factors of up to about 15. Faster methods, e.g. FeatureTrees, achieve similar enrichment factors, but deliver molecules more similar to known binding ligands and do not detect such a diverse range of binding molecules.

Besides the savings in time and resources, another advantage of virtual screening of drug libraries over high-throughput screening in the laboratory is that the computer models produced can be inspected in order to find out why certain drugs bind tightly to the protein and others do not. This insight can be a fruitful source of further pharmaceutical innovation.

6   Summary

Computational biology is an extremely interesting and demanding branch of applied computer science. This article could only touch upon a few research topics in this complex field. Computational biology is a very young field. The biological systems under study are not very well understood yet. Models are rough, data are voluminous but often noisy. This limits the accuracy of computational biology predictions. However, the analyses improve quickly, due to improvements on the algorithmic and statistical side and to the accessibility to more and better data. Nevertheless, computational biology can be expected to be a major challenge for some time to come.

New kinds of experimental data are already in sight. These include protein data from mass spectrometry experiments. With these data, we can look at proteins not only on the transcription (mRNA) level but also, as they have been modified after their synthesis. These protein modifications, which include glycosylation (attachment of large tree-like sugar molecules to the protein surface) and phosphorylation (attachment of phosphate groups that often activate proteins), greatly influence protein function. Computational biology has not really advanced to analyzing these post-translational modifications of proteins yet.

Another emerging area is that of studying the genomic differences between individuals in a species, e.g., between different people. Of course, the susceptibility to disease and the receptivity to drug treatment finally is based on these dif-

ferences. Genomic differences between people amount to single nucleotide poly-
morphisms (SNPs), i.e., changes in single bases in the genomic text every few
thousand letters. These variations have to be analyzed with computational biology methods. Eventually, these activities will result in today's computational
biology merging with the field of statistical genetics and epidemiology.

One important point that we want to stress in the end is this. Computational
biology is a field which needs much basic research. At the same time, however,
the pressure for immediate biological innovation is tremendous. In this situation,
the impact of computational biology research critically depends on an accurate
understanding of the biological process which is to be empowered by the computational biology contribution. It is critical to ask the right questions, and often
modeling takes priority over optimization. Lastly, we need people that understand and love both computer science and biology to bring the field forward.
Fortunately, it seems that a growing number of people discover their interest in
both disciplines that make up computational biology.

References

1. M.D. Adams et al., The Genome Sequence of Drosophila Melanogaster,
   Science 287 (2000) 2185-2195.
   http://www.celera.com/celerascience/index.cfm
   http://flybase.bio.indiana.edu/ 9
2. A. Bateman et al., The Pfam Protein Families Database, Nucleic Acids
   Research 28 (2000) 263-266.
   http://pfam.wustl.edu/ 12
3. Proteins: Structure, Function and Genetics, Suppl 1 (1997).
   http://PredictionCenter.llnl.gov/casp2/Casp2.html 14
4. Proteins: Structure, Function and Genetics, Suppl: Third Meeting on the
   Critical Assessment of Techniques for Protein Structure Prediction (1999).
   http://PredictionCenter.llnl.gov/casp3/Casp3.html 14
5. The C. elegans Sequencing Consortium, Genome Sequence of the Nematode
   C. elegans: A Platform for Investigating Biology, Science 282 (1998) 2012-
   2018.
   http://www.wormbase.org/ 9
6. J. L. DeRisi, V.R. Iyer, P.O. Brown, Exploring the metabolic and genetic
   control of gene expression on a genomic scale, Science 278 (1997) 680-685.
   http://cmgm.stanford.edu/pbrown/explore/ 9
7. S.R. Eddy, Profile hidden Markov Models, Bioinformatics 14,9 (1998) 755-
   763. 13
8. R.D. Fleischmann, et al., Whole-genome random sequencing and assembly
   of Haemophilus influenzae Rd., Science 269 (1995) 496-512.
   http://www.tigr.org/tdb/mdb/mdb.html 9
9. A. Goffeau et al., Life with 6000 genes, Science 274 (1996) 546-567.
   http://www.mips.biochem.mpg.de/proj/yeast/ 9
10. T. Jaakola, M. Diekhans, D. Haussler, Using the Fisher kernel method to
    detect remote protein homologies, In Proceedings of the Seventh Interna-
    tional Conference on Intelligent Systems for Molecular Biology (ISMB99),
    AAAI Press (1999) 149-158. 13

11. A. Krogh, M. Brown, I. S. Mian, K. Sjölander und D. Haussler (1994). Hidden Markov Models in computational biology: application to protein modeling. Journal of Molecular Biology 235, 1501–1531. 13

12. E.M. Marcotte et al., A combined algorithm for genome-wide prediction of protein function, Nature 402 (1999) 83-86. 13

13. H. Nielsen, S. Brunak, G. von Heijne, Review. Machine learning approaches to the prediction of signal peptides and other protein sorting signals, Protein Engineering 12 (1999) 3-9. 13

14. M. Pellegrini et al., Assigning protein functions by comparative genome analysis: protein phylogenetic profiles, Proc. Natl. Acad. Sci. USA 96 (1999) 4285-4288. 13

References from the GMD Group

A list of references by the GMD group follows. More references can be found on the internet at http://www.gmd.de/SCAI/.

15. N.N. Alexandrov, R. Nussinov, R.M. Zimmer, Fast Protein Fold Recognition via Sequence to Structure Alignment and Contact Capacity Potentials, Proceedings of the First Pacific Symposium on Biocomputing, World Scientific Publ. (1996) 53-72.
http://cartan.gmd.de/ToPLign.html – Presents a dynamic programming method for protein folding. The key of the contribution is a novel and very effective cost function. A frequently used protein folding program on the internet. 12

16. I. Koch, T. Lengauer, E. Wanke, An Algorithm for Finding all Common Subtopologies in a Set of Protein Structures, Journal of Computational Biology 3,2 (1996) 289-306. – Presents an algorithm based on clique enumeration that compares 3D protein structures

17. B. Kramer, M. Rarey, T. Lengauer, Evaluation of the FlexX incremental construction algorithm for protein-ligand docking, PROTEINS: Structure, Function and Genetics 37 (1999) 228-241. – Detailed evaluation of FlexX on a set of 200 protein-ligand complexes. 16

18. C. Lemmen, C. Hiller, T. Lengauer, RigFit: A New Approach to Superimposing Ligand Molecules, Journal of Computer-Aided Molecular Design 12,5 (1998) 491-502. – Fast rigid molecular superpositioning method based on algorithmic approach taken from computer-based crystallography. 15

19. C. Lemmen, T. Lengauer, G. Klebe, FlexS: A Method for Fast Flexible Ligand Superposition, Journal of Medicinal Chemistry 41,23 (1998) 4502-4520. http://cartan.gmd.de/flex-bin/FlexS – Presents a method for superposing two drug molecules in 3D space in order to ascertain whether they have similar biochemical function. This algorithm offers an answer to the protein-ligand docking problem in the common case that the 3D structure of the protein is not available. FlexS is a commercialized software product that is a successor development of FlexX (see below). 15

20. T. Lengauer, Molekulare Bioinformatik: Eine interdisziplinäre Herausforderung. In Highlights aus der Informatik (I. Wegener ed.), Springer Verlag, Heidelberg (1996) 83-111. – An earlier review paper on algorithmic aspects of computational biology, somewhat more detailed. 14

21. M. Rarey, J.S. Dixon, Feature Trees: A new molecular similarity measure based on tree matching, Journal of Comput. Aided Mol. Design, 12 (1998) 471-490.
http://cartan.gmd.de/ftrees/ – Presents a method for very fast structural molecular comparison of drug molecules. 15

22. M. Rarey, B. Kramer, T. Lengauer, G. Klebe, A Fast Flexible Docking Method Using an Incremental Construction Algorithm. Journal of Molecular Biology 261,3 (1996) 470-489.
http://cartan.gmd.de/flexx//html/flexx-intro.html – Presents a fast algorithm for docking drug molecules into the binding sites of target proteins. Thus the program takes the structures of the two involved molecules and computes the structure of the complex (drug bound to protein). The program also returns a estimate of the binding energy, in order to discriminate between tightly and loosely binding drugs. The structural flexibility of the drug molecule is treated algorithmically; the protein is considered rigid. This algorithm is the basis of a successfully commercialized software product FlexX. 15

23. M. Rarey, T. Lengauer, A Recursive Algorithm for Efficient Combinatorial Library Docking, to appear in Drug Discovery and Design (2000). – Presents a version of FlexX that treats combinatorial libraries. 15

24. R. Thiele, R. Zimmer, T. Lengauer, Protein Threading by Recursive Dynamic Programming, Journal of Molecular Biology 290 (1999) 757-779.
http://cartan.gmd.de/ToPLign.html – introduces a combinatorial method that heuristically solves an NP-hard formulation of the protein folding problem. 12

25. A. Zien, G. Rätsch, S. Mika, B. Schölkopf, C. Lemmen, A. Smola, T. Lengauer, K. Müller, Engineering support vector machines kernels that recognize translation initiation sites, to appear in Bioinformatics (2000) – Uses support vector machines to solve an important classification problem in computational biology 13

26. A. Zien, R. Küffner, R. Zimmer, T. Lengauer, Analysis of Gene Expression Data With Pathway Scores, International Conference on Intelligent Systems for Molecular Biology (ISMB'00), AAAI Press (2000) – Presents a statistical method for validating a biological hypothesis that assigns a protein under study to a certain biochemical pathway. The method uses correlation studies of differential expression data to arrive at its conclusion. 13

27. A. Zien, R. Zimmer, T. Lengauer, A simple iterative approach to parameter optimization, Proceedings of the Fourth Annual Conference on Research in Computational Molecular Biology (RECOMB'00), ACM Press (2000) – Offers a combinatorial method for optimizing hard-to-set parameters in protein folding programs. 14

28. R. Zimmer, T. Lengauer, Fast and Numerically Stable Parametric Alignment of Biosequences. Proceedings of the First Annual Conference on Research in Computational Molecular Biology (RECOMB'97) (1997) 344-353. – Solves an algorithmically challenging parametric version of the pairwise sequence alignment problem. 14

Polygon Decomposition for Efficient
Construction of Minkowski Sums*

Pankaj K. Agarwal[1], Eyal Flato[2], and Dan Halperin[2]

[1] Department of Computer Science, Duke University, Durham, NC 27708-0129
pankaj@cs.duke.edu
[2] Department of Computer Science, Tel Aviv University, Tel-Aviv 69978, Israel
{flato, halperin}@math.tau.ac.il

Abstract. Several algorithms for computing the Minkowski sum of two
polygons in the plane begin by decomposing each polygon into convex
subpolygons. We examine different methods for decomposing polygons by
their suitability for efficient construction of Minkowski sums. We study
and experiment with various well-known decompositions as well as with
several new decomposition schemes. We report on our experiments with
the various decompositions and different input polygons. Among our
findings are that in general: (i) triangulations are too costly (ii) what
constitutes a good decomposition for one of the input polygons depends
on the other input polygon—consequently, we develop a procedure for
simultaneously decomposing the two polygons such that a "mixed" objective function is minimized, (iii) there are optimal decomposition algorithms that significantly expedite the Minkowski-sum computation, but
the decomposition itself is expensive to compute — in such cases simple heuristics that approximate the optimal decomposition perform very
well.

1   Introduction

Given two sets P and Q in $\mathbb{R}^d$, their *Minkowski sum* (or vector sum), denoted
by $P \oplus Q$, is the set $\{p + q \mid p \in P, q \in Q\}$. Minkowski sums are used in a wide
range of applications, including robot motion planning [19], assembly planning
[11], computer-aided design and manufacturing (CAD/CAM) [6], and geographic
information systems.

* P.A. is supported by Army Research Office MURI grant DAAH04-96-1-0013, by
a Sloan fellowship, by NSF grants EIA–9870724, EIA–997287, and CCR–9732787
and by a grant from the U.S.-Israeli Binational Science Foundation. D.H. and E.F.
have been supported in part by ESPRIT IV LTR Projects No. 21957 (CGAL)
and No. 28155 (GALIA), and by a Franco-Israeli research grant (monitored by
AFIRST/France and The Israeli Ministry of Science). D.H. has also been supported
by a grant from the U.S.-Israeli Binational Science Foundation, by The Israel Science Foundation founded by the Israel Academy of Sciences and Humanities (Center
for Geometric Computing and its Applications), and by the Hermann Minkowski –
Minerva Center for Geometry at Tel Aviv University.

M. Paterson (Ed.): ESA 2000, LNCS 1879, pp. 20–31, 2000.

Motivated by these applications, there has been much work on obtaining sharp bounds on the size of the Minkowski sum of two sets in two and three dimensions, and on developing fast algorithms for computing Minkowski sums. It is well known that if P is a polygonal set with m vertices and Q is another polygonal set with n vertices, then $P \oplus Q$ is a portion of the arrangement of mn segments, where each segment is the Minkowski sum of a vertex of P and an edge of Q, or vice-versa. Therefore the size of $P \oplus Q$ is $O(m^2 n^2)$ and it can be computed within that time; this bound is tight in the worst case [14]. If P is convex, then a result of Kedem et al. [15] implies that $P \oplus Q$ has $\Theta(mn)$ vertices, and it can be computed in $O(mn \log(mn))$ time [20]. If both P and Q are convex, then $P \oplus Q$ is a convex polygon with at most $m + n$ vertices, and it can be computed in $O(m + n)$ time [19]. In motion-planning applications, one is often interested in computing only a single connected component of the complement of $P \oplus Q$. Faster algorithms are known for computing a single connected component of the complement of $P \oplus Q$ [12,23]. A summary of the known results on computing the Minkowski sum of two sets in three and higher dimensions can be found in a recent survey by Agarwal and Sharir [2].

We devised and implemented several algorithms for computing the Minkowski sum of two simple polygons[1] based on the CGAL software library [1]. Our main goal was to produce a *robust* and exact implementation. This goal was achieved by employing the CGAL *planar map* package [9] while using exact number types. We are currently using our software to solve translational motion planning problems in the plane. We are able to compute collision-free paths even in environments cluttered with obstacles, where the robot could only reach a destination placement by moving through tight passages, practically moving in contact with the obstacle boundaries. This is in contrast with most existing motion planning software for which tight or narrow passages constitute a significant hurdle. For more details on our algorithms and implementation see [8]; we briefly describe the algorithms in Section 2 below.

The robustness and exactness of our implementation come at a cost: they slow down the running time of the algorithms in comparison with a more standard implementation that uses floating point arithmetic. This makes it especially necessary to try and speed up the algorithms in other ways. All our algorithms start with decomposing the input polygons into convex subpolygons. We discovered that not only the number of subpolygons in the decomposition of the input polygons but also their shapes had dramatic effect on the running time of the Minkowski-sum algorithms; see Figure 1 for an example.

In this paper we examine different methods for decomposing polygons by their suitability for efficient construction of Minkowski sums. In the theoretical study of Minkowski-sum computation (e.g., [15]), the choice of decomposition is often irrelevant (as long as we decompose the polygons into convex subpolygons) because it does not affect the worst-case *asymptotic* running time of the

---

[1] Our Minkowski sum software works for general complex polygonal sets including polygons with holes. Although not all the decomposition methods discussed in this paper extend to polygons with holes, many of them are easy to extend.

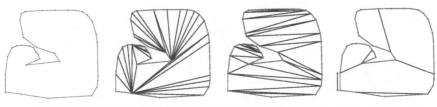

	naïve triang.	min Σd_i^2 triang.	min convex
Σd_i^2	754	530	192
# of convex subpolygons in P	33	33	6
time (mSec) to compute $P \oplus Q$	2133	1603	120

Fig. 1. Different decomposition methods applied to the polygon P (leftmost in the figure), from left to right: naïve triangulation, minimum Σd_i^2 triangulation and minimum convex decomposition (the details are given in Section 3). We can see in the table for each decomposition the sum of squares of degrees, the number of convex subpolygons, and the time in milliseconds to compute the Minkowski sum of the polygon with a small convex polygon, Q, with 4 vertices.

algorithms. In practice however, as we already mentioned above, different decompositions can induce a large difference in running time of the Minkowski-sum algorithms. The decomposition can affect the running time of algorithms for computing Minkowski sums in several ways: some of them are global to all algorithms that decompose the input polygons into convex polygons, while some others are specific to certain algorithms or even to specific implementations. We examine these various factors and report our findings below.

Polygon decomposition has been extensively studied in computational geometry; it is beyond the scope of this paper to give a survey of results in this area and we refer the reader to the survey papers by Keil [18] and Bern [3], and the references therein. As we proceed, we will provide details on specific decomposition methods that we will be using.

In the next section we survey the Minkowski sum algorithms that we have implemented. In Section 3 we describe the different decomposition algorithms we have implemented. We present a first set of experimental results in Section 4 and filter out the methods that turn out to be inefficient. In Section 5 we focus on the decomposition schemes that are not only fast to compute but also help to compute the Minkowski sum efficiently.

2   Minkowski Sum Algorithms

Given a collection $\mathcal{C}$ of curves in the plane, the *arrangement* $\mathcal{A}(\mathcal{C})$ is the subdivision of the plane into vertices, edges and faces induced by the curves in $\mathcal{C}$. *Planar maps* are arrangements where the curves are pairwise interior disjoint. Our algorithms for computing Minkowski sums rely on arrangements and planar maps, and in the discussion below we assume some familiarity with these structures, and with a refinement thereof called the *vertical decomposition*; we refer the reader to [2,10,25] for information on arrangements and vertical decomposition, and to [9] for a detailed description of the planar map package in CGAL on which our algorithms are based.

The input to our algorithms are two simple polygons P and Q, with m and n vertices respectively. Our algorithms consist of the following three steps:

Step 1: Decompose P into the convex subpolygons $P_1, P_2, \ldots, P_s$ and Q into the convex subpolygons $Q_1, Q_2, \ldots, Q_t$.

Step 2: For each $i \in [1..s]$ and for each $j \in [1..t]$ compute the Minkowski *subsum* $P_i \oplus Q_j$ which we denote by R_{ij}. We denote by R the set $\{R_{i,j} \mid i \in [1..s], j \in [1..t]\}$.

Step 3: Construct the union of all the polygons in R, computed in Step 2; the output is represented as a planar map.

The Minkowski sum of P and Q is the union of the polygons in R. Each R_{ij} is a convex polygon, and it can easily be computed in time that is linear in the sizes of P_i and Q_j [19]. Let k denote the overall number of edges of the polygons in R, and let I denote the overall number of intersections between (edges of) polygons in R

We briefly present two different algorithms for performing Step 3, computing the union of the polygons in R, which we refer to as the *arrangement* algorithm and the *incremental union* algorithm. A detailed description of these algorithms is given in [8].

Arrangement algorithm. The algorithm constructs the arrangement $\mathcal{A}(R)$ induced by the polygons in R (we refer to this arrangement as the *underlying arrangement* of the Minkowski sum) by adding the (boundaries of the) polygons of R one by one in a random order and by maintaining the vertical decomposition the arrangement of the polygons added so far; each polygon is chosen with equal probability at each step. Once we have constructed the arrangement, we efficiently traverse all its cells (vertices, edges, or faces) and we mark a cell as belonging to the Minkowski sum if it is contained inside at least one polygon of R. The construction of the arrangement takes randomized expected time $O(I + k \log k)$ [22]. The traversal stage takes $O(I + k)$ time.

Incremental union algorithm. In this algorithm we incrementally construct the union of the polygons in R by adding the polygons one after the other in random order. We maintain the planar map representing the partial union of polygons in R. For each $r \in R$ we insert the edges of r into the map and then remove redundant edges from the map. All these operations can be carried out efficiently using the planar map package. We can only give a naïve bound $O(k^2 \log^2 k)$ on the running time of this algorithm, which in the worst case is higher than the worst-case running time of the arrangement algorithm. Practically however the incremental union algorithm works much better on most problem instances.

Remarks. (1) We also implemented a union algorithm using a divide-and-conquer approach but since it mostly behaves worse than the incremental algorithm we do not describe it here. The full details are given in [8]. **(2)** Our planar map package provides full support for maintaining the vertical decomposition, and for efficient point location in a map. However, using simple point location strategies (naïve, walk-along-a-line) is often faster in practice [9]. Therefore we ran the tests reported below without maintaining the vertical decomposition.

3   The Decomposition Algorithms

We briefly describe here different algorithms that we have implemented for decomposing the input polygons into convex subpolygons. We used both decomposition with or without Steiner points. Some of the techniques are optimal and some use heuristics to optimize certain objective functions. The running time of the decomposition stage is significant only when we search for the optimal solution and use dynamic programming; in all other cases the running time of this stage is negligible even when we implemented a naïve solution. Therefore we only mention the running time for the "heavy" decomposition algorithms. In what follows P is a polygon with n vertices $p_1, \ldots, p_n$, r of which are reflex.

3.1   Triangulation

Greedy triangulation. This procedure searches for a pair of vertices p_i, p_j such that the segment $p_i p_j$ is a diagonal, namely it lies inside the polygon. It adds such a diagonal, splits the polygon into two subpolygons by this diagonal, and triangulates each subpolygon recursively. The procedure stops when the polygon becomes a triangle. See Figure 1 for an illustration.

In some of the following decompositions we are concerned with the degrees of vertices in the decomposition (namely the number of diagonals incident to a vertex). Our motivation for considering the degree comes from an observation on the way our planar map structures perform in practice: we noted that the existence of high degree vertices in the decomposition results in high degree vertices in the arrangement, which makes the maintenance of union of polygons slower (see the full version for details).

Optimal triangulation—minimizing the maximum degree. Using dynamic programming we compute a triangulation of the polygon where the maximum degree of a vertex is minimal. The algorithm is described in [13], and runs in $O(n^3)$ time.

Optimal triangulation—minimizing Σd_i^2. We adapted the minimal-maximum-degree algorithm to find the triangulation with minimum Σd_i^2 where d_i is the degree of vertex v_i of the polygon. See Figure 1.

3.2   Convex Decomposition without Steiner Points

Greedy convex decomposition. The same as the greedy triangulation algorithm except that it stops as soon as the polygon does not have a reflex vertex.
Minimum number of convex subpolygons (min-convex). We apply the algorithm of Keil [16] which computes a decomposition of a polygon into the minimum number convex subpolygons without introducing new vertices (Steiner

Fig. 2. From left to right: Slab decomposition, angle bisector (AB) decomposition, and KD decomposition

points). The running time of the algorithm is $O(nr^2 \log n)$. This algorithm uses dynamic programming. See Figure 1. This result was recently improved to $O(n + r^2 \min\{r^2, n\})$ [17].

Minimum Σd_i^2 convex decomposition. We modified Keil's algorithm so that it will compute decompositions that minimize Σd_i^2, the sum of squares of vertex degree.

3.3 Convex Decomposition with Steiner Points

Slab decomposition. Given a direction e, from each reflex vertex of the polygon we extend a segment in directions e and $-e$ inside the polygon until it hits the polygon boundary. The result is a decomposition of the polygon into convex slabs. If e is vertical then this is the well-known vertical decomposition of the polygon. See Figure 2. The obvious advantage of this decomposition is its simplicity.

Angle bisector decomposition (AB). In this algorithm we extend the internal angle bisector from each reflex vertex until we first hit the polygon's boundary or a diagonal that we have already extended from another vertex. This decomposition gives a 2-approximation to the optimal convex decomposition as described in [4]. See Figure 2.

KD decomposition. This algorithm is inspired by the KD-tree method to partition a set of points in the plane [5]. See Figure 2. By this method we try to lower the *stabbing number* of the subdivision (namely, the maximum number of subpolygons in the subdivision intersected by any line)—the detailed discussion can be found at [8].

4   A First Round of Experiments

Our implementation of the Minkowski sum package is based on the CGAL (version 2.0) [7] and LEDA (version 4.0) [21] libraries. Our package works with Linux (g++ compiler) as well as with WinNT (Visual C++ 6.0 compiler). The tests were performed under WinNT workstation on an (unloaded) 500 MHz PentiumIII machine with 128 Mb of RAM. We measured the running times for the various algorithms with different input data.

The input data that we present here is just a small representative sample of the polygonal sets on which tests were performed. As mentioned in the Introduction, the complexity of the Minkowski sum can range from $\Theta(m + n)$ (e.g., two

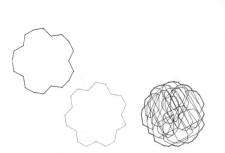
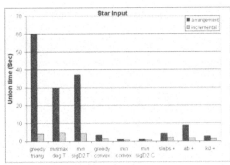

Fig. 3. Star input: The input (on the left-hand side) consists of two star-shaped polygons. The underlying arrangement of the Minkowski sum is shown in the middle. Running times in seconds for different decomposition methods (for two star polygons with 20 vertices each) are presented in the graph on the right-hand side.

convex polygons) to $\Theta(m^2 n^2)$ (e.g. the fork input—Figure 4). Tests on two convex polygons are meaningless in our context. The "intermediate" inputs (star shaped polygons—Figure 3, random looking polygons) are interesting in that there are many different ways to decompose them.

4.1   Results and discussion

We ran both union algorithms (arrangement and incremental-union) with all nine decomposition methods on the input data. The running times for the computation of the Minkowski sum for two input examples are summarized in Figures 3 and 4. Additional experimental results can be found at [8] and http://www.math.tau.ac.il/~flato/TriminkWeb/.

It is obvious from the experimental results that triangulations result in poor union running times (the left three pairs of columns in the histograms of Figures 3 and 4). By triangulating the polygons, we create $(n-1)(m-1)$ hexagons in R with potentially $\Omega(m^2 n^2)$ intersections between the edges of these polygons. We get those poor results since the performance of the union algorithms strongly depends on the number of vertices in the arrangement of the hexagon edges. Minimizing the maximum degree or the sum of squares of degrees in a triangulation is a slow computation that results in better union performance (compared to the naïve triangulation) but is still much worse than other simple convex-decomposition techniques.

In most cases the arrangement union algorithm runs much slower than the incremental union approach. By removing redundant edges from the partial sum during the insertion of polygons, we reduce the number of intersections of new polygons and the current planar map features. The fork input is an exception since the complexity of the union is roughly the same as the complexity of the underlying arrangement and the edges that we remove in the incremental algorithm do not significantly reduce the complexity of the planar map; see Figure 4.

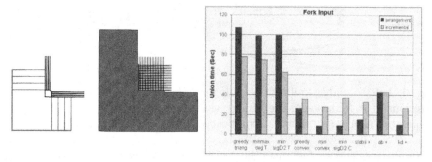

Fig. 4. Fork input: the input (on the left-hand side) consists of two orthogonal fork polygons. The Minkowski sum in the middle has complexity $\Theta(m^2 n^2)$. The running times in seconds for different decomposition methods (for two fork polygons with 8 teeth each) are shown in the graph on the right-hand side.

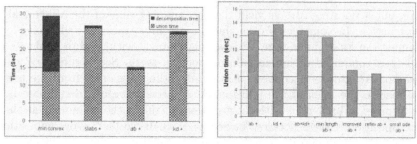

Fig. 5. On the left: when using the min-convex decomposition the union computation time is the smallest but it becomes inefficient when counting the decomposition time as well (running times for two star polygons with 100 vertices each). On the right: average union running times for star inputs with the improved decomposition algorithms.

The min-convex algorithm almost always gives the best union computation time but constructing this optimal decomposition may be expensive—see Figure 5. Minimizing the sum of squares of degrees in a convex decomposition rarely results in a decomposition that is different from the min-convex decomposition.

This first round of experiments helped us to filter out inefficient methods. In the next section we focus on the better decomposition algorithms (i.e., min convex, slab, angle bisector), we further study them and attempt to improve their performance.

5   Revisiting the More Efficient Algorithms

In this section we focus our attention on the algorithms that were found to be efficient in the first round of experiments. We present an experiment that shows that, contrary to the impression that the first round of results may give, minimizing the number of convex subpolygons in the decomposition does not always lead to better Minkowski-sum computation time. We also show in this

Fig. 6. Left to right: knife input polygons (P above Q), two types of decompositions of P (enlarged), the Minkowski sum and the underlying arrangement (generated using the "short" decomposition).

section that in certain instances the decision how to decompose the input polygon P may change depending on the other polygon Q.

5.1   Nonoptimality of Min-convex Decompositions

Minimizing the number of convex parts of P and Q can be expensive to compute, but it does not always yield the best running time of the Minkowski sum construction. In some cases other factors are important as well. Consider for example the knife input data. P is a long triangle with j teeth along its base and Q is composed of horizontal and vertical teeth. See Figure 6. P can be decomposed into $j + 1$ convex parts by extending diagonals from the teeth in the base to the apex of the polygon. Alternatively, we can decompose it into $j + 2$ convex subpolygons with short diagonals (this is the "minimal length AB" decomposition described below in Section 5.3). If we fix the decomposition of Q, the latter decomposition of P results in considerably faster Minkowski-sum running time, despite having more subpolygons, because the Minkowski sum of the long subpolygons in the first decomposition with the subpolygons of Q results in many intersections between the edges of polygons in R. In the first decomposition we have $j + 1$ long subpolygons while in the latter we have only one "long" subpolygon and $j + 1$ small subpolygons.

5.2   Mixed objective function

Good decomposition techniques that handle P and Q separately might not be sufficient because what constitutes a good decomposition of P depends on Q. We measured the running time for computing the Minkowski sum of a knife polygon P (Figure 6) and a random looking polygon Q. We scaled Q differently in each test. We fixed the decomposition of Q and decomposed the knife polygon P once with the short $j + 2$ "minimal length AB" decomposition and then with the long $j + 1$ minimum convex decomposition. The results are presented in Figure 7. We can see that for small Q's the short decomposition of the knife P with more subpolygons performs better but as Q grows the long decomposition of P with fewer subpolygons wins.

These experiments imply that a more careful strategy would be to simultaneously decompose the two input polygons, or at least take into consideration properties of one polygon when decomposing the other.

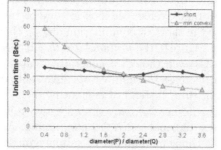

Fig. 7. Minkowski sum of a knife, P, with 22 vertices and a random polygon, Q, with 40 vertices using the arrangement union algorithm. On the left-hand side the underlying arrangement of the sum with the smallest random polygon and on the right-hand side the underlying arrangement of the sum with the largest random polygon. As Q grows, the number of vertices I in the underlying arrangement is dropping from (about) 15000 to 5000 for the "long" decomposition of P, and from 10000 to 8000 for the "short" decomposition.

The running time of the arrangement union algorithm is $O(I + k \log k)$, where k is the number of edges of the polygons in R and I is the overall number of inter-sections between (edges of) polygons in R (see Section 2). The value of k depends on the complexity of the convex decompositions of P and Q. Hence, we want to keep this complexity small. It is harder to optimize the value of I. Intuitively, we want each edge of R to intersect as few polygons of R as possible. Considering the standard rigid-motion invariant measure on lines in the plane [24], we developed the following cost function $c(D_{P,Q})$ of a simultaneous convex decomposition of P and Q. Let π_P, k_P and Δ_P be the perimeter, number of convex subpolygons and the sum of the lengths of the diagonals of P, respectively. Similarly define π_Q, k_Q and Δ_Q. Then $c(D_{P,Q}) = k_Q(2\Delta_P + \pi_P) + k_P(2\Delta_Q + \pi_Q)$, which is the sum of the perimeters of the polygons in R.

If we do not allow Steiner points, we can modify the dynamic-programming algorithm by Keil [16] to compute a decomposition that minimizes this cost function in $O(m^2 r_P^4 + n^2 r_Q^4)$ time, where r_P (r_Q) is the number of reflex vertices of P (resp. Q). We did not make any serious attempt to improve the running time. Because of lack of space we omit the details; see further information in [8].

If we allow Steiner points, then it is an open question whether an optimal de-composition can be computed in polynomial time. Currently, we do not even have a constant-factor approximation algorithm. The difficulty arises because unlike the minimum-size decomposition for which an optimal algorithm is known [4], no constant-factor approximation is known for minimum-length convex decom-position of a simple polygon if Steiner points are allowed.

5.3   Improving the AB method

Most of the tests suggest that in general the AB decomposition algorithm works better than the other heuristics. We next describe our attempts to improve this algorithm.

Minimal length angle bisector decomposition. In each step we handle one reflex vertex. For a reflex vertex we look for one or two diagonals that will eliminate it. We choose the shortest combination among the eliminators we have found. As we can see in Figure 5, the *minimal length AB* decomposition performs better than the naïve AB even though it generally creates more subpolygons.

We tried to further decrease the number of convex subpolygons generated by the decomposition algorithm. Instead of emanating a diagonal from any reflex vertex, we first tested whether we can eliminate two reflex vertices with one diagonal (we call such a diagonal a *2-reflex eliminator*). All the methods listed below generate at most the same number of subpolygons generated by the AB algorithm but practically the number is likely to be smaller.

Improved angle bisector decomposition. At each step, we choose a reflex vertex v and search for a 2-reflex eliminator incident upon v. If we cannot find such a diagonal, we continue as in the standard AB algorithm.

Reflex angle bisector decomposition. At each step, we check all pairs of reflex vertices to find a 2-reflex eliminator diagonal. When there are no more 2-reflex eliminators, we continue with the standard AB algorithm on the rest of the reflex vertices.

Small side angle bisector decomposition. As in the *reflex AB* decomposition, we are looking for 2-reflex eliminators. Such an eliminator decomposes the polygon into two parts, one on each side. Among the candidate eliminators we choose the one that has the minimal number of reflex vertices on one of its sides. By this we are trying to "block" the minimal number of reflex vertices from being connected (and eliminated) by another 2-reflex eliminator diagonal.

Experimental results are shown in Figure 5. These latter improvements to the AB decomposition seem to have the largest effect on the union running time, while keeping the decomposition method very simple to understand and implement. Note that the *small side AB* heuristic results in 20% faster union time than the *improved AB* and *reflex AB* decompositions, and 50% faster than the standard *angle bisector* method.

References

1. *The CGAL User Manual, Version 2.0*, 1999. http://www.cs.ruu.nl/CGAL. 21
2. P. K. Agarwal and M. Sharir. Arrangements. In J.-R. Sack and J. Urrutia, editors, *Handbook of Computational Geometry*, pages 49–119. Elsevier Science Publishers B.V. North-Holland, Amsterdam, 1999. 21, 23
3. M. Bern. Triangulations. In J. E. Goodman and J. O'Rourke, editors, *Handbook of Discrete and Computational Geometry*, chapter 22, pages 413–428. CRC Press LLC, Boca Raton, FL, 1997. 22
4. B. Chazelle and D. P. Dobkin. Optimal convex decompositions. In G. T. Toussaint, editor, *Computational Geometry*, pages 63–133. North-Holland, Amsterdam, Netherlands, 1985. 25, 29
5. M. de Berg, M. van Kreveld, M. Overmars, and O. Schwarzkopf. *Computational Geometry: Algorithms and Applications*. Springer-Verlag, Berlin, 1997. 25
6. G. Elber and M.-S. Kim, editors. *Special Issue of Computer Aided Design: Offsets, Sweeps and Minkowski Sums*, volume 31. 1999. 20

7. A. Fabri, G. Giezeman, L. Kettner, S. Schirra, and S. Schönherr. On the design of CGAL, the Computational Geometry Algorithms Library. Technical Report MPI-I-98-1-007, MPI Inform., 1998. To appear in *Software—Practice and Experience*. 25

8. E. Flato. Robust and efficient construction of planar Minkowski sums. Master's thesis, Dept. Comput. Sci., Tel-Aviv Univ., 2000. Forthcoming. http://www.math.tau.ac.il/ flato/thesis.ps.gz. 21, 23, 25, 26, 29

9. E. Flato, D. Halperin, I. Hanniel, and O. Nechushtan. The design and implementation of planar maps in CGAL. In J. Vitter and C. Zaroliagis, editors, *Proceedings of the 3rd Workshop on Algorithm Engineering*, volume 1148 of *Lecture Notes Comput. Sci.*, pages 154–168. Springer-Verlag, 1999. Full version: http://www.math.tau.ac.il/ flato/WaeHtml/index.htm. 21, 23, 24

10. D. Halperin. Arrangements. In J. E. Goodman and J. O'Rourke, editors, *Handbook of Discrete and Computational Geometry*, chapter 21, pages 389–412. CRC Press LLC, Boca Raton, FL, 1997. 23

11. D. Halperin, J.-C. Latombe, and R. H. Wilson. A general framework for assembly planning: The motion space approach. *Algorithmica*, 26:577–601, 2000. 20

12. S. Har-Peled, T. M. Chan, B. Aronov, D. Halperin, and J. Snoeyink. The complexity of a single face of a Minkowski sum. In *Proc. 7th Canad. Conf. Comput. Geom.*, pages 91–96, 1995. 21

13. G. Kant and H. L. Bodlaender. Triangulating planar graphs while minimizing the maximum degree. In *Proc. 3rd Scand. Workshop Algorithm Theory*, volume 621 of *Lecture Notes Comput. Sci.*, pages 258–271. Springer-Verlag, 1992. 24

14. A. Kaul, M. A. O'Connor, and V. Srinivasan. Computing Minkowski sums of regular polygons. In *Proc. 3rd Canad. Conf. Comput. Geom.*, pages 74–77, Aug. 1991. 21

15. K. Kedem, R. Livne, J. Pach, and M. Sharir. On the union of Jordan regions and collision-free translational motion amidst polygonal obstacles. *Discrete Comput. Geom.*, 1:59–71, 1986. 21

16. J. M. Keil. Decomposing a polygon into simpler components. *SIAM J. Comput.*, 14:799–817, 1985. 24, 29

17. J. M. Keil and J. Snoeyink. On the time bound for convex decomposition of simple polygons. In *Proc. 10th Canad. Conf. Comput. Geom.*, 1998. 25

18. M. Keil. Polygon decomposition. In J.-R. Sack and J. Urrutia, editors, *Handbook of Computational Geometry*. Elsevier Science Publishers B.V. North-Holland, Amsterdam, 1999. 22

19. J.-C. Latombe. *Robot Motion Planning*. Kluwer Academic Publishers, Boston, 1991. 20, 21, 23

20. D. Leven and M. Sharir. Planning a purely translational motion for a convex object in two-dimensional space using generalized Voronoi diagrams. *Discrete Comput. Geom.*, 2:9–31, 1987. 21

21. K. Melhorn and S. Näher. *The LEDA Platform of Combinatorial and Geometric Computing*. Cambridge University Press, 1999. 25

22. K. Mulmuley. *Computational Geometry: An Introduction Through Randomized Algorithms*. Prentice Hall, Englewood Cliffs, NJ, 1994. 23

23. R. Pollack, M. Sharir, and S. Sifrony. Separating two simple polygons by a sequence of translations. *Discrete Comput. Geom.*, 3:123–136, 1988. 21

24. L. Santaló. *Integral Probability and Geometric Probability*, volume 1 of *Encyclopedia of Mathematics and its Applications*. Addison-Wesley, 1979. 29

25. M. Sharir and P. K. Agarwal. *Davenport-Schinzel Sequences and Their Geometric Applications*. Cambridge University Press, 1995. 23

An Approximation Algorithm for Hypergraph Max k-Cut with Given Sizes of Parts

Alexander A. Ageev[1]* and Maxim I. Sviridenko[2]**

[1] Sobolev Institute of Mathematics, pr. Koptyuga 4, 630090, Novosibirsk, Russia
ageev@math.nsc.ru
[2] BRICS, University of Aarhus, Aarhus, Denmark
sviri@brics.dk

Abstract. An instance of Hypergraph Max k-Cut with given sizes of parts (or HYP MAX k-CUT WITH GSP) consists of a hypergraph $H = (V, E)$, nonnegative weights w_S defined on its edges $S \in E$, and k positive integers $p_1, \dots, p_k$ such that $\sum_{i=1}^{k} p_i = |V|$. It is required to partition the vertex set V into k parts $X_1, \dots, X_k$, with each part X_i having size p_i, so as to maximize the total weight of edges not lying entirely in any part of the partition. The version of the problem in which $|X_i|$ may be arbitrary is known to be approximable within a factor of 0.72 of the optimum (Andersson and Engebretsen, 1998). The authors (1999) designed an 0.5-approximation for the special case when the hypergraph is a graph. The main result of this paper is that HYP MAX k-CUT WITH GSP can be approximated within a factor of $\min\{\lambda_{|S|} : S \in E\}$ of the optimum, where $\lambda_r = 1 - (1 - 1/r)^r - (1/r)^r$.

1 Introduction

The last decade has been marked by striking breakthroughs in designing approximation algorithms with provable performance guarantees. Most of them to all appearances are due to using novel methods of rounding polynomially solvable fractional relaxations. Applicability of a rounding method is highly dependent on the type of constraints in the relaxation. In [1] the authors presented a new rounding technique (the pipage rounding) especially oriented to tackle some NP-hard problems, which can be formulated as integer programs with assignment-type constraints. The paper [1] contains four approximation results demonstrating the efficiency of this technique. One of the problems treated in [1] is Max k-Cut with given sizes of parts (or MAX k-CUT WITH GSP). An instance of this problem consists of an undirected graph $G = (V, E)$, nonnegative edge weights $w_e, e \in E$, and k positive integers $p_1, p_2, \dots, p_k$ such that $\sum_{i=1}^{k} p_i = |V|$. It is required to find a partition of V into k parts $V_1, V_2, \dots, V_k$ with each part V_i having size p_i, so as to maximize the total weight of edges whose ends lie in different parts of the partition. The paper [1] gives an 0.5-approximation for

---

* Supported in part by the Russian Foundation for Basic Research, grant 99-01-00601.
** Supported in part by the Russian Foundation for Basic Research, grant 99-01-00510.

this problem. Very recently, Feige and Langberg [7] by a combination of the method of [1] with the semidefinite programming technique designed an $0.5 + \varepsilon$-approximation for MAX 2-CUT WITH GSP, where ε is some unspecified small positive number.

The MAX CUT and MAX k-CUT problems are classical in combinatorial optimization and have been extensively studied in the absence of any restrictions on the sizes of parts. The best known approximation algorithm for MAX CUT is due to Goemans and Williamson [9] and has a performance guarantee of 0.878. Frieze and Jerrum [8] extended the technique of Goemans and Williamson to MAX k-CUT and designed a $(1 - 1/k + 2\ln k/k^2)$-approximation algorithm. On the other hand, as it was shown by Kann et al. [10], no approximation algorithm for MAX k-CUT can have a performance guarantee better than $1 - 1/34k$, unless P=NP.

Approximation results for some special cases of MAX k-CUT WITH GSP have been also established. In particular, Frieze and Jerrum [8] present an 0.65-approximation algorithm for MAX BISECTION (the special case of MAX k-CUT WITH GSP where $k = 2$ and $p_1 = p_2 = |V|/2$). Very recently, Ye [11] announced an algorithm with a better performance guarantee of 0.699. The best known approximation algorithm for MAX k-SECTION (the case where $p_1 = \cdots = p_k = |V|/k$) is due to Andersson [2] and has a performance guarantee of $1 - 1/k + \Omega(1/k^3)$.

In this paper we consider a hypergraph generalization of MAX k-CUT WITH GSP — Hypergraph Max k-Cut with given sizes of parts or, for short, HYP MAX k-CUT WITH GSP. An instance of HYP MAX k-CUT WITH GSP consists of a hypergraph $H = (V, E)$, nonnegative weights w_S on its edges S, and k positive integers $p_1, \ldots, p_k$ such that $\sum_{i=1}^{k} p_i = |V|$. It is required to partition the vertex set V into k parts $X_1, X_2, \ldots, X_k$, with each part X_i having size p_i, so as to maximize the total weight of the edges of H, not lying entirely in any part of the partition (i.e., to maximize the total weight of $S \in E$ satisfying $S \nsubseteq X_i$ for all i).

Several closely related versions of HYP MAX k-CUT WITH GSP were studied in the literature but few results have been obtained. Andersson and Engebretsen [3] presented an 0.72-approximation algorithm for the ordinary HYP MAX CUT problem (i.e., for the version without any restrictions on the sizes of parts). Arora, Karger, and Karpinski [4] designed a PTAS for dense instances of this problem or, more precisely, for the case when the hypergraph H is restricted to have $\Theta(|V|^d)$ edges, under the assumption that $|S| \leq d$ for each edge S and some constant d.

In this paper, by applying the pipage rounding method, we prove that HYP MAX k-CUT WITH GSP can be approximated within a factor of $\min\{\lambda_{|S|} : S \in E\}$ of the optimum, where $\lambda_r = 1 - (1 - 1/r)^r - (1/r)^r$. By direct calculations it easy to get some specific values of λ_r: $\lambda_2 = 1/2 = 0.5$, $\lambda_3 = 2/3 \approx 0.666$, $\lambda_4 = 87/128 \approx 0.679$, $\lambda_5 = 84/125 = 0.672$, $\lambda_6 \approx 0.665$ and so on. It is clear that λ_r tends to $1 - e^{-1} \approx 0.632$ as $r \to \infty$. A bit less trivial fact is that $\lambda_r > 1 - e^{-1}$ for each $r \geq 3$ (Lemma 2 in this paper). Summing up we arrive at the following conclusion: our algorithm finds a feasible cut of weight within a factor of 0.5 on general hypergraphs, i.e., in the case when each edge of the hypergraph has size at least 2, and within a factor of $1 - e^{-1} \approx 0.632$ in the case when each edge has size at least 3. Note that the first bound coincides with that obtained in [1] for the case of graphs. In this paper we also show that in the case of hypergraphs with each edge of size at least 3 the bound of $1 - e^{-1}$ cannot be improved, unless P=NP.

2　The Pipage Rounding: A General Scheme

We begin with a description of the pipage rounding [1] for the case of a slightly more general constraints.

Assume that a problem P can be reformulated as the following nonlinear binary program:

$$\max \quad F(x_{11}, \ldots, x_{nk}) \tag{1}$$

$$\text{s. t.} \quad \sum_{i=1}^{n} x_{it} = p_t, \quad t = 1, \ldots, k, \tag{2}$$

$$\sum_{t=1}^{k} x_{it} = 1, \quad i = 1, \ldots, n, \tag{3}$$

$$x_{it} \in \{0, 1\}, \quad t = 1, \ldots, k, \quad i = 1, \ldots, n, \tag{4}$$

where $p_1, p_2, \ldots, p_k$ are positive integers such that $\sum_t p_t = n$, $F(x)$ is a function defined on the rational points $x = (x_{it})$ of the $n \times k$-dimensional cube $[0, 1]^{n \times k}$ and computable in polynomial time. Assume further that one can associate with $F(x)$ another function $L(x)$ that is defined and polynomially computable on the same set, coincides with $F(x)$ on binary x satisfying (2)–(3), and such that the program

$$\max \quad L(x) \tag{5}$$

$$\text{s. t.} \quad \sum_{i=1}^{n} x_{it} = p_t, \quad t = 1, \ldots, k, \tag{6}$$

$$\sum_{t=1}^{k} x_{it} = 1, \quad i = 1, \ldots, n, \tag{7}$$

$$0 \leq x_{it} \leq 1, \quad t = 1, \ldots, k, \quad i = 1, \ldots, n \tag{8}$$

(henceforth called the *nice relaxation*) is polynomially solvable. Assume next that the following two main conditions hold. The first—*F/L lower bound condition*—states: there exists $C > 0$ such that $F(x)/L(x) \geq C$ for each $x \in [0, 1]^{n \times k}$. To formulate the second—*ε-convexity condition*—we need a description of the so-called pipage step.

Let x be a feasible solution to (5)–(8). Define the bipartite graph H with the bipartition $(\{1, \ldots, n\}, \{1, \ldots, k\})$ so that $jt \in E(H)$ if and only if x_{jt} is non-integral. Note that (6) and (7) imply that each vertex of H is either isolated or has degree at least 2. Assume that x has fractional components. Since H is bipartite it follows that H has a cycle D of even length. Let M_1 and M_2 be the matchings of H whose union is the cycle D. Define a new solution $x(\varepsilon)$ by the following rule: if jt is not an edge of D, then $x_{jt}(\varepsilon)$ coincides with x_{jt}, otherwise, $x_{jt}(\varepsilon) = x_{jt} + \varepsilon$ if $jt \in M_1$, and $x_{jt}(\varepsilon) = x_{jt} - \varepsilon$ if $jt \in M_2$.

By definition $x(\varepsilon)$ is a feasible solution to the linear relaxation of (5)–(8) for all $\varepsilon \in [-\varepsilon_1, \varepsilon_2]$ where

$$\varepsilon_1 = \min\{ \min_{jt \in M_1} x_{jt}, \ \min_{jt \in M_2} (1 - x_{jt})\}$$

and

$$\varepsilon_2 = \min\{ \min_{jt \in M_1} (1 - x_{jt}), \ \min_{jt \in M_2} x_{jt}\}.$$

The ε-convexity condition states that for every feasible x and every cycle D in the graph H, $\varphi(\varepsilon) = F(x(\varepsilon))$ is a convex function on the above interval.

Under the above assumptions we claim that there exists a polynomial-time C-approximation algorithm for solving P. Indeed, since the function $\varphi(\varepsilon) = F(x(\varepsilon))$ is convex,

$$F(x(\varepsilon^*)) \geq F(x) \geq CL(x)$$

for some $\varepsilon^* \in \{-\varepsilon_1, \varepsilon_2\}$. The new solution $x(\varepsilon^*)$, being feasible for (5)–(8), has a smaller number of fractional components. Set $x' = x(\varepsilon^*)$ and, if x' has fractional components, apply to x' the above described pipage step and so on. Ultimately, after at most nk steps, we arrive at a solution $\tilde{x}$ which is feasible for (1)–(4) and satisfies

$$F(\tilde{x}) \geq CL(x) \geq CF^*$$

where F^* is an optimal value of (1)–(4) (and of the original problem P). The rounding procedure described (and henceforth called the *pipage rounding*) can be clearly implemented in polynomial time.

Thus we obtain a C-approximation algorithm for P. It consists of two phases: the first phase is to find a feasible (fractional) solution to (5)–(8), and the second is to round off this solution by using the pipage rounding.

3   The Pipage Rounding: An Application to the Problem

It is easy to see that an instance of HYP MAX k-CUT WITH GSP can be equivalently formulated as the following (nonlinear) integer program:

$$\max \quad F(x) = \sum_{S \in E} w_S \left(1 - \sum_{t=1}^{k} \prod_{i \in S} x_{it}\right) \tag{9}$$

$$\text{s. t.} \quad \sum_{t=1}^{k} x_{it} = 1 \text{ for all } i, \tag{10}$$

$$\sum_{i=1}^{n} x_{it} = p_t \quad \text{for all } t, \tag{11}$$

$$x_{it} \in \{0,1\} \text{ for all } i \text{ and } t. \tag{12}$$

The equivalence is shown by the one-to-one correspondence between optimal solutions to the above program and optimal k-cuts $\{X_1,\ldots,X_k\}$ of instance of HYP MAX k-CUT WITH GSP defined by the relation "$x_{it} = 1$ if and only if $i \in X_t$".

As a nice relaxation we consider the following linear program:

$$\max \quad \sum_{S \in E} w_S z_S \tag{13}$$

$$\text{s. t.} \quad z_S \le |S| - \sum_{i \in S} x_{it} \quad \text{for all } S \in E, \tag{14}$$

$$\sum_{t=1}^{k} x_{it} = 1 \quad \text{for all } i, \tag{15}$$

$$\sum_{i=1}^{n} x_{it} = p_t \quad \text{for all } t, \tag{16}$$

$$0 \le x_{it} \le 1 \quad \text{for all } i \text{ and } t, \tag{17}$$

$$0 \le z_S \le 1 \quad \text{for all } S \in E. \tag{18}$$

It is easy to see that, given a feasible matrix x, the optimal values of z_S in the above program can be uniquely determined by simple formulas. Using this observation we can exclude the variables z_S and rewrite (13)–(18) in the following equivalent way:

$$\max \quad L(x) = \sum_{S \in E} w_S \min\{1, \min_t(|S| - \sum_{i \in S} x_{it})\} \tag{19}$$

subject to (15)–(17). Note that $F(x) = L(x)$ for each x satisfying (10)–(12).

We claim that, for every feasible x and every cycle D in the graph H (for definitions, see Section 2), the function $\varphi(\varepsilon) = F(x(\varepsilon))$ is a quadratic polynomial with a nonnegative leading coefficient. Indeed, observe that each product $\prod_{i \in S} x_{it}(\varepsilon)$ contains at most two modified variables. Assume that a product $\prod_{i \in S} x_{it}(\varepsilon)$ contains exactly two such variables $x_{i_1 t}(\varepsilon)$ and $x_{i_2 t}(\varepsilon)$. Then they can have only one of the following forms: either $x_{i_1 t} + \varepsilon$ and $x_{i_2 t} - \varepsilon$ or $x_{i_1 t} - \varepsilon$ and $x_{i_2 t} + \varepsilon$, respectively. In either case ε^2 has a nonnegative coefficient in the term corresponding to the product. This proves that the ε-convexity condition does hold.

For any $r \ge 1$, set $\lambda_r = 1 - (1 - 1/r)^r - (1/r)^r$.

Lemma 1. *Let $x = (x_{it})$ be a feasible solution to* (19),(15)–(17) *and $S \in E$. Then*

$$\left(1 - \sum_{t=1}^{k} \prod_{i \in S} x_{it}\right) \ge \lambda_{|S|} \min\{1, \min_t(|S| - \sum_{i \in S} x_{it})\}.$$

Proof. Let $z_S = \min\{1, \min_t(|S| - \sum_{i \in S} x_{it})\}$. Define q_S and t' by the equalities

$$q_S = \max_t \sum_{i \in S} x_{it} = \sum_{i \in S} x_{it'}.$$

Note that

$$z_S = \min\{1, |S| - q_S\}. \tag{20}$$

Using the arithmetic-geometric mean inequality and the fact that

$$\sum_{t=1}^{k} \sum_{i \in S} x_{it} = |S|$$

we obtain that

$$
\begin{aligned}
1 - \sum_{t=1}^{k} \prod_{i \in S} x_{it} &= 1 - \prod_{i \in S} x_{it'} - \sum_{t \neq t'} \prod_{i \in S} x_{it} \\
&\geq 1 - \left(\frac{\sum_{i \in S} x_{it'}}{|S|}\right)^{|S|} - \sum_{t \neq t'} \left(\frac{\sum_{i \in S} x_{it}}{|S|}\right)^{|S|} \\
&\geq 1 - \left(\frac{q_S}{|S|}\right)^{|S|} - \left(\frac{\sum_{t \neq t'} \sum_{i \in S} x_{it}}{|S|}\right)^{|S|} \\
&= 1 - \left(\frac{q_S}{|S|}\right)^{|S|} - \left(\frac{|S| - \sum_{i \in S} x_{it'}}{|S|}\right)^{|S|} \\
&= 1 - \left(\frac{q_S}{|S|}\right)^{|S|} - \left(1 - \frac{q_S}{|S|}\right)^{|S|}.
\end{aligned}
\tag{21}
$$

Let $\psi(y) = 1 - \left(1 - \frac{y}{|S|}\right)^{|S|} - \left(\frac{y}{|S|}\right)^{|S|}$.

Case 1. $|S| - 1 \leq q_S \leq |S|$. Then by (20), $z_S = |S| - q_S$, and hence by (21),

$$1 - \sum_{t=1}^{k} \prod_{i \in S} x_{it} \geq 1 - \left(1 - \frac{z_S}{|S|}\right)^{|S|} - \left(\frac{z_S}{|S|}\right)^{|S|} = \psi(z_S).$$

Since the function ψ is concave and $\psi(0) = 0$, $\psi(1) = \lambda_{|S|}$, it follows that

$$1 - \sum_{t=1}^{k} \prod_{i \in S} x_{it} \geq \lambda_{|S|} z_S.$$

Case 2. $1 \leq q_S \leq |S| - 1$. Here $z_S = 1$. Since $\psi(y)$ is concave and $\psi(1) = \psi(|S| - 1) = \lambda_{|S|}$,

$$1 - \sum_{t=1}^{k} \prod_{i \in S} x_{it} \geq \lambda_{|S|}.$$

Case 3. $0 \leq q_S \leq 1$. Again, $z_S = 1$. For every t, set $\mu_t = \sum_{i \in S} x_{it}$. Note that, by the assumption of the case,

$$0 \leq \mu_t \leq 1, \tag{22}$$

and, moreover,

$$\sum_{t=1}^{k} \mu_t = |S|. \tag{23}$$

By the arithmetic-geometric mean inequality it follows that

$$\sum_{t=1}^{k} \prod_{i \in S} x_{it} \leq \sum_{t=1}^{k} \left(\frac{\mu_t}{|S|} \right)^{|S|}$$

$$(\text{by } (22)) \quad \leq |S|^{-|S|} \sum_{t=1}^{k} \mu_t$$

$$(\text{by } (23)) \quad = |S|^{-|S|} |S|.$$

Consequently,

$$1 - \sum_{t=1}^{k} \prod_{i \in S} x_{it} \geq 1 - |S| \left(\frac{1}{|S|} \right)^{|S|}$$

$$= 1 - \left(\frac{1}{|S|} \right)^{|S|} - (|S| - 1) \left(\frac{1}{|S|} \right)^{|S|}$$

$$\geq 1 - \left(\frac{1}{|S|} \right)^{|S|} - (|S| - 1)^{|S|} \left(\frac{1}{|S|} \right)^{|S|}$$

$$= \lambda_{|S|}.$$

$\square$

Corollary 1. *Let* $x = (x_{it})$ *be a feasible solution to* (19),(15)–(17). *Then*

$$F(x) \geq (\min_{S \in E} \lambda_{|S|}) L(x).$$

$\square$

The corollary states that the F/L lower bound condition holds with

$$C = \min_{S \in E} \lambda_{|S|}.$$

Hence the pipage rounding provides an algorithm that finds a feasible k-cut whose weight is within a factor of $\min_{S \in E} \lambda_{|S|}$ of the optimum.

Note that $\lambda_2 = 1/2$. We now establish a lower bound on λ_r for all $r \geq 3$.

Lemma 2. *For any* $r \geq 3$,

$$\lambda_r > 1 - e^{-1}.$$

Proof. We first deduce it from the following stronger inequality:

$$\left(1 - \frac{1}{r}\right)^r < e^{-1}\left(1 - \frac{1}{2r}\right) \text{ for all } r \geq 1. \tag{24}$$

Indeed, for any $r \geq 3$,

$$\lambda_r = 1 - \frac{1}{r^r} - \left(1 - \frac{1}{r}\right)^r$$

$$> 1 - \frac{1}{r^r} - e^{-1}\left(1 - \frac{1}{2r}\right)$$

$$= 1 - e^{-1} + \frac{1}{r}\left(\frac{e^{-1}}{2} - \frac{1}{r^{r-1}}\right)$$

$$> 1 - e^{-1}.$$

To prove (24), by taking natural logarithm of both sides of (24) rewrite it in the following equivalent form:

$$1 + r\ln(1 - \frac{1}{r}) < \ln(1 - \frac{1}{2r}) \text{ for all } r \geq 1.$$

Using the Taylor series expansion

$$\ln(1 - \sigma) = -\sum_{i=1}^{\infty} \frac{\sigma^i}{i}$$

we obtain that for each $r = 1, 2, \ldots$,

$$1 + r\ln\left(1 - \frac{1}{r}\right) = 1 + r\left(-\frac{1}{r} - \frac{1}{2r^2} - \frac{1}{3r^3} - \cdots\right)$$

$$= -\frac{1}{2r} - \frac{1}{3r^2} - \frac{1}{4r^3} \cdots$$

$$< -\frac{1}{2r} - \frac{1}{2(2r)^2} - \frac{1}{3(2r)^3} \cdots$$

$$= \ln(1 - \frac{1}{2r}),$$

as required.   □

We now show that in the case of r-uniform hypergraphs the integrality gap for the relaxation (13)–(18) can be arbitrarily close to λ_r. It follows that no other rounding of this relaxation can provide an algorithm with a better performance guarantee.

Indeed, consider the following instance: the complete r-uniform hypergraph on $n = rq$ vertices, $k = 2$, $w_S = 1$ for all $S \in E$, $p_1 = q$ and $p_2 = n - q$. It is clear that any feasible cut in this hypergraph has weight

$$C_n^r - C_q^r - C_{n-q}^r.$$

Consider the feasible solution to (15)–(18) in which

$$x_{i1} = 1/r \text{ and } x_{i2} = 1 - 1/r \text{ for each } i.$$

The weight of this solution is equal to C_n^r, since for each edge S we have

$$r - \sum_{i \in S} x_{i1} \geq r - \sum_{i \in S} x_{i2} = 1$$

and therefore $z_S = 1$ for all $S \in E$. Thus the integrality gap for this instance is at most

$$
\begin{aligned}
\frac{C_n^r - C_q^r - C_{n-q}^r}{C_n^r} &= 1 - \frac{q!(n-r)!}{(q-r)!n!} - \frac{(n-q)!(n-r)!}{(n-q-r)!n!} \\
&\leq 1 - \frac{q!}{(q-r)!n^r} - \frac{(n-q)!}{(n-q-r)!n^r} \\
&\leq 1 - \frac{(q-r)^r}{n^r} - \frac{(n-q-r)^r}{n^r} \\
&= 1 - \left(\frac{1}{r} - \frac{1}{q}\right)^r - \left(1 - \frac{1}{r} - \frac{1}{q}\right)^r,
\end{aligned}
$$

which tends to λ_r as $q \to \infty$.

We conclude the paper with a proof that the performance bound of $1 - e^{-1}$, our algorithm provides on hypergraphs with each edge of size at least 3, cannot be improved, unless $P = NP$.

In the Maximum Coverage problem (MAXIMUM COVERAGE for short), given a family $\mathcal{F} = \{S_j : j \in J\}$ of subsets of a set $I = \{1, \ldots, n\}$ with associated nonnegative weights w_j and a positive integer p, it is required to find a subset $X \subseteq I$ (called *coverage*) with $|X| = p$ so as to maximize the total weight of the sets in $\mathcal{F}$ having nonempty intersections with X. It is well known that a simple greedy algorithm solves MAXIMUM COVERAGE approximately within a factor of $1 - e^{-1}$ of the optimum (Cornuejols, Fisher and Nemhauser [5]). Feige [6] proved that no polynomial algorithm can have better performance guarantee, unless P=NP.

Our proof consists in constructing an approximation preserving reduction from MAXIMUM COVERAGE to HYP MAX k-CUT WITH GSP. Let a set I, a collection $S_1, \ldots, S_m \subseteq I$, nonnegative weights (w_j), and a positive number p form an instance A of MAXIMUM COVERAGE. Construct an instance B of HYP MAX k-CUT WITH GSP as follows: $I' = I \cup \{u_1, \ldots, u_m\}$ (assuming that $I \cap \{u_1, \ldots, u_m\} = \emptyset$), $(S_1' = S_1 \cup \{u_1\}, \ldots, S_m' = S_m \cup \{u_m\})$, the same weights w_j, and $p_1 = p$, $p_2 = |I'| - p$. Let $(X, I' \setminus X)$ be a maximum weight cut in B with the sizes of parts p_1 and p_2. It is clear that its weight is at least the weight of a maximum coverage in A. Thus it remains to transform $(X, I' \setminus X)$ into a coverage of A with the same weight. If $X \subseteq I$, we are done. Assume that X contains u_j for some j. Then successively, for each such j, replace u_j in X by an arbitrary element in S_j that is not a member of X, or if $S_j \subseteq X$, by an arbitrary element of I that is not a member of X. After this transformation and after possibly including a few more elements from I to get exactly p elements, we arrive at a coverage $Y \subseteq I$ in A whose weight is at least the weight of the cut $(X, I' \setminus X)$ in B, as desired.

References

1. A. A. Ageev and M. I. Sviridenko, Approximation algorithms for Maximum Coverage and Max Cut with given sizes of parts, *Lecture Notes in Computer Science (Proceedings of IPCO'99)* **1610** (1999), 17–30. 32, 33, 34
2. G. Andersson, An approximation algorithm for Max p-Section, *Lecture Notes in Computer Science (Proceedings of STACS'99)* **1563** (1999), 237–247. 33
3. G. Andersson and L. Engebretsen, Better approximation algorithms for Set splitting and Not-All-Equal SAT, *Inform. Process. Letters* **65** (1998), 305–311. 33
4. S. Arora, D. Karger, and M. Karpinski, Polynomial Time Approximation Schemes for Dense Instances of NP-Hard Problems, *J. of Comput. and Syst. Sci.* **58** (1999), 193–210. 33
5. G. Cornuejols, M. L. Fisher, and G. L. Nemhauser, Location of bank accounts to optimize float: an analytic study exact and approximate algorithms, *Management Science* **23** (1977), 789–810. 40
6. U. Feige, A threshold of $\ln n$ for approximating set cover, *J. of ACM* **45** (1998), 634–652. 40
7. U. Feige and M. Langberg, Approximation algorithms for maximization problems arising in graph partitioning, manuscript, 1999. 33
8. A. Frieze and M. Jerrum, Improved approximation algorithms for MAX k-CUT and MAX BISECTION, *Algorithmica* **18** (1997), 67–81. 33
9. M. X. Goemans and D. P. Williamson, Improved Approximation Algorithms for Maximum Cut and Satisfiability Problems Using Semidefinite Programming, *J. of ACM* **42** (1995), 1115–1145. 33
10. V. Kann, S. Khanna, J. Lagergren, and A. Panconesi, On the hardness of approximating Max k-Cut and its dual, *Chic. J. Theor. Comput. Sci.* 1997, Article no. 2 (1997). 33
11. Y. Ye, An 0.699-approximation algorithm for Max-Bisection, manuscript, 1999. 33

Offline List Update is NP-hard

Christoph Ambühl

Institute for Theoretical Computer Science, ETH Zürich, 8092 Zürich, Switzerland.
ambuehl@inf.ethz.ch

Abstract. In the offline list update problem, we maintain an unsorted linear list used as a dictionary. Accessing the item at position i in the list costs i units. In order to reduce access cost, we are allowed to update the list at any time by transposing consecutive items at a cost of one unit. Given a sequence σ of requests one has to serve in turn, we are interested in the minimal cost needed to serve all requests. Little is known about this problem. The best algorithm so far needs exponential time in the number of items in the list. We show that there is no polynomial algorithm unless $P = NP$.

Keywords: On-line algorithms, competitive analysis, list-update, NP.

1 Introduction

The *list update problem* is a classical online problem in the area of self-organizing data structures [16,4,8]. In this paper, we will be concerned with the *offline* version of that problem (*OLUP*). In both versions, requests to items in an unsorted linear list must be served by accessing the requested item. If the item is at position i in the list, the access incurs a cost of i units in the *full* cost model and $i - 1$ units in the *partial* cost model [10,11]. The goal is to keep access cost small by rearranging the items in the list. We have so-called *free exchanges* (the requested item can be moved closer to the list at no charge just after the request) and *paid exchanges* (any two consecutive items can be transposed at cost one).

An *online* algorithm must serve the sequence σ of requests one item at a time, without seeing future requests. An *offline* algorithm knows the entire sequence σ in advance. A feasible (but not necessarily optimal) solution for an instance is called a *schedule*.

To measure the performance of an online algorithm A on some sequence σ, we compare its overall cost $A(\sigma)$ with the cost $OPT(\sigma)$ incurred by an optimal offline algorithm, and we call A c-competitive, if for a suitable b,

$$A(\sigma) \leq c \cdot OPT(\sigma) + b, \tag{1}$$

for all sequences σ. If A is randomized, $A(\sigma)$ denotes the expected cost incurred by serving σ.

M. Paterson (Ed.): ESA 2000, LNCS 1879, pp. 42–51, 2000.

It is known that the best randomized online algorithm satisfies $1.5 \leq c \leq 1.6$ in the full cost model [2,17]. For the partial cost model, a lower bound of $c \geq 1.50084$ can be shown [5].

Because of (1), the analysis of an online algorithm requires some understanding of the optimal offline algorithm *OPT*. Most algorithms analyzed so far have a property called *projectivity* [7]. For a projective algorithm A, a proof that A is c-competitive on lists with only two items generalizes automatically to longer lists. Fortunately on two items, *OPT* is fully understood. Recently it was shown that the upper bound of 1.6 cannot be further improved by projective algorithms [6]. This means that understanding *OPT* has become an even more important issue.

A better understanding of *OPT* could be helpful in two respects for online list update. First and most obviously, it may lead to better lower bounds on $OPT(\sigma)$ to plug into (1); second, it might reveal structural properties that turn out to be useful in devising better online algorithms. Properties of *OPT* have already been studied in the past [14,15,2,3].

The size of an *OLUP* instance on n items and m requests is $\Theta(\log(n) \cdot m)$. But we can assume $m \geq n$. Therefore, an algorithm is still polynomial if its runtime is polynomial in n.

The currently best algorithm for *OLUP* runs in $O(2^n n! m)$ [14]. It is based on a straightforward dynamic programming algorithm for metrical task systems [13] which works as follows.

Let $d(i, L)$ be the minimal cost needed to serve the first i requests of the request sequence σ and end up in the list state L. We define $d(0, L) = 0$ if L is the initial list state and $d(0, L) = \infty$ for all other L. We can compute $d(i, L)$ using

$$d(i, L) = \min_{L'}\big(d(i - 1, L') + trans(L', L) + acc(i, L)\big). \qquad (2)$$

Here, $trans(L', L)$ denotes the minimal cost to move from state L' to L and $acc(i, L)$ denotes the cost for accessing σ_i in L.

Note that there is always an optimal algorithm which does no free exchanges, therefore $trans(L, L')$ is easy to compute. The time needed to compute all $d(i, L)$ is $O((n!)^2 m)$.

This runtime can be reduced to $O(2^n n! m)$ by using the fact that there is an optimal algorithm which uses only so-called *subset transfers* [14]. In a subset transfer, one moves a subset of the items preceding the requested item x just behind x without changing their relative orders. Only $O(2^n)$ among the $n!$ possible transformations are subset transfers.

In this paper, we show that it is NP-hard to decide whether a sequence σ can be served with cost less or equal k. Therefore, there cannot be a polynomial time algorithm in n and m unless $P = NP$. This solves a longstanding open question ([8], *page 22*). One interpretation of this result is that the structure of *OPT* is not easy to understand; consequently, it might be very difficult to develop (and analyze) non-projective list update algorithms that improve the current upper bound of 1.6.

As an important part of the proof, we introduce a generalization of *OLUP* called *weighted list update problem* (*WLUP*). Here the items have a weight that influences access and transposition cost. A version of *WLUP* was considered already in [9], but our definitions and applications are different.

Throughout this paper, we assume the partial cost model. This is simpler to analyze than the original full cost model. It is easy to obtain the value of $OPT(\sigma)$ in the full cost model by adding $|\sigma|$ to the optimal cost in the partial cost model. Therefore, the proof certainly holds for the full cost model as well.

2 The Weighted List Update Problem

In this section, we generalize offline list update to items with *weights*. These weights have to be positive integers. We denote weighted items by capital letters.

An instance (σ, L, W) of $(WLUP)$ consists of a request sequence σ and an initial list L over a set of weighted items X_i with weights w_i. Let W be the vector whose ith entry w_i is the weight of item X_i.

The cost incurred by operating on weighted items are the following. In order to transpose two items X_i and X_j with weights w_i and w_j respectively, we pay $w_i \cdot w_j$ units. The access cost for an item X_i with weight w_i are the following. Let S be the set of items in front of X_i, then accessing X_i costs

$$\binom{w_i}{2} + \sum_{l:X_l \in S} w_i \cdot w_l \tag{3}$$

This setting is motivated by the following interpretation of *WLUP* in terms of *OLUP*. Think of a weighted item X_i as a vector of items $[x_{i,1} \ldots x_{i,w_i}]$. Accessing X_i means to access every item in X_i's vector in turn. In order to access $x_{i,j}$, one has to pass all $x_{l,k}$ with $X_l \in S$ plus all $x_{i,k}$ with $k < j$. Summing up over all items in X_i, we get (3). If two weighted items X_i and X_j are transposed, every item in X_i's vector has to pass every item in X_j's vector. In total, we need $w_i \cdot w_j$ transpositions.

The next lemma shows that the optimal schedules for the *WLUP* instance and its interpretation as an *OLUP* instance have the same cost. The lemma also serves as a proof for the reduction

$$WLUP \leq_p OLUP. \tag{4}$$

Note that the reduction is polynomial only for instances where the weights of the items are polynomial in the number of items or requests.

Lemma 1. *Let (σ, L, W) be an instance of WLUP and let (σ', L') be the transformation of (σ, L, W) to OLUP by replacing the items of (σ, L) by vectors of items. Then the costs of their optimal schedules are the same.*

Proof. Let us assume that the items of (σ, L, W) are denoted by X_i, and let w_i be the weight of X_i. In order to obtain (σ', L'), we replace the X_i by $x_{i,1} \ldots x_{i,w_i}$ in L and σ. As an example, $\sigma = X_2 X_2 X_1$, $L = [X_1 X_2 X_3]$, and $W = [3, 2, 2]$ transforms to

$$\sigma' = x_{2,1} x_{2,2}\, x_{2,1} x_{2,2}\, x_{1,1} x_{1,2} x_{1,3} \text{ and } L' = [x_{1,1} x_{1,2} x_{1,3}\, x_{2,1} x_{2,2}\, x_{3,1} x_{3,2}].$$

Clearly, the optimal cost of (σ', L') are not larger than those of (σ, L, W), because we can serve (σ', L') according to the interpretation given above. It remains to show the the optimal cost for (σ, L, W) are not larger than those of (σ', L').

From (σ', L'), we can produce an instance of WLUP called $(\sigma'', L'', 1)$ which has the same optimal value as (σ', L'): In a first step, we replace the $x_{i,j}$ by weighted items $Y_{i,j}$ with weight $w_{i,j}$. Note that for the moment, we do not set values to the $w_{i,j}$. Let us refer to this instance by (σ'', L'', W''). In order to obtain $(\sigma'', L'', 1)$, we further assign all weights to $w_{i,j} = 1$. An optimal schedule for (σ', L') is clearly also optimal for $(\sigma'', L'', 1)$. (The cost for a pair of items with weight one in WLUP are the same as for a pair of items in OLUP.) Therefore, the cost of an optimal schedule is the same for both $(\sigma'', L'', 1)$ and (σ', L').

We can apply the optimal schedule for $(\sigma'', L'', 1)$ also to (σ'', L'', W'') with any setting of W''. Note that the schedule remains the same by changing the weights. Only the cost for serving the schedule change. The cost for serving this schedule can be expressed as a function in the $w_{i,j}$.

In order to do so, we define constants $f_{(i,j),(k,l)}$ for each pair $\{Y_{i,j}, Y_{k,l}\}$. In $f_{(i,j),(k,l)}$, we sum up the number of times $Y_{i,j}$ and $Y_{k,l}$ are transposed in the schedule plus the number of times $Y_{i,j}$ is in front of $Y_{k,l}$ when $Y_{k,l}$ is requested and vice versa. So these constants are independent of $w_{i,j}$ and $w_{k,l}$. The total cost of the optimal schedule for $(\sigma'', L'', 1)$ applied to (σ'', L'', W'') then is

$$C = \sum_{i \le k, j, l} f_{(i,j),(k,l)} \cdot w_{i,j} w_{k,l} + \sum_{i,j} \binom{w_{i,j}}{2} \cdot |\sigma''_{Y_{i,j}}|. \tag{5}$$

Here $|\sigma''_{Y_{i,j}}|$ denotes the number or requests to $Y_{i,j}$ in σ''. The second sum comes from the binomial coefficients in (3). The second term of (3) is part of the $f_{(i,j),(k,l)}$.

For any choice of the $w_{i,j}$, equation (5) upper bounds the cost of the optimal schedule for (σ'', L'', W''). This holds because (5) denotes the cost of a legal (but not necessary optimal) schedule for (σ'', L'', W'').

In order to prove $(\sigma'', L'', 1) \ge (\sigma, L, W)$, we proceed as follows. Starting with the initial values $w_{i,j} = 1$ for all i and j, we will change the values of the $w_{i,j}$ step by step in such a way that the value of (5) does not increase and we will end up with an instance which is equivalent to (σ, L, W).

This is how we change the weights. As long as there is an i with $w_{i,j} > 0$ and $w_{i,k} > 0$ for some $i \ne k$, we set one of them to the sum of the two. The other one is set to zero. Note that an item with weight zero incurs neither access nor transposition cost.

Let us write (5) such that we can more easily detect how its value changes if we change $w_{i,j}$ and $w_{i,k}$.

$$C = C_0 + C_1 \cdot w_{i,j} + C_2 \cdot w_{i,k} + f_{(i,j),(i,k)} \cdot w_{i,j} w_{i,k}$$
$$+ \binom{w_{i,j}}{2} |\sigma''_{Y_{i,j}}| + \binom{w_{i,k}}{2} |\sigma''_{Y_{i,k}}| \tag{6}$$

Here, C_0 denotes all costs independent of both $w_{i,j}$ and $w_{i,k}$. By $C_1 \cdot w_{i,j}$ and $C_2 \cdot w_{i,k}$, we denote cost depending linearly only on one of the two.

Assume w.l.o.g. $C_1 \leq C_2$. If we set the new value of $w_{i,j}$ to $w_{i,j} + w_{i,k}$ and set $w_{i,k}$ to zero, the value of (5) does not increase. This holds because C_0 does not change and

$$C_1(w_{i,j} + w_{i,k}) + C_2 \cdot 0 \leq C_1 \cdot w_{i,j} + C_2 \cdot w_{i,k}, \tag{7}$$

and furthermore

$$f_{(i,j),(i,k)} \cdot w_{i,j} w_{i,k} + \binom{w_{i,j}}{2} |\sigma''_{Y_{i,j}}| + \binom{w_{i,k}}{2} |\sigma''_{Y_{i,k}}| \geq$$
$$f_{(i,j),(i,k)} \cdot (w_{i,j} + w_{i,k}) \cdot 0 + \binom{w_{i,j} + w_{i,k}}{2} |\sigma''_{Y_{i,j}}| + \binom{0}{2} |\sigma''_{Y_{i,k}}|. \tag{8}$$

Using $|\sigma''_{Y_{i,j}}| = |\sigma''_{Y_{i,k}}|$ and $f_{(i,j),(i,k)} \geq |\sigma''_{Y_{i,j}}|$, inequality (8) is straightforward.

This reweighting process must terminate because in each step, the number of $w_{i,j}$ set to zero increases by one. What we end up with is an instance where for each i we have exactly one $w_{i,j}$ with $w_{i,j} = w_i$, and all the other $w_{i,j}$ are zero. This instance is equivalent to (σ, L, W) we started with. Just rename the $Y_{i,j}$ with $w_{i,j} = w_i$ to X_i and forget about the $Y_{i,j}$ which have weight zero. Because we did not increase the value of (5) by changing the weights, we have a schedule for (σ, L, W) which has the same cost as the optimal schedule for (σ', L'). □

3   A Lower Bound

A total ordering of n items in a list L is defined by the relative ordering of all $\binom{n}{2}$ pairs of items in L. For example, the list state $L' = [X_2 X_3 X_1]$ corresponds uniquely to the set

$$\{X_2 < X_1, X_3 < X_1, X_2 < X_3\}.$$

Instead of one list containing n items, we can represent the list state by two-item lists $L_{X_i X_j}$, for each pair of items in L. The state formed by these $\binom{n}{2}$ pair lists is legal if and only if we can combine the relative orders induced by the pair lists to a total ordering. Thus, the state

$$L_{X_1 X_2} = [X_1 X_2], \ L_{X_2 X_3} = [X_2 X_3], \ L_{X_1 X_3} = [X_3 X_1]$$

is not legal. But if we changed the ordering of the items in $L_{X_2 X_3}$, it would correspond to $[X_3 X_1 X_2]$.

We will serve our requests on the pair lists. In order to satisfy a request to an item X_i, we must access this item in all pair list it occurs. There are $n - 1$ of them. Hence, on a pair list $L_{X_i X_j}$, we serve only those requests in σ that go to X_i or X_j. We denote this subsequence by $\sigma_{X_i X_j}$ and call it the *projection* of σ to X_i and X_j. At any point in time, we are allowed to transpose the items in a pair list, paying $w_i \cdot w_j$ units. But we have to make sure that this leads to a legal state. If we access X_j in the list $[X_iX_j]$, we pay $w_i \cdot w_j$ units, but in $[X_jX_i]$ we pay noting. We call these access and transposition cost *pair costs*, because we can assign them uniquely to a pair if items, namely $\{X_i, X_j\}$. In addition to the access cost we already mentioned, we pay b units for each request, where $b = \binom{w}{2}$ if the requested item has weight w. We refer to these cost as *item cost*. Note that the item cost depend only on the request sequence, but not on the schedule. It can be easily seen from the definitions of the two models that they are equivalent.

From the pair list based model, we can derive a lower bound for $WOPT$, to which we will refer to as $\overline{WOPT}$ [2]. In order to do so, we forget about the condition that the state defined by the pair lists must correspond to a total ordering of all n items. So in $\overline{WOPT}$, any state of the pair lists is legal. This means that for each pair of items $\{X_i, X_j\}$, we can choose their relative order independently of the other pairs. In order to get the optimal schedule for the pair $\{X_i, X_j\}$, we just serve σ_{X_i, X_j} on L_{X_I, X_j} optimally. Thus, to derive $\overline{WOPT}$, we just have to solve $\binom{n}{2}$ $WLUP$ instances on two items each. Fortunately, the algorithm described in the introduction runs in linear time if the number of items is fixed. Thus, $\overline{WOPT}$ is computable in polynomial time.

Note as well that the optimal schedule for an instance on two items does not depend on the weights, at least if both weights are strictly positive. This holds because of two reasons. First, the item cost are constant for every schedule of a given request sequence. Second, the pair cost are of the form $k \cdot (w_i \cdot w_j)$, where k does not depend on the weights.

Thus, the hardness of the problem comes from the total ordering that must hold at any time in the schedule of $WOPT$. Most of the time, the actual state $\overline{WOPT}$ is in will not be a legal one for $WOPT$. Therefore, some pairs will have to be served non-optimally in $WOPT$.

4   The Reduction

Lemma 1 allows us to show a reduction from an NP-complete problem to $WLUP$ in order to prove NP-hardness of $OLUP$.

The *minimum feedback arc set problem* (*MINFAS*) [12] will serve well for that purpose. Its decision variant $MINFAS(G, k)$ is defined as follows. Given a directed graph $G = (V, E)$ and $k \in \mathbf{N}$, we want to decide whether there is a subset $E' \subseteq E$ with $|E'| \leq k$ such that the Graph $G' = (V, E - E')$ is acyclic.

There is a second interpretation which is more natural for our purpose. We interpret an arc pointing from v_i to v_j as a constraint "v_i should be ahead of v_j". What we want to decide is whether there exists a total ordering of the vertices such that less than k constraints are unsatisfied.

We show a reduction from $MINFAS(G, k)$ to the decision version of $WLUP$, denoted by $WLUP(\sigma, L, W, k')$. Here we want to decide whether there is a schedule which serves σ from the initial state L at maximal cost k'. More precisely, we give a function f that takes G and k as arguments and returns (σ, L, W) such that

$$MINFAS(G, k) \Leftrightarrow WLUP(\sigma, L, W, \overline{WOPT}(\sigma, L, W) + k). \tag{9}$$

From now on, we abbreviate the cost $\overline{WOPT}(\sigma, L, W)$ by $\overline{wopt}$. For the following section, it is important to understand how $\overline{WOPT}$ behaves in different situations. As $\overline{WOPT}$ treats all pairs of items independently, we have to investigate how sequences on two items are served optimally. Remember that in the two items case, the behavior does not depend on the weights, therefore it behaves like $\overline{OPT}$.

We consider a list containing only the items a and b. In order to describe $\overline{WOPT}$, we must find out in which cases one must, can or must not swap the two items. The following table gives the answer for a few cases, depending on how the remaining part of the request sequence looks like. We will encounter these cases later in this section and then refer to them by their number in $\langle \cdot \rangle$. The notation is analogous to the one for regular expressions. Thus, $(ba)^*$ denotes the empty sequence or any number of repetitions of ba. The sequence λ can be any sequence on a and b. If there is no λ at the end of the sequence, we assume that this is the end of the sequence.

must not swap:	$\langle 0 \rangle$	[ab]	$a\lambda$
	$\langle 1 \rangle$	[ab]	$baa\lambda$
	$\langle 2 \rangle$	[ab]	$(ba)^*$
can swap:	$\langle 3 \rangle$	[ab]	$babb\lambda$
	$\langle 4 \rangle$	[ab]	$(ba)^*b$
must swap:	$\langle 5 \rangle$	[ab]	$bbb\lambda$

We now describe the function f which transforms a *MINFAS* instance into a *WLUP* instance. For every vertex v_i of $G = (V, E)$, we have a weighted item V_i with weight $k + 1$. We call them *vertex items* and define $n := |V|$. Additionally, we have two items c and d both with weight one.

These are all the items we need. Let us check briefly that the weights are not too large in order to make (4) work. Clearly, the hard *MINFAS* instances obey $k < |E| \le |V|^2$. Hence in those cases, the weights of the items are polynomial in the number of items. Thus, the reduction from *WLUP* to *OLUP* is polynomial.

We set $L = [V_1 V_2 V_3 \ldots V_n c\, d]$. The sequence σ is basically of the form

$$(V_1 V_2 V_3 \ldots V_n)^*,$$

with additional requests to c and d. It consists of two parts σ' and σ''. The first part is

$$\sigma' = V_1 V_2 V_3 \ldots V_n.$$

The second part consists of so called *arc gadgets*. An arc gadget γ for $(v_i, v_j) \in E$ looks as follows. Basically, it consists of 6 repetitions of σ', with additional requests to c and d. If $i < j$ the gadget looks as follows.

$$\gamma = V_1...V_{i-1}c\,V_i...V_j dV_{j+1}...V_i cc V_{i+1}...V_{j-1}dddcccV_j...V_n\sigma'\sigma'\sigma'\sigma' \qquad (10)$$

For $j < i$, we have

$$\gamma = V_1...V_j...V_{i-1}c\,V_i...V_j dV_{j+1}...V_i cc V_{i+1}...V_{j-1}dddcccV_j...V_i...V_n\sigma'\sigma'\sigma' \qquad (11)$$

The idea of γ is to make *WOPT* pay one unit extra if and only if V_j is behind V_i at the first request to d in the gadget.

Lets partition the set of arcs in G into two subsets. E^+ contains the arcs (v_i, v_j) with $i > j$, whereas E^- contains those with $i < j$. In σ'', we have one arc gadget for each arc in G, with the additional restriction that all the arc gadgets of the arcs in E^+ precede those in E^-.

In order to prove (9), we first look at some properties of this instance. We have seen in section 3 that the cost spent by $\overline{WOPT}$ on a pair of items lower bounds the cost *WOPT* needs for this pair. Therefore, in a schedule that costs no more than $\overline{wopt} + k$ units every pair of items involving a vertex item must be served according to $\overline{WOPT}$. This holds because the pair cost of such a pair are a multiple of $(k+1)$. Therefore, any non-optimal step involving a vertex item costs at least $k + 1$ additional units.

Let us first look at pairs consisting of two vertex items V_i and V_j, $i < j$. In the initial state, V_i is in front of V_j. Therefore, $\overline{WOPT}$ has to serve the following request sequence from initial state $[V_i V_j]$:

$$V_i V_j V_i V_j \ldots V_i V_j \tag{12}$$

One way of optimally serving this instance is to do nothing at all. But there are even more optimal schedules for this sequence: In order to stay optimal, we are allowed to swap the two items exactly once (check $\langle 0 \rangle$, $\langle 2 \rangle$, and $\langle 4 \rangle$). Because one should never move V_j in front of V_i when the next request goes to V_i $\langle 0 \rangle$, this swap has to take place before a request to V_j.

It is easy to see that in an optimal schedule which pays less than $\overline{wopt} + k$ units, the list state before and after every gadget must have the sublist $[cd]$ at the end of the list state: Because of the three repetitions of σ' at the end of the edge gadgets and because of $\langle 5 \rangle$, the items c and d must surely be at the end of the list. If we now have a closer look at the requests to c and d only, we notice that a gadget ends up with three requests to c and starts with another one to c. Therefore, in order to be optimal, c has to be in front of d.

To see how *WOPT* serves the gadget for (v_i, v_j), we have to look at two cases. If V_i is in front of V_j, we can serve it with the same amount $\overline{WOPT}$ would pay. This is easy to check by showing that there is a schedule which serves each pair of items like $\overline{WOPT}$ would do: The crucial point is that in $\overline{WOPT}$, when the first request to d takes place, d must still be behind c $\langle 1 \rangle$, while c must be behind V_i $\langle 1 \rangle$ and d must be in front of V_j $\langle 5 \rangle$. Only if V_i is in front of V_j, we can fulfill these conditions. Note that c and d can pass V_k, $k \notin \{i, j\}$, but they do not have to $\langle 3 \rangle$. At the next request to c, we move c to the front of the list $\langle 3, 5 \rangle$. Later, at the first request of the request triple to d, we move d to the front as well $\langle 5 \rangle$, but c will have to pass d again later $\langle 5 \rangle$. Because of the additional σ'^3 finishing γ, both c and d must be moved behind all vertex items at the end without changing the relative order of c and d $\langle 5 \rangle$.

If V_i is behind V_j, not all the conditions mentioned in the previous paragraph can be fulfilled at the first request to d. The only way to fix this without paying more than k extra units is to move d in front of c at the first request to d and thus pay one unit more than $\overline{WOPT}$.

Now we are ready to prove (9). The easier part is the $\Rightarrow$ direction. If we can get an acyclic graph G' by removing only k arcs, we sort the vertices of G' topologically. The schedule which costs at most $\overline{wopt} + k$ looks as follows. We use the initial sequence σ' to rearrange the items V_i according to the topological order $\langle 4 \rangle$. For the rest of σ, we do not change the ordering of the vertex items anymore. Thus, we serve all vertex pairs optimally.

Concerning the arc gadgets, we can serve all those corresponding to the arcs in G' like $\overline{WOPT}$ would do it. For each arc we removed from G, we have to pay one unit extra. As there are at most k of them, we pay at most $\overline{wopt} + k$ units to serve σ.

It remains to prove the $\Leftarrow$ direction of (9). There are at most k gadgets whose cost where higher than those of $\overline{WOPT}$. We say a gadget is served *well* if we payed no more than $\overline{WOPT}$ to serve it. We will show that if we remove the arcs corresponding to those gadgets, the resulting graph will be acyclic.

Let the arcs of $C \subseteq E$ form a cycle in G. We have to prove that there is at least one arc gadget belonging to arcs in C which is not served well. For any arc $e = (v_i, v_j)$ and a list state L, we say e is *open* if we have V_i in front of V_j in L and *closed* otherwise. The arcs in $C \subseteq E^+$ are those which are closed in the initial list. In order to serve their gadget well, it has to be open when their gadget is served, but we cannot close them anymore $\langle 2 \rangle$. The arcs in $C \subseteq E^-$ are open in the initial list. If we want to serve them well, we can not close them before their gadget is served because we cannot reopen them $\langle 2 \rangle$.

Let us have a look at the list just after we served all arc gadgets for E^+ in σ. In order to serve all gadgets belonging to C well, all of them must be open at this time. This means for any arc $e = (v_i, v_j)$ in C that the item V_i must be in front of V_j in the current list. Because C forms a cycle, at least one of them must be closed and hence was not (if it belongs to E^+) or will not be (if it belongs to E^-) served well. This concludes the proof.

5   Conclusions

We have shown that the offline list update problem is NP-complete using a reduction from minimum feedback arc set to weighted list update. By changing the gadgets slightly, one can prove the same result also for the case where only free exchanges are allowed in order to update the list.

As an open question, it remains to show whether it can be decided in polynomial time whether

$$OPT(\sigma) = \overline{OPT}(\sigma).$$

Acknowledgements

The author likes to thank Bernd Gärtner, Bernhard von Stengel, and Rolf Möhring for helpful discussions.

References

1. S. Albers (1998), Improved randomized on-line algorithms for the list update problem. *SIAM J. Comput.* 27, no. 3, 682–693 (electronic). Preliminary version in *Proc. 6th Annual ACM-SIAM Symp. on Discrete Algorithms* (1995), 412–419.
2. S. Albers, B. von Stengel, and R. Werchner (1995), A combined BIT and TIMESTAMP algorithm for the list update problem. *Inform. Process. Lett.* 56, 135–139. 43, 47
   S. Albers (1995), Improved randomized on-line algorithms for the list update problem, *Proc. 6th Annual ACM-SIAM Symposium on Discrete Algorithms*, 412–419.
3. S. Albers (1998), A competitive analysis of the list update problem with lookahead. *Theoret. Comput. Sci.* 197, no. 1-2, 95–109. 43
4. S. Albers and J. Westbrook (1998), Self Organizing Data Structures. In A. Fiat, G. J. Woeginger, "Online Algorithms: The State of the Art", *Lecture Notes in Comput. Sci.*, 1442, Springer, Berlin, 13–51. 42
5. C. Ambühl, B. Gärtner, and B. von Stengel (2000), A new lower bound for the list update problem in the partial cost model. To appear in *Theoret. Comput. Sci.* 43
6. C. Ambühl, B. Gärtner, and B. von Stengel (2000), Optimal Projective Bounds for the List Update Problem. To appear in *Proc. 27th ICALP.* 43
7. J. L. Bentley, and C. C. McGeoch (1985), Amortized analyses of self-organizing sequential search heuristics. *Comm. ACM* 28, 404–411. 43
8. A. Borodin and R. El-Yaniv (1998), Online Computation and Competitive Analysis. Cambridge Univ. Press, Cambridge. 42, 43
9. F. d'Amore, A. Marchetti-Spaccamela, and U. Nanni (1993), The weighted list update problem and the lazy adversary. *Theoret. Comput. Sci.* 108, no. 2, 371–384 44
10. S. Irani (1991), Two results on the list update problem. *Inform. Process. Lett.* 38, 301–306. 42
11. S. Irani (1996), Corrected version of the SPLIT 42
12. M. R. Garey, D. S. Johnson (1979), Computers and intractability. W. H. Freeman and Co, San Francisco. 47
13. M. Manasse, L. A. McGeoch, and D. Sleator (1988), Competitive algorithms for online problems. *Proc. 20nd STOC* (1988), 322-333. 43
14. N. Reingold, and J. Westbrook (1996), Off-line algorithms for the list update problem. *Inform. Process. Lett.* 60, no. 2, 75–80. Technical Report YALEU/DCS/TR-805, Yale University. 43
15. N. Reingold, J. Westbrook, and D. D. Sleator (1994), Randomized competitive algorithms for the list update problem. *Algorithmica* 11, 15–32. 43
16. D. D. Sleator, and R. E. Tarjan (1985), Amortized efficiency of list update and paging rules. *Comm. ACM* 28, 202–208. 42
17. B. Teia (1993), A lower bound for randomized list update algorithms, *Inform. Process. Lett.* 47, 5–9. 43

Computing Largest Common Point Sets under Approximate Congruence

Christoph Ambühl[1], Samarjit Chakraborty[2], and Bernd Gärtner[1]

[1] Institut für Theoretische Informatik
[2] Institut für Technische Informatik und Kommunikationsnetze
ETH Zürich, ETH-Zentrum, CH-8092 Zürich, Switzerland
ambuehl@inf.ethz.ch, samarjit@tik.ee.ethz.ch, gaertner@inf.ethz.ch

Abstract. The problem of computing a *largest common point set* (LCP) between two point sets under ε-congruence with the bottleneck matching metric has recently been a subject of extensive study. Although polynomial time solutions are known for the planar case and for restricted sets of transformations and metrics (like translations and the Hausdorff-metric under L_∞-norm), no complexity results are formally known for the general problem. In this paper we give polynomial time algorithms for this problem under different classes of transformations and metrics for any fixed dimension, and establish NP-hardness for unbounded dimensions. Any solution to this (or related) problem, especially in higher dimensions, is generally believed to involve implementation difficulties because they rely on the computation of intersections between algebraic surfaces. We show that (contrary to intuitive expectations) this problem can be solved under a rational arithmetic model in a straightforward manner if the set of transformations is *extended* to general affine transformations under the L_∞-norm (difficulty of this problem is generally expected to be in the order: translations < rotation < isometry < more general). To the best of our knowledge this is also the first paper which deals with the LCP-problem under such a general class of transformations.

1 Introduction

Let $\varepsilon \geq 0$ be a real number, $\mathcal{G}$ be a transformation group (such as translations, rotations, isometry, or linear transformations), and A and B be two d-dimensional point sets. A *largest common point set* (LCP) [2,3] between A and B under ε-congruence is a subset A' of A, having the largest possible cardinality, for which there exists a transformation $g \in \mathcal{G}$ such that the *distance* between the sets $g(A')$ and B' is less than ε, where B' is some subset of B, and the distance is measured using some appropriate metric. Related geometric problems for determining the similarity between two point sets have been extensively studied. Two commonly used metrics for quantifying the notion of similarity have been the *Hausdorff distance* [11,12,18] which is defined as the maximum distance between a point in one set and its nearest neighbor in the other set, and the *bottleneck matching metric* [14] seeks a perfect bipartite matching between two equal cardinality

M. Paterson (Ed.): ESA 2000, LNCS 1879, pp. 52–64, 2000.

point sets such that the maximum distance between any two matched points is minimized, and it returns this distance.

A systematic study of these problems was initiated by Alt *et al.* [5]. They presented algorithms for several versions of the problem for planar point sets under the bottleneck matching metric. In particular, they proposed an $O(n^8)$ decision algorithm to determine if there exists an isometric transformation using which two equal cardinality planar point sets can be brought to within ε distance of each other under the Euclidean bottleneck matching metric. Although they do not mention it explicitly, it is straightforward to adapt this algorithm to compute the LCP of two planar point sets (not necessarily of equal cardinality), without incurring any increase in running time.

Problems involving the *exact matching metric*, where two points are said to match only when the underlying geometric transformation takes one of them exactly on to the other have been studied in [1,3,13,21]. Given two point sets, the problem of computing the minimum Hausdorff distance between them has been studied with different parameters such as only translation and general Euclidean motion, planar and d-dimensional point sets, and the underlying metrics being L_∞, L_1 and L_2 [11,12,18]. In an effort to improve the running time, various approximation algorithms for either the Hausdorff or the bottleneck metric for point sets in two, three, and in general d-dimensions have been presented in [9,10,15,16,17,19,20,22].

Pattern matching using bottleneck metric It should be noted that most of the known exact algorithms, especially those involving three and higher dimensional point sets, are restricted to either the exact or the Hausdorff metric. While the exact metric is ill-posed for many practical applications, many problems also demand a one-to-one matching between the two point sets, thereby rendering the Hausdorff metric unsuitable in such situations [14]. This motivates the study of the problem using the bottleneck matching metric. However, in contrast to the Hausdorff metric where the distance between two point sets is in essence determined by the distance between two points each of which belongs to one of the sets, the existence of a bottleneck matching is a global property of the point sets and therefore complicates the problem. As a result, it is not apparent how the algorithms concerned with the Hausdorff metric can be adapted for computing the bottleneck matching. Neither do the algorithms of [5] extend from the planar case to work in three or higher dimensions in any simple way. Very recently, a new paradigm for point set pattern matching based on algebraic convolutions was proposed in [9] and [19]. This reduced the complexity of the problem under Hausdorff metric to nearly quadratic time. However, as noted in [20], "the one-to-one restriction imposed by bottleneck matching distance seems not to fit well within the rigid framework of algebraic convolutions".

Our results In this paper we present polynomial time exact (in contrast to approximation) algorithms for computing the LCP between two d-dimensional point sets under ε-congruence with the bottleneck matching metric under the L_2- and the L_∞-norms and various classes of transformations. All of our algorithms are based on the general framework of traversing an arrangement of surfaces or hyperplanes in some high dimensional space, based on the dimensionality of the point sets. This can be considered as a generalization of the concepts in the original algorithms for bottleneck matching presented by Alt *et al.* in [5] and in those proposed more recently for the Hausdorff metric by Chew *et al.* in [11] and [12].

We prove that for unbounded dimensions the problem is NP-hard. All previous hardness results pertain to the exact matching metric and show that *subset matching* [3] is NP-hard for unbounded dimensions [1] and computing the LCP of an unbounded number of point sets even in one dimension is hard [2]. Polynomial time algorithms for d-dimensional point sets (fixed d) are known either for the exact metric [1] or for Hausdorff metric under the L_∞-norm and restricted set of transformations [11].

Many geometric pattern matching algorithms tend to suffer from implementation difficulties because of their reliance on algebraic techniques. As noted by Alt and Guibas [4] these algorithms "are probably difficult to implement and numerically unstable due to the necessary computation of intersection points of algebraic surfaces". We show that (surprisingly) if we extend our transformation group to include general affine transformations (isometries are a special case of this) then under the L_∞-norm this problem can be solved using a realistic rational arithmetic model. To the best of our knowledge this is the first time that the LCP-problem is being considered under such general transformations. All of the previous algorithms have considered either translations, rotations, isometries, or at most scaling [21]. It was claimed that the algorithms in [21] generalize to broader classes of transformations, but these were not dealt with explicitly.

Before proceeding further we first give a formal definition of our problem. A point set S is ε-congruent under the bottleneck matching metric and a transformation group $\mathcal{G} : \mathbb{R}^d \to \mathbb{R}^d$, to a point set S' if there exists a transformation $g \in \mathcal{G}$ and a bijective mapping $l : S \to S'$ such that for each point $s \in S$, $\delta(g(s), l(s)) < \varepsilon$, where $\delta(\cdot, \cdot)$ denotes some metric such as L_2 or L_∞ (we will only treat these two in the sequel). Given two point sets A and B, and a real number ε, the LCP between A and B is the maximum cardinality subset $A' \subseteq A$ which is ε-congruent to some subset of B. In what follows, ε-congruence is always assumed to be under the bottleneck matching metric.

We outline our basic method in Section 2. In Section 3 we present the two-dimensional realization of the general scheme, followed by the d-dimensional case in Section 4. In Section 5 we establish the NP-hardness result.

2   Computing the LCP – the General Scheme

Let $A = \{a_1, \ldots, a_n\}$ and $B = \{b_1, \ldots, b_m\}$ be two point sets in d-dimensional real space $\mathbb{R}^d$, and fix any metric δ on $\mathbb{R}^d$. For any transformation $L : \mathbb{R}^d \to \mathbb{R}^d$ and given indices $i \in [n], j \in [m]$, we define $\mathcal{T}_{ij}(L)$ as the set of translations that map the point $L(a_i)$ into the ε-neighborhood of b_j under the metric δ. Identifying a translation with its defining vector v, we can write this set as

$$\mathcal{T}_{ij}(L) := \{v \in \mathbb{R}^d \mid \delta(L(a_i) + v, b_j) < \varepsilon\}.$$

If both, the transformation L and the translation vector v are fixed, then the *distance* between the point sets $L(A) + v$ and B can be computed by constructing a bipartite graph $G = (A \cup B, E)$ and computing the maximum matching in it. Here, for $a_i \in A$ and $b_j \in B$, the edge $(a_i, b_j) \in E$ if $\delta(L(a_i) + v, b_j) < \varepsilon$. Computing the LCP between A and B under ε-congruence essentially amounts to finding the optimal transformation L and the translation vector v under which the graph G has the largest maximum matching. For isometric transformations, we would restrict the transformation group from which L is chosen to consist of only pure rotation. This restricted definition of an isometry does not result in any loss of generality because isometry including mirror image just increases the computation time of any of our algorithm by only a constant factor. An algorithm now has to be run once with A and B, and then with A and a mirror image of B on a plane that can be chosen arbitrarily.

LCP under translation Here, the transformation L is fixed, and we are looking for the LCP between the point sets $L(A) = \{L(a_i) \mid i \in [n]\}$ and B under some translation v. The 'overlay' of all the possible sets of translations $\mathcal{T}_{ij}(L)$ for all indices $i \in [n]$ and $j \in [m]$ forms an *arrangement* $\mathcal{A}(L)$, inducing a decomposition of the space $\mathbb{R}^d$ into a number of *faces*. A face of $\mathcal{A}(L)$ is a maximal set of vectors with the property that any two v, v' in the set have the same relation to all $\mathcal{T}_{ij}(L)$, meaning that v lies inside (on the boundary of, outside) $\mathcal{T}_{ij}(L)$ if and only if the same is true for v'. Cells are faces not contained in any boundary of some $\mathcal{T}_{ij}(L)$.

If the metric δ is induced by the L_∞-norm or the L_2-norm, the sets $\mathcal{T}_{ij}(L)$ are balls of radius ε which are cubes in the former case and Euclidean-balls in the later. In the L_∞-case, the cells of the arrangement $\mathcal{A}(L)$ are then simply axis-aligned boxes, while they are bounded by spherical surface patches in the L_2-case. Also, the following property clearly holds.

Property 1. Let $v_{opt} \in \mathbb{R}^d$ be some translation that enables in computing the LCP between $L(A)$ and B. Then v_{opt} lies in some cell of $\mathcal{A}(L)$, and all vectors v' in that cell are optimal translations as well, meaning that they could also be used in computing the LCP.

This property results in the formulation of a discrete version of the problem since the process of finding v_{opt} now reduces to examining just one vector in every cell of $\mathcal{A}(L)$ and computing the maximum bipartite matching in the resulting graph. Any vector in the cell for which the graph has the largest matching serves as an optimal translation.

LCP under general transformations Now we turn to the more general problem of computing the LCP between A and B under any transformation L from some group (such as rotation, scaling, or linear transformations), followed by a translation v. It is clear that as L varies, the LCP between $L(A)$ and B under translation remains invariant as long as the arrangement $\mathcal{A}(L)$ does not undergo any combinatorial change, meaning that a cell does not disappear and no new cells appear. The different possible combinatorial structures arising from the arrangement $\mathcal{A}(L)$ as L varies, partitions the space $\mathcal{L}$ of all transformations L under consideration into a number of cells. Each cell in this case is defined as the maximal set of transformations generating combinatorially equivalent arrangements $\mathcal{A}(L)$. Therefore the LCP-problem is reduced to a cell enumeration problem in an arrangement in the space $\mathcal{L}$. We are required to find a transformation L in every cell of this arrangement, compute $\mathcal{A}(L)$ and compute an optimal translation for that given L by solving the LCP-problem under translation.

3   LCP in 2-d under Isometry with the L_2-Norm

Let us assume without any loss of generality that the point set A is being rotated about the origin followed by a translation of the rotated set. Since the point sets involved are planar, the angle of rotation can be parametrized by a single parameter $\theta \in [0, 2\pi)$, the transformation L as described in Section 2 is therefore rotations and the space $\mathcal{L}$ is $[0, 2\pi)$. Since the underlying norm is L_2, for each pair of points $a_i \in A$ and $b_j \in B$, and a fixed angle θ, the set $\mathcal{T}_{ij}(\theta)$ of translations that take $R_\theta(a_i)$ (which is the point resulting out of rotating a_i about the origin by θ) into the ε-ball around the point b_j is a circular disk of radius ε in the space of possible translations. Overlaying all such $O(mn)$ circular disks for $i \in [n]$ and $j \in [m]$ forms the arrangement $\mathcal{A}(\theta)$. Corresponding to each cell c of this arrangement we can construct a bipartite graph $G_c = (A \cup B, E)$ where for each pair of points $a_i \in A$ and $b_j \in B$, the edge $(a_i, b_j) \in E$ if the cell c lies within the disk $\mathcal{T}_{ij}(\theta)$. We are interested in finding the graph G_c with the largest maximum bipartite matching over all the possible graphs corresponding to the different cells of $\mathcal{A}(\theta)$ arising from all possible values of θ.

As θ varies, the combinatorial structure of the arrangement changes as any of the following two conditions are fulfilled: two circular disks of $\mathcal{A}(\theta)$ touching and three disks meeting at a point. The first condition results in a linear equation in $\sin \theta$ and $\cos \theta$ and therefore can be transformed into a quadratic equation in either $\sin \theta$ or $\cos \theta$. The second condition results in a cubic equation in $\sin \theta$ and $\cos \theta$ and can be transformed into an algebraic equation of degree six in either of them. Since there are $O(mn)$ disks in the arrangement $\mathcal{A}(\theta)$, it gives rise to a collection of $O(m^3 n^3)$ univariate algebraic equations of at most degree six.

The solutions to the equations result in partitioning the space $[0, 2\pi)$ into $O(m^3 n^3)$ intervals and each interval corresponds to a set of combinatorially equivalent arrangements arising out of the angles θ lying within the interval. It may be observed that as we move from one interval to its adjacent one, the combinatorial structure of an arrangement $\mathcal{A}(\theta)$ (for any θ belonging to the former interval) changes with the introduction of a new cell or the disappearance of an existing one.

Our algorithm traverses the $O(m^3n^3)$ intervals of $[0, 2\pi)$ and at each interval if a new cell appears in the arrangement of disks then it constructs the bipartite graph corresponding to this cell and computes the maximum matching in it, which takes $O(mn\sqrt{m+n})$ time. The worst case overall running time of the algorithm is therefore $O(m^4n^4\sqrt{m+n})$.

4   LCP in d Dimensions

This section gives polynomial-time algorithms for computing the LCP of two d-dimensional point sets, under isometries and general affine transformations, and under the L_2- as well as the L_∞-norm. The case of affine transformations with the L_∞-norm is special in the sense that it can be solved with rational arithmetic in a straightforward manner. In the other cases, one needs to resort to algebraic techniques even in two dimensions [5].

4.1   Affine Transformations and the L_∞-Norm

Following the general scheme given in Section 2, we consider the arrangement $\mathcal{A}(L)$ for any fixed linear transformation L. For any set $\mathcal{T}_{ij}(L) = (x_1, x_1 + \varepsilon) \times \ldots \times (x_d, x_d + \varepsilon)$, the values $x_t, x_t + \varepsilon$ are the first and the second *coordinate* of $\mathcal{T}_{ij}(L)$ in *direction* t. As L varies, the combinatorial structure of $\mathcal{A}(L)$ can only change at points where for some i, j, k, ℓ, $\mathcal{T}_{ij}(L)$ and $\mathcal{T}_{k\ell}(L)$ share a coordinate in some direction t. It is easy to see that

$$\mathcal{T}_{ij}(L) = \mathcal{T}_{ij}(id) + a_i - L(a_i) := \{v + a_i - L(a_i) \mid v \in \mathcal{T}_{ij}(id)\}, \qquad (1)$$

where id denotes the identity transformation. Let x_t^{ij} and $x_t^{k\ell}$ be the coordinates of $\mathcal{T}_{ij}(id)$ and $\mathcal{T}_{k\ell}(id)$ in direction t.

According to (1), $\mathcal{T}_{ij}(L)$ and $\mathcal{T}_{k\ell}(L)$ then share a coordinate in direction t if and only if

$$x_t^{ij} - x_t^{k\ell} + (a_i - a_k - L(a_i - a_k))_t \in \{-\varepsilon, 0, \varepsilon\}. \qquad (2)$$

Considering L as a $d \times d$-matrix with variable coefficients $x_{11}, \ldots x_{dd}$, we see that conditions (2) are linear equality constraints involving the variables $x_{t1}, \ldots x_{td}$.

In other words, all combinatorial changes in $\mathcal{A}(L)$ can be characterized by hyperplanes in d^2-dimensional space. Any cell in the arrangement $\mathcal{H}$ of those hyperplanes corresponds to a set of linear transformations L that generate combinatorially equivalent arrangements $\mathcal{A}(L)$.

To find the LCP of A and B under affine transformations, we therefore proceed as follows.

1. Traverse all cells of $\mathcal{H}$. Because $\mathcal{H}$ is defined by $O(m^2n^2)$ hyperplanes in dimension d^2, the number of cells is $(mn)^{2d^2}$ in the worst case, and all cells can be traversed in time proportional to their number, using *reverse search* with $O(m^2n^2)$ space [6,23].

2. In each cell of $\mathcal{H}$, choose a representative transformation L (the reverse search algorithms [6,23] used to traverse $\mathcal{H}$ actually generate such 'points' L, so there is no extra effort involved here). Traverse all cells in the arrangement $\mathcal{A}(L)$, which is an arrangement of unit cubes. The reverse search paradigm can also be adapted to this case within the space bounds of step 1 and with time $(mn)^{O(d)}$. For this, one can consider $\mathcal{A}(L)$ as a subset of cells of the arrangement induced by the facet-defining hyperplanes of all the cubes contributing to $\mathcal{A}(L)$.
3. In each cell of $\mathcal{A}(L)$, perform the graph matching. This can be done in $O((mn)^{O(1)})$ time.
4. Among all the matchings that have been computed, return the one with maximum cardinality.

The overall runtime of the algorithm is $(mn)^{2d^2+O(d)}$ which is polynomial for any fixed d. The space complexity is $O(m^2n^2)$, thus independent from d.

Note that step 3 can be improved by only considering the cells of $\mathcal{A}(L)$ which have been created in the combinatorial change encountered by going from $\mathcal{A}(L')$ to $\mathcal{A}(L)$, where L' is a representative transformation in some neighboring cell of $\mathcal{H}$ that has previously been processed. Typically, there are only few 'new' cells and they are easier to compute than the whole arrangement $\mathcal{A}(L)$. This approach is possible, because the reverse search traverses $\mathcal{H}$ along a tree of cells whose edges correspond to neighboring cells. However, as the complexity of this step is dominated by step 1 of the algorithm, we will not elaborate on this.

4.2   Incorporating Isometries and the L_2-Norm

First let us consider the case of general affine transformations under the L_2-norm. In this case, $\mathcal{A}(L)$ is an arrangement of Euclidean ε-balls, whose combinatorial structure changes if and only if some circumscribed ball of k ball centers ($2 \leq k \leq d+1$) attains radius ε. If $q_1, \ldots, q_k$ are those ball centers, it can be shown that their circumcenter c is given by $c = q_1 + \sum_{\ell=2}^{k} \lambda_\ell \mathbf{q}_\ell$, where $\mathbf{q}_\ell = q_\ell - q_1$, and the λ_ℓ are obtained from solving the system of linear equations

$$M \begin{pmatrix} \lambda_2 \\ \vdots \\ \lambda_k \end{pmatrix} = \begin{pmatrix} \mathbf{q}_2^T \mathbf{q}_2 \\ \vdots \\ \mathbf{q}_k^T \mathbf{q}_k \end{pmatrix}, \quad M := \begin{pmatrix} 2\mathbf{q}_2^T \mathbf{q}_2 \cdots 2\mathbf{q}_2^T \mathbf{q}_k \\ \vdots \qquad \vdots \\ 2\mathbf{q}_2^T \mathbf{q}_k \cdots 2\mathbf{q}_k^T \mathbf{q}_k \end{pmatrix}. \tag{3}$$

The ball centers $\mathbf{q}_\ell$ depend linearly on the entries of L (cf. equation (1)); Cramer's rule then implies that the entries of $\det(M)M^{-1}$ are polynomials in the entries of L ($\det(M)$ itself is a polynomial too, of course). This again means that the condition

$$\| \sum_{\ell=2}^{k} \lambda_\ell \mathbf{q}_\ell \|^2 = \varepsilon^2 \tag{4}$$

for a combinatorial change can be written in the form $f(x) = 0$, where f is a constant-degree polynomial (the constant being in $O(d)$) in the entries

$x_{11}, \ldots, x_{dd}$ of the matrix L. Our strategy will be as before: we consider the set of all the $(mn)^{O(d)}$ polynomials f coming from the conditions on any k ($2 \leq k \leq d+1$) out of the mn ε-balls for the pairs of points in $A \times B$; these polynomials define an arrangement $\mathcal{H}$, whose cells correspond to combinatorial equivalence classes of arrangements $\mathcal{A}(L)$. By traversing the cells of $\mathcal{H}$, we can generate all the combinatorial types that $\mathcal{A}(L)$ may assume; if we perform for each type the graph matching in all the cells, the LCP will be found. We use the following result to 'compute' $\mathcal{H}$.

Theorem 1 (Basu, Pollack, Roy [7]). *Let $\mathcal{P} = \{f_1, \ldots, f_r\}$ be a set of p-variate polynomials with rational coefficients and maximum algebraic degree s ($p < r$). Two points $q, q' \in \mathbb{R}^p$ are equivalent if $\mathrm{sign}(f_\ell(q)) = \mathrm{sign}(f_\ell(q'))$ for $\ell = 1, \ldots, r$. The vector $(\mathrm{sign}(f_1(q)), \ldots, \mathrm{sign}(f_\ell(q)))$ is a sign condition of $\mathcal{P}$. In $O(r(r/p)^p s^{O(p)})$ arithmetic operations, one can compute all sign conditions determined by $\mathcal{P}$.*

Applied to our setting with $r = (mn)^{O(d)}, p = d^2$, and $s = O(d)$, we can obtain all sign conditions in $(mn)^{O(d^3)}$ time. The 'full-dimensional' sign conditions (the ones containing no zero sign) correspond to the combinatorial equivalence classes of linear transformations L we are interested in. Therefore, this solves our problem, once we can obtain a representative transformation L from every such class.

Although full-dimensional sign conditions are always attained by suitable transformations L with rational coordinates, such transformations might be difficult to find. However, all we need is the combinatorial structure of $\mathcal{A}(L)$, and this can be derived directly from the sign condition. More precisely, for every subset $\mathcal{B}$ of ε-balls, we can check whether $\mathcal{A}(L)$ contains a point in all the balls of $\mathcal{B}$, by checking whether all $d+1$-element subsets of $\mathcal{B}$ have a common point (Helly's Theorem). The latter information is given by the sign condition, though. Processing sets $\mathcal{B}$ by increasing size in a dynamic programming fashion, we can construct all sets $\mathcal{B}$ that define cells of $\mathcal{A}(L)$ in time proportional to their number (which is $(mn)^{O(d)}$), incurring only some polynomial overhead.

The overall runtime (including all graph matchings) is then bounded by $(mn)^{O(d^3)}$

In order to handle isometries, we exploit the fact that L defines an isometry if and only if $L^{-1} = L^T$ (and $\det(L) = 1$, in case we want to limit ourselves to rotations). These conditions give rise to $O(d^2)$ additional polynomial equations which we add to the ones already obtained from the combinatorial change conditions. As before, Theorem 1 can be used to deal with this case in $(mn)^{O(d^3)}$ time.

4.3 L_2-Norm and Rotations in 3-Space

The goal of this subsection is to give a concrete bound on the exponent entering the runtime in case of a three-dimensional LCP problem. Here it is crucial to use the fact that a rotation in 3-space has only 3 degrees of freedom. More precisely, a rotation in 3-space can be parametrized by three parameters $\phi_1, \phi_2, \phi_3,$

and the resulting transformation matrix L has entries which are polynomials in $\sin\phi_i, \cos\phi_i, i = 1, 2, 3$. The combinatorial change conditions of type (4) mentioned in Section 4.2, resulting from all 2-, 3- and 4-tuples of balls give rise to $O(m^4 n^4)$ polynomial equations of some constant maximum degree in six variables, say r_i and s_i, $i = 1, 2, 3$, where $r_i = \sin\phi_i$ and $s_i = \cos\phi_i$.

Consider a family of r polynomials $\mathcal{P}$ in p variables, each of degree at most s, and an algebraic variety $\mathcal{V}$ of real dimension p' which is defined as the zero set of a polynomial of degree at most s. Using $r^{p'+1} s^{O(p)}$ arithmetic operations it is possible to compute a set of points in each non-empty semi-algebraically connected component of $\mathcal{P}$ over $\mathcal{V}$, along with the signs of all the polynomials of $\mathcal{P}$ at each of these points [8]. The number of such points is $r^{p'} s^{O(p)}$. Applied to our case with the set $\mathcal{P}$ as the $O(m^4 n^4)$ polynomials in the six variables r_i and s_i, $i = 1, 2, 3$, and the variety $\mathcal{V}$ as the zero set of the polynomial $\sum_{i=1,2,3}(r_i^2 + s_i^2 - 1)^2$ we can obtain in $O(m^{16} n^{16})$ time the sign conditions of $O(m^{12} n^{12})$ cells of the arrangement defined by $\mathcal{P}$. Given a fixed sign condition the corresponding arrangement $\mathcal{A}(L)$ consists of $O(m^3 n^3)$ cells, (combinatorial descriptions of) which can be computed by dynamic programming as indicated before in $O(m^4 n^4)$ time (for each cell we get an overhead which is at most linear in the number of balls). The graph matching for each cell takes $O(mn\sqrt{m+n})$ time, therefore resulting in an algorithm with total complexity $O(m^{16} n^{16}\sqrt{m+n})$.

5   NP-Hardness of LCP

In this section we prove that the LCP problem under approximate congruence is (not surprisingly) NP-hard, even when the transformation group is restricted to only translations.[1] However, the proof establishes hardness also if we allow in addition general linear transformations, or isometries. The proof is a reduction from SAT. Suppose we are given a SAT formula ϕ with m clauses over n variables.

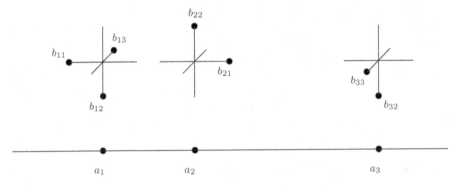

Fig. 1. Point sets for $\phi = (\bar{x}_1 \vee \bar{x}_2 \vee \bar{x}_3) \wedge (x_1 \vee x_2) \wedge (\bar{x}_2 \vee x_3)$

---

[1] David Eppstein (personal communication) independently found another proof of this.

We transform ϕ into a pair of point sets $A = A(\phi)$, $B = B(\phi) \subseteq \mathbb{R}^n$ such that $|A| = m$, $|B| = k$, $k \geq m$ the number of literals in ϕ, and with the property that ϕ is satisfiable if and only if the LCP of A and B has the largest possible size m. This shows that LCP is NP-hard.

For every clause C_i, $i < m$ we create a point $a_i := (i, 0, \ldots, 0)$. For C_m, we define $a_m := (m + 1, \ldots, 0)$. The collection of all the a_is forms the set A.

To get the set B, we proceed as follows. For every variable x_j occurring in clause C_i (in form of x_j or $\bar{x}_j$) we introduce a point b_{ij}, with coordinates

$$b_{ij} = a_i \pm \varepsilon' e_j,$$

where we choose the plus sign if the variable occurs non-negated, and the minus sign otherwise. e_j is the j-th unit vector, and ε' is chosen slightly larger than ε (we will see below what that exactly means). The collection of all the b_{ij}s forms the set B. Fig. 1 gives an example for $m = n = 3$ (for better readability, the set B has been shifted vertically; this does not change the LCP, of course, if translations are among the allowed transformations).

Theorem 2. *The formula ϕ is satisfiable if and only if the ε-LCP of $A(\phi)$ and $B(\phi)$ has size m.*

Proof. Let us first assume ϕ is satisfiable, and let $\tilde{x}_i, i = 1 \ldots, n$ with $\tilde{x}_i \in \{\texttt{true}, \texttt{false}\}$ be a satisfying assignment. Define an n-vector v by

$$v_i = \begin{cases} \varepsilon'/n, & \text{if } \tilde{x}_i = \texttt{true} \\ -\varepsilon'/n, & \text{otherwise.} \end{cases}$$

The claim is that v defines a translation which maps every a_i into the ε-neighborhood of some b_{ij}, thus giving rise to a ε-LCP of size m. To see this, consider a fixed a_i and let x_j be a variable whose value $\tilde{x}_j$ is responsible for satisfying clause C_i. If x_j occurs in C_i unnegated, we have

$$a_i + v - b_{ij} = v - \varepsilon' e_j = \begin{pmatrix} \pm\varepsilon'/n \\ \vdots \\ \pm\varepsilon'/n \\ -(n-1)\varepsilon'/n \\ \pm\varepsilon'/n \\ \vdots \\ \pm\varepsilon'/n \end{pmatrix}.$$

This implies

$$\|a_i + v - b_{ij}\|_2 = \varepsilon'\sqrt{1 - \frac{1}{n}} \leq \varepsilon,$$

if ε' is suitably chosen, so a_i is indeed approximately matched with b_{ij}. The case where x_j occurs negated in C_i is similar.

This argument actually holds for any L_p-norm with $p \geq 2$, in particular for the L_∞-norm.

For the other direction, let us assume that a ε-LCP of size m exists, i.e. all a_i are matched. We first observe that for any i, a_i must have been matched to some b_{ij}. This holds by construction, if ε is sufficiently small ($\varepsilon \ll 1$), which we assume for the rest of the proof. Under translations, this is quite obvious, and under general affine transformations, the gap between a_{n-1} and a_n prevents the only alternative matching $a_i \leftrightarrow b_{n+1-i,j}, i = 1 \ldots n$.

Now we define an assignment of truth values to variables as follows: we set x_j to true (false) if some a_i is matched to b_{ij}, where x_j occurs non-negated (negated) in C_i. If none of the two cases occur, the value of x_j is arbitrary. This is well-defined, because if x_j occurs negated in C_i and non-negated in C_k, then $\|(a_i - a_k) - (b_{ij} - b_{kj})\| = 2\varepsilon'$, which means that one of $\|a_i - b_{ij}\|$ and $\|a_k - b_{kj}\|$ must be at least $\varepsilon' > \varepsilon$. This shows that if some a_i is matched to a point b_{ij} coming from a non-negated (negated) instance of x_j, then this holds for all a_k that are matched to b_{kj}. This also easily implies that the defined truth-assignment satisfies one literal in every clause.

6 Concluding Remarks

The main aim of this paper was to establish polynomial time bounds for solving the LCP-problem in bounded dimensions under bottleneck matching. As a consequence the time bounds presented here are not the tightest possible ones. We believe that there is scope for reasonable improvements in this direction. For example, the algorithm for planar point sets presented in Section 3, which is a straightforward realization of our general scheme, runs in time $O(n^{8.5})$ compared to $O(n^8)$ of [5]. Both the algorithms require solutions to algebraic equations of degree six. However, the number of events in which three disks touch at a point can be bounded by $O(m^2 n^3)$ instead of $O(m^3 n^3)$, using the fact that the dynamic Voronoi diagram of k sets of rigidly moving points on a plane with each set having m points has a complexity $O(m^2 k^2 \log^* k)$ [12]. This reduces the worst case time complexity of our algorithm to $O(n^{7.5})$. For realistic point sets it would be faster since the number of graph matchings depend on the number of combinatorial changes in the arrangement of disks as the angle of rotation varies, which in general would be much smaller.

References

1. T. Akutsu. On determining the congruence of point sets in d dimensions. *Computational Geometry: Theory and Applications*, 9:247–256, 1998. 53, 54
2. T. Akutsu and M.M. Halldórsson. On approximation of largest common subtrees and largest common point sets. *Theoretical Computer Science*, 233(1-2):33–50, 2000. 52, 54
3. T. Akutsu, H. Tamaki, and T. Tokuyama. Distribution of distances and triangles in a point set and algorithms for computing the largest common point sets. *Discrete and Computational Geometry*, 20:307–331, 1998. 52, 53, 54

4. H. Alt and L. Guibas. Discrete geometric shapes: Matching, interpolation, and approximation. In J.-R. Sack and J. Urrutia, editors, *Handbook of Computational Geometry*, pages 121–153. Elsevier Science Publishers B.V. North-Holland, 1999. 54

5. H. Alt, K. Mehlhorn, H. Wagener, and E. Welzl. Congruence, similarity, and symmetries of geometric objects. *Discrete and Computational Geometry*, 3:237–256, 1988. 53, 54, 57, 62

6. D. Avis and K. Fukuda. Reverse search for enumeration. *Discrete and Applied Mathematics*, 65:21–46, 1996. 57, 58

7. S. Basu, R. Pollack, and M.-F. Roy. A new algorithm to find a point in every cell defined by a family of polynomials. In B.F. Caviness and J. Johnson, editors, *Proc. Symp. on Quantifier Elimination and Cylindrical Algebraic Decomposition*. Springer Verlag, 1995. 59

8. S. Basu, R. Pollack, and M.-F. Roy. On computing a set of points meeting every semi-algebraically connected component of a family of polynomials on a variety. *Journal of Complexity*, 13:28–37, 1997. 60

9. D.E. Cardoze and L.J. Schulman. Pattern matching for spatial point sets. In *Proc. 39th Annual Symposium on Foundations of Computer Science*, pages 156–165, 1998. 53

10. S. Chakraborty and S. Biswas. Approximation algorithms for 3-D common substructure identification in drug and protein molecules. In *Proc. 6th. International Workshop on Algorithms and Data Structures*, LNCS 1663, pages 253–264, 1999. 53

11. L.P. Chew, D. Dor, A. Efrat, and K. Kedem. Geometric pattern matching in d-dimensional space. *Discrete and Computational Geometry*, 21:257–274, 1999. 52, 53, 54

12. P. Chew, M. Goodrich, D. Huttenlocher, K. Kedem, J. Kleinberg, and D. Kravets. Geometric pattern matching under eucledian motion. *Computational Geometry: Theory and Applications*, 7:113–124, 1997. 52, 53, 54, 62

13. P.J. de Rezende and D.T. Lee. Point set pattern matching in d-dimensions. *Algorithmica*, 13:387–404, 1995. 53

14. A. Efrat, A. Itai, and M. Katz. Geometry helps in bottleneck matching and related problems. To appear in Algorithmica. 52, 53

15. P.J. Heffernan. The translation square map and approximate congruence. *Information Processing Letters*, 39:153–159, 1991. 53

16. P.J. Heffernan. Generalized approximate algorithms for point set congruence. In *Proc. 3rd. Workshop on Algorithms and Data Structures*, LNCS 709, pages 373–384, Montréal, Canada, 1993. 53

17. P.J. Heffernan and S. Schirra. Approximate decision algorithms for point set congruence. In *Proc. 8th. Annual ACM Symp. on Computational Geometry*, pages 93–101, 1992. 53

18. D.P. Huttenlocher, K. Kedem, and M. Sharir. The upper envelope of Voronoi surfaces and its applications. *Discrete and Computational Geometry*, 9:267–291, 1993. 52, 53

19. P. Indyk, R. Motwani, and S. Venkatasubramanian. Geometric matching under noise: Combinatorial bounds and algorithms. In *Proc. 10th. Annual ACM-SIAM Symp. on Discrete Algorithms*, pages 457–465, 1999. 53

20. P. Indyk and S. Venkatasubramanian. Approximate congruence in nearly linear time. In *Proc. 11th. Annual ACM-SIAM Symp. on Discrete Algorithms*, 2000. 53

21. S. Irani and P. Raghavan. Combinatorial and experimental results for randomized point matching algorithms. *Computational Geometry: Theory and Applications*, 12:17–31, 1999. 53, 54
22. S. Schirra. Approximate decision algorithms for approximate congruence. *Information Processing Letters*, 43:29–34, 1992. 53
23. N. Sleumer. Output-sensitive cell enumeration in hyperplane arrangements. In *Proc. 6th. Scandinavian Workshop on Algorithm Theory*, LNCS 1432, pages 300–309, 1998. 57, 58

Online Algorithms for Caching Multimedia Streams

Matthew Andrews[*] and Kamesh Munagala[**]

[1] Bell Laboratories, Murray Hill NJ 07974.
[2] Computer Science Department, Stanford University, Stanford CA94305.

Abstract. We consider the problem of caching multimedia streams in the internet. We use the dynamic caching framework of Dan et al. and Hofmann et al.. We define a novel performance metric based on the maximum number of simultaneous cache misses, and present near-optimal on-line algorithms for determining which parts of the streams should be cached at any point in time for the case of a single server and single cache. We extend this model to case of a single cache with different per-client connection costs, and give an 8-competitive algorithm in this setting. Finally, we propose a model for multiple caches in a network and present an algorithm that is $O(K)$-competitive if we increase the cache sizes by $O(K)$. Here K is the number of caches in the network.

1 Introduction

In the classical caching problem, requests are made on-line for *pages*. There is a *cache* that can hold a small number of pages. Whenever a request is made for a page we have a *cache hit* if that page is currently in the cache, and a *cache miss* if the page is not in the cache. The goal is to maintain a set of pages in the cache so that the total number of cache misses is minimized.

However, for the problem of caching multimedia streams in the Internet, a different model is more appropriate for two reasons. First, the size of some streams (e.g. video streams) can be extremely large which means that it is infeasible to fit entire streams in a cache. Second, one of the main reasons for using caches is to reduce the usage of bandwidth in the network. Hence, we are more concerned with the maximum number of *simultaneous* cache misses rather than the total number of cache misses.

To address the first issue we shall consider the framework of *dynamic caching* as proposed by Dan and Sitaram [3] and Hofmann et al. [5]. We restrict our attention to a single cache that can access streams from a single server. Suppose that there is a request for some data stream at time t_1 and another request for the same data stream at time $t_2 = t_1 + \Delta$. (In the terminology of [5], Δ is the *temporal distance* between the two requests.) Suppose also that there is enough

---

[*] Email: andrews@research.bell-labs.com.
[**] Email: kamesh@cs.stanford.edu. Supported by ONR N00014-98-1-0589. Part of this work was done while the author was visiting Bell Labs.

M. Paterson (Ed.): ESA 2000, LNCS 1879, pp. 64–75, 2000.

space in the cache to store Δ time units of the stream. Then, we can serve *both* requests using only *one* connection to the server (see Figure 1). We always cache the last Δ time units of the stream that were seen by the first request. The second request can always obtain from the cache the current stream data that it needs.

In this paper we consider the problem of determining which parts of the streams to maintain in the cache so as to minimize the number of simultaneous connections to the server. We hence minimize the bandwidth required on the link between the cache and the server. We also consider the case of multiple caches in a network and propose offline and online algorithms.

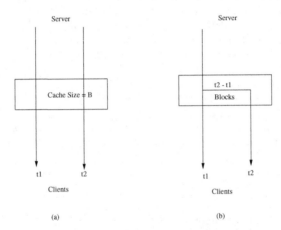

Fig. 1. Dynamic Caching: (a) The two requests before caching. (b) After caching the interval $t_2 - t_1$, the bandwidth to the server reduces by one.

1.1   The Models

Single cache. We begin by considering a cache and a server connected by a single link. There is a data stream (e.g. a video clip) of duration T stored in the server. We call one time unit of the clip a *block*, and denote by $b_1, b_2, \ldots, b_T$ the T blocks of the clip.

Requests for the clip arrive in an online fashion. We denote by t_i the time at which client i makes a request for the *entire* clip. Without loss of generality, $t_1 < t_2 < \ldots$. We assume that the client requests must be serviced *without delay*, i.e., client i must receive the block b_j at time $t_i + j - 1$.

The cache has a buffer that can store B blocks of the clip. At time $t_i + j - 1$, if block j is in the cache, then client i can obtain it without cost (we have a *Cache Hit.*) If block b_j is not in the cache then it must be obtained from the server at a cost of one unit of bandwidth on the cache-server link (we have a *Cache Miss.*) In the latter case, when block b_j is obtained from the server, we are allowed to place it in the cache, subject to the condition that we never cache more than B blocks.

Our goal is to determine which blocks of the clip should be cached at any point in time so as to minimize the maximum bandwidth that is ever used on the link. That is, we wish to minimize the maximum number of *simultaneous* cache misses. More specifically, let M_t^i be an indicator variable that equals 1 if cache i has a cache miss at time t and 0 otherwise. The bandwidth used at time t is $\sum_i M_t^i$. We wish to minimize $\max_t \sum_i M_t^i$.

Throughout this paper we shall focus on *interval strategies*. We define the interval i to be the set of blocks lying between the block currently required by request $i - 1$ and the block currently required by request i. Hence at time t interval i consists of blocks $b_{t-t_i+1}, \ldots, b_{t-t_{i-1}}$. An interval strategy always tries to cache entire intervals.

We emphasize that our algorithms must be *online* in the sense that they cannot know about the request from client i until time t_i. However, since the request is for the entire stream of blocks, the algorithm will know at time t_i that block b_j will be needed at time $t_i + j - 1$. We shall use the framework of competitive analysis [13] and compare our candidate algorithms against the optimal *offline* algorithm that knows all the requests in advance.

Single cache with per-client connection costs. We next consider a more general model in which we have per-client costs. We assume that for each client there is a cost d_i for connecting the client to the cache, and a cost s_i for connecting the request directly to the server. These costs reflect the bandwidth required by these connections. There is also a cost c for connecting the cache to the server. For each block required by the client, if the block is present in the cache it can be obtained at a cost of d_i. Otherwise, the block can either be obtained via the direct connection to the server at a cost of s_i or else it can be obtained from the server and brought into the cache at a cost of $c + d_i$. Our goal is to minimize the maximum cost incurred at any time instant.

Multiple caches In our most general model we have many caches, each of which could be placed in a different part of the network. Cache j has size B and can be connected to the server at a cost of c_j. Client i can obtain a block from cache j at a cost of d_{ij}. If the block is not cached anywhere then client i can obtain it from the server and bring it into cache j at a cost of $c_j + d_{ij}$.[1]

Multiple clips Although the results we derive assume that the server has just one clip, the discussion can be easily extended to the case where the server has many clips.

1.2   Results

- In Section 2 we consider the single cache problem where we have a cache-server link and we wish to minimize the required bandwidth on this link. Let

---

[1] Note that we do not need to consider the cost of connecting to the server directly since we can assume that there is a cache at the server with $B_j = 0$ and $c_j = 0$.

U be the bandwidth of the optimal offline algorithm (that is not necessarily an interval strategy). We first present an interval strategy in which we always cache the smallest intervals that will fit into the buffer. We show that this strategy requires bandwidth at most $U + 1$. We also consider how to carry out this strategy when the requests arrive online. We first show how to do this when we have *auxiliary* buffer of size B in which we cache the first B blocks of the clip. We next show how to dispense with the auxiliary buffer at the expense of two additional connections to the server, i.e. the bandwidth required is $U + 3$.

- In Section 3 we analyze the single cache problem where we have per-client connection costs. We present an online algorithm that is 8-competitive with respect to the optimal offline interval strategy.
- In Section 4, we study the multiple cache problem. Let K be the number of caches. We first present an integer programming formulation for the problem of choosing the best set of intervals to store at each cache. We show that the solution of the LP-relaxation can be rounded to give an integral solution in which all caches are increased by a factor of $4K$ and the connection costs are increased by a factor $4K$. We also show that if the size of cache j is increased once more to $12KB$ then we can obtain an online algorithm that is $12K$-competitive with respect to the optimal offline interval strategy.

1.3 Previous Work

There has been much work on the classical caching problem in which requests are for arbitrary blocks and we wish to minimize the aggregate number of cache misses [1,2,4,9,13]. For example, Sleator and Tarjan [13] showed that the Least-Recently-Used (LRU) protocol is B-competitive (i.e. the number of cache misses is at most B times the number of misses due to the optimal offline algorithm). However, the problem of caching streaming data in a network has received less attention.

Dan and Sitaram presented the framework of dynamic caching in [3]. They propose an interval strategy similar to the algorithm we present in Section 2 in which we aim to cache the intervals that require the smallest buffer space. However, Dan and Sitaram do not present a theoretical analysis of this approach and they do not consider how to carry out the algorithm in the online setting.

Hofmann et al. [5] present heuristics for dynamic caching when there are many co-operating caches in a network. The caches store different sections of the stream and control processes called helpers decide which cache(s) to use for any particular client so that connection cost is minimized.

Sen et al. [11] study the benefits of caching as a method to minimize the delay associated with the playback of a multimedia stream. They advocate caching the initial parts of a clip so that the playback can be started earlier.

More generally, the problem of caching in distributed environments such as the Internet is receiving increasing attention. Li et al. [8] present algorithms for the placement of caches. Karger et al. [7] and Plaxton and Rajaraman [10]

give methods for the placement of data and the assignment of client requests to
caches.

2   Single Cache

We begin by considering the single cache model in which we wish to minimize
the required bandwidth on the cache-server link. Let $\Delta_i = t_{i+1} - t_i$. We call the
We include the interval between the first request and the most recently expired
request in our list of intervals. We denote the length of this interval by Δ_0.

At any time instant t, let $\Delta_{i_1} \le \Delta_{i_2} \le \ldots \Delta_{i_n}$, and let $N_t = max\{n|$
$\sum_{j=1}^{n} \Delta_{i_j} \le B\}$. Note that N_t changes with time as new intervals get added
and old ones cease to exist.

We first assume that the cache has an *auxiliary buffer* of size B, in addition
to its regular buffer. We present a caching strategy that we call the GREEDY-
INTERVAL-STRATEGY. It is similar to the caching scheme proposed in [3], and
can be described as follows:

(1) **Basic Strategy:** Assume the current time instant is t. Identify the N_t
    smallest intervals in the current set of requests. Merge adjacent intervals
    in this set to create *super-intervals*. For a super-interval of length L, connect
    the request that came earliest in time to the server, and allocate L blocks
    of the cache to this super-interval. Store the last L blocks of the connected
    request in this space. Service all other requests in that super-interval from
    the cached blocks.
(2) **Dying streams:** When a request i expires, we say that the interval between
    request $i - 1$ and i has *died*. This may cause the set of N_t cached intervals
    to change. To maintain this set online, we do the following. After request
    $i - 1$ expires, if the dying interval is one of the N_t smallest intervals then we
    only cache that part of it which is required by the request i. If the current
    time is t, then we cache only the last $t_i + T - t$ blocks of the dying interval
    in the cache. Note that if the dying interval is evicted from the cache by a
    newly arriving interval, it will never get re-inserted, as re-insertions occur
    only when an interval dies.
(3) **Vacant Space:** At any instant of time, let δ be the number of blocks of
    cache that are not used by any interval. We cache the first δ part the smallest
    uncached interval in this space. We will show below that this can be done
    online.
(4) **New Requests:** When a request arrives, it creates a new interval. We re-
    compute N_t and find the N_t new smallest intervals and store these in the
    cache. Note that except the new interval, we do not add any interval to the
    existing set of intervals. Since we have the first B blocks of the clip in the
    auxiliary buffer, it is trivial to add the new interval.

The following theorem shows that GREEDY-INTERVAL-STRATEGY is a valid
online strategy.

Theorem 1. *When an interval is added to the set of cached intervals, it is already present in the cache.*

Proof. We prove this by induction on time. Assume the claim is true at time t, *i.e.*, the N_t smallest intervals are present in the cache. The only events that change the set of cached intervals are listed below. We will show in each case that the new set of N_t smallest intervals are already present in the cache.

1. A new interval arrives. Since we cache the first B blocks of clip on the auxiliary buffer, we have this interval in cache when it arrives.
2. An interval expires when both streams corresponding to its end-points die. This may cause the interval i_{N_t+1} to get cached. But in this case, by step **(3)** of the algorithm, we would have cached this interval completely in the vacant space.

Note further that once a dying interval is removed from the cache by a newly arriving interval, it will never be re-inserted.

2.1   Competitive Analysis

The performance metric we use is the maximum bandwidth used by the algorithm. We compare it with an optimum offline strategy, which we call OPT. Let us assume that OPT uses U units of bandwidth on a sequence of requests. We can explicitly characterize OPT. It looks at all the X requested blocks at at any instant of time t, and caches those $X - U$ blocks for which the previous requests were closest in time. Note that these blocks must have been cached at some time in the past, so the algorithm is effectively looking into the future to decide which blocks to cache. The algorithm is *successful* if and only if it caches at most B blocks at any instant of time. We state the following results without proof.

Lemma 1. *Given any sequence of requests, if it is possible to cache the blocks in such a way that the maximum bandwidth required is U, then OPT, as described above is successful.*

Lemma 2. *If OPT caches the block required by request i_j at time t then it must also cache the block required by i'_j for all $j' < j$.*

We now compare the performance of GREEDY-INTERVAL-STRATEGY with the bandwidth-optimal strategy OPT.

Theorem 2. *If GREEDY-INTERVAL-STRATEGY uses bandwidth U at time t, then there exists time $t' \geq t$ at which OPT uses bandwidth $U - 1$.*

Proof. Suppose that OPT uses bandwidth $U - 2$ throughout the interval $[t, t + \Delta_{i_{N_t+1}})$.

There are two cases to consider.

- **Case 1** All but one of the streams that are *not* cached by the GREEDY-INTERVAL-STRATEGY at time t are still in the system at time $t + \Delta_{i_{N_t+1}}$. Then, since OPT uses bandwidth $U - 2$ during the time period $[t, t + \Delta_{i_{N_t+1}})$, one of these streams must be cached by OPT throughout the time period $[t, t + \Delta_{i_{N_t+1}})$. By Lemma 2, this means that OPT caches $N_t + 1$ intervals at time t. But, if we consider the $N_t + 1$ smallest intervals at time t, we have $\sum_{j=1}^{N_t+1} \Delta_{i_j} > B$ (by definition of GREEDY-INTERVAL-STRATEGY). This is a contradiction.
- **Case 2** Two or more streams that are not cached by the GREEDY-INTERVAL-STRATEGY at time t die before time $t + \Delta_{i_{N_t+1}}$. However, by the definition of the GREEDY-INTERVAL-STRATEGY, the gap between these streams must be at least $\Delta_{i_{N_t+1}}$. This is a contradiction.

Note that if a stream is not cached at time t and dies during $t + \Delta_{i_{N_t+1}}$ then it must be the earliest stream in the system at time t.

Corollary 1. *The* GREEDY-INTERVAL-STRATEGY *requires at most one more unit of bandwidth than* OPT.

2.2   Removing the Auxiliary Buffer

We now present a scheme LOOKAHEAD-GREEDY that removes the auxiliary buffer of size B. First of all we modify GREEDY-INTERVAL-STRATEGY so that we only begin to cache an interval when there is enough free buffer space to cache the entire interval. This modification creates one extra connection to the server.

Suppose that a new request arrives at time t. If the new interval (of length δ, say) needs to be cached, we start caching the stream corresponding to the previous request starting from time t. By the above comment there is free buffer space equal to δ at time t. This space is completely filled up only at time $t + \delta$. Therefore, at time $t + \gamma$, there is unused space of size $\delta - \gamma$. We cache as many blocks of the beginning of the new stream in this space as possible. Note that the cached section of the new stream increases from 0 to $\frac{\delta}{2}$ blocks, but then decreases to 0, as the free space decreases.

Suppose that the next interval that must be cached arrives at time $t + \mu$. The size of this interval is at most μ. There are three cases depending on the value of μ:

1. If $\mu < \frac{\delta}{2}$, we have the beginning μ blocks in the cache, and we can serve the new stream directly from the cache, as in GREEDY-INTERVAL-STRATEGY.
2. If $\delta > \mu > \frac{\delta}{2}$, we will have the first $\delta - \mu$ blocks in the cache. This means that we can serve the new stream from the cache for the interval $[t + \mu, t + \delta]$. At time $t + \delta$ the first interval will be fully cached.
3. If $\mu > \delta$, the new stream does not arrive until the first interval is fully cached.

This implies that LOOKAHEAD-GREEDY requires at most one more connection to the server than the modified version of GREEDY-INTERVAL-STRATEGY.

Recall that the modified version of GREEDY-INTERVAL-STRATEGY requires one more connection than the original version. We have the following result.

Theorem 3. *If* OPT *requires maximum bandwidth* U *on any sequence of requests,* LOOKAHEAD-GREEDY *requires bandwidth at most* $U + 3$.

3   Single Cache with Connection Costs

We will now consider online algorithms in a more general cost model than the one used above. If some request is not part of any super-interval then it is not necessary for that request to be routed via the cache.

In this new model any client can either connect to the server directly or can connect via the cache. We assume that the connection cost per unit bandwidth from the cache to the server is c. (see Figure 2). Furthermore, for each request i we pay cost d_i if we connect the request to the cache. We pay s_i if we connect the request directly to the server. Without loss of generality we assume that $d_i \leq s_i \leq c + d_i$.

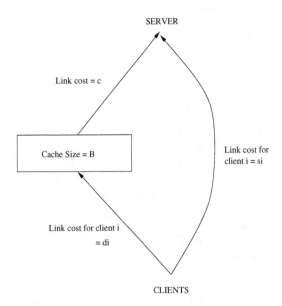

Fig. 2. The cost model for a single server and single cache. Note that $d_i \leq s_i \leq c + d_i$.

We shall compare our online scheme only against schemes which cache complete intervals from those formed by the active requests, at any instant of time. We call such schemes *Interval Caching* schemes. We do not consider offline algorithms that are not interval strategies. This does not matter as long as we assume that the caching policy at a cache is *oblivious* to the algorithm that assigns requests to caches. More precisely, the caching protocol assumes that the requests currently assigned to the cache will remain assigned to that cache until they die.

The cost of an interval caching scheme at any time instant is simply the total cost of the bandwidth used in the cost model described above. Our objective is to derive an online strategy that is competitive (with respect to the maximum cost needed by any interval caching scheme) on a given sequence of requests.

As before, at any time instant, we number the requests $1, 2, \ldots, n$ according to the order in which they arrive. A *super-interval* is a contiguous set of cached intervals, each of which is serviced by one connection to the server. Note that the cost of connecting one super-interval to the server is c, irrespective of the number of requests cached within it.

If there are n active requests at any instant of time, we can compute the optimum interval caching strategy[2] for that time instant using dynamic programming. The running time of this dynamic program is $O(n^2 \sum s_i)$. However, we can obtain a polynomial time approximation scheme for this problem by scaling all the costs.

3.1   An Online Algorithm

We now show how to obtain a 8-competitive online algorithm with respect to the optimum interval caching strategy for this problem. The algorithm is very simple to describe. We periodically compute the optimum, and try to cache the intervals corresponding to it. We call this strategy the GREEDY-PREFETCH strategy.

Compute Optimum For the given set of requests, we compute the optimum solution. This solution may differ in some super-intervals from the previous optimum. We remove all super-intervals in the old optimum but not in the new one from the cache, and start caching the new super-intervals.

Greedy Caching For a new super-interval p, let L_p be the total uncached length. Note that this length need not be contiguous. Let m_p be the point on this super-interval where the length of the uncached portion is exactly $\lceil \frac{L_p}{2} \rceil$. Let the two halves be X_p and Y_p. Let S_p be the current total cost of connecting the requests in super-interval p to the server. Among X_p and Y_p we find that half with the larger current connection cost, and start caching that half of the super-interval with one connection to the server.

Prefetch When we compute the optimum and start caching the new super-intervals, the time it would take for the solution to stabilize is the size of the largest uncached part of any super-interval. Let us denote this amount of space as L. In this space, we cache the beginning of the clip with one extra connection to the server. We can cache at most $\lfloor \frac{L}{2} \rfloor$ blocks of the clip before this space starts to diminish in size. For any new request arriving within time $\lfloor \frac{L}{2} \rfloor$ of computing the optimum, we serve it from the cache. Note that the

---

[2] This problem is NP-hard, with a reduction from the Knapsack problem [6].

request arriving at time $\lceil \frac{L}{2} \rceil$ after computing optimum must be connected directly to the server. We can do this at an extra cost of c.

Recompute Optimum We recompute the optimum after time $\lceil \frac{L}{2} \rceil$ of the previous computation, and repeat the whole process. Note that if the optimal solution does not cache anything, we recompute the solution at the next request arrival.

3.2   Competitive Analysis

We will now show that GREEDY-PREFETCH is competitive with respect to the maximum optimum interval caching cost on any sequence of requests. Denote the times at which we compute the optimum solution by $x_1, x_2, \ldots$. Let the cost of the optimum interval caching strategy at time x_i be $M(i)$, and the cost of GREEDY-PREFETCH be $G(i)$. Also, let $S(i)$ be the total excess cost of connecting requests within a new super-interval to the server at time x_i. We state the following results without proof.

Lemma 3. *For the functions G, S and M as defined above, the following are true:*

1. $S(i+1) \leq 3M(i) + \lfloor \frac{S(i)}{2} \rfloor$.
2. $G(i) \leq 2M(i) + S(i)$.
3. *The cost G attains its maximum at one of the times x_i.*

Theorem 4. *Given any sequence of requests, if the optimum interval caching strategy can service them within a cost of C, GREEDY-PREFETCH requires cost at most $8C$ to service them.*

4   Multiple Caches

The most general version of this problem can be stated as follows. We have a single server, K caches and n clients in a network. The caches have capacity B and the connection cost per unit bandwidth to the server from cache j is c_j. The cost for connecting client i to cache j is d_{ij}. We first present an offline $O(K)$-approximation algorithm[3] for finding the optimum set of intervals to cache if we are allowed to increase each buffer size by a factor $O(K)$. We then show how to convert the offline scheme into an online scheme if we are allowed a further increase in the cache size.

4.1   Offline Approximation Algorithm

The server stores a single clip. Client i requests the clip at time t_i, and we assume that $t_1 \leq t_2 \leq \cdots \leq t_n$. Let $\Delta_i = t_i - t_{i-1}$. The objective is to compute

---

[3] The offline problem is NP-hard, with a reduction from the Generalized Assignment Problem [12].

the optimum set of intervals to store in each cache so that the total cost of the bandwidth used is minimized.

The above problem can be formulated as an IP as follows. Let x_{ij} be equal to 1 if request i is served via cache j, and 0 otherwise. Let y_{ij} be equal to 1 if request i is connected to the server via cache j. Let z_{ij} be equal to 1 if the interval $[t_{i-1}, t_i]$ is present in cache j. Note that request i need not be served by this interval[4].

$$\text{Minimize} \quad \sum_{j=1}^{K} \sum_{i=1}^{n} c_j y_{ij} + d_{ij} x_{ij}$$

$$\begin{aligned}
\sum_{j=1}^{K} x_{ij} &= 1 & \forall i \in \{1, 2, \ldots, n\} \\
\sum_{i=1}^{n} \Delta_i z_{ij} &\leq B & \forall j \in \{1, 2, \ldots, K\} \\
z_{ij} &\leq z_{i-1j} + y_{i-1j} & \forall i, j \\
x_{ij} &\leq z_{ij} + y_{ij} & \forall i, j \\
x_{ij}, y_{ij}, z_{ij} &\in \{0, 1\}
\end{aligned}$$

Definition 1. *A solution to the above IP is a (α, β)-**approximation** if the cost of the solution is at most α times the cost of the optimal fractional solution (obtained by relaxing the IP), and the required buffer space cache j used is at most $\beta \cdot B_j$ for all j.*

We first show a lower bound on the approximability via this formulation, without proof.

Theorem 5. *If the above IP is (α, β)-approximable by rounding its LP relaxation, then $\alpha\beta \geq K$.*

We will now show how to obtain a $(4K, 4K)$-approximation by rounding the relaxation of the IP described above. For each cache j, look at the variables in increasing order of i, applying the following rules. We denote the rounded x_{ij} as X_{ij}, y_{ij} as Y_{ij}, and z_{ij} as Z_{ij}.

1. $x_{ij} \geq \frac{1}{K} \Rightarrow X_{ij} = 1$, else $X_{ij} = 0$.
2. $z_{ij} \geq \frac{1}{2K}$, or $z_{ij} \geq \frac{1}{4K}$ and $Z_{i-1j} = 1 \Rightarrow Z_{ij} = 1$, else $Z_{ij} = 0$.
3. $y_{ij} \geq \frac{1}{2K}$, or $Z_{i+1j} = 1$ and $Z_{ij} = 0 \Rightarrow Y_{ij} = 1$, else $Y_{ij} = 0$.

Theorem 6. *The rounding scheme described above gives a $(4K, 4K)$-approximation.*

4.2   Online Scheme

We now show how to convert the offline algorithm into an online scheme. Suppose that we increase the buffer size at cache j to $12KB$. We compute the approximate offline solution every $4KB$ time steps. We use the additional buffer space as follows. At each cache we store the first $4KB$ blocks of the clip. Any new request i is served from the cache j with the smallest d_{ij} until the optimum is re-computed.

---

[4] For any super-interval, the first and last requests will be served by that super-interval. However, intermediate requests may not be served.

5   Open Problems

The most interesting question is whether it is possible to improve the approximation factor in the multiple cache offline problem. Another interesting question is whether it is possible to obtain online schemes in the multiple cache setting without blowing up the cache size. Finally, it would be interesting to see if it is possible to extend these schemes to the case where the requests are not for the complete clip, but for parts of it[5].

References

1. S. Albers, S. Arora and S. Khanna. Page replacement for general caching problems. *Proceedings of the 10th Annual Symposium on Discrete Algorithms*, 31–40, 1999. 67
2. L. Belady. A study of replacement algorithms for virtual storage computers. *IBM System Journal*, 5:78–101, 1966. 67
3. A. Dan and D. Sitaram. A generalized interval caching policy for mixed interactive and long video environments. *Multimedia Computing and Networking*, January 1996. 64, 67, 68
4. A. Fiat, R. Karp, M. Luby, L. McGeoch, D. Sleator and N. Young. Competitive paging algorithms. *Journal of Algorithms*, 12:685–699, 1991. 67
5. M. Hofmann, T.S.E. Ng, K. Guo, S. Paul and H. Zhang. Caching techniques for streaming multimedia over the internet. *Bell Laboratories Technical Memorandum*, May 1999. 64, 67
6. D. Hochbaum. Various notions of approximations: Good, better, best, and more. In *Approximation Algorithms for NP-Hard Problems*, D. Hochbaum, Ed. *PWS Publishing Company*, 1995. 72
7. D. Karger, E. Lehman, T. Leighton, M. Levine, D. Lewin and R. Panigrahy. Consistent hashing and random trees: Distributed caching protocols for relieving hot spots on the World Wide Web. *Proceedings of the 29th Annual ACM Symposium on Theory of Computing*, 654–663, 1997. 67
8. B. Li, M. Golin, G. Italiano, X. Deng and K. Sohraby. On the optimal placement of web proxies in the internet. *Proceedings of INFOCOM '99*, 1282–1290, 1999. 67
9. L. McGeoch and D. Sleator. A strongly competitive randomized paging algorithm. *Algorithmica*, 6:816–825, 1991. 67
10. G. Plaxton and R. Rajaraman. Fast fault-tolerant concurrent access to shared objects. *Proceedings of the 37th Annual Symposium on Foundations of Computer Science*, 570–579, 1996. 67
11. S. Sen, J. Rexford and D. Towsley. Proxy prefix caching for multimedia streams. *Proceedings of INFOCOM '99*, 1310–1319, 1999. 67
12. D. Shmoys and É. Tardos. An approximation algorithm for the generalized assignment problem. *Mathematical Programming*, 62:461–474, 1993. 73
13. D. Sleator and R. Tarjan. Amortized efficiency of list update and paging rules. *Communications of the ACM*, 28(2):202–208, 1985. 66, 67, 75

---

[5] If the requests are for arbitrary single blocks of the clip, then this reduces to the classical caching problem [13].

On Recognizing Cayley Graphs*

Lali Barrière[1], Pierre Fraigniaud[2], Cyril Gavoille[3], Bernard Mans[4], and John M. Robson[3]

[1] Departament de Matemàtica Aplicada i Telemàtica, *Univ. Politècnica de Catalunya*, Campus Nord, Edifici C3, c/ Jordi Girona 1-2, 08034 Barcelona, Spain.
[2] Laboratoire de Recherche en Informatique, Bât. 490, *Univ. Paris-Sud*, 91405 Orsay cedex, France.
[3] Laboratoire Bordelais de Recherche en Informatique, *Univ. Bordeaux I*, 33405 Talence cedex, France.
[4] Department of Computing, *Macquarie Univ.*, Sydney, NSW 2109, Australia.

Abstract. Given a class $\mathcal{C}$ of Cayley graphs, and given an edge-colored graph G of n vertices and m edges, we are interested in the problem of checking whether there exists an isomorphism ϕ preserving the colors such that G is isomorphic by ϕ to a graph in $\mathcal{C}$ colored by the elements of its generating set. In this paper, we give an $O(m \log n)$-time algorithm to check whether G is color-isomorphic to a Cayley graph, improving a previous $O(n^{4.752} \log n)$ algorithm. In the case where $\mathcal{C}$ is the class of the Cayley graphs defined on Abelian groups, we give an optimal $O(m)$-time algorithm. This algorithm can be extended to check color-isomorphism with Cayley graphs on Abelian groups of given rank. Finally, we propose an optimal $O(m)$-time algorithm that tests color-isomorphism between two Cayley graphs on $\mathbb{Z}_n$, i.e., between two circulant graphs. This latter algorithm is extended to an optimal $O(n)$-time algorithm that tests color-isomorphism between two Abelian Cayley graphs of bounded degree.

1 Introduction

Checking whether two input graphs with n vertices and m edges are isomorphic is known as the *graph-isomorphism* problem and is a famous combinatoric problem. For general graphs the graph-isomorphism problem is known to be in NP, not known to be in P, and probably is not NP-complete (see Section 6 of [3]). A relaxed problem is to recognize classes of graphs, that is to check whether a given graph G is isomorphic to an element of a specified class of graphs (e.g., interval graphs, chordal graphs, planar graphs, etc.). Depending on the class of graphs, this problem can be easy (e.g., checking planarity). However, for many classes, checking whether a graph belongs to that class remains difficult. This is particularly true for Cayley graphs [12].

* This work is supported by an Australian-French cooperation granted by CNRS and ARC. Part of this work has been done while the three first authors were visiting the Department of Computing of Macquarie University, Sydney. Additional support by Spanish Research Council (CICYT) under project TIC97-0963, by the CNRS, and by the Aquitaine Region project #98024002.

M. Paterson (Ed.): ESA 2000, LNCS 1879, pp. 76–87, 2000.

A graph $G = (V, E)$ is *Cayley* if there exists a group Γ and a generating set S of Γ such that V is the set of the elements of Γ, and there is an arc from x to y in G, that is $(x, y) \in E$, if and only if there exists $s \in S$ such that $y = s * x$, or, in other words, if and only if $y * x^{-1} \in S$. (In the following, we will use the additive notation for the operation $*$ of Γ). If G is Cayley, then G is denoted by the pair (Γ, S). If S is involutive, that is if $(s \in S \Rightarrow s^{-1} \in S)$ holds, then G is a symmetrically-oriented directed graph which can be seen as an undirected graph. The regularity and the underlying algebraic structure of Cayley graphs make them good candidates for interconnecting nodes of a network [11]. For instance, Cycles, Tori, Hypercubes, Cube-Connected Cycles (CCC), Butterflies with wraparound, Star-graphs, Bubble Sort graphs, Pancake graphs, Chordal rings (i.e., circulant graphs) are all Cayley graphs. Some of them are defined on Abelian groups (e.g., hypercubes and tori), whereas some others are defined on non-Abelian groups (e.g., Star-graphs, and Pancake graphs).

The complexity of graph-isomorphism for Cayley graphs is a challenging problem [12,14]. Currently, the problem is still open even if we restrict our attention to circulant graphs [2,7], i.e., Cayley graphs defined on $\mathbb{Z}_n$. The problem is known to be polynomial only for specific classes of circulant graphs, e.g., those with a prime number of vertices [6,15]. In this paper, we study a variant of the graph-isomorphism problem. Given two (properly or not) arc-colored digraphs $G_1 = (V_1, E_1)$ and $G_2 = (V_2, E_2)$, these two digraphs are said *color-isomorphic* if there exists an isomorphism from G_1 to G_2 that preserves the colors. More formally, let C_1 be the set of colors of G_1, and C_2 be the set of colors of G_2. G_1 and G_2 are color-isomorphic if there exist a one-to-one mapping $\phi : V_1 \to V_2$ and a one-to-one mapping $\psi : C_1 \to C_2$ such that for every $c \in C_1$, and for every $(x, y) \in V_1 \times V_1$:

$$(x, y) \in E_1 \text{ and is of color } c \iff (\phi(x), \phi(y)) \in E_2 \text{ and is of color } \psi(c) .$$

Graph-isomorphism is the particular case of the color-isomorphism problem obtained by choosing $C_1 = C_2 = \{c\}$. Conversely, one can construct a polynomial reduction from color-isomorphism to graph-isomorphism[1]. Therefore, the two problems are equivalent. That is why we will focus on *specific* colorings of graphs to solve the color-isomorphism problem. In particular, the generating set of a Cayley graph (Γ, S) induces a coloring of the edges, simply by coloring the edge $(x, s + x)$ by the element $s \in S$. Such a coloring is called the *natural* edge-coloring of a Cayley graph. As we show in this paper, the color-isomorphism problem between naturally colored Cayley graphs is apparently simpler than graph-isomorphism, or color-isomorphism, between arbitrary colored Cayley graphs.

The notion of color-isomorphism applied to naturally colored Cayley graphs was introduced to characterize edge-labeled regular graphs that have so-called *minimal sense of direction* [8]. In [9] it is shown that such regular graphs are

---

[1] In fact, we will show in the full version of the paper that there is a polynomial reduction from color-isomorphism of Cayley graphs to Cayley isomorphism.

exactly those that are color-isomorphic to a Cayley graph whose edges are colored by the elements of its generating set. Boldi and Vigna proposed in [5] an $O(n^{4.752} \log n)$-time algorithm to check whether a colored graph is color-isomorphic to a naturally colored Cayley graph. A parallel version of this algorithm runs in $O(\log n)$ time with n^6 processors.

Note that there are major differences between, on one hand, the problem of checking whether a given graph is isomorphic (or color-isomorphic) to a Cayley graph of a specified class, and, on the other hand, the problem of checking whether a given group is isomorphic to a group of a specified class [13]. For instance, an $O(n)$-time algorithm exists for Abelian group isomorphism [16], but the addition table of the group is given as input of the problem. In the color-isomorphism problem, the algebraic structure of the graph is hidden, and it may not be possible to perform elementary group operations in constant time. In this paper we show how to extract the structure of the underlying group of colored Cayley graphs. This is done by a pure graph and algorithmic approach.

Results and structure of the paper

1) In Section 3, we present an $O(m \min\{k, \log n\}) = O(m \log n)$-time algorithm that checks whether a colored graph of degree k given as input is color-isomorphic to a naturally colored Cayley graph, improving a previous $O(n^{4.752} \log n)$ algorithm.

2) In Section 4, we give an $O(m)$-time algorithm that checks whether a colored graph given as input is color-isomorphic to a naturally colored Cayley graph defined on an Abelian group. This algorithm trivially extends to an algorithm that, given any integer r and any graph G, checks whether G is color-isomorphic to a naturally colored Cayley graph on an Abelian group of rank r. As a consequence, we can check in $O(m)$-time whether a colored graph given as input is color-isomorphic to a naturally colored circulant graph ($r = 1$). The decrease of the complexity from $O(m \log n)$ for arbitrary Cayley graphs to $O(m)$ for Abelian Cayley graphs is obtained by a sophisticated technique used to perform arithmetic computations in $\mathbb{Z}_{n_1} \times \cdots \times \mathbb{Z}_{n_r}$, $\prod_{i=1}^{r} n_i = n$, with the standard $O(\log n)$-word RAM model [10].

3) Finally, in Section 5, we give an $O(m)$-time algorithm that checks color-isomorphism between two naturally colored circulant graphs given as input. This algorithm can be generalized to produce an $O(n)$-time algorithm that checks color-isomorphism between two bounded degree naturally colored Abelian Cayley graphs.

Note that our $O(m)$-time algorithms are optimal since, in the worst-case, we have to check the m edges of the input graph. As far as the sense of direction theory is concerned, our results have the following consequence:

Corollary 1. *There is an $O(m \min\{k, \log n\})$-time algorithm which, given any regular graph and any edge-coloring of this graph, returns whether this coloring gives G a minimal sense of direction.*

The next section gives preliminary results that are helpful for solving our problems.

2   Notation and Preliminary Results

The model. In each of our problems, we assume that the input graph is a digraph (that is a strongly connected directed graph). Again, a graph is a particular case of this setting by viewing the graph as a symmetrically-oriented directed graph. The input arc-colored digraph of order n is described by its adjacency list $\mathcal{L}$, including the colors. The vertices are labeled from 1 to n, and, for every vertex x, $\mathcal{L}[x]$ is the list $((y_1, c_1), \ldots, (y_k, c_k))$ of the k out-neighbors of x together with the colors of the arcs linking x to these out-neighbors. Each color c_i is an integer taken in the set of colors $C = \{1, \ldots, k\}$. Actually, for k-regular digraphs (that is where each vertex is of in- and out-degree k), the list can be replaced in $O(nk)$ time by a $n \times k$ *adjacency table* $\mathcal{T}$ so that, for every vertex x and for every $i \in \{1, \ldots, k\}$, $\mathcal{T}[x, c_i]$ returns in constant time the i-th neighbor y_i of x such that the arc (x, y_i) is colored c_i. Therefore, for a k-regular digraph of n vertices, the size of the input is nk integers taken from the set $\{1, \ldots, n\}$.

Note that one can check in $O(nk)$ time whether a digraph is k-regular, and whether a digraph is properly colored by k colors (that is every vertex has its k incoming arcs colored with k different colors, and its k outgoing arcs colored with the same k different colors). Therefore, unless specified otherwise, we will always assume in the following that the input digraph is k-regular, and properly colored with k colors.

Notation. Let x_0 be an arbitrary vertex of a properly colored digraph $G = (V, E)$. Since we are checking whether G is isomorphic to a subclass of Cayley digraphs, one can assume that G is vertex-transitive. Indeed, if this assumption happens to be false, it will create a contradiction further at some stage of the algorithm, and therefore it will be detected. Thus, we can assume, w.l.o.g., that x_0 is the zero of the underlying group. As stated previously, the elementary group operations are not directly available to be performed in constant time given the digraph G only. Indeed, the (supposed) algebraic structure of the digraph is hidden in its adjacency table. So, let us redefine standard group operations applied to digraphs.

The concatenation of two words w and w' whose letters are taken in the set of colors C, that of two words w and w' of C^*, is denoted by $w \otimes w'$. Let $\mathcal{S}$ be any directed tree spanning G, and rooted at x_0. We can label each vertex x of G by a sequence of colors according to the color of the arcs encountered in the path from x_0 to x in $\mathcal{S}$. More precisely x_0 is labeled $\ell(x_0) = \epsilon$ (the empty sequence), and every vertex $x \neq x_0$ is labeled $\ell(x) = \ell(y) \otimes c$, i.e., the concatenation of the label of the father y of x in $\mathcal{S}$, and of the color c of the arc (y, x). Once this labeling is fixed, one can define the *orbit* of a vertex $x \neq x_0$ as the ordered sequence of vertices $y_0, y_1, y_2, \ldots$ where $y_0 = x_0$, and, for $i > 0$, y_i is the vertex reached from y_{i-1} by following the sequence of colored arcs given by the sequence of colors of $\ell(x)$. In particular $y_1 = x$. G being properly colored, one can then define the *order* of a vertex $x \neq x_0$ as the smallest integer $t > 0$ for which $y_t = x_0$. The order of x is denoted by $\mathbf{o}(x)$, and $\mathbf{o}(x_0)$ is set to 1.

By extension, the order of a color c, denoted by $\mathbf{o}(c)$, is defined as the order of the out-neighbor x_c of x_0 such that the arc (x_0, x_c) is colored c. The orbit of c is the orbit of x_c. Note that the orbit of a color is an elementary cycle in G, whereas the orbit of an arbitrary vertex x may not be a cycle, in particular if x is not an out-neighbor of x_0. Finally, we define the *addition* of vertices. Given two vertices x and y, $x + y$ is defined as the vertex reached from x by following the path $\ell(y)$. The orbit of a vertex x *passing through* $y \in V$ is simply the sequence $y, y + x, y + 2x, \ldots$

Correspondence tables. Computing the order of a color c can be done in $O(\mathbf{o}(c))$ time by simply following from x_0 the path composed of arcs of color c. Adding a color c to any vertex y takes $O(1)$ time using the adjacency table $\mathcal{T}$, because the vertex $y + c$ is given by $\mathcal{T}[y, c]$. However, computing the order of an arbitrary vertex x looks harder since following the path of color $\ell(x)$ from a vertex y requires $|\ell(x)|$ accesses to the adjacency table, where $|\ell(x)|$ denotes the length of sequence $\ell(x)$. Nevertheless, this complexity can be reduced if we restrict our study to the case $\ell(x) = c \otimes \cdots \otimes c$, with $|\ell(x)| = \alpha$. One can compute the vertex $y + \alpha c$ in $O(\alpha)$ time using $\mathcal{T}$, but one can do better as shown in the following lemma:

Lemma 1. *There is an $O(nk)$-time and $O(nk)$-space algorithm that, given a properly colored k-regular digraph G of order n described by its adjacency table $\mathcal{T}$, either returns* FALSE, *and then G is not a Cayley digraph, or returns a data structure such that, if G is color-isomorphic to a naturally colored Cayley digraph (Γ, S), then it enables computation of vertex $x + \alpha s$ in constant time, for any $x \in \Gamma$, any integer $\alpha \geq 1$ and any $s \in S$.*

Proof. Let us first assume that G is a naturally colored Cayley digraph (Γ, S) of order n given by its adjacency table $\mathcal{T}$. First, let us describe the structure, and show how to compute $z = y + \alpha c$ in constant time using that structure, that is a vertex reached from y by following the path $\ell(\alpha c) = c \otimes \cdots \otimes c$ composed of α times the arc colored by c. Then we will show how this structure can be constructed in $O(nk)$ time. For every $c \in C$, that is for every color of the input digraph, G can be viewed as $n_c = n/\mathbf{o}(c)$ c-colored orbits of length $\mathbf{o}(c)$ with arcs of the other colors between the orbits. Let us concentrate on the n_c orbits of c. Arbitrarily order these orbits from 1 to n_c, and, inside each orbit, order the vertices of this orbit from 0 to $\mathbf{o}(c) - 1$ by starting at any vertex, and by respecting the natural order induced by the orbit. We get an $(n \times k)$-table T, and k tables T_c, one for each color $c \in C$, of respective size $n_c \times \mathbf{o}(c)$. $T_c[i, j]$ is the vertex x at the j-th position in the i-th orbit of c. Conversely, $T[x, c]$ returns the pair (i, j). Computing $z = y + \alpha c$ can be done in constant time using the table T and the tables T_c, $c \in C$, as follows:

- $T[y, c]$ specifies that y is the j-th element of the i-th orbit of c;
- $z = T_c[i, (j + \alpha) \bmod \mathbf{o}(c)]$.

The construction of T and T_c, $c \in C$, can be done in $O(nk)$ time as follows. There is one phase for each color $c \in C$. Once c is fixed, we visit all the vertices and construct the n_c orbits of c (the visit of the first orbit allows to set $\mathbf{o}(c)$). This is done in n_c steps, one step for each orbit. At each step, say at step i, an arbitrary not yet visited vertex x is chosen. $T[x, c]$ is set to $(i, 0)$, and $T_c[i, 0]$ is set to x. Visiting the $\mathbf{o}(c) - 1$ remaining vertices $x_1, \ldots, x_{\mathbf{o}(c)-1}$ of the current orbit allows to set $T[x_j, c]$ to (i, j), and $T_c[i, j]$ to x_j, for all j, $1 \le j < \mathbf{o}(c)$. The total cost of this construction is $O(nk)$ because there are $k = |C|$ phases, and, at each phase, every vertex is visited exactly once because G is a Cayley digraph.

Given an arbitrary properly colored k-regular digraph G, one can apply the previous construction. This construction may fail if G does not have enough symmetry, and, conversely, it may succeed even if G is not a Cayley digraph. Nevertheless, for any digraph such that, for every color c, the set of vertices can be decomposed in n_c c-colored orbits of the same length, the construction will succeed. Since G is properly colored, the only thing that must be checked is whether all orbits of the same color have the same length. The algorithm returns FALSE if it fails. This checking can be done on the fly, during the construction of the tables, with no additional cost. □

Definition The tables resulting of the proof of Lemma 1 are called the *correspondence tables* of the colored digraph G.

3 Recognizing Cayley Colored Digraphs

Recall that $C = \{1, \ldots, k\}$ is the set of colors. For any vertex x, we define the equivalence relation $\equiv_x$ on words in C^* by $w_1 \equiv_x w_2$ if and only if $x + w_1 = x + w_2$. Then we define the equivalence relation $\sim$ on vertices by $x_1 \sim x_2$ if and only if $\equiv_{x_1}$ and $\equiv_{x_2}$ are identical.

Lemma 2. *A k-regular connected properly colored digraph $G = (V, E)$ is a Cayley digraph if and only if $v_1 \sim v_2$ holds for every pair (v_1, v_2) of vertices of V.*

Proof. The "only if" part is trivial: if the graph is Cayley, then for every vertex v, the relation $\equiv_v$ is exactly the relation under which words are equivalent if and only if their values in the group are equal. We write $\equiv$ for the relation $\equiv_v$ of every vertex $v \in V$. Choose a vertex $x_0 \in V$, choose a spanning tree rooted at x_0, and define $\ell(x)$ for every vertex $x \in V$ as in Section 2. Recall the definition of $x_1 + x_2$ as $x_1 + \ell(x_2)$ in Section 2. It is easy to see that x_0 is an identity for this $+$ and that if w is the label of a path from x_1 to x_0, then $x_0 + w$ is an inverse of x_1. We have that $\ell(x_1) \otimes \ell(x_2) \equiv \ell(x_1 + x_2)$ since both are the colors of paths from x_0 to $x_1 + x_2$. Now the operation $+$ is associative since $\ell(x_1 + (x_2 + x_3)) \equiv \ell(x_1) \otimes (\ell(x_2) \otimes \ell(x_3))$ and $\ell((x_1 + x_2) + x_3) \equiv (\ell(x_1) \otimes \ell(x_2)) \otimes \ell(x_3)$. □

Lemma 3. *If $x_1 \sim x_2$, then $x_1 + c \sim x_2 + c$.*

Proof. Suppose that $x_1 \sim x_2$. If $w_1 \equiv_{x_1+c} w_2$, then $(x_1 + c) + w_1 = (x_1 + c) + w_2$ so $x_1 + (c \otimes w_1) = x_1 + (c \otimes w_2)$. In other words, $c \otimes w_1 \equiv_{x_1} c \otimes w_2$ and so $c \otimes w_1 \equiv_{x_2} c \otimes w_2$ giving $x_2 + (c \otimes w_1) = x_2 + (c \otimes w_2)$ and finally $(x_2 + c) + w_1 = (x_2 + c) + w_2$ so that $w_1 \equiv_{x_2+c} w_2$. □

The following $O(m)$-time algorithm tests for two vertices x and y whether $x \sim y$:

Testing equivalence algorithm

1. Choose a spanning tree rooted at x and label the vertices $x = x_0, x_1, \ldots, x_{n-1}$ according to a pre-order traversal. Traverse the similar tree rooted at y (i.e., using the same colors as in the tree rooted at x), checking that no vertex is encountered twice and labeling the vertices $y = y_0, y_1, \ldots, y_{n-1}$ again in pre-order.
2. For every $i < n$ and every color $c \in C$, consider $x_i + c = x_j$ and $y_i + c = y_{j'}$; if in any case $j \neq j'$ then return FALSE, else return TRUE.

Proof. Let w_i be the word in C^* labeling the path in the spanning tree from x to x_i. If a vertex is encountered twice in traversing the tree rooted at y, say in the positions corresponding to x_i and x_j then we have that $w_i \equiv_y w_j$ but $w_i \not\equiv_x w_j$; if we ever found $x_i + c = x_j$, $y_i + c = y_{j'}$ with $j \neq j'$, then $w_i \otimes c \equiv_x w_j$ but $w_i \otimes c \not\equiv_y w_j$. On the other hand, if we always have $j = j'$, we have a way of reducing any word to a unique equivalent w_i which is valid for both $\equiv_x$ and $\equiv_y$. $\square$

Given Lemma 3, we will know that all vertices are equivalent to x_0 if we find a set $X = \{c_1, \ldots, c_t\}$ of colors such that $x_0 \sim x_0 + c_i$ ($1 \leq i \leq t$) and every vertex is reachable from x_0 by a path labeled by colors from X. If the graph is Cayley, each new color added to X takes us from a subgroup to a larger subgroup and so at least doubles the number of reachable vertices. Hence at most $\min\{k, \log n\}$ colors can be added before either exhausting V or finding that the graph is not Cayley.

Theorem 1. *There is an $O(m \min\{k, \log n\})$-time algorithm which, given a properly colored k-regular digraph G, checks whether G is color-isomorphic to a naturally colored Cayley digraph.*

Proof. Starting with a set S of vertices including only x_0, and a set of colors X initially empty, we consider each color $c \in C$ in turn. If $x_0 + c$ is already in S we do nothing. Otherwise we test that $x_0 + c \sim x_0$, rejecting if not, add c to X and then add to S all vertices reachable from existing elements of S by a path starting with color c and then using any colors from X. If this exhausts V accept and otherwise if the set of colors is exhausted reject.

Given that we stop the recursive process of adding vertices to S when we encounter a vertex already in S, any edge is only considered once and so the time is dominated by the at most $\min\{k, \log n\}$ tests for $x_0 + c \sim x_0$ each of which takes time $O(m)$ giving the total time bound of $O(m \min\{k, \log n\})$.

Note that we do not need to test that $|S|$ has at least doubled after each incremental operation because the subgraph consisting of the vertices in S and the edges between them colored by colors in X is always a Cayley graph. $\square$

We were not able to prove or disprove the optimality of our algorithm. It is obviously optimal for bounded degree graphs, but there is a $O(\log n)$ gap between its worst-case complexity and the trivial lower bound $\Omega(m)$ for arbitrary graphs. In the next section, we give an optimal algorithm for the class of Abelian Cayley graphs.

4   Recognizing Colored Cayley Digraphs on Abelian Groups

Recall that the class of Abelian Cayley graphs contains networks as popular as the hypercubes, the tori, and the circulant graphs. Let $n = \prod_{i=1}^{t} p_i^{m_i}$ be the prime-decomposition of n. An Abelian group Γ of order n is determined by a specific decomposition of the m_i's in $m_i = \sum_{j=1}^{r_i} \alpha_{i,j}$, $\alpha_{i,1} \geq \ldots \geq \alpha_{i,r_i} \geq 1$. We then get (see for instance [4]):

$$\Gamma \simeq (\mathbb{Z}_{p_1^{\alpha_{1,1}}} \times \cdots \times \mathbb{Z}_{p_1^{\alpha_{1,r_1}}}) \times (\mathbb{Z}_{p_2^{\alpha_{2,1}}} \times \cdots \times \mathbb{Z}_{p_2^{\alpha_{2,r_2}}}) \times \cdots \times (\mathbb{Z}_{p_t^{\alpha_{t,1}}} \times \cdots \times \mathbb{Z}_{p_t^{\alpha_{t,r_t}}}) \tag{1}$$

By ordering the $\alpha_{i,j}$'s so that $\alpha_{i,j} \geq \alpha_{i,j+1}$, and the p_i's so that $p_i > p_{i+1}$, let

$$n_j = \prod_{i=1}^{t} p_i^{\alpha_{i,j}}, \; j = 1, \ldots, r,$$

where $r = \max_i r_i$, and $\alpha_{i,j} = 0$ for $j > r_i$. The rank of Γ is then equal to r. Note that, for any Cayley digraph $G = (\Gamma, S)$ to be connected, we necessarily have $r \leq k = |S|$. From the definition of the n_j's, one can express Γ in the form $\mathbb{Z}_{n_1} \times \cdots \times \mathbb{Z}_{n_r}$, $n_i \geq n_{i+1}$. (Note that n_{i+1} divides n_i.) This latter form is called the *standard form* of Γ.

Our recognition algorithm performs in three phases. The first finds a basis of the unknown $\mathbb{Z}$-module Γ corresponding to the input digraph G. Once this basis has been determined, that is once the group Γ has been identified, we try, during the second phase, to label the vertices, and to compute a candidate for S such that $G = (\Gamma, S)$. After that, a one-to-one mapping between the vertex-sets of the two graphs has been set. Therefore, the third phase checks whether or not the colors of the edges of G and (Γ, S) correspond by a one-to-one mapping. We actually do not label directly the vertices of G by r-tuples because it is not computationally cheap to find elements of order $n_1, \ldots, n_r$. Rather, we still first make use of Γ in its original form depicted by Eq. (1). This will yield labels of $\sum_{i=1}^{t} r_i$ coordinates. We will refine this construction later, to finally get labels on r coordinates.

Notation. We will denote by $\langle x_1, \ldots, x_t \rangle$ the *orbit* of $x_1, \ldots, x_t$, that is the set of the linear combinations of $x_1, \ldots, x_t$, no matter if the x_i's are vertices of a digraph, or elements of a group.

   In the statement of the following lemma, x_0 refers to the root of the tree $\mathcal{S}$ defined in Section 2.

Lemma 4. *There is an $O(k\sqrt{n} + n)$-time algorithm which, given a properly colored k-regular digraph G and its correspondence tables, returns*

1. *a t-tuple $(r_1, \ldots, r_t)$;*
2. *t r_i-tuples $(\alpha_{i,1}, \ldots, \alpha_{i,r_i})$, $i \in \{1, \ldots, t\}$; and*
3. *$\sum_{i=1}^{t} r_i$ vertices $g_{i,j}$, $i \in \{1, \ldots, t\}$, $j \in \{1, \ldots, r_i\}$;*

such that, if G is color-isomorphic to a naturally colored Abelian Cayley digraph on Γ, then

- $\mathbf{o}(g_{i,j}) = p_i^{\alpha_{i,j}}$ *for every i and j;*
- $\langle g_{i,j} \rangle \cap \langle g_{i,1}, \ldots, g_{i,j-1} \rangle = \{x_0\}$ *for every i, and every $j > 1$;*
- $\langle g_{i,1}, \ldots, g_{i,r_i} \rangle \cap \langle g_{i',1}, \ldots, g_{i',r_{i'}} \rangle = \{x_0\}$ *for every $i \neq i'$; and*
- $\Gamma \simeq (\mathbb{Z}_{p_1^{\alpha_{1,1}}} \times \cdots \times \mathbb{Z}_{p_1^{\alpha_{1,r_1}}}) \times \cdots \times (\mathbb{Z}_{p_t^{\alpha_{t,1}}} \times \cdots \times \mathbb{Z}_{p_t^{\alpha_{t,r_t}}})$.

The proof of Lemma 4 will be given in the full version of the paper.

Theorem 2. *There is an $O(m)$-time and $O(m)$-space algorithm which, given a colored digraph G of m arcs, checks whether G is color-isomorphic to a naturally colored Cayley digraph on an Abelian group. In the affirmative, the algorithm returns the corresponding Cayley digraph (Γ, S), Γ given in the standard form $\Gamma = \mathbb{Z}_{n_1} \times \cdots \times \mathbb{Z}_{n_r}$, and the two one-to-one mappings $\phi : V \to \Gamma$ and $\psi : C \to S$.*

Proof. As said before, we first check in $O(m)$ time that G is properly colored and regular. If it succeeds we apply Lemma 1 and Lemma 4, otherwise return FALSE and stop. To label the vertices, we will not directly use the $g_{i,j}$'s derived in Lemma 4, but we will group some of them together to get a system of r coordinates rather than $\sum_{i=1}^{t} r_i$ coordinates. Let

$$h_j = \sum_{i=1}^{t} g_{i,j}, \quad j = 1, \ldots, r$$

where $g_{i,j} = 0$ if $j > r_i$. We get $\mathbf{o}(h_j) = n_j$ for every j, and we label the vertices in $\Gamma = \mathbb{Z}_{n_1} \times \cdots \times \mathbb{Z}_{n_r}$ as follows.

Vertex x_0 is labeled by the r-tuple $(0, 0, \ldots, 0)$. Vertices s_i's are simply represented by the neighbors of x_0. The β_1-th vertex of the orbit of h_1 is labeled the r-tuple $(\beta_1, 0, \ldots, 0)$. The β_2-th vertex of the orbit of h_2 passing through $(\beta_1, 0, \ldots, 0)$ (starting from $(\beta_1, 0, \ldots, 0)$) is labeled $(\beta_1, \beta_2, 0, \ldots, 0)$. And so on. Thanks to Lemma 1, each jump along the orbit of h_1 (starting at x_0) can be computed in $O(\min\{k, t\})$ time since h_1 is the sum of multiples of at most $\min\{k, t\}$ colors. Similarly, each jump along the orbit of h_{i+1}, starting at $(\beta_1, \beta_2, \ldots, \beta_i, 0, \ldots, 0)$ takes a time $O(\min\{k, it\})$ because h_{i+1} is the sum of multiples of at most $O(\min\{k, it\})$ colors. Therefore, the whole labeling can be performed in $O(n \min\{k, rt\})$ time, that is in at most $O(m)$ time. Meanwhile, the sequence $(n_1, \ldots, n_r)$ has been computed in at most $O(m)$ time.

Once the labeling has been performed, it remains to check that, for every $i = 1, \ldots, k$, all arcs

$$(x, y) = ((x_1, \ldots, x_r), (y_1, \ldots, y_r))$$

colored c_i satisfy

$$\begin{cases} y_1 - x_1 = \psi_1(c_i) & (\mathrm{mod}\ n_1); \\ y_2 - x_2 = \psi_2(c_i) & (\mathrm{mod}\ n_2); \\ \quad\vdots \\ y_r - x_r = \psi_r(c_i) & (\mathrm{mod}\ n_r); \end{cases} \tag{2}$$

where $\psi(c) = (\psi_1(c), \ldots, \psi_r(c))$ is a r-tuple of $\mathbb{Z}^r$ depending only on $c \in \{c_1, \ldots, c_k\}$. If these equalities are satisfied by every color c_i, then G is color-isomorphic to (Γ, S) where $S = \{\psi(c_1), \ldots, \psi(c_k)\}$. Otherwise, G is not color-isomorphic to a naturally colored Cayley digraph on an Abelian group (and we return FALSE and stop). Indeed, the map $\phi : \Gamma \to \Gamma$ such that $\phi(a_1 h_1 + \ldots + a_r h_r) = (a_1, \ldots, a_r)$ is a group-automorphism, and therefore, if $G = (\Gamma, \{s_1, \ldots, s_k\})$, then $\psi(c_i)$ is actually the labeling of s_i, that is $\psi(c_i) = \phi(s_i) = \phi(v^{(i)})$ where $v^{(1)}, \ldots, v^{(k)}$ are the k neighbors of x_0 such that, for every $i \in \{1, \ldots, k\}$, $v^{(i)} = s_i + x_0$. A naive component-wise approach to check Eq. (2) yields an $O(nkr)$-time algorithm with a $O(nk)$-space. As detailed in the full version of the paper Eq. (2) can be checked in $O(1)$ time for each arc by using pre-computed $O(nk)$-space tables requiring $O(nk)$ time to be set. This can be done with the standard $O(\log n)$-word RAM model [10]. □

Recognizing colored Abelian Cayley digraphs of a given rank r can be done by a simple variant of the algorithm of Theorem 2. Indeed, once the $g_{i,j}$'s have been computed in $O(m)$ time (see Lemma 4), the rank of the group Γ such that G is possibly color-isomorphic to a Cayley digraph on Γ is determined. Therefore, one can stop the algorithm if this rank is different from r. The algorithm will perform in $O(m)$ time as shown in Theorem 2. As a consequence:

Corollary 2. *For any integer $r > 0$, there is an $O(m)$-time and $O(m)$-space algorithm which, given a colored digraph G, checks whether G is color-isomorphic to a naturally colored Cayley digraph on an Abelian group of rank r. In the affirmative, the algorithm returns*

- *the corresponding Cayley digraph (Γ, S) where Γ given in its standard form;*
- *the labeling function $\phi : V \to \Gamma$; and*
- *the coloring function $\psi : C \to S$.*

5   Color-Isomorphism for Circulant and Abelian Cayley Digraphs

Circulant graphs have attracted a lot of attention in the literature from both a theoretical and practical point of view. Indeed, the algebraic structure of circulant graphs is simple but rich, and their regularity make them suitable candidates for connecting nodes of a network. Let us show how one can check whether two given naturally colored circulant graphs are color-isomorphic. The elements of the generating set S of a circulant graph $(\mathbb{Z}_n, S)$ are called *chords*. We say that two subsets T and S of $\mathbb{Z}_n$ are *proportional* if there is an integer λ such that $\gcd(\lambda, n) = 1$ and $T = \lambda S$. Two circulant digraphs are then said to be *Ádám-isomorphic* if the sets of chords are proportional [1]. One can easily show that Ádám-isomorphism and color-isomorphism between naturally colored circulant digraphs are two equivalent notions, that is:

Lemma 5. *Two naturally colored circulant digraphs are color-isomorphic if and only if they are Ádám-isomorphic.*

Therefore, we get:

Theorem 3. *There is an $O(m)$-time algorithm that determines, for any pair G and H of naturally colored circulant digraphs, whether G is color-isomorphic to H.*

Proof. Let G and H be the two input digraphs, both of order n. Let $S, T \subseteq \mathbb{Z}_n$ be the sets of chords returned by the algorithm of Corollary 2 (with $r = 1$) on G and H respectively. From Lemma 5, G and H are color-isomorphic if and only if there exists an invertible constant λ in $\mathbb{Z}_n$ such that $T = \lambda S$. Let $S = \{s_1, \ldots, s_k\}$, and $T = \{t_1, \ldots, t_k\}$. There are at most k candidates for λ: the solutions of $t_i = \lambda s_1$, $i = 1, \ldots, k$, if exists. Each candidate can be computed in $O(\log n)$ time by Euclid's algorithm. For every candidate, one checks whether $T = \lambda S$ (that can be performed in $O(k)$ time assuming a pre-computed n bitmap encoding the set T). All these operations take $O(n + k(\log n + k)) = O(n + k^2)$ time, which is bounded by the time required by the algorithm of Corollary 2.   □

The previous result can be generalized to Abelian Cayley digraphs by noticing that, as a generalization of Lemma 5, two naturally colored Abelian Cayley digraphs (Γ, S) and (Γ, T) are color-isomorphic if and only if there exists an automorphism ϕ of Γ such that $T = \phi(S)$, and for any basis B of the $\mathbb{Z}$-module Γ, $\phi(B)$ is also a basis. Therefore, given two naturally colored Abelian Cayley digraphs G_1 and G_2, one can apply Theorem 2 to compute their respective groups Γ_1 and Γ_2, and their respective generating sets S and T. If $\Gamma_1 \neq \Gamma_2$, then G_1 and G_2 are not color-isomorphic. (This latter checking can be done easily be comparing the two sequences $n_1, \ldots, n_r$ of Γ_1 and Γ_2 since our algorithm returns the group in its standard form.) Otherwise, one can test all the one-to-one functions mapping S to T. Computing one generating subset $S' \subseteq S$ of Γ (which can be done in $O(m)$ time), one can show that there are only $k!/(k - |S'|!) \leq k^r$ mappings to test. Choosing a suitable base B, one can compute $\phi(B)$ and check whether $\phi(B)$ is a basis in $O(m)$ time. In total, it gives rise to an $O(k^r m)$-time algorithm.

Corollary 3. *There is an $O(k^r m)$-time algorithm that determines, for any pair G, H of naturally colored Cayley digraphs on Abelian groups of rank r, whether G is color-isomorphic to H.*

Our algorithm is linear for bounded degree and polynomial for bounded rank. We leave as an open problem the following:

Problem 1. Design a polynomial-time algorithm that checks color-isomorphism between two naturally colored Cayley graphs on Abelian groups of arbitrary rank.

Acknowledgments. The authors are thankful to Nicola Santoro who suggested that they study the color-isomorphism problem. The authors also want to express their deep thanks to Sebastiano Vigna for his comments, and Charles Delorme who gave them many hints on how to attack the $\mathbb{Z}_n$-isomorphism problem.

References

1. A. Ádám. Research problem 2-10. *J. Combin. Theory*, 2:393, 1967. 86
2. B. Alspach and T. Parsons. Isomorphism of circulant graphs and digraphs. *Discrete Mathematics*, 25:97–108, 1979. 77
3. L. Babai. Automorphism groups, isomorphism, reconstruction. In R. Graham, M. Grötschel, and L. Lovász, editors, *Handbook of Combinatorics*, vol. 2. Elsevier and MIT Press, 1995. 76
4. G. Birkhoff and S. MacLane. *Algebra*. Macmillan, 1967. 83
5. P. Boldi and S. Vigna. Complexity of deciding Sense of Direction. *SIAM Journal on Computing*, 29(3):779–789, 2000. 78
6. B. Codenotti, I. Gerace and S. Vigna. Hardness Results and Spectral Techniques for Combinatorial Problems on Circulant Graphs. *Linear Algebra Appl.*, 285(1–3):123–142, 1998. 77
7. B. Elspas and J. Turner. Graphs with circulant adjacency matrices. *J. Comb. Theory*, 9:229–240, 1970. 77
8. P. Flocchini, B. Mans, and N. Santoro. Sense of direction in distributed computing. In *Proceedings of DISC '98*, LNCS vol. 1499, Springer-Verlag, S. Kutten (Ed.), pages 1–15, 1998. 77
9. P. Flocchini, A. Roncato, and N. Santoro. Symmetries and sense of direction in labeled graphs. Discrete Applied Mathematics, 87:99-115, 1998. 77
10. T. Hagerup. Sorting and searching on the word RAM. In *Proceedings of STACS '98*, LNCS vol. 1379, Springer-Verlag, M. Morvan and C. Meinel (Eds.), pp. 366–398, 1998. 78, 85
11. T. Leighton Introduction to Parallel Algorithms and Architectures: Arrays, Trees, Hypercubes. Morgan Kaufmann, 1992. 77
12. C. Li, C. Praeger, and M. Xu. On finite groups with the Cayley isomorphism property. *Journal of Graph Theory*, 27:21–31, 1998. 76, 77
13. G. Miller. On the $n^{\log n}$ isomorphism technique. In *Proceedings Tenth Annual ACM Symposium on Theory of Computing (STOC)*, pp. 51–58, 1978. 78
14. J. Morris. Isomorphic Cayley graphs on nonisomorphic groups. *Journal of Graph Theory*, 31:345–362, 1999. 77
15. M. Muzychuk and G. Tinhoffer. Recognizing circulant graphs of prime order in polynomial time. *The electronic journal of combinatorics*, 3, 1998. 77
16. N. Vikas. An $O(n)$ algorithm for Abelian p-group isomorphism and an $O(n \log n)$ algorithm for Abelian group isomorphism. *Journal of Computer and System Sciences*, 53:1–9, 1996. 78

Fast Algorithms for Even/Odd Minimum Cuts and Generalizations

András A. Benczúr[1*] and Ottilia Fülöp[2**]

[1] Computer and Automation Institute, Hungarian Academy of Sciences, and
Department of Operations Research, Eötvös University, Budapest
[2] Institute of Mathematics, Technical University, Budapest
{benczur,otti}@cs.elte.hu

Abstract. We give algorithms for the directed minimum odd or even cut problem and certain generalizations. Our algorithms improve on the previous best ones of Goemans and Ramakrishnan by a factor of $O(n)$ (here n is the size of the ground vertex set). Our improvements apply among others to the minimum directed T-odd or T-even cut and to the directed minimum Steiner cut problems. The (slightly more general) result of Goemans and Ramakrishnan shows that a collection of *minimal* minimizers of a submodular function (i.e. minimum cuts) contains the odd minimizers. In contrast our algorithm selects an n-times smaller class of *not necessarily minimal* minimizers and out of these sets we construct the odd minimizer. If $M(n,m)$ denotes the time of a u–v minimum cut computation in a directed graph with n vertices and m edges, then we may find a directed minimum

- odd or T-odd cut with V (or T) even in $O(n^2m + n \cdot M(n,m))$ time;
- even or T-even cut in $O(n^3m + n^2 \cdot M(n,m))$ time.

The key of our construction is a so-called *parity uncrossing* step that, given an arbitrary set system with odd intersection, finds an odd set with value not more than the maximum of the initial system.

1 Introduction

The notion of minimum cuts in graphs and its generalization, the minimum of a submodular function, plays a key role in combinatorial optimization. See [15] for a survey of the application of minimum cuts and [13] for that of submodular functions.

In this paper we consider minimum cuts or minimizers of submodular functions by restricting minimization to either the even or the odd sets. More generally we will minimize in so-called triple families ([9] and [6]), set systems that satisfy certain properties of even or odd (for the definition see Section 1.3). The fact that the minimum odd cut or T-odd cut (requiring odd intersection with a prescribed set T) problems can be solved in polynomial time by maximum flow computations is first shown in [14]; the same for even cuts as well as its generalizations for matroids is shown in [2].

---

* Supported from grants OTKA T-30132 and T-29772; AKP 98-19; FKFP 0206/1997
  URL: http://www.cs.elte.hu/~benczur
** Supported from OTKA grant T-29772

M. Paterson (Ed.): ESA 2000, LNCS 1879, pp. 88–99, 2000.

1.1 Minimum Cut Problems.

The cut of the graph is a bipartition of its vertex set V into $C \subset V$ and its complement; the value of the cut $f(C)$ in a directed graph is the number or the total capacity of the edges leaving C. We consider the following optimization problems related to cuts in directed graphs:

- The minimum odd (even) cut problem asks for a cut $C \neq \emptyset, V$ such that $|C|$ is odd (even) with $f(C)$ minimum.
- The minimum T-odd (T-even) cut problem asks for a cut $C \neq \emptyset, V$ such that $|C \cap T|$ is odd (even) with $f(C)$ minimum.
- For a given integer p, we ask for the minimizer C of the cut value function f with $|C|$ (or for a given set T, the intersection $|C \cap T|$) not divisible by p.
- The generalized directed Steiner cut problem asks for a cut C with minimum value $f(C)$ in a family of cuts defined as follows. Given sets $T_1, \ldots, T_k \subseteq V$, a cut C is a *generalized Steiner cut* if for some $i \leq k$ we have $\emptyset \neq C \cap T_i \neq T_i$. (For the notion and use of *undirected* Steiner cuts see [7].)

1.2 Submodular Functions.

A function f over all subsets of a ground set V is called *submodular* if all $X, Y \subseteq V$ satisfy $f(X) + f(Y) \geq f(X \cap Y) + f(X \cup Y)$.

An example of a submodular function is the cut value function f seen in the previous subsection. In this paper we generalize the cut problems of Section 1.1 to arbitrary submodular functions; for example we will consider the problem of finding the set $Y \subseteq V$ with $f(Y)$ minimum such that $|Y|$ is odd or even.

1.3 Triple Families: A Generalization of Even/Odd.

Grötschel et al. [9,10,11] generalize the notion of an odd set and define a *triple family* as follows.

> A family of subsets of ground set V forms a *triple family* if for all members X and Y whenever three of the four sets $X, Y, X \cap Y$ and $X \cup Y$ are not in the triple family, then so is the fourth.

Notice that each example in Section 1.1 asks for the minimum value cut in a certain triple family; thus all these problems form special cases of the algorithms described in this paper. This fact is immediate for cuts with cardinality odd, even, or not divisible by p. It is also easy to see that directed generalized Steiner cuts form a triple family. For the undirected version this follows by the results of [7] and [6]; we obtain the same for the directed version since the definition of a Steiner cut is symmetric.

To simplify the notation we will call members of the triple family **odd** while other sets **even**. Notice that this notion of even and odd is not related to the actual cardinality of a set C; in fact when we optimize for even sets, an odd set is a set with even cardinality. Also note that the empty set may, in this sense, be odd as well.

In the discussion we will only use the next two key properties of triple families. Strictly speaking, the second property is more restrictive than the definition of Grötschel et al. [9,10,11] since they also consider triple families of lattices where we cannot take the complement of a set.

($*$) if X and Y are even (non-members) and $X \cap Y$ (or $X \cup Y$) is odd, then so is $X \cup Y$ (or $X \cap Y$).

($**$) if $\emptyset$ is even and $X_1 \subset X_2$ with both X_1 and X_2 even, then $X_2 - X_1$ is even.

1.4   Previous Results.   The previous best result of Goemans and Ramakrishnan [6] finds the minimum f-value odd set over a ground set with $|V| = n$ by $O(n^2)$ calls to a submodular minimization oracle. Their algorithm also applies to a slight generalization of triple families called *parity families*, a notion defined in Section 4.4. The minimizer they find may both be the empty set as well as the entire ground set V. If one wants to exclude these possibilities (and $\emptyset$ or V belongs to the triple family), then the algorithm must be repeated $O(n)$ times, thus yielding $O(n^3)$ oracle calls.

Goemans and Ramakrishnan [6] prove that a minimal odd minimizer of a submodular function f is either $\emptyset$ or V or it is among the (inclusionwise) minimal minimizers X of function f with the property that X has $u \in X$ and $v \notin X$, where u and v takes all possible values from V. Hence for a minimum cut problem they must use a flow/cut algorithm that returns the *minimal* minimum cut, but for the price of this additional restriction they are able obtain a minimal minimizer.

As for earlier results, the first one related to even/odd minimization are those of Padberg and Rao [14] who prove that the odd cut and T-odd cut problems can be solved in polynomial time and those of Barahona and Conforti [2] who prove the same for even cuts. Grötschel et al. [9] improve their results both in that they only use $O(n^3)$ minimization oracle calls and in that their result applies to all triple families. The algorithm of Goemans and Ramakrishnan [6] mentioned above further improve both the efficiency and the scope of applicability of these even/odd mincut results.

1.5   New Results and Organization.   We give algorithms that find the minimum of a submodular function f in a triple family over a ground set V with $|V| = n$. All these algorithms are faster than the Goemans–Ramakrishnan algorithm [6] by a factor of n. None of our algorithms requires *minimal* minimizers, just any minimizers of the function f. This might be an advantage for example for cuts in graphs in that arbitrary max-flow algorithms may be used. While we cannot get a speedup for the more general scenario of Goemans and Ramakrishnan [6] (parity families), we also sketch an algorithm (Section 4.4) that matches the asymptotic efficiency of their algorithm.

The main technique of our algorithms is a specific uncrossing procedure that we call *parity uncrossing*. All of our algorithms follow the generic framework of Section 2 by starting with a collection of (not necessarily odd) submodular minimizers and apply the parity uncrossing procedure of Section 2.1. As input, this procedure gets a collection of sets whose intersection is odd; in this collection it repeatedly picks sets and replaces them by either of the union or the intersection with smaller f-value. The procedure terminates when an odd set is found; these odd sets become the minimizer candidates.

We give four similar algorithms that minimize submodular functions over triple families depending on whether the empty set $\emptyset$ or the ground set V are odd (belong to the family). We need such a distinction when minimizing submodular functions, since in general we want to avoid obtaining these sets as minimizers. For instance in case of even cardinality cuts in a graph $f(\emptyset) = f(V) = 0$ while $f(S) \geq 0$ for all other sets; the minimizer we get should be different from these two trivial ones.

Our first algorithm in Section 3 solves the simplest case when neither the empty set nor the ground set is in the family. We use $O(n)$ calls to a minimization oracle (max-flow computation) and another $O(n^2)$ to an evaluation oracle (one that tells the value of

$f(C)$ for a set C). This algorithm may for example find the minimum odd cardinality cut in $\tilde{O}(n^2 m)$ time for $|V| = 2k$.

The next two algorithms drop the restriction that $\emptyset$ and V must be even by keeping the asymptotic running time of the first one; however these algorithms may return both $\emptyset$ and V as the odd minimizer of f. The second algorithm in Section 4.1 drops the restriction on the ground set V (the empty set still must be even); the third one in Section 4.2 drops both restrictions.

The fourth and strongest algorithm in Section 4.3 applies to all triple families and finds the odd minimizer of f different from $\emptyset$ and V. The algorithm, just like that of Goemans and Ramakrishnan [6] for this case, makes $O(n)$ calls to the algorithm that may return $\emptyset$ or V as minimizers. Thus we spend the time of $O(n^2)$ calls to a minimization oracle and another $O(n^3)$ to an evaluation oracle. This algorithm may for example find a minimum even cardinality cut in $\tilde{O}(n^3 m)$ time.

2   The Main Algorithmic Idea: Parity Uncrossing

Our generic submodular minimization algorithm over triple families (the *odd sets* as defined in Section 1.3) begins with a collection of certain minimizers of the submodular function f among all (not necessarily odd or even) subsets. Two earlier algorithms also follow this line: the algorithm of Goemans and Ramakrishnan [6] considers the $n(n-1)$ minimizers separating all possible pairs of the ground set $|V| = n$; and the algorithms for symmetric submodular functions such as in [5] arrange $n - 1$ minimizers in a so-called Gomory–Hu tree [8] to select the odd minimizer.

We aim to achieve the same efficiency for minimization as in the second type of algorithms that use Gomory–Hu trees. Although Gomory–Hu trees do not exist in the asymmetric case (as pointed out by [3,16]), we can still use the abstraction of *uncrossing* in the Gomory–Hu construction [8] to reduce the number of minimizations from $O(n^2)$ to $O(n)$ for odd minimization of asymmetric submodular functions.

Definition 1.   Two sets X and Y *cross* if neither of $X \cap Y$, $X - Y$, $Y - X$ and $V - (X \cup Y)$ is empty. For two sets X and Y, *uncrossing* means that we replace one of X and Y with larger f-value by one of $X \cap Y$ and $X \cup Y$ with smaller f-value.

Note that submodularity $f(X \cap Y) + f(X \cup Y) \leq f(X) + f(Y)$ implies

$$\min\{f(X \cap Y), f(X \cup Y)\} \leq \max\{f(X), f(Y)\}$$

and thus uncrossing will not increase the minimum of $f(X)$ in a set system $\mathcal{X}$.

The key idea of our algorithm is *parity uncrossing*, a particular way of uncrossing that applies to asymmetric functions and can be related to the properties of triple families. Given that we start out with a set system $\mathcal{X}$, we replace elements X of $\mathcal{X}$ one by one by odd sets (members of the triple family) Y with $f(Y) \leq f(X)$. All we require for this procedure is a subcollection $\mathcal{X}' \subseteq \mathcal{X}$ with $\bigcap\{X \in \mathcal{X}'\}$ odd. By using this odd set to guide uncrossing, we may ensure an odd set as the outcome of uncrossing.

Now we give Algorithm GENERIC-MINIMIZE-ODDSETS, the framework of minimization based on the parity uncrossing subroutine described in detail in the next subsection. In this algorithm we proceed by reducing the size of $\mathcal{X}$ by using the odd sets

Y found by parity uncrossing. Given such a set Y, we will discard all $X \in \mathcal{X}$ with $f(X) \geq f(Y)$. In order to find Y, we require a subcollection of $\mathcal{X}$ with odd intersection; the minimization algorithm terminates when there is none. Then our particular implementations will complete the task of odd minimization by using the candidate minimizer Y_0 and the remaining collection $\mathcal{X}$.

GENERIC-MINIMIZE-ODDSETS($\mathcal{X}$)

$\quad \mathcal{Y} \leftarrow \emptyset \qquad \triangleright \mathcal{Y}$ contains the candidate odd sets that minimize f
$\quad$ **while** $\exists \, \mathcal{X}' \subset \mathcal{X}$ with $\bigcap \{X : X \in \mathcal{X}'\}$ odd **do**
$\quad\quad$ **obtain** an odd set Y with $f(Y) \leq \max\{f(X) : X \in \mathcal{X}'\}$
$\quad\quad$ $\mathcal{X} = \{X \in \mathcal{X} : f(X) < f(Y)\}$
$\quad\quad$ $\mathcal{Y} = \mathcal{Y} + Y$
$\quad$ **output** $\mathcal{X}$ and $Y_0 \in \mathcal{Y}$ with $f(Y)$ minimum

---

2.1 Subroutine PARITY-UNCROSS.

We introduce the technique of *parity uncrossing* that, given a collection $\mathcal{X}$ with $\bigcap\{X \in \mathcal{X}\}$ odd, finds an odd set Y with $f(Y) \leq f(X)$ for some $X \in \mathcal{X}$. The family of odd sets must satisfy $(*)$ but not necessarily $(**)$. We give a divide-and-conquer parity uncrossing procedure that makes $O(n)$ calls to the *evaluation* oracle. Notice that in case of graph cuts the evaluation oracle is $\tilde{\Theta}(n)$ times faster than the minimization oracle, hence in order not to exceed the time for finding $O(n)$ minimum cuts, we may make a total of $O(n^2)$ evaluation calls in the minimization algorithm. Our procedure stays within this bound as long as the algorithm calls parity uncrossing $O(n)$ times. However we may not use a more straightforward non-divide-and-conquer approach for parity uncrossing since that would cost a total of $O(n^3)$ evaluation calls.

We start the description of Algorithm PARITY-UNCROSS with the easy cases. If we have a single set X_1 in the input, it must be odd by the requirement on the intersection and thus we may immediately return this $Y = X_1$. For two sets X_1, X_2 we proceed by selecting either X_1 or X_2 if one of them is odd. If this is not the case, by Property $(*)$ $X_1 \cup X_2$ must be odd if (as assumed) $X_1 \cap X_2$ is odd. Thus we may select the odd set $X_1 \cup X_2$ or $X_1 \cap X_2$ with smaller f-value; by submodularity this value cannot be more than $\max\{f(X_1), f(X_2)\}$. Notice that here we do not require X and Y cross; in fact if $X \cap Y = \emptyset$, then $\emptyset$ is odd and if $V - X \cup Y = \emptyset$, then V is odd. In general we want to avoid these two trivial sets as minimizers but we cannot do it right at this step.

For a general input $\mathcal{X} = \{X_1, \ldots, X_k\}$ with $k > 2$ we use divide-and-conquer: we recurse on the two halves of the set $\mathcal{X}$; then on the outputs Y_1 and Y_2 of the recursive calls we perform the steps for case $k = 2$. Unfortunately we rely on the fact that the intersection of $\mathcal{X}$ is odd, but in a recursive call we may not expect this for either half of $\mathcal{X}$. Hence we must extend the input by a *restriction set* R; instead of requiring certain sets to be odd, we only require their intersection with R be odd. Thus initially $R = V$ and at a general recursive call it is a set with $R \cap \bigcap_i X_i$ odd.

We are ready with the description of Algorithm PARITY-UNCROSS if we describe the selection of the restriction set R. For the first half $\{X_1, \ldots, X_{\lceil k/2 \rceil}\}$ we may add the natural choice R as the intersection of all remaining X_i currently not processed. However for the second call we may simplify the proof if, instead of taking the corresponding choice by exchanging the role of the two halves of the input $\{X_i\}$, we take R as the output Y_1 of the first recursive call. Notice that Y_1 has an odd intersection with the previous $R = X_{\lceil k/2 \rceil + 1} \cap \ldots \cap X_k$ by the definition of parity uncrossing; this is exactly the requirement for the new $R = Y_1$ for the second recursive call. Finally the output Y_2 of this second call, again by the definition of parity uncrossing, has odd intersection with Y_1 and thus we may perform the steps of case $k = 2$ for Y_1 and Y_2.

PARITY-UNCROSS$(X_1, \ldots, X_k, R)$

   if $k = 1$ then
1   return $Y = X_1$
     $Y_1 \leftarrow$ PARITY-UNCROSS$(X_1, \ldots, X_{\lceil k/2 \rceil}, R \cap \bigcap_{i > \lceil k/2 \rceil} X_i)$
     $Y_2 \leftarrow$ PARITY-UNCROSS$(X_{\lceil k/2 \rceil + 1}, \ldots, X_k, R \cap Y_1)$
     if $Y_i \cap R$ $(i = 1$ or $2)$ is odd then
2   return $Y = Y_i$
     else
3   return either $Y = Y_1 \cup Y_2$ or $Y = Y_1 \cap Y_2$ with smaller f-value

---

Lemma 1. *If $X_1 \cap \ldots \cap X_k \cap R$ is odd, then the set Y returned by Algorithm PARITY-UNCROSS has $Y \cap R$ odd and $f(Y) \leq \max\{X_1, \ldots, X_k\}$.*

Proof. By induction on the recursive calls. First of all we show that the recursive calls are valid, i.e. their input satisfies that

$$X_1 \cap \ldots \cap X_{\lceil k/2 \rceil} \cap R \cap \bigcap_{i > \lceil k/2 \rceil} X_i \quad \text{and} \quad X_{\lceil k/2 \rceil + 1} \cap \ldots \cap X_k \cap R \cap Y_1$$

are odd. For the first intersection this follows since we have $X_1 \cap \ldots \cap X_k \cap R$ odd by the requirement on the input of Algorithm PARITY-UNCROSS. Next we apply the inductive hypothesis for the first recursive call to obtain $Y_1 \cap R \cap \bigcap_{i > \lceil k/2 \rceil} X_i$ odd. However this is exactly the requirement for the input of the second recursive call.

Now we show the first part of the claim stating $Y \cap R$ odd for the output of the algorithm. The claim is trivial for $k = 1$ (line 1) and for $Y_i \cap R$ odd (line 2). For the remaining case (line 3) we notice that the inductive hypothesis gives $Y_1 \cap Y_2 \cap R = (Y_1 \cap R) \cap (Y_2 \cap R)$ odd for the output of the second recursive call. Since we are not exiting in line 2, we must have $Y_i \cap R$ even for $i = 1, 2$. But then

$$R \cap (Y_1 \cup Y_2) = (R \cap Y_1) \cup (R \cap Y_2)$$

must also be odd by Property $(*)$.

We prove the second part again separate for the three output types in lines 1, 2 and 3 by induction. The claim is trivial for $k = 1$ (line 1). For the output $Y = Y_1$ at line 2 we get

$$f(Y_1) \leq \max\{f(X_1), \ldots, f(X_{\lceil k/2 \rceil})\}$$
$$\leq \max\{f(X_1), \ldots, f(X_k)\}$$

where the first inequality follows by induction. A similar proof can be obtained for $Y = Y_2$:

$$f(Y_2) \le \max\{f(X_1), \ldots, f(X_k)\}.$$

Finally if we exit at line 3,

$$\min\{f(Y_1 \cap Y_2), f(Y_1 \cup Y_2)\} \le \max\{f(Y_1), f(Y_2)\}$$
$$\le \max\{f(X_i) : i \le k\}$$

where the first inequality follows by submodularity while the second by the previous two inequalities for $f(Y_1)$ and $f(Y_2)$. $\square$

Lemma 2. *Algorithm* PARITY-UNCROSS *runs with $O(k)$ calls to the evaluation oracle.*

Proof. The number of calls to find $f(X)$ for $X \subset V$ satisfies the recurrence

$$T(k) = 2T(k/2) + O(1). \square$$

3   The Simplest Case: When $\emptyset$ and V Are not in the Triple Family

Our first minimization algorithm applies when the empty set and the ground set are even (not in the triple family). This assumption is not only necessary for ensuring an outcome different from the trivial minimizers $\emptyset$ and V but also technically necessary for the correctness of our first algorithm below.

The algorithm proceeds as follows. First it computes a collection $\mathcal{X}$ of $n - 1 = |V| - 1$ sets with the property that

> for each pair $u, v \in V$ there exists an $X \in \mathcal{X}$ such that $f(X)$ is minimum over all sets X that contain exactly one of u and v.

In terms of cuts in graphs, X is the smaller of a minimum u–v or a minimum v–u cut. The collection $\mathcal{X}$ can be computed by $2n - 2$ minimization oracle calls as proved in [4] or [1, Chapter 8.7]. Note that this fact is a weakening of the result of Gomory and Hu [8] since the collection $\mathcal{X}$ cannot be cross-free in general.

Given the above collection $\mathcal{X}$, we proceed by discarding its members by Algorithm GENERIC-MINIMIZE-ODDSETS. If a discarded set X minimizes f over sets containing exactly one of u, v, then no odd set S with value less than Y may separate u and v since $f(Y) \le f(X) \le f(S)$ for all sets S separating u and v. Hence we may contract u and v for the further steps, or instead we may consider the partition given by the remaining elements of $\mathcal{X}$ defined below.

Definition 2. *For a set system $\mathcal{X}$, let $\mathcal{P}(\mathcal{X})$ be the partition of the ground set V defined with both $u, v \in P$ for some $P \in \mathcal{P}$ if and only if either $u, v \in X$ or $u, v \notin X$ for all $X \in \mathcal{X}$.*

The correctness of the algorithm immediately follows by the next two theorems. The first theorem will also give a way of finding, at each iteration of the algorithm, a subcollection of $\mathcal{X}$ with odd intersection. The description of the algorithm is completed with this additional step in Section 3.2.

Theorem 1. *Let us consider a family of odd sets satisfying* $(*)$, $(**)$ *with* $\emptyset$ *and* V *even. Let* $\mathcal{X}$ *be an arbitrary collection of sets such that all non-empty subcollections* $\mathcal{X}' \subseteq \mathcal{X}$ *have* $\bigcap\{X \in \mathcal{X}'\}$ *even. Then* $\mathcal{P}(\mathcal{X})$ *contains no odd sets.*

Theorem 2. *Assume the family of odd sets satisfy* $(*)$ *with* $\emptyset$ *even. Let us run Algorithm* GENERIC-MINIMIZE-ODDSETS *with a collection* $\mathcal{X}$ *containing, for each pair* $u, v \in V$, *a minimizer of* f *over all sets* X *that contain exactly one of* u *and* v. *If* $\mathcal{P}(\mathcal{X})$ *contains no odd sets, then* $\mathcal{Y}$ *contains the minimizer of* f *over the odd sets.*

3.1   Proofs.   First we prove Theorem 2. The proof requires odd sets satisfy $(*)$ with $\emptyset$ even; neither $(**)$ nor the fact that V is even need to be used. These facts are used only for Theorem 1 that we prove later via a sequence of lemmas. In these lemmas we explicitly show where we use the requirement that V is even; we in fact get a weaker result that also applies when V is odd.

Proof (Theorem 2). Let S be an odd minimizer of f. First we show that S subdivides at least one set $P \in \mathcal{P}(\mathcal{X})$. This follows since the partition consists exclusively of even sets. The union of disjoint even sets is even by the repeated use of Property $(*)$ and the fact that $\emptyset$ is even. Thus the odd set S may not arise as the union of some sets in $\mathcal{P}(\mathcal{X})$.

Next consider two elements $u, v \in V$ with $u \in P \cap S$ and $v \in P - S$. The initial set system $\mathcal{X}$ contains a set X minimizing f over all sets that contain exactly one of u and v. This set must, at some stage of the algorithm, be discarded from $\mathcal{X}$; at the same time, a set Y is added to $\mathcal{Y}$ with $f(Y) \leq f(X)$.

Now we are ready with the proof. Since S is a set containing exactly one of u and v, we must have $f(S) \geq f(X)$, thus $f(S) \geq f(Y)$. On the other hand the algorithm returns an odd set $Y_0 \in \mathcal{Y}$ with $f(Y_0) \leq f(Y)$. In conclusion the odd set Y_0 returned at termination minimizes f over odd sets since $f(Y_0) \leq f(S)$. $\square$

Lemma 3. *Consider an arbitrary classification of subsets of* V *as even and odd and an arbitrary set system* $\mathcal{X}$. *Let* $P \in \mathcal{P}(\mathcal{X})$ *be odd with the maximum number of sets* $X \in \mathcal{X}$ *with* $P \subseteq X$. *Then all* $P' \in \mathcal{P}(\mathcal{X})$ *with* $P' \neq P$ *and* $P' \subseteq \bigcap\{X \in \mathcal{X} : P \subseteq X\}$ *are even. Here the intersection of an empty collection* $\{X \in \mathcal{X} : P \subseteq X\}$ *is defined to be the entire* V.

Proof. Assume in contrary that some $P' \neq P$ as above is odd. Consider some $X_0 \in \mathcal{X}$ that separate P and P', i.e. have $P \not\subseteq X_0$ and $P' \subseteq X_0$. Then P' is contained by one more set of $\mathcal{X}$ than P, a contradiction. $\square$

Lemma 4. *Let odd sets satisfy* $(*)$ *and* $(**)$. *Let* $\emptyset$ *be even. Let* $\mathcal{X}$ *and* P *be as in Lemma 3. Then* $\bigcap\{X \in \mathcal{X} : P \subseteq X\}$ *is odd.*

Proof. Let $Z = \bigcap\{X \in \mathcal{X} : P \subseteq X\}$. By Lemma 3 all $P' \in \mathcal{P}(\mathcal{X})$ with $P' \neq P$, $P' \subset Z$ are even. By repeatedly applying $(*)$ we get that $\bigcup\{P' \in \mathcal{P}(\mathcal{X}) : P' \subset Z, P' \neq P\} = Z - P$ is even. The proof is completed by applying $(**)$ to Z and $Z - P$. $\square$

The lemma almost implies Theorem 1: we may parity uncross $\{X \in \mathcal{X} : P \subseteq X\}$ whenever $P \in \mathcal{P}(\mathcal{X})$ is odd, provided the subcollection of $\mathcal{X}$ is *non-empty*. However for the possible element $P_0 = V - \bigcup\{X \in \mathcal{X}\}$ of $\mathcal{P}(\mathcal{X})$ we intersect the empty subcollection; while that is V by our definition and it could be odd, we cannot proceed by parity uncrossing. The final lemma shows that this may happen only if V is odd.

Lemma 5. *Let odd sets satisfy* $(*)$ *and* $(**)$. *Let* $\emptyset$ *and* V *be even. For a set system* $\mathcal{X}$, *let* $P_0 = V - \bigcup\{X \in \mathcal{X}\}$. *If all* $P \neq P_0$, $P \in \mathcal{P}(\mathcal{X})$ *are even, then so is* P_0.

Proof. By repeatedly applying $(*)$ we get that $\bigcup\{P \in \mathcal{P}(\mathcal{X}) : P \neq P_0\} = V - P_0$ is even. Then we are done by applying $(**)$ to V and $V - P_0$.   □

3.2   Implementation.   In Algorithm MINIMIZE-ODDSETS we give an efficient implementation for the algorithm of Section 3. Notice that currently we use an unspecified selection rule for $\mathcal{X}'$ in Algorithm GENERIC-MINIMIZE-ODDSETS that could take exponential time. Our procedure below, in contrast, requires only $O(n^2)$ time in addition to parity uncrossing.

MINIMIZE-ODDSETS

    $\mathcal{X} \leftarrow$ a family of sets that, for all $u, v \in V$ with $u \neq v$, contains a
    minimizer of $f(X)$ with exactly one of u and v in X
    $\mathcal{Y} \leftarrow \emptyset$
    while $\exists\, P \in \mathcal{P}(\mathcal{X})$ odd **do**
        $P \leftarrow$ odd element $P \in \mathcal{P}(\mathcal{X})$ with $|\{X : P \subseteq X\}|$ maximum
        $Y \leftarrow$ PARITY-UNCROSS$(\{X : P \subseteq X\}, V)$
        $\mathcal{X} = \{X \in \mathcal{X} : f(X) < f(Y)\}$
        $\mathcal{Y} = \mathcal{Y} + Y$
    output $Y_0 \in \mathcal{Y}$ with $f(Y_0)$ minimum

In our implementation we use Lemma 3 by selecting $P \in \mathcal{P}(\mathcal{X})$ odd with the maximum number of sets in $\mathcal{X}$ containing P; these elements of $\mathcal{X}$ will have odd intersection then. For this selection we must maintain the partition $\mathcal{P}(\mathcal{X})$ as well as counters for each member P of the partition with the number of sets in $\mathcal{X}$ containing P.

First we build the initial partition $\mathcal{P}(\mathcal{X})$. We do this by adding sets X in the order of increasing value $f(X)$. After each addition we maintain the partition $\mathcal{P}$ as a linked list; an update of the list as well as the counters takes $O(n)$ time for a set X.

The implementation of removals from $\mathcal{X}$ is easy, given the way we build the initial partition. Since we increment $\mathcal{X}$ with sets of increasing f-value and we always delete sets of value at least a certain $f(Y)$, we may perform deletion by simply moving back in history to the stage when the first set of f-value at least $f(Y)$ was added. The history may be preserved in the linked list of $\mathcal{P}(\mathcal{X})$ as follows: whenever a set X is inserted and it subdivides some $P \in \mathcal{P}$ into two sets P_1 and P_2, we insert P_2 in the list as an immediate successor of P_1. If we mark the link from P_1 to P_2 by set X, we may remove X from $\mathcal{X}$ in $O(n)$ time by traversing the linked list once.

4   Algorithms

In this section we describe four algorithms based on the parity uncrossing technique. In two steps we extend the algorithm of Section 3 first to the case when V (Section 4.1), then to the case when both $\emptyset$ and V belong to the parity family (Section 4.2). These algorithms may return both $\emptyset$ and V as minimizer and thus they are useless for example in the case of graph cuts. However their running time remains within the asymptotic bound of the algorithm of Section 3. The third algorithm is guaranteed to output a minimizer different from $\emptyset$ and V for the price of using $O(n)$ times more steps (Section 4.3).

The final Section 4.4 gives us a way of applying parity uncrossing to parity families (or parity subfamilies of lattices), a more general problem class introduced by Goemans and Ramakrishnan [6] where odd sets satisfy only $(*)$. Unfortunately our algorithm is $O(n)$ times slower than its counterpart for triple families and thus only achieves the same efficiency as that of Goemans and Ramakrishnan.

4.1   The Case $\emptyset$ Even, V Odd.   We consider a triple family satisfying $(*)$ and $(**)$ where V belongs to the family (V is odd) while $\emptyset$ does not ($\emptyset$ even). Recall that in this case Algorithm MINIMIZE-ODDSETS may terminate with a collection $\mathcal{X}$ with

$$P_0 = V - \bigcap \{X \in \mathcal{X}\} \text{ odd.}$$

We can easily modify the proof of Theorem 2 to get that if the odd minimizer S is not Y_0 returned by Algorithm MINIMIZE-ODDSETS, then it must arise as the union of certain members of partition $\mathcal{P}(\mathcal{X})$ and must contain P_0.

We will continue minimization hence by finding the minimal odd set containing P_0 in a modified ground set with all members of partition $\mathcal{P}(\mathcal{X})$ contracted to a single vertex. In this ground set V' a single element $v \in V'$ corresponding to P_0 has $\{v\}$ odd, while all other single-point sets are even.

Now we apply Algorithm GENERIC-MINIMIZE-ODDSETS in the following implementation. We start with a different

$$\mathcal{X} = \{X_u : f(X_u) \text{ is minimum among all sets } X \text{ with } v \in X, u \notin X\}.$$

Now we call Algorithm PARITY-UNCROSS always with the *entire* $\mathcal{X}$. Hence the algorithm terminates with a collection $\mathcal{X}$ such that $Z = \bigcap \{X \in \mathcal{X}\}$ is even. Notice that all sets containing Z are even since they arise as the union of Z and single-point sets other than $\{v\}$. The union of disjoint even sets is however even by $(*)$ and the fact that $\emptyset$ is even. We are thus done by the next lemma:

Lemma 6. *Let us run Algorithm* GENERIC-MINIMIZE-ODDSETS *with a family of odd sets satisfying* $(*)$, *with* $\mathcal{X} = \{X_u\}$ *as above, and by always parity uncrossing over the entire remaining collection* $\mathcal{X}$. *Let* $f(Y_0)$ *be minimum in* $\mathcal{Y}$ *and let* $Z = \bigcap \mathcal{X}$ *for the output* $\mathcal{X}$ *and* $\mathcal{Y}$. *Then if an odd set* S *containing* v *has* $f(S) < f(Y_0)$, *then* $S \supset Z$.

Proof. By contradiction let $u \in Z - S$. Since $u \notin X_u$, set X_u must have been discarded from $\mathcal{X}$ to obtain the intersection Z with $u \in Z$. Thus the output Y_0 of the algorithm has $f(Y_0) \le f(X_u)$. On the other hand S has $v \in S$ and $u \notin S$; by the minimality of $f(X_u)$ we have $f(X_u) \le f(S)$. We get the contradiction $f(Y_0) \le f(S)$. $\square$

4.2   A General Algorithm that May Return $\emptyset$ or V.   The previous algorithms may give an incorrect result if $\emptyset$ is odd: the proofs rely on the fact that the union of disjoint even sets is even and this fails if $\emptyset$ is odd. We may, in case V is even, turn to the dual (either complement every set or exchange the notions of union and intersection) to obtain a correct algorithm. However we need additional steps if both $\emptyset$ and V are odd.

Our algorithm builds on Lemma 6. First we make two subproblems by selecting a vertex v and finding the odd minimizer containing and not containing v separately. Since the second subproblem is the dual of the other, we only discuss the first one. Now by the lemma we may immediately get an even set Z such that the minimizer is either already found or else it contains Z. The second scenario can now be handled by the algorithm of Section 4.1 by considering a new ground set $V' = V - Z$ where set $X \subseteq V'$ is odd iff $X \cup Z$ is (originally) odd; over V' we define a new submodular function $f'(X) = f(X \cup Z)$. Since $\emptyset$ is even in the new family, we may obtain the odd minimizer Y of f'. Set Y is precisely the odd minimizer of f among sets containing Z.

4.3   Minimization with Sets $\emptyset$ and V Excluded.   The final algorithm finds the minimum f-value odd set different from $\emptyset$ and V even if $\emptyset$ or V is in the triple family. In order to avoid the possible trivial minimizers $\emptyset$ and V, we pay the price of calling the algorithm of the previous section $O(n)$ times. Note that the algorithm of Goemans and Ramakrishnan [6] also has this drawback and hence we are still faster by a factor of $O(n)$ in this case.

Our algorithm is based on a very simple reduction. As we saw in the previous section, we may restrict attention to sets containing an arbitrarily prescribed $v \in V$. Now we find the odd minimizer S of f containing v by $n - 1$ calls to the algorithm of Section 4.2 as follows. Since $S = V$ as minimizer is excluded, there exists $w \in V$ with $w \notin S$. For all $n - 1$ possible such w we proceed by finding the minimizer over $V' = V - \{v, w\}$ with $f'(X) = f(X + v)$ by the algorithm of Section 4.2. The result is an odd minimizer of f containing v but not containing w; we take the minimum over all choices of w.

Note that in case of even cardinality T-cuts C, we may also find the minimizer with $C \cap T \neq \emptyset, T$ by selecting both vertices u and v from T.

4.4   Parity Families.   A parity family is defined as a collection of subsets of V satisfying only $(*)$, i.e. if neither X nor Y belongs to the family, then $X \cap Y$ and $X \cup Y$ must either both belong to the family or both be outside the family. We may use the terminology of even and odd sets just like in the case of triple families.

For parity families Algorithm MINIMIZE-ODDSETS fails in Theorem 1 since parity families do not satisfy Property $(**)$. Thus it may happen that the partition $\mathcal{P}(\mathcal{X})$ given by the f-minimizer sets $\mathcal{X}$ contains an odd set P while, unlike in Lemma 4, the intersection $Z = \bigcap\{X \in \mathcal{X} : P \subseteq X\}$ is even. In this case the minimizer may be a set containing P but not the entire Z. We may apply Lemma 6 to find the minimizer for this case; however in the worst case we may have $\Omega(n)$ such odd P and use $\Omega(n^2)$ minimization oracle calls. In this algorithm and proof of correctness we never take complements of sets and thus we can also apply it to parity subfamilies of lattices.

5   Conclusion

We have given improved algorithms for the minimum odd or even cut problems and their generalizations. The algorithms achieve a factor n speedup over the previous best ones. While we believe that an algorithm with $O(n)$ flow computations is optimal for example for the directed minimum odd cut problem, we conjecture that some of the remaining open questions may have a positive answer:

- Can one solve the *undirected* minimum odd (even) cut problem by pipelining the $O(n)$ flow computations needed to a single one, in the flavor of the Hao–Orlin algorithm [12]?
- Can one optimize over parity families that are not triple families (cf. [6]) with only $O(n)$ minimization oracle calls (max-flows)?
- Can one improve the algorithm in the case $\emptyset$ or V is odd (such as the minimum even cut problem)?

6 Acknowledgement

The second author would like to thank András Frank for fruitful discussions.

References

1. Ahuja, R.K., T.L. Magnanti and J.B. Orlin, *Network flows: Theory, Algorithms and Applications*. Prentice Hall (1993). 94
2. Barahona, F. and M. Conforti, A construction for binary matroids. *Discr. Math.* **66**:213–218 (1987) 88, 90
3. Benczúr, A. A., Counterexamples for directed and node capacitated cut-trees. *Siam J. Comp.* **24**(3):505–510 (1995) 91
4. Cheng, C. K. and T. C. Hu, Ancestor trees for arbitrary multi-terminal cut functions. *Ann. Op. Res.* **33**:199–213 (1991) 94
5. Gabow, H. N., M. X. Goemans and D. P. Williamson, An efficient approximation algorithm for the survivable network design problem. *Proc. 3rd IPCO*, pp. 57–74 (1993) 91
6. Goemans, M. X. and V. S. Ramakrishnan, Minimizing submodular functions over families of sets. *Combinatorica* **15**(4):499–513 (1995) 88, 89, 90, 91, 97, 98, 99
7. Goemans, M. X. and D. P. Williamson, The Primal-Dual Method for Approximation Algorithms and its Application to Network Design Problems, Chapter 4 in *Approximation Algorithms*, D. Hochbaum, Ed. (1997) 89
8. Gomory, R. E. and T. C. Hu, Multi-terminal network flows, *J. SIAM* **9**:551–570 (1961) 91, 94
9. Grötschel, M., L. Lovász and A. Schrijver, The ellipsoid method and its consequences in combinatorial optimization. *Combinatorica* **1**:169–197 (1981) 88, 89, 90
10. Grötschel, M., L. Lovász and A. Schrijver, Corrigendum to our paper The ellipsoid method and its consequences in combinatorial optimization. *Combinatorica* **4**:291–295 (1984) 89
11. Grötschel, M., L. Lovász and A. Schrijver, *Geometric Algorithms and Combinatorial Optimization*. Springer-Verlag, Berlin, 1988. 89
12. Hao, J. and J. B. Orlin, A faster algorithm for finding the minimum cut in a graph, *Proc. of 3rd SODA*, pp. 165–174 (1992) 99
13. Lovász, L., Submodular functions and convexity, in: A. Bachem, M. Grötschel and B. Korte (eds.), *Mathematical Programming: The State of the Art, Bonn, 1982*. Springer, Berlin, pp. 235–257 (1983) 88
14. Padberg, M. W. and M. R. Rao, Odd minimum cut-sets and *b*-matchings. *Math. Op. Res.* **7**:67–80 (1982) 88, 90
15. Picard, J. C. and M. Queyranne. Selected applications of minimum cuts in networks, *I.N.F.O.R: Canadian Journal of Operations Research and Information Processing* **20**:394–422 (1982) 88
16. Rizzi, R., Excluding a simple good pair approach to directed cuts. *Graphs and Combinatorics*, to appear 91

Efficient Algorithms for Centers and Medians in Interval and Circular-Arc Graphs

S. Bespamyatnikh[1], B. Bhattacharya[2], J. Mark Keil[3],
D. Kirkpatrick[1], and M. Segal[1]

[1] Dept. of Computer Science, University of British Columbia, Vancouver, B.C.
Canada, V6T 1Z4   {besp,kirk,msegal}@cs.ubc.ca
[2] School of Computing Science, Simon Fraser University, Burnaby, B.C.,
Canada, V5A 1S6   binay@cs.sfu.ca
[3] Dept. of Computer Science, University of Saskatchewan, 57 Campus Dr.,
Saskatoon, Sask.,
Canada, S7N 5A9   keil@cs.usask.ca

Abstract. The p-center problem is to locate p facilities on a network so as to minimize the largest distance from a demand point to its nearest facility. The p-median problem is to locate p facilities on a network so as to minimize the average distance from one of the n demand points to one of the p facilities. We provide, given the interval model of an n vertex interval graph, an $O(n)$ time algorithm for the 1-median problem on the interval graph. We also show how to solve the p-median problem, for arbitrary p, on an interval graph in $O(pn \log n)$ time and on an circular-arc graph in $O(pn^2 \log n)$ time. Other than for trees, no polynomial time algorithm for p-median problem has been reported for any large class of graphs. We introduce a *spring model* of computation and show how to solve the p-center problem on an circular-arc graph in $O(pn)$ time, assuming that the arc endpoints are sorted.

1   Introduction

The p-center problem is to locate p facilities on a network so as to minimize the largest distance between the n demand points and p facilities. In the p-median problem we endeavour to locate p facilities on a network so as to minimize the average distance from one of n demand points to one of the p facilities. These problems are central to the field of location theory and has been researched extensively [4,8,11,15,16,18,22,23]. Applications include the location of industrial plants, warehouses, and public service facilities on transportation networks as well as the location of various service facilities in telecommunication networks [4,11,16,18,23].

   We model the network with a graph $G = (V, E)$ on n vertices and assume that the demand points coincide with the vertices. We also restrict the facilities to vertices as Hakimi [11] has shown that for the p-median problem the possible sites for the facilities can always be restricted to the set of vertices without increasing the cost. The p-median problem then becomes that of finding a set $X \subset V$ such that $|X| = p$ and for which $\sum_{i=1}^{n} d(v_i, X)$ is minimum.

M. Paterson (Ed.): ESA 2000, LNCS 1879, pp. 100–112, 2000.
© Springer-Verlag Berlin Heidelberg 2000

If $p = 1$ the problem is known as the 1-median problem and Hakimi [11] optimally solved it in a general network in $O(n^3)$ time. In a tree network Goldman [9] and Kariv and Hakimi [15] derive $O(n)$ time algorithms for the 1-median problem. Auletta et al. [1] gave a linear time algorithm for the 2-median problem on a tree. One can also find the 2-median of a general network in $O(n^3)$ time by considering all possible pairs of vertices as medians. For tree networks Gavish and Sridhar present an $O(n \log n)$ time algorithm [8].

For general p, Kariv and Hakimi [15] showed that the p-median problem is NP-complete. They were, however, able to produce an $O(p^2n^2)$ time algorithm for the case of tree networks [15]. The tree network algorithm was recently improved to $O(pn^2)$ by Tamir [22]. Other than for trees, no polynomial time algorithm for the p-median has been reported for any large class of graphs. In this paper we provide algorithms for the p-median problem on interval graphs.

The p-center problem for a given graph $G = (V, E)$ (restricting the facilities to be a subset of V) is to find a set $C \subset V$ such that $|C| = p$ and $\max_{i=1}^n d(v_i, C)$ is minimized. Regarding this problem Olariu provides an $O(n)$ time algorithm for locating a single central facility that minimizes the maximum distance to a demand point [20]. Frederickson [6] showed how to solve this problem for trees in optimal linear time (not necessarily restricting the facilities to be the vertices of the tree) using parametric search. The work of Kariv and Hakimi [14] presents results for general graphs; however since the problem is known to be NP-complete they were able to give only an $O(n^{2p+1} \log n/(p-1)!)$ runtime algorithm. Some work also has been done for approximating the p-center solution, see e.g, [2].

A graph $G(\mathcal{S}) = (V, E)$ ($G(\mathcal{A}) = (V, E)$) is an interval (circular-arc) graph if there exists a set $\mathcal{S}$ ($\mathcal{A}$) of intervals (arcs) on the real line (unit circle) such that there is a one-to-one correspondence between vertices $v_i \in V$ and intervals (arcs) $I_i \in \mathcal{S}$ such that an edge $(v_i, v_j) \in E$ if and only if $I_i \cap I_j \neq \emptyset$. The set $\mathcal{S}$ ($\mathcal{A}$) is called the interval (circular-arc) model for G. Interval and circular-arc graphs are important tools in many application areas, including scheduling and VLSI layout [10,19].

The problem of recognizing interval and circular-arc graphs is known to be solved in $O(|V|+|E|)$ time (see e.g, [3]), and we assume an interval (circular-arc) model $\mathcal{S}$ ($\mathcal{A}$) for G is available.

In the next section we review some relevant results on interval and circular-arc graphs. Section 3 provides an $O(n)$ time algorithm for the 1-median problem in interval graphs. In Section 4 we generalize the result for arbitrary p and give an $O(pn \log n)$ time algorithm. We show how to apply this result in order to solve the p-median problem in circular-arc graphs in Section 5. Section 6 presents an $O(pn)$ runtime solution for the p-center problem.

2   Preliminaries

Let $\mathcal{S}$ ($\mathcal{A}$) be the set of n intervals (arcs) in the interval (circular-arc) model for G. Without loss of generality, we assume that all the interval (arc) endpoints are distinct. We define each interval (arc) $I_i \in \mathcal{S}$ ($I_i \in \mathcal{A}$) by its left endpoint a_i and its right endpoint b_i. Once the endpoints are sorted we can replace the real value of an endpoint by its rank in the sorted order. Thus we use the integers from 1 to $2n$ as coordinates for the endpoints of the intervals (arcs) in S (A). That is, each integer $j \in [1 \ldots 2n]$ is a_i (or b_i) for some interval (arc) $I_i \in S$ ($I_i \in A$).

From the sorted list of interval endpoints, in $O(n)$ time, we compute the numbers of a's and b's to the left or right of every point q. In particular, let $\#aL(q)$ (likewise $\#bL(q)$) be the number of left (likewise right) endpoints of intervals in S that lie to the left of integer q, for $q \in [1 \ldots 2n]$. Similarly define $\#aR(q)$ (and $\#bR(q)$) to be the number of left (or right) endpoints of intervals in S that lie to the right of q. We use these quantities to quickly compute some of the structure of S. For example, the number of intervals left of interval I_i is $\#bL(a_i)$.

Chen et al [5] (see also [21]) define a successor function on intervals (arcs) in their paper on solving the all pairs shortest path problem on interval and circular-arc graphs. We use their idea to define a right successor and a left successor of an integer q. We say $RSUC(q) = I_i \in S$ if and only if $b_i = max\{b_j | I_j$ contains $q\}$ and $LSUC(q) = I_i \in S$ if and only if $a_i = min\{a_j | I_j$ contains $q\}$. For an interval I_i, $RSUC(I_i) = RSUC(b_i)$, and $LSUC(I_i) = LSUC(a_i)$.

We also define for an integer q, its ith iterated right successor $RSUC(q, i)$ to be $RSUC(RSUC(\ldots RSUC(q, 0)))$ where $RSUC$ appears i times. Define $RSUC(q, 0)$ to be q. Similarly we define $LSUC(q, i)$. For any interval I_j, we define $LSUC(I_j, i)$ to be $LSUC(a_j, i)$ and $RSUC(I_j, i)$ to be $RSUC(b_j, i)$.

Using a tree data structure based on the successor function, Chen et al. were able to compute iterated successors in constant time [5] (the same holds for the circular arcs as well).

Lemma 1. *After $O(n)$ time preprocessing, given integers $q \in [1 \ldots 2n]$ and $i \in [1 \ldots n]$ $RSUC(q, i)$ or $LSUC(q, i)$ can be computed in constant time.*

Chen et al. [5] make further use of their tree structure to attain the following.

Lemma 2. *After $O(n)$ time preprocessing, given two intervals (arcs) $I \in S$ and $J \in S$ ($I \in A$ and $J \in A$) the distance between I and J in G can be computed in constant time.*

3   1-Median

We will first consider the problem of locating one facility at a vertex (interval) of an interval graph to minimize the sum of the distances to the remaining vertices. For a candidate median interval I, $cost(I) = \sum_{J \in S} d(I, J)$.

We say an interval in S is maximal if it is not contained within any other interval in S. We need not consider a non-maximal interval I_i as a candidate for a median as any interval I_j containing I_i can replace I_i as median without increasing the cost.

For a candidate median (maximal interval) I_i, the cost of servicing the other intervals can be broken down into two parts according to whether the right endpoint b_j of an interval $I_j \in S$ lies left of b_i or right of b_i. Thus $Cost(I_i) = LSUM(b_i) + RSUM(b_i)$ where for an endpoint i of a maximal interval I_i we define $LSUM(i) = \sum_{I_j \in S | b_j < i} d(I_j, I_i)$ and $RSUM(i) = \sum_{I_j \in S | b_j > i} d(I_j, I_i)$ Here i in $LSUM(i)$ or $RSUM(i)$ could be a_i or b_i.

Then to compute the 1-median it suffices to compute $LSUM(i)$, and $RSUM(i)$ for each endpoint i of a maximal interval and let the median be the maximal interval I_k for which $LSUM(b_k) + RSUM(b_k)$ is minimum.

Let us turn to the problem of computing $LSUM(i)$ for each maximal interval endpoint i. If $\#bL(i) = 0$ then $LSUM(i) = 0$. In general, once the $LSUM$ values of all maximal interval endpoints left of i are computed, $LSUM(i)$ is computed in constant time using the formula in the following lemma.

Lemma 3. *If endpoint i is the left endpoint of a maximal interval I_i then*
$LSUM(i) = LSUM(a_{LSUC(i)}) + \#bL(a_{LSUC(i)}) + 2 * (\#bL(i) - \#bLa_{LSUC(i)})$.
If i is the right endpoint of maximal interval I_i then $LSUM(i) = LSUM(a_i) + \#bL(i) - \#bL(a_i)$.

Proof. To prove the first part we note that the contribution to $LSUM(i)$ of intervals whose right endpoints are left of the left endpoint of the maximal interval $LSUC(i)$ is given by $LSUM(a_{LSUC(i)}) + \#bL(a_{LSUC(i)})$. The intervals whose right endpoints lie between $a_{LSUC(i)}$ and i each contribute 2 to $LSUM(i)$.

To prove the second part, notice that the intervals whose right endpoints lie in I_i contribute 1 to $LSUM(i)$, and the contribution of other intervals is captured by $LSUM(a_i)$. ∎

Similarly $RSUM(i)$ can be computed in constant time once the $RSUM$ values of maximal interval endpoints right of i have been computed. The following lemma gives the formula.

Lemma 4. *If i is the left endpoint of interval I_i then $RSUM(i) = RSUM(b_i) + \#bL(b_i) - \#bL(a_i)$. If i is the right endpoint of interval I_i then $RSUM(i) = RSUM(b_{RSUC(i)}) + \#bR(i) + \#aR(i) - \#bR(b_{RSUC(i)})$.*

The formulae in the previous lemmas allow the computation of $RSUM(i)$ and $LSUM(i)$ for all i such that i is an endpoint of a maximal interval in $O(n)$ time. The 1-median is the maximal interval $I_i \in S$ for which $RSUM(b_i) + LSUM(b_i)$ is minimum. Thus

Theorem 1. *The 1-median of a set of intervals whose endpoints have been sorted can be computed in $O(n)$ time.*

4 p-Median in Interval Graphs

In this section we consider how best to locate p facilities in intervals to minimize the total distance from the remaining intervals to their nearest facility. When locating more than one median we need to be able to quickly determine the nearest median for an interval.

If there are no more than p maximal intervals in $\mathcal{S}$ the required set of p medians will consist of the maximal intervals in $\mathcal{S}$ plus an arbitrary selection of the remaining required number of intervals. Thus in the following we assume that there are more than p maximal intervals in S and we can also again restrict our candidate medians to maximal intervals.

Consider two candidate non-adjacent medians M_1 and M_2 in a possible solution to the p-median problem, $p \geq 2$ such that $b_1 < b_2$ and no other median I_i in the candidate solution exists such that $b_1 < b_i < b_2$. For an interval I_j such that $a_1 < a_j$ and $b_j < b_2$, we want to determine whether I_j will be serviced by M_1 or M_2. How we do this depends upon the distance between M_1 and M_2.

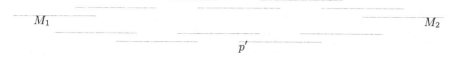

Fig. 1. The distance between M_1 and M_2 is even (six).

[**Even Case**] First we consider the case where the distance d between M_1 and M_2 is even. See the example in the figure 1. Let p' be the left endpoint of $LSUC(M_2, \frac{d}{2} - 1)$.

Claim (R,R split). Any interval with right endpoint right of p' is as close to M_2 as M_1 and any interval with right endpoint left of p' is as close to M_1 as M_2.

All intervals with right endpoint right of p' are at distance $\frac{d}{2}$ or less from M_2. They are at distance $\frac{d}{2}$ or more from M_1, or the fact that d is the minimum distance from M_1 to M_2 would be contradicted. Likewise all intervals with right endpoint left of p' are at distance at least $\frac{d}{2} + 1$ from M_2 and at distance at most $\frac{d}{2} + 1$ from M_1.

[**Odd Case**] Now we consider the case where the distance between M_1 and M_2 is odd and equal to $d + 1$. See the example in Figure 2. For this case let p be the right endpoint of $RSUC(M, \frac{d}{2})$. The following claim can be proven analogously to the corresponding claim in the even case.

Claim (R,R split). Any interval with right endpoint left of p is as close to M_1 as M_2 and any interval with right endpoint right of p is as close to M_2 as M_1.

We have then that the set of intervals to be serviced by one median I of a group of p medians can be determined by the proximity of the right endpoints of the intervals with respect to I. In order to account for the costs of these intervals to a solution we generalize $LSUM$ and $RSUM$ as follows. For an endpoint j of a maximal interval I_j and another maximal endpoint i such that $i < j$, define $LSUM(i,j) = \sum_{I_k \in S | i \leq b_k < j} d(I_k, I_j)$. Likewise for endpoint i of maximal interval I_i and another maximal endpoint j such that $i < j$, define $RSUM(i,j) = \sum_{I_k \in S | i < b_k \leq j} d(I_i, I_k)$.

To be able to efficiently compute $LSUM(i,j)$ and $RSUM(i,j)$ we relate these quantities to $LSUM(j)$ and $RSUM(i)$.

Fig. 2. The distance between M_1 and M_2 is odd (seven).

Lemma 5. *Let i and j, $i < j$, be endpoints of maximal intervals I_i and I_j respectively. Let I_p be the leftmost left iterated successor of I_j whose left endpoint a_p lies right of i, then*

$$LSUM(i,j) = LSUM(j) - LSUM(a_p) - \#bL(a_p) * d(I_p, I_j)$$
$$+(\#bL(a_p) - \#bL(i)) * (2 + d(I_p, I_j))$$

Proof: To compute $LSUM(i,j)$ the sum of the distances to I_j of the intervals whose right endpoints lie in $[i,j)$, we start with $LSUM(j)$ (the sum of the distances to I_j for intervals whose right endpoints lie left of j). Next we subtract away the contributions of intervals lying left of a_p, finally we add back in the contributions of intervals whose right endpoints lie in $[i,p)$.■.

The following lemma gives the analogously derived formula for $RSUM(i,j)$.

Lemma 6. *Let i and j, $i < j$, be endpoints of maximal intervals I_i and I_j respectively. Let I_p be the rightmost right iterated successor of I_i whose right endpoint b_p lies left of j, then*

$$RSUM(i,j) = RSUM(i) - RSUM(b_p) - \#bR(b_p) * d(I_i, I_p)$$
$$+(2 + d(I_i, I_p)) * \{ \text{ The \# of intervals contained in } [p,j) = \#In[p,j]\}$$
$$+(1 + d(I_i, I_p)) * \{ \text{ The \# of intervals containing point } b_p \text{ but not}$$
$$\text{point } j \text{ which is } \#bL(j) - \#bL(b_p) - \#In[b_p, j]\}.$$

All of the needed quantities can be easily computed except $\#In[p, p']$, the number of intervals contained in the interval $[p, p']$. The interval $[p, p']$ is not an interval in S and we need to be able to quickly compute the number of intervals of S contained in a query interval such as $[p, p']$. To do this we represent the intervals in S as points in the plane where interval I_i is represented by point (a_i, b_i). Then intervals in $[p, p']$ have the x coordinates of their corresponding points in the interval $[p, p']$ and the y coordinate of their corresponding points in the interval $[0, p']$. Thus the number of intervals of S contained in $[p, p']$ can be computed by range search counting query, using the priority search tree of McCreight in $O(\log n)$ query time [17].

The previous lemmas allow us to conclude that $LSUM(i,j)$ and $RSUM(i,j)$ can be efficiently computed.

Lemma 7. *For $i < j$ and i and j endpoints of maximal intervals, $LSUM(i,j)$ or $RSUM(i,j)$ can be computed in $O(\log n)$ time.*

With $LSUM(i,j)$ and $RSUM(i,j)$ available we now turn to the dynamic programming approach of Hassin and Tamir [12] for computing medians of points on the real line.

For j the endpoint of a maximal interval in S and $1 \le q \le p$ we define $F^q(j)$ to be the size of the optimal q median solution where each of the q medians must have at least one endpoint left of or equal to j and the intervals to be serviced have right endpoints less than or equal to j.

For j the right endpoint of a maximal interval in S and $1 \le q \le p$ we define $G^q(j)$ to be the size of the solution of the same subproblem that defines $F^q(j)$ except that the interval I_j, one of whose endpoints is j, is included in the solution as one of the q medians.

Let R be the set of right endpoints of maximal intervals in S and let M be the set of all endpoints of maximal intervals in S. Then for each endpoint j of a maximal interval

$$F^q(j) = min_{\{i<j \mid i\in R\}}\{G^q(i) + RSUM(i,j)\} \qquad (1)$$

and for each right endpoint j of a maximal interval

$$G^q(j) = min_{\{i<j \mid i\in M\}}\{F^{q-1}(i) + LSUM(i,j)\} \qquad (2)$$

with boundary conditions $F^q(1) = 0$ and $G^1(j) = LSUM(1,j)$.

Hassin and Tamir [12] exploit the quadrangle inequality in order to apply the fast dynamic programming algorithms of [7,13] to solve recurrences similar to $F^q(j)$ and $G^q(j)$. Here we have the quadrangle inequality for $LSUM(i,j)$ and $RSUM(i,j)$.

Lemma 8. *Let j, k, l and m be endpoints of maximal intervals such that $1 \le j \le k \le l \le m$. Furthermore restrict l and m to be right endpoints of maximal intervals. Then*

$$LSUM(j,m) - LSUM(j,l) \ge LSUM(k,m) - LSUM(k,l)$$

Proof: Let $M = LSUM(j,m) - LSUM(k,m)$. Then $M = \sum_{I_t \in S \mid j \le b_t < k} d(I_t, I_m)$. Let $JK = \{I_t \in S \mid j \le b_t < k\}$. Then $M = \sum_{I_t \in JK} d(I_t, I_m)$. Let $L = LSUM(j,l) - LSUM(k,l)$. Then $L = \sum_{I_t \in JK} d(I_t, I_l)$. Since m and l are right endpoints of maximal intervals and $1 \le j \le k \le l \le m$, we have that $M \ge L$ and the lemma follows. ∎

We can similarly show that the quadrangle inequality holds for $RSUM(i,j)$.

Lemma 9. *Let j, k, l and m be endpoints of maximal intervals such that $1 \le j \le k \le l \le m$. Furthermore restrict j and k to be right endpoints of maximal intervals. Then $RSUM(j,m) - RSUM(j,l) \ge RSUM(k,m) - RSUM(k,l)$.*

If $LSUM(i,j)$ and $RSUM(i,j)$ could be computed in constant time, the quadrangle inequality would allow us to follow Hassin and Tamir [12] in applying the fast dynamic programming algorithms in [7,13] and use equations (1) and (2) to compute $G^q(j)$ and $F^q(j)$ for all $1 \le q \le p$ and all relevant j (j an endpoint of a maximal interval for $F^q(j)$ and j the right endpoint of a maximal interval for $G^q(j)$) in $O(pn)$ total time. However the best time we have for computing each $LSUM(i,j)$ or $RSUM(i,j)$ from lemma 4.5 is $O(\log n)$. Thus we conclude

Theorem 2. *The p-median problem for a given interval graph can be computed in $O(pn \log n)$ time.*

5 p-Median in Circular Arc Graphs

Recall that we are given a set $\mathcal{A} = \{I_1, ..., I_n\}$ of n arcs on the unit circle centered at the origin. For the circular arc graph $G(\mathcal{A})$ we wish to find a set M of p nodes (medians) such that the sum of the distances from all nodes in $G(\mathcal{A})$ to their nearest nodes in M is minimized. We show how to use the algorithm for finding p medians for an interval graph in order to solve the same problem for the circular arc graphs.

The main result can be stated as the following theorem.

Theorem 3. *The p-median problem for a circular arc graph with n nodes can be solved in $O(nT(n))$ time, where $T(n)$ is the running time of an algorithm that solves the p-median problem for an interval graph with n nodes.*

Proof: Starting from the left endpoint of some arc $I_j \in \mathcal{A}$ we sort all the arcs in the clockwise order of their left endpoints. We assume that the indices of arcs in $\mathcal{A}$ already correspond to the sorted order of the arcs. We allow the use of any integer values for indices of arcs, using modulo n computation. For example, $I_1 = I_{1-n} = I_{n+1} = I_{2n+1} = \ldots$. Consider the set $M = \{I_{i_1}, I_{i_2}, \ldots, I_{i_p}\}$ of arcs that represent an optimal solution. Let C be the set of arcs whose left endpoints are between i_1 and i_2 in the clockwise order. Note that C is equal to $\mathcal{A}$ for $p = 1$. The set C is divided into two disjoint subsets C_1 and C_2 such that the nearest median for all arcs in C_1 (C_2) is I_{i_1} (I_{i_2}). In the case when some arc in C has the same distances to I_{i_1} and I_{i_2}, we put this arc into C_1. If $p = 1$ then instead of considering distances to I_{i_1} and I_{i_2} we consider distances computed in the clockwise and counterclockwise directions. Obviously, the indices of the arcs in C_1 (C_2) form a consecutive sequence of integers starting at i_1 (ending at i_2). Let I_k be the arc with the largest index in C_1. We shoot a ray which passes through I_k (to the left of left endpoint of I_{k+1}) in order to unroll the circle and obtain collection of the intervals, see Figure 3. Each arc from C that is intersected by a shooting ray is divided into two intervals. We discard the left intervals as depicted in Figure 3. The optimal solution for the resulted interval graph corresponds to the optimal solution for $G(\mathcal{A})$. Thus, in order to solve the p-median problem for a given circular arc graph $G(\mathcal{A})$ we apply n times the algorithm for an interval arc graph obtained by unrolling $G(\mathcal{A})$ using a ray through every endpoint. ∎

The following corollary follows immediately from the theorem.

Corollary 1. *The p-median problem for a given circular arc graph with n nodes can be solved in $O(pn^2 \log n)$ time.*

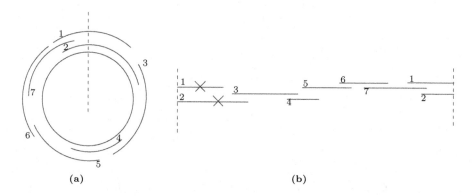

Fig. 3. (a) The initial circular arc graph; (b) Unrolling the circular arc graph into an interval graph, $k = 2$; discarded intervals are marked by a cross.

6 p-Center in Circular Arc Graphs

We wish to find a set M of p nodes (centers) in $G(\mathcal{A})$ such that the maximum distance from the nodes of the graph $G(\mathcal{A})$ to the nearest node in M is minimized.

The algorithm below is based on a *spring model* of computation. We first describe the idea of the algorithm and then provide technical details.

6.1 Spring Model

Starting from the left endpoint of some arc $I_j \in \mathcal{A}$ we sort all the arcs in the clockwise order of their left endpoints. We assume that the indices of arcs in $\mathcal{A}$ already correspond to the sorted order of arcs. As before, we allow the use of any integer values for indices of arcs, using modulo n computation. For each pair of arcs $I_j, I_k \in \mathcal{A}$ we define a set $T_{j,k}$ of arcs such that their left endpoints are between a_j and a_k in the clockwise order. This set $T_{j,k}$ can be divided into two subsets $T^1_{j,k}$ and $T^2_{j,k}$, such that the distance from every arc in $T^1_{j,k}$ $(T^2_{j,k})$ to I_j is smaller or equal (greater) than the distance to I_k. At any execution time we maintain a temporary set $C = \{I_{i_1}, \ldots, I_{i_p}\}$ of p centers such that $1 \le i_1 < i_2 < \ldots < i_p \le n$. For a sake of clarity we use C to denote centers in $G(\mathcal{A})$ and corresponding arcs as well. Initially, $C = \{I_1, \ldots, I_p\}$. For each arc $I_{i_j} \in C$ we define two *spring forces*, left force and right force as following.
- Left force $LFORCE(i_j)$ is defined as the largest distance between arcs in $T^2_{i_{j-1}, i_j}$ and I_{i_j}.
- Right force $RFORCE(i_j)$ is defined as the largest distance between arcs in $T^1_{i_j, i_{j+1}}$ and I_{i_j}.

In order to obtain a new set of centers we find some center I_{i_j} in the current set C and change it to I_{i_j+1}, see Figure 4. Note, that the centers in C always move in the clockwise order. However, not all the centers are allowed to move in the current moment. Let $C' \subseteq C$ be a set of centers that are allowed to move. The set C' is defined by the following rules.

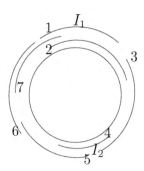

Fig. 4. For the 2-center problem the arcs I_1 and I_2 with the endpoints numbered 1 and 4, respectively, are the current centers. According to the definition $LFORCE(1) = RFORCE(1) = LFORCE(2) = 1$ and $RFORCE(2) = 2$. The arc I_2 is moved to be the arc with the endpoint numbered at 5.

Rule 1. $I_{i_j} \in C$ cannot move if $i_{j+1} = i_j + 1$, i.e. $I_{i_j} \notin C'$.

Rule 2. If the last center I_{i_p} is equal to I_p for the second time it is not allowed to move at future.

At any time we have $2p$ computed forces. The maximal force defines the movement of centers. If the maximal force is the right force for some center then this center moves to the right. If the maximal force is the left force for some center then the left neighbor of this center moves to the right. Formally, the moving center I_{i_j} is defined as an arc in C' with the largest value $\max(RFORCE(i_j), LFORCE(i_{j+1}))$. and in the case of tie we can choose any of them. The algorithm stops when $C' = \emptyset$. At the end of the algorithm's execution the last set of centers C is returned.

Theorem 4. *The algorithm above correctly computes p centers for a given circular arc graph $G(\mathcal{A})$.*

Proof: Let $\{I_{c_1}, \ldots, I_{c_p}\}$ be the optimal solution. Assume, to the contrary, that our algorithm misses the correct solution. Let us consider the last step of the algorithm where $C = \{I_{i_1}, \ldots, I_{i_p}\}$ and $i_1 \leq c_1, i_2 \leq c_2, \ldots, i_p \leq c_p$. The next step of the algorithm makes the wrong movement of center $I_{i_j}, 1 \leq j \leq p$, which changes to I_{i_j+1}. In other words, $i_j + 1 > c_j$. It follows that $i_j = c_j$. Therefore, $T_{i_j, i_{j+1}} \subseteq T_{c_j, c_{j+1}}$ and $T^1_{i_j, i_{j+1}} \subseteq T^1_{c_j, c_{j+1}}$, $T^2_{i_j, i_{j+1}} \subseteq T^2_{c_j, c_{j+1}}$ We obtain that $\max(RFORCE(i_j), LFORCE(i_{j+1})) \leq \max(RFORCE(c_j), LFORCE(c_{j+1}))$ and, thus, a set C presents an optimal solution. ∎

6.2   Implementation

In order to find forces efficiently we use the data structure proposed by Chen et al. [5]. Their data structure can be constructed in $O(n)$ time and $O(n)$ space. Using this data structure the query on the length of the shortest path between any two arcs can be answered in $O(1)$ time. This data structure also supports computing iterated right and left successors for a given integer value in constant time. We use Lemma 1 and Lemma 2 which provide an efficient way to maintain the spring model.

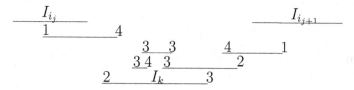

Fig. 5. Odd case in Theorem 5. The numbers on the arcs denote the distances to corresponding centers.

We distinguish between two cases : even and odd distance d. If d is odd then there is an arc I_k at distance $\frac{d-1}{2}$ and $\frac{d+1}{2}$ from the centers I_{i_j} and $I_{i_{j+1}}$. We claim that the maximum force is equal to $\frac{d+1}{2}$, i.e. there is no arc at distance greater than $\lceil \frac{d}{2} \rceil$ from both I_{i_j} and $I_{i_{j+1}}$. Assume, to the contrary, that such an arc I_l exists. If arcs I_k and I_l intersect then the distance of I_l to one of the centers is at most $\frac{d-1}{2} + 1 = \frac{d+1}{2}$. Otherwise, the arc I_l is either to the left of I_k or to the right. The distance from I_l to the corresponding center is less or equal to the distance from I_k to the same center. This contradicts the definition of I_l.

It remains to show how to deal with the even case. Let I_k be the arc at distance $\frac{d}{2}$ from both centers I_{i_j} and $I_{i_{j+1}}$. We define a vertical slab for I_k whose left side passes through the right endpoint of arc $RSUC(i_j, d/2 - 1)$ and whose right side passes though the left endpoint of arc $LSUC(i_{j+1}, d/2 - 1)$. We claim that the maximum force is equal to $d/2$ if and only if this slab contains no arcs. Otherwise the maximum force equals $d/2 + 1$. Note that all the arcs containing some point to the left of the slab (including the left side) are at distance at most $d/2$ from I_{i_j}. Similarly, all the arcs containing a point to the right of the slab (including the right side) are at distance at most $d/2$ from $I_{i_{j+1}}$. Thus, if the slab does not contain any arc, clearly all arcs are within the distance $d/2$ from centers; otherwise the arcs inside the slab are at distance $d/2 + 1$ from each center.

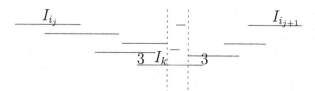

Fig. 6. Even case in Theorem 5. The dashed lines show the slab.

The computation of the length between two arcs as well as the computation of the right and left iterated successors can be done in constant time [5]. Determining whether the slab contains any arc also can be done in constant time. We define the array T whose indices corresponds to the left endpoints of the arcs and for any given index $i, 1 \leq i \leq n$, $T[i]$ keeps the rightmost left endpoint of the arcs with right endpoint less than i. This array T can be precomputed in linear time after once the endpoints of the arcs are sorted. The slab contains some arc if and only if the value of element of T corresponding to the right side of slab is larger than the left side of the slab.

We conclude by the following theorem.

Theorem 5. *The algorithm computes p centers for a given circular arc graph $G(\mathcal{A})$ in $O(pn)$ time if the endpoints of the arcs are sorted.*

Remark. Applying the spring model strategy to the p-median problem in a circular-arc graph leads to the similar result obtained in Section 5.

References

1. V. Auletta, D. Parente and G. Persiano, "Dynamic and static algorithms for the optimal placement of resources in a tree", *Theoretical Computer Science*, 165 (1996), 441–461. 101
2. J. Bar-Ilan and D. Peleg, "Approximation algorithms for selecting network centers", In *Proc. Workshop on Algorithms and Data Structures'91*, 1991, 343–354. 101
3. K.S. Booth and G.S. Leuker, "Testing for the consecutive ones property, interval graphs and graph planarity using PQ-tree algorithms", *Journal of Computer and System Sciences*, 13 (1976), 335 – 379. 101
4. M.L. Brandeau and S.S. Chiu, "An overview of representative problems in location research" *Management Science*, 35 (1989), 645-674. 100
5. D. Chen , D.T. Lee, R. Sridhar and C. Sekharam, "Solving the All-Pair Shortest Path Query Problem on Interval and Circular-Arc Graphs", *Networks*, 31(4), 1998, 249–258. 102, 109, 110
6. G. Frederickson, "Parametric search and locating supply centers in trees", In *Proc. Workshop on Algorithms and Data Structures'91*, 1991, 299–319. 101
7. Z. Galil and K. Park, "A linear-time algorithm for concave one-dimensional dynamic programming", *Information Processing Letters*, 33 (1990), 309–311. 106
8. B. Gavish and S. Sridhar, "Computing the 2-Median on Tree Networks in $O(n \log n)$ Time", *Networks*, 26 (1995), 305 – 317. 100, 101
9. A. J. Goldman, "Optimal center location in simple networks", *Transportation Science*, 5 (1971), 212 – 221. 101
10. M.C. Golumbic, *Algorithmic Graph Theory and Perfect Graphs*, Academic Press, New York, 1980. 101
11. S. L. Hakimi, "Optimal locations of switching centers and the absolute centers and medians of a graph", *Operations Research*, 12 (1964), 450 – 459. 100, 101
12. R. Hassin and A. Tamir, "Improved complexity bounds for location problems on the real line", *Operations Research Letters*, 10 (1991), 395–402. 106
13. M. Klawe, "A simple linear time algorithm for concave one-dimensional dynamic programming", Technical Report 89-16, University of British Columbia, Vancouver, 1989. 106
14. O. Kariv and S. L. Hakimi, "An algorithmic approach to network location problems I: The p-centers", *SIAM Journal on Applied Mathematics*, 37 (1979), 514 – 538. 101

15. O. Kariv and S. L. Hakimi, "An algorithmic approach to network location problems II: The p-medians", *SIAM Journal on Applied Mathematics*, 37 (1979), 539 – 560. 100, 101

16. M. Labbe, D. Peeters and J.F. Thisse, "Location on Networks", in *Handbooks in Operations Research and Management Science*, 8, Ball et al editors, Elsevier Science, 1995. 100

17. E. M. McCreight, "Priority Search Trees", *SIAM Journal on Computing*, 14 (1985), 257 – 276. 105

18. P. Mirchandani, "The p-median problem and generalizations", in *Discrete Location Theory*, Mirchandani and Francis editors, Wiley, 1990, 55 – 117. 100

19. R. Mohring, "Graph problems related to gate matrix layout and PLA folding", *Computational Graph Theory*, Tinhofer et al editors, Springer-Verlag, 1990, 17 – 51. 101

20. S. Olariu, "A simple linear-time algorithm for computing the center of an interval graph", *International Journal of Computer Mathematics*, 24 (1990), 121-128. 101

21. R. Ravi, M. Marathe and C. Rangan, "An optimal algorithm to solve the all-pair shortest path problem on interval graphs", Networks, 22 (1992), 21–35. 102

22. A. Tamir, "An $O(pn^2)$ algorithm for the p-median and related problems on tree graphs", *Operations Research Letters*, 19 (1996), 59 – 64. 100, 101

23. B.C. Tansel, R. L. Francis and T. J. Lowe, "Location on Networks: A Survey - Part I: The p-center and p-median problems", *Management Science*, 29 (1983), 482 – 497. 100

Exact Point Pattern Matching and the Number of Congruent Triangles in a Three-Dimensional Pointset

Peter Brass

Free University Berlin, Institute of Computer Science,
Takustrasse 9, D-14195 Berlin, Germany,
brass@inf.fu-berlin.de

Abstract. In this paper we study the complexity of the problem of finding all subsets of an n-point set B in threedimensional euclidean space that are congruent to a given m-point set A. We obtain a randomized $O(mn^{\frac{7}{4}} \log n\beta(n))$-algorithm for this problem, improving on previous $O(mn^{\frac{5}{2}})$ and $O(mn^2)$-algorithms of Boxer. By the same method we prove an $O(n^{\frac{7}{4}}\beta(n))$ upper bound on the number of triangles congruent to a given one among n points in threedimensional space, improving an $O(n^{\frac{9}{5}})$-bound of Akutsu et al. The corresponding lower bound for both problems is $\Omega(n^{\frac{4}{3}})$.

1 Introduction

Geometric pattern matching is an important practical as well as an interesting theoretical class of problems (see [4] for a survey); depending on the type of patterns to be matched (e.g. point sets, line drawings, greyscale images) it has developed connections to a variety of mathematical subjects. In this paper we will study pattern matching for finite point sets in euclidean space, looking for exact matches. This problem class is related to some Erdős-type questions in combinatorial geometry.

In this model, the most important questions, in order of difficulty, are

1. Are pattern A and B the same? (i.e. does there exist a rigid motion μ with $\mu(A) = B$?)
2. Does pattern A occur in B ? (i.e. does there exist a rigid motion μ with $\mu(A) \subset B$?)
3. Do patterns A and B have a large common subpattern? (i.e. what is the largest subset $A' \subset A$ for which there is a rigid motion μ with $\mu(A') \subset B$?)

The first of these problems, the congruence test problem, has been satisfactorily solved in dimensions two and three, with $O(n \log n)$ algorithms in the plane [20,17] and the space [3,5,6,22]. Higher-dimensional congruence test seems to be related to the unsolved problem of classifying finite symmetry groups, the currently best algorithm in d-dimensional space has running-time $O(n^{\lceil \frac{1}{3}d \rceil} \log n)$ [9].

M. Paterson (Ed.): ESA 2000, LNCS 1879, pp. 112–119, 2000.

The second of these problems has in the plane a simple solution in time $O(mn^{\frac{4}{3}}\log n)$: choose a point pair $a_1 a_2 \in A$, find all pairs $b_1 b_2$ with $d(b_1, b_2) = d(a_1, a_2)$, and try for each the motion defined by $a_i \mapsto b_i$. The complexity of this algorithm is determined by the number of possible matches to be tried, i.e. the number of occurences of the distance $d(a_1, a_2)$ in B. This is the famous difficult 'number of unit distances' problem in combinatorial geometry, the current best upper bound is $O(n^{\frac{4}{3}})$, and no improvement on the pattern matching problem seems possible without an improvement in the bound of this combinatorial geometry problem.

The three-dimensional case of the second problem was treated by Rezende and Lee [21] ($O(mn^d)$ in d dimensions) and Boxer [7] ($O(mn^{\frac{5}{2}}\log n)$) and [8] (claims $O(mn^2 \log n)$, analysis incorrect, also lower bound incorrectly cited). In this paper we obtain a randomized $O(mn^{\frac{7}{4}}\log n\beta(n))$ algorithm (where $\beta(n)$ is an extremely slow growing function) and establish a connection to the combinatorial geometry problem of the maximum number of congruent triangles [15], for which we find the same upper bound. This improves a previous $O(n^{1.8})$ bound of Akutsu et al. [2] to $O(n^{1.75}\beta(n))$.

For higherdimensional detection of congruent subsets the only result is the $O(mn^d)$ algorithm of Rezende and Lee [21], which is a very small improvement on the trivial algorithm in $O(mn^d \log n)$ (for a fixed d-tupel in A, check all possible image d-tuples in B, whether they define the correct congruence mapping). The correct order is probably $O(mn^{\frac{1}{2}d}\log n)$ (for even $d \geq 4$), since $cn^{\frac{d}{2}}$ congruent subsets are possible, but it is again an unsolved problem by Erdős and Purdy [16] whether this number is maximal.

The third problem, detection of common congruent subsets, was treated up to now only by Akutsu et al [2], who found algorithms of complexity $O(n^{1.43}m^{1.77} + n^2)$ for the twodimensional case, $O(n^{1.89}m^{2.8} + n^3)$ in three, $O(n^{2.87}m^4 + n^4)$ in four, and $O(n^{d-1}m^d + n^d)$ in d dimensions.

Exact point pattern matching is by its strong connections to problems of combinatorial geometry an interesting theoretical problem, but it should also be noted that it is at least related to some practical problems which are not satisfactorily solved in some critical applications. As an example one should consider the 'starmatching' problem: a satellite has to deduce its orientation in space from a picture of some part of the sky (giving positions of some stars, where a high accuracy is possible) and its internal starmap [23]. This is a critical component, permanent loss of orientation leads to loss of the satellite, and even temporary loss may make a mission useless. Thus the satellite 'Deep Space 1' had in summer 1999 a close encounter with the asteroid 1992KD 'Braille', but took only pictures of empty sky, since it made a false matching and had therefore the wrong orientation (later it recovered its orientation, but then the encounter was past) [11]. Indeed, point pattern matching occurs in a number of situations, where variously defined 'feature points' are matched.

Of course, as it is, exact point patten matching is a purely theoretical problem, since real-world data is always subject to measurement errors (although stars are a quite good approximation for points, and their angular distance can be measured with good accuracy). But the obvious alternative, Hausdorff-approximate pattern matching, suffers from a high complexity which also makes it quite impractical: enumerating all ε-approximately congruent matches of an m-point planar set in an n-point set has already an output complexity of $\left(\frac{n}{m}\right)^m$, and the best known algorithm for the for finding the optimum Hausdorff-approximately congruent set has a complexity of $O((n+m)^6 \log(nm))$ [18,4] even in the planar case, which is also quite impractical. Better bounds can be obtained only by introducing additional restrictions on the sets, some of which (like bounds on the ratio of diameter to smallest distance) sound rather artificial. So this should be considered as a subject in which we still try various models to find one which is at the same time sufficiently realistic, mathematically interesting and computationally well-behaved.

2   The Result

Theorem 1. *A set of n points in threedimensional space contains at most $O(n^{\frac{7}{4}}\beta(n))$ triangles congruent to a given triangle.*
They can be found by a randomized algorithm in $O(n^{\frac{7}{4}} \log n\beta(n))$ time.

The lower bound for the number of congruent triangles is $cn^{\frac{4}{3}}$ using a construction of Erdős, Hickerson and Pach [14], which gives a set of $n-1$ points on the sphere which give together with the sphere center this number of isosceles right-angled triangles. For other triangles no such construction is known.

In the usual way we can select any triangle $a_1a_2a_3$ in A, find by this algorithm the congruent triangles in B construct for each of these possible image triangles the rigid motion μ and check for it whether $\mu(A) \subset B$. This gives

Theorem 2. *For given threedimensional sets A and B of m and n points, respectively, all subsets of B congruent to A an be found by a randomized algorithm in $O(mn^{\frac{7}{4}} \log n\beta(n))$ expected time*

The algorithm to find the congruent triangles is very simple (with the exception of the distance listing operations, which are discussed in section 3). Given a triangle $a_1a_2a_3$ and a set B of n points in threedimensional space, we do the following:

1. Construct the graph of point pairs with distance $d_1 = d(a_1a_2)$ in B
2. Construct the graph of point pairs with distance $d_2 = d(a_1a_3)$ in B
3. For each point $b_i \in B$
   (a) List the d_1-neighbours and the d_2-neighbours of b_i
   (b) Find the $d_3 = d(a_2, a_3)$-distances between these two sets.
       Each d_3-distance pair b_jb_k gives together with b_i a triangle congruent to $a_1a_2a_3$.

We will show in section 3 that the point pairs with a given distance among n points in threedimensional space can be found in $O(n^{\frac{17}{11}+\varepsilon})$ (certainly not optimal) time, and the occurences of a given distance between two sets of x and

y points on two concentric spheres can be found in $O((x^{\frac{2}{3}}y^{\frac{2}{3}}+x+y)\log(x+y))$ time, thus the key step, dominating the complexity, is step 3(b). Let x_i be the number of d_1-neighbours of point b_i, and y_i be the number of its d_2-neighbours, then the total complexity is

$$O(n^{\frac{17}{11}+\varepsilon}) + \sum_{i=1}^{n} O(x_i^{\frac{2}{3}}y_i^{\frac{2}{3}} + x_i + y_i)\log n .$$

Since $\frac{1}{2}\sum x_i$ and $\frac{1}{2}\sum y_i$ are the number of d_1 and d_2-distances in B, and it is known that any fixed distance (unit distance) occurs at most $O(n^{\frac{3}{2}}\beta(n))$ times among n points in three-dimensional space [13], we only have to bound the sum $\sum_{i=1}^{n} x_i^{\frac{2}{3}}y_i^{\frac{2}{3}}$. Here we first use the Cauchy-Schwarz inequality to separate both distances and reduce it to the case of a single distance:

$$\sum_{i=1}^{n} x_i^{\frac{2}{3}}y_i^{\frac{2}{3}} \leq \left(\sum_{i=1}^{n} x_i^{\frac{4}{3}}\right)^{\frac{1}{2}} \left(\sum_{i=1}^{n} y_i^{\frac{4}{3}}\right)^{\frac{1}{2}}$$

$$\leq \max_{\substack{\text{unit distance graphs} \\ \text{in threedimensional space}}} \sum_{i=1}^{n} \deg(v_i)^{\frac{4}{3}} ,$$

where the last maximum is taken over all unit-distance graphs with n vertices in threedimensional space (so especially the d_1 and d_2-distance graphs of B).

Now for threedimensional unit distance graphs we have $\sum_{i=1}^{n} \deg(v_i) = O(n^{\frac{3}{2}}\beta(n))$ (number of unit distances [13]), but this is not enough to obtain a bound for $\sum_{i=1}^{n} \deg(v_i)^{\frac{4}{3}}$ (this is where the analysis of Boxer [8] fails). But we can obtain a bound on $\sum_{i=1}^{n} \deg(v_i)^3$ using the property that a threedimensional unit distance graph does not contain a subgraph $K_{3,3}$ (using the technique of [19]). For all common unit-distance neighbours of three points lie on the three unit spheres around these points, which have at most two triple intersection points. Now any graph contains $\sum_{i=1}^{n} \binom{\deg(v_i)}{3}$ subgraphs $K_{1,3}$, but in a $K_{3,3}$-free graph, each point triple can belong to at most two $K_{1,3}$s, so we have for our unit distance graph $\sum_{i=1}^{n} \binom{\deg(v_i)}{3} \leq 2\binom{n}{3}$. Thus we have the two inequalities

$$\sum_{i=1}^{n} \deg(v_i) \leq c_1 n^{\frac{3}{2}}\beta(n) \quad \text{and} \quad \sum_{i=1}^{n} \deg(v_i)^3 \leq c_2 n^3 .$$

We combine these using the Hölder inequality

$$\sum_{i=1}^{n} \deg(v_i)^{\frac{4}{3}} = \sum_{i=1}^{n} \left(\deg(v_i)^{\frac{5}{6}}\right)\left(\deg(v_i)^{\frac{1}{2}}\right)$$

$$\leq \left(\sum_{i=1}^{n} \left(\deg(v_i)^{\frac{5}{6}}\right)^{\frac{6}{5}}\right)^{\frac{5}{6}} \left(\sum_{i=1}^{n} \left(\deg(v_i)^{\frac{1}{2}}\right)^{6}\right)^{\frac{1}{6}}$$

$$= \left(\sum_{i=1}^{n} \deg(v_i)\right)^{\frac{5}{6}} \left(\sum_{i=1}^{n} \deg(v_i)^3\right)^{\frac{1}{6}}$$

$$\leq \left(c_1 n^{\frac{3}{2}} \beta(n)\right)^{\frac{5}{6}} \left(c_2 n^3\right)^{\frac{1}{6}}$$

$$\leq c_3 n^{\frac{7}{4}} \beta(n) \,.$$

This proves the claimed bound of Theorem 1.

3   Listing Distance Pairs

There remains the problem of listing all unit distance pairs among n points in threedimensional space, or between two sets of n_1, n_2 points on two concentric spheres (which can be replaced by a projection with sets the same sphere, for a different distance). There are already a number of papers on a very similar looking question [1,10], finding the k-th smallest distance pair among n points, with some implicit ordering of the pairs in case of equal distances. This of course allows to find some unit distance fast, and indeed to find the number of unit distances (first and last unit distance in that implicit ordering) by binary search, but it would be inefficient to use this algorithm again and again to find all unit distance pairs.

Instead we can reuse the technique of [13] (which is indeed the only point where randomisation comes in), which we sketch below. Further components we need are

- an algorithm for twodimensional point location in an arrangement of n unit circles on a sphere with $O(n^2 \log n)$ preprocessing $O(\log n)$ query time: any algorithm for line arrangements can be adapted to this setting.
- an algorithm of [12] for threedimensional point location in an arrangement of n spheres, with $O(n^{4+\varepsilon})$ preprocessing, $O(\log n)$ query time. This is certainly not optimal (the setting of that algorithm is much more general), but this step is not critical for our time bound

To find the unit distance pairs among n points in threedimensional space, we take for each point the unit sphere around that point and look for incident pairs (point, unit sphere). We randomly select a subset of $n^{\frac{1}{4}}$ of these spheres, they decompose space in $O(n^{\frac{3}{4}})$ cells. Each point belongs to exactly one cell, so the average number of points per cell is $n^{\frac{1}{4}}$. Each sphere intersects $O(n^{\frac{1}{2}})$ cells, since the $n^{\frac{1}{4}}$ selected spheres cut each sphere in $O(n^{\frac{1}{2}})$ twodimensional cells. Thus the total number of pairs 'sphere intersecting cell' is $O(n^{\frac{3}{4}})$, and the average number of spheres intersecting each cell is $O(n^{\frac{3}{4}})$. For each cell (on the average containing $O(n^{\frac{1}{4}})$ points, and intersected by $O(n^{\frac{3}{4}})$ spheres) we replace each intersecting sphere by its center, and each point by the unit sphere around that point. This preserves incidences and gives us instead for each cell a new 'local' arrangement of $O(n^{\frac{1}{4}})$ new spheres and $O(n^{\frac{3}{4}})$ new points, and there are at most $O(n^{\frac{3}{4}})$ incidences in this 'local' arrangement, since any three spheres are incident to at most two common points. Taking the sum over all cells of the incidences in each 'local' arrangement we get $O(n^{\frac{3}{2}})$ incidences, assuming the averages behave nicely. Working out the details, Clarkson et al. [13] obtained an upper bound $O(n^{\frac{3}{2}}\beta(n))$.

To obtain an enumeration algorithm from this counting method, a slight change in exponents is necessary, since an arrangement of n spheres has complexity $O(n^3)$, but the point location algorithm in [12] takes $O(n^{4+\varepsilon})$, resulting in a slightly worse total time bound of $O(n^{\frac{17}{11}+\varepsilon})$ for the enumeration algorithm.

We construct that arrangement of $n^{\frac{3}{11}}$ randomly selected 'special' spheres, and a point-location structure for it, in $O(n^{\frac{12}{11}+\varepsilon})$ time. For all points we find their cells in $O(n\log n)$ time. Then for each of the n spheres we construct the (twodimensional) arrangement induced by the $n^{\frac{3}{11}}$ 'special' spheres on that sphere in $O(n^{\frac{6}{11}}\log n)$ time, choose from each of the $O(n^{\frac{6}{11}})$ cells of this arrangement a point and perform for it a point location in the threedimensional arrangement. Thus we find the cells that are intersected by this sphere in $O(n^{\frac{6}{11}}\log n)$ time. Now we have for each cell the list of points contained in that cell $(O(n^{\frac{2}{11}}))$, and the list of spheres intersecting that cell $(O(n^{\frac{8}{11}}))$. For each cell $(O(n^{\frac{3}{11}})$ times) we again replace spheres by centerpoints, and points by new unit spheres, and construct the arrangement of the new spheres in $O(n^{\frac{8}{11}+\varepsilon})$ time, then perform a point location for each of the $O(n^{\frac{8}{11}})$ new points in $O(\log n)$ time to find its incident spheres. Thus we find all point-sphere incidences (unit distances) in $O(n^{\frac{17}{11}+\varepsilon})$ time.

The same method works even simpler for unit distances between two sets X,Y of n and m points on a sphere: we replace one set of points by unit circles on the sphere around these points and count incidences point-circle. If one set is much smaller $(n < m^{\frac{1}{2}}$ or $n > m^2)$, take the smaller set (e.g. X, so $n < m^{\frac{1}{2}})$, replace each point by the unit circle around that point, construct in $O(n^2 \log n)$ time a point location structure for that arrangement and look up for each point of the larger set (Y) to which cell it belongs $(O(m\log n))$. This takes a total time of $O(n^2 \log n + m\log n)$, which is $O((n^{\frac{2}{3}}m^{\frac{2}{3}} + n + m)\log(n+m))$ for $n < m^{\frac{1}{2}}$. The same works for $n > m^2$.

If $m^{\frac{1}{2}} < n < m^2$, replace each point of Y by the radius one circle around that point of Y, giving m circles. We randomly select $s = n^{\frac{2}{3}}m^{-\frac{1}{3}}$ 'special' circles, by which we decompose the sphere in $O(s^2)$ cells, and construct a point location structure for that arrangement in $O(s^2 \log s) = O(n^{\frac{4}{3}}m^{-\frac{2}{3}}\log(n+m))$ time. Then for each point we find in which cell it is (total time $O(n\log s)$), and for each circle which cells it intersects (each circle intersects $2s$ cells, total time $O(ms\log s) = O(n^{\frac{2}{3}}m^{\frac{2}{3}}\log(n+m))$. The average cell contains $O(\frac{n}{s^2})$ points and is intersected by $O(\frac{m}{s})$ circles. Again for each cell we replace circles by centerpoints and points by unit circles around them, construct (for each cell) a point location structure for this new arrangement in $O(\frac{n^2}{s^4}\log\frac{n}{s^2}) = O(n^{-\frac{2}{3}}m^{\frac{4}{3}}\log(n+m))$ time, and look up for each point the cell (the incident circle) in $O(\frac{m}{s}\log(n+m)) = O(n^{-\frac{2}{3}}m^{\frac{4}{3}}\log(n+m))$ time. This gives a total complexity of $O(n^{-\frac{2}{3}}m^{\frac{4}{3}}\log(n+m)) + O(n^{\frac{4}{3}}m^{-\frac{2}{3}}\log(n+m)) + O(n^{\frac{2}{3}}m^{\frac{2}{3}}\log(n+m))$, which is (for $m^{\frac{1}{2}} < n < m^2$) again $O((n^{\frac{2}{3}}m^{\frac{2}{3}} + n + m)\log(n+m))$, as claimed.

Again we suppressed the details of why a random selection generates sufficiently 'average' cells; for this see [13].

References

1. Agarwal, P.K., Aronov, B., Sharir, M., Suri, S.: Selecting distances in the plane. Algorithmica **9** (1993) 495–514 116
2. Akutsu, T., Tamaki, H., Tokuyama, T.: Distribution of distances and triangles in a point set and algorithms for computing the largest common point set. Discrete Comput. Geom. **20** (1998) 307–331 113
3. Alt, H., Mehlhorn, K., Wagener, H., Welzl, E.: Congruence, similarity, and symmetries of geometric objects. Discrete Comput. Geom. **3** (1988) 237–256 112
4. Alt, H., Guibas, L.: Resemblance of geometric objects. In J.-R. Sack, J. Urrutia (Eds.): Handbook of Computational Geometry Elsevier 1999, 121–153 112, 114
5. Atallah, M.J.: On symmetry detection. IEEE Trans. Comput. **34** (1985) 663–666 112
6. Atkinson, M.D.: An optimal algorithm for geometrical congruence. J. Algorithms **8** (1987) 159–172 112
7. Boxer, L.: Point set pattern matching in 3-D. Pattern Recog. Letters **17** (1996) 1293–1297 113
8. Boxer, L.: Faster point set pattern matching in 3-D. Pattern Recog. Letters **19** (1998) 1235–1240 113, 115
9. Braß, P., Knauer, C.: Testing the congruence of d-dimensional point sets. To appear in ACM Symposium on Comput. Geom. 2000 112
10. Chan, T.M.: On enumerating and selecting distances. ACM Symposium on Comput. Geom. 1998, 279–286 116
11. Chapman, C.R.: News and Reviews. The Planetary Report **19** No. 5 (1999), p.18 113
12. Chazelle, B., Edelsbrunner, H., Guibas, L.J., Sharir, M.: A singly-exponential stratification scheme for real semi-algebraic varieties and its applications. Theoretical Comput. Sci. **84** (1991) 77–105 116, 117
13. Clarkson, K., Edelsbrunner, H., Guibas, L.J., Sharir, M., Welzl, E.: Combinatorial complexity bounds for arrangements of curves and spheres. Discrete Comput. Geom. **5** (1990) 99–160 115, 116, 117, 118
14. Erdős, P., Hickerson, D., Pach, J.: A problem of Leo Moser about repeated distances on the sphere. Amer. Math. Mon. **96** (1989) 569–575 114
15. Erdős, P., Purdy, G.: Some extremal problems in geometry. J. Comb. Theory **10** (1971) 246–252 113
16. Erdős, P., Purdy, G.: Some extremal problems in geometry IV. Proc. 7th South-Eastern Conf. Combinatorics, Graph Theory, and Computing 1976, 307–322 113
17. Highnam, P.T.: Optimal algorithms for finding the symmetries of a planar point set. Inf. Proc. Letters **22** (1986) 219–222 112
18. Huttenlocher, D.P., Kedem, K., Kleinberg, J.M.: On dynamic Voronoi diagrams and the minimum Hausdorff distance for point sets under Euclidean motion in the plane. Proc. 8th Annual ACM Symposium Comput. Geom. (1992) 110–120 114

19. Kővári, T., Turán, P., Sós, V.T.: On a problem of K. Zarankiewicz Colloq. Math. **3** (1954) 50–57   115

20. Manacher, G.K.: An application of pattern matching to a problem in geometrical complexity. Inf. Proc. Letters **5** (1976) 6–7   112

21. de Rezende, P.J., Lee, D.T.: Point set pattern matching in d-dimensions. Algorithmica **13** (1995) 387–404   113

22. Sugihara, K.: An $n \log n$ algorithm for determining the congruity of polyhedra. J. Comput. Syst. Sci. **29** (1984) 36–47   112

23. Weber, G., Knipping, L., Alt, H.: An Application of Point Pattern Matching in Astronautics. Journal of Symbolic Computation **17** (1994) 321–340   113

Range Searching over Tree Cross Products

Adam L. Buchsbaum[1], Michael T. Goodrich[2*], and Jeffery R. Westbrook[3**]

[1] AT&T Labs, Shannon Laboratory, 180 Park Ave., Florham Park, NJ 07932,
alb@research.att.com, http://www.research.att.com/info/alb
[2] Dept. of Computer Science, Johns Hopkins University, Baltimore, MD 21218,
goodrich@jhu.edu, http://www.cs.jhu.edu/~goodrich/home.html
[3] 20th Century Television, Los Angeles, CA 90025,
jwestbrook@acm.org

Abstract. We introduce the *tree cross-product problem*, which abstracts a data structure common to applications in graph visualization, string matching, and software analysis. We design solutions with a variety of tradeoffs, yielding improvements and new results for these applications.

1 Introduction

Range searching is a classic problem in data structure design [8,10,12,23,29]. It is typically defined on a *ground set*, S, of ordered tuples $(x_1, \ldots, x_d)$ that are to be queried by *ranges*, which are specified by intersections of tuples of intervals on the coordinates. A *simple query* asks if such a range R contains any point of S, and a *reporting query* asks for all points of S inside R.

We introduce and study an interesting variant of range searching, which we call *range searching over tree cross products*. Given a priori are d rooted trees, $T_1, \ldots, T_d$, and a ground set E of tuples $(x_1, \ldots, x_d)$, such that $x_i \in T_i$, $1 \le i \le d$. The nodes of the trees and the set E define a d-partite hypergraph G. We wish to perform several possible queries on tuples of nodes $\boldsymbol{u} = (u_1, \ldots, u_d)$, such that $u_i \in T_i$, $1 \le i \le d$. Analogous to a simple range searching query, an *edge query* determines if there is a hyperedge in G that connects descendents of all the u_i nodes. We say that such a hyperedge *induces* $\boldsymbol{u}$. Likewise, a *reporting query* determines all hyperedges in G that induce $\boldsymbol{u}$. An *expansion query* determines, for each y that is a child of some designated u_i, whether the tuple formed by replacing u_i with y in $\boldsymbol{u}$ is induced by a hyperedge in G. The goal is to preprocess the trees and hypergraph so that these queries can be performed efficiently on-line. Dynamic variants admit updates to the trees and hypergraph.

1.1 Motivating Applications. We explore four applications of tree cross products. (We define the applications formally in Section 5.) First is the *hierarchical graph-view maintenance problem*, which has applications in graph visualization [7,9]. We are given a rooted tree, T, and a graph, G, of edges superimposed between leaves of T. At any point, there exists a *view*, U, of G, which

---

* Work supported by NSF Grant CCR-9732300 and ARO Grant DAAH04-96-1-0013.
** Work completed while a member of AT&T Labs.

M. Paterson (Ed.): ESA 2000, LNCS 1879, pp. 120–131, 2000.

is a set of nodes of T such that each leaf of T has precisely one ancestor in U. The *induced graph*, G/U, is obtained by contracting each vertex in G into its representative in U, removing multiple and loop edges. The problem is to maintain G/U as nodes in U are expanded (replaced by their children) or contracted (replace their children). Buchsbaum and Westbrook [7] show how to expand and contract nodes in U in optimal time for an unweighted graph G: linear in the number of changes to G/U. The tree cross-product problem generalizes the graph-view problem. Our approach is orthogonal to that of [7], however.

Two more applications involve string matching. In *text indexing with one error*, we want to preprocess a string T of length n so that we can determine on-line all the occurrences of a pattern P of length m with one error (using Hamming or Levenshtein [19] distance). This problem has applications in password security [20] and text filtering [3,4,22,24]. We improve on the work of Ferragina, Muthukrishnan, and de Berg [11] and Amir et al. [2], whose solution traverses two suffix trees in tandem, periodically performing analogous reporting queries. Grossi and Vitter [15] use a similar strategy to report contiguous entries in compressed suffix arrays, and we also improve their bounds.

Finally, we consider finding *hammocks* in directed graphs. These are regions of nodes that are separated from designated source and sink nodes by the equivalent of articulation points. Hammocks have been studied in the context of compiler control-flow analysis [17], in which solutions to the problem of finding all hammocks off-line are known. We present the first results for finding hammocks on-line, with an application to software system analysis [5].

1.2   Our Results. We contribute a formal definition of tree cross-product operations, and we relate them to range searching. Rather than use classical range search techniques, we exploit the structure of the input to devise a framework based upon simpler search structures. Let $n = \sum_{i=1}^{d} |T_i|$, $m = |E|$, and k be the number of edges reported by a reporting or expansion query. Using $O(m \log^{\frac{d-1}{2}} n)$ (rsp., $O(m(\log \log n)^{d-1}))$ space, we can perform edge queries in $O(2^{d-1} \log n / \log \log n)$ time (rsp., $O(\log n (\log \log n)^{d-2}))$ time; reporting queries add $O(k)$ to the edge-query time, and expansion queries multiply the edge-query time by $O(k)$. In the dynamic case, we can perform insertions and deletions of hyperedges in G in $O(\log^{\frac{d+1}{2}} n / \log \log n)$ (rsp., $O(\log n (\log \log n)^{d-2}))$ time. In the two-dimensional case, note that we can achieve logarithmic query **and** update times in almost-linear space. No classical range searching result provides the same bounds. In the static case, the query times improve by $\log n / (\log \log n)^2$ factors, and the preprocessing time equals the space. All are deterministic, worst-case bounds. Our framework allows simple implementations for practical cases, using nothing more sophisticated than balanced search trees.

Applied to graph views, we present a dynamic solution that improves the results of Buchsbaum and Westbrook [7] by factors of $\log^2 n / \log \log n$ in space (to $O(m \log \log n)$) and $\log^2 n$ in update time (to $O(\log n)$). The cost is a $\log n$ factor penalty in expand time; contract remains optimal. For the static problem, the space (and preprocessing time) reduction is the same, while the expand penalty is only $(\log \log n)^2$. All of our bounds are deterministic, worst-case, whereas the

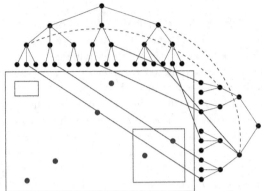

Fig. 1. Tree cross products for $d = 2$. Graph edges are shown as solid arcs and corresponding points; two potential induced edges are shown as dashed arcs and corresponding rectangular ranges.

prior update and preprocessing time bounds [7] are expected. For text indexing with one error, we improve all the bounds of Amir et al. [2] by $\log n$ factors. For compressed suffix-array traversal, we improve the additional reporting time of Grossi and Vitter [15] by a factor of $\log n/(\log \log n)^2$. For on-line hammock finding, we give a data structure that uses $O(n\sqrt{\log n})$ (rsp., $O(n \log \log n)$) space and preprocessing time and returns hammocks of size k in $O(\log \log n + k)$ (rsp., $O((\log \log n)^2 + k)$) time each; these are the first such results.

Table 1 in Section 4 summarizes our results.

2   Tree Cross Products

Let $G = (V(G), E(G))$ denote a (hyper)graph with vertex set $V(G)$ and edge set $E(G)$. Let T be a rooted tree. For $v \in V(T)$: let $children(v)$ be the set of children of v, and let $desc(v)$ be the set of descendents of v. Given d rooted trees, $T_1, \ldots, T_d$, consider some d-partite hypergraph, G, such that $V(G) = \bigcup_{i=1}^{d} V(T_i)$ and $E(G) \subseteq V(T_1) \times \cdots \times V(T_d)$, which we also write $\prod_{i=1}^{d} V(T_i)$. G is a subset of the *cross product of* $T_1, \ldots, T_d$. Denote by $\boldsymbol{u}$ an element $(u_1, \ldots, u_d) \in \prod_{i=1}^{d} V(T_i)$. Define $E(\boldsymbol{u})$ to be the set $\{\boldsymbol{x} \in E(G) : x_i \in desc(u_i), 1 \le i \le d\}$, i.e., the set of hyperedges of G connecting descendents of each component of $\boldsymbol{u}$.

Let $\mathcal{I}$ be the hypergraph with vertex set $\bigcup_{i=1}^{d} V(T_i)$ and edge set $\{\boldsymbol{u} : |E(\boldsymbol{u})| > 0\}$. We call $\mathcal{I}$ an *induced hypergraph* and hyperedges of $\mathcal{I}$ *induced hyperedges*, because they are induced by the cross-product subset G of the T_is. (See Fig. 1.) If each hyperedge $\boldsymbol{e} \in E(G)$ has a weight, $w(\boldsymbol{e}) \in \mathcal{G}$, from a given abelian group, $(\mathcal{G}, \odot)$, then the *weight* of $\boldsymbol{u} \in E(\mathcal{I})$ is defined to be $w(\boldsymbol{u}) = \bigodot \{w(\boldsymbol{x}) : \boldsymbol{x} \in E(\boldsymbol{u})\}$.

Given $T_1, \ldots, T_d$, and G, the *tree cross-product problem* is to perform on-line the following *tree cross-product operations* on tuples $\boldsymbol{u} \in \prod_{i=1}^{d} V(T_i)$.
edgeQuery($\boldsymbol{u}$*).* Determine if $\boldsymbol{u} \in E(\mathcal{I})$.
edgeReport($\boldsymbol{u}$*).* Determine $E(\boldsymbol{u})$.
edgeWeight($\boldsymbol{u}$*).* Determine $w(\boldsymbol{u})$ (undefined if $\boldsymbol{u} \notin E(\mathcal{I})$).
edgeExpand($\boldsymbol{u}, j$*),* where $1 \le j \le d$. For each $x \in children(u_j)$, determine if
   $(u_1, \ldots, u_{j-1}, x, u_{j+1}, \ldots, u_d) \in E(\mathcal{I})$.

We consider the static version of the problem and also one in which hyperedges can be added and deleted. Although we can implement $edgeExpand(\boldsymbol{u}, j)$ as an appropriate collection of $edgeQuery(\cdot)$ operations, we demonstrate that considering $edgeExpand(\boldsymbol{u}, j)$ as a primitive can lead to more efficient solutions.

We denote by T the forest containing $T_1, \ldots, T_d$, and define $n = |V(G)| = |V(T)|$, and $m = |E(G)|$. For $v \in V(T)$: let $depth(v)$ be the depth of v; let $p(v)$ be the parent of v; and let $leaves(v)$ be the set of leaf descendents of v. Let $depth(T) = \max_{v \in V(T)} depth(v)$. We assume T is given explicitly, although our methods extend when it is implicitly defined by $depth(\cdot)$, $p(\cdot)$, $children(\cdot)$, etc.

In an $O(n)$-time preprocessing phase, we perform postorder traversals of $T_1, \ldots, T_d$ so that each node is assigned its postorder number (ordinally from 1 for each tree) and depth. We abuse notation and use u itself to denote the postorder number of node u in its respective tree. For any node $u \in V(T)$, define $\min(u) = \min\{x : x \in desc(u)\}$ and $\max(u) = \max\{x : x \in desc(u)\}$.

3   A Range Searching Framework

We first assume $d = 2$ (and discuss G in terms of a graph with edges of the form (u, v)) and later extend our solutions to higher dimensions. Determining if $(u, v) \in E(\mathcal{I})$ can be viewed as a two-dimensional range query. Consider a $|V(T_1)| \times |V(T_2)|$ integral grid, with a point (x, y) for each edge $(x, y) \in E(G)$, as shown in Fig. 1. $E(u, v)$ corresponds precisely to the points in the range $([\min(u), \max(u)], [\min(v), \max(v)])$. A straightforward solution applies classical range searching results, but this ignores the structure imposed by T.

Each of T_1 and T_2 defines only $O(n)$ one-dimensional subranges that can participate in any two-dimensional query. To exploit this structure, we store a set $S(u)$ with each node $u \in V(T)$, which contains the far endpoints of all graphs edges incident on vertices in $desc(u)$. For $u \in V(T)$, we maintain the invariant that $S(u) = \{y : (x, y) \in E(G) \vee (y, x) \in E(G), \ x \in desc(u)\}$. Each $y \in S(u)$ is thus in the tree other than that containing u. The operations on $S(u)$ are $insert(S(u), y)$, $delete(S(u), y)$, and $succ(S(u), y)$. Operation $succ(S(u), y)$ returns the successor of y in $S(u)$, or ∞ if there is none.

Thus, (u, v) is an induced edge of $\mathcal{I}$ if and only if (1) u and v are in separate trees, and (2) $succ(S(u), \min(v) - 1) \leq \max(v)$. By the invariant that defines $S(u)$, test (2) succeeds if there is an edge from a descendent of u to one of v. Equivalently, test (2) can be replaced by (2$'$) $succ(S(v), \min(u) - 1) \leq \max(u)$.

To implement $edgeExpand((u, v), u)$ (sym., $edgeExpand((u, v), v)$), we iteratively perform $succ(S(v), \min(x) - 1)$ on the children x of u, in left-to-right order, using the intermediate results to skip children with no induced edges. This is more efficient than performing $edgeQuery(x, v)$ for each $x \in children(u)$.

To insert edge (x, y) into $E(G)$, we perform $insert(S(u), y)$ for each node u on the x-to-$root(T_1)$ path and $insert(S(v), x)$ for each node v on the y-to-$root(T_2)$ path. Deletion of (x, y) substitutes $delete(\cdot, \cdot)$ for $insert(\cdot, \cdot)$.

Theorem 1. *Let $D = depth(T)$. With $O(mD \log \log n)$ space, we can insert or delete an edge into G in $O(D \log \log n)$ time and perform $edgeQuery(\cdot)$ in*

$O(\log \log n)$ *time, edgeExpand*$(\cdot, \cdot)$ *in* $O(k \log \log n)$ *time, and edgeReport*$(\cdot)$ *in* $O(\log \log n + k)$ *time, where k is the number of edges reported.*

Proof (Sketch): We maintain each set $S(u)$ as a contracted stratified tree (CST) [25], linking the leaves to facilitate *edgeReport*$(\cdot)$. Each edge in $E(G)$ appears in at most $2D$ such sets. The number of *succ*$(\cdot, \cdot)$ operations for *edgeExpand*$(\cdot, \cdot)$ is one plus the number of induced edges returned, because each *succ*$(\cdot, \cdot)$ operation except the last returns a distinct induced edge. □

Ferragina, Muthukrishnan, and de Berg [11] similarly solve the related *point enclosure problem* on a multi-dimensional grid. They maintain a recursive search structure on the grid, however, whereas we exploit the structure of T.

4 Decompositions and Higher Dimensions

4.1 Compressed Trees and Three-Sided Range Queries. When $D = \omega(\log n / \log \log n)$, we can improve the space bound using compressed trees [13,16,27]. Call tree edge $(v, p(v))$ *light* if $2|desc(v)| \leq |desc(p(v))|$, and *heavy* otherwise. Each node has at most one heavy edge to a child, so deletion of the light edges produces a collection of node-disjoint *heavy paths*.

The *compressed forest*, $C(T)$, is constructed (in $O(n)$ time) by contracting each heavy path in T into a single node. Each tree edge in $C(T)$ corresponds to a light edge of T. Since there are $O(\log n)$ light edges on the path from any node to the root of T, $C(T)$ has depth $O(\log n)$. Let $h(\nu)$, $\nu \in C(T)$, denote the heavy path of T that generates node ν. Define $h^{-1}(v) = \nu$ for all $v \in h(\nu)$.

Consider node $u \in T$ and the corresponding node $\mu = h^{-1}(u) \in C(T)$. Number the nodes in the heavy path $h(\mu)$ top-down $(u_1, \ldots, u_\ell)$. For some $1 \leq i \leq \ell$, $u = u_i$. Associated with u are the corresponding node $\mu \in C(T)$, the value $index(u) = i$ and a pointer, $t(u)$, to u_1 (which is a node in T). We also define $t(\mu) = u_1$, the top of the heavy path $h(\mu)$. For a node $\mu \in C(T)$ and vertex $x \in desc(t(\mu))$, we define $entry(\mu, x)$ to be the maximum i such that $x \in desc(u_i)$, where u_i is the ith node in $h(\mu)$.

We now maintain the sets $S(\cdot)$ on nodes in $C(T)$, not T. For $\mu \in V(C(T))$, $S(\mu) = \{(y, entry(\mu, x)) : (x, y) \in E(G) \vee (y, x) \in E(G), \ x \in desc(t(\mu))\}$. Consider $u \in V(T_1)$, $v \in V(T_2)$, $\mu = h^{-1}(u)$, and $\nu = h^{-1}(v)$. (u, v) is an induced edge of $\mathcal{I}$ if and only if there exists some $(y, d) \in S(\mu)$ such that (a) $\min(v) \leq y \leq \max(v)$ and (b) $index(u) \leq d$. (a) implies that y is a descendent of v; (b) implies that y is adjacent to a descendent of a node at least as deep as u on $h(\mu)$ and thus to a descendent of u. We can also query $S(\nu)$ symmetrically.

The update operations become $insert(S(\mu), (i, j))$ and $delete(S(\mu), (i, j))$. The query operations are: $tsrQuery(S(\mu), (x_1, x_2), y)$, which returns an arbitrary pair $(i, j) \in S(\mu)$ such that $x_1 \leq i \leq x_2$ and $j \geq y$, or $\emptyset$ if none exists; and $tsrReport(S(\mu), (x_1, x_2), y)$, which returns the set of such pairs. An $edgeQuery(u, v)$ is effected by $tsrQuery(S(\mu), (\min(v), \max(v)), index(u))$, and an $edgeReport(u, v)$ by $tsrReport(S(\mu), (\min(v), \max(v)), index(u))$ (or symmetrically on $S(\nu)$). These queries are sometimes called *three-sided range queries*.

We implement $edgeExpand((u,v), \cdot)$ iteratively, as in Section 3. To update $E(G)$, we use the $t(\cdot)$ values to navigate up T and the $h^{-1}(\cdot)$ and $index(\cdot)$ values to create the proper arguments to $insert(\cdot, \cdot)$ and $delete(\cdot, \cdot)$.

Theorem 2. *Let* $p = \min\{depth(T), \log n\}$. *With* $O(mp)$ *space, we can insert or delete an edge into* G *in* $O(p \log n / \log \log n)$ *time and perform* $edgeQuery(\cdot)$ *in* $O(\log n / \log \log n)$ *time,* $edgeExpand(\cdot, \cdot)$ *in* $O(k \log n / \log \log n)$ *time, and* $edgeReport(\cdot)$ *in* $O(\log n / \log \log n + k)$ *time;* k *is the number of edges reported.*

Proof (Sketch): We maintain each $S(\mu)$ as a separate priority search tree (PST) [29]. Each edge $(x, y) \in E(G)$ appears in at most $2p$ such sets. During an $edgeExpand(\cdot, \cdot)$, each $tsrQuery(\cdot, \cdot, \cdot)$ either engenders a new induced edge or else terminates the procedure $\square$

Theorem 3. *Let* $p = \min\{depth(T), \log n\}$. *With* $O(mp)$ *preprocessing time and space, we can build a data structure that performs* $edgeQuery(\cdot)$ *in* $O(\log \log n)$ *time,* $edgeExpand(\cdot, \cdot)$ *in* $O(k \log \log n)$ *time, and* $edgeReport(\cdot)$ *in* $O(\log \log n + k)$ *time, where* k *is the number of edges reported.*

Proof (Sketch): We use a static three-sided range query data structure [12,23]. To provide access to the leaves of the underlying data structures without the high overhead (e.g., perfect hashing) of previous solutions [12,23], we use one array of size n. Each element i points to a CST that contains pointers to the leaf representing i in each structure, indexed by structure, which we number ordinally 1 to $|V(C(T))|$. Each leaf in each underlying structure appears in one such CST, so the total extra space and preprocessing time is $O(n + m \log \log n)$. The initial access to a leaf requires an $O(\log \log n)$-time CST look-up. $\square$

4.2 Stratification. We further reduce the space by stratifying T into $\sqrt{D}$ strata of $\sqrt{D}$ levels each, where $D = depth(T)$. Entries for an edge (u, v) are made in set $S(x)$ only for each x that is in ancestor of u (sym., v) in the same stratum. Each node x at the *top* of a stratum (such that $p(x)$ is in a higher stratum) maintains a set $S'(x)$ containing corresponding entries for edges incident on descendents in deeper strata. Thus, each edge occurs in only $O(\sqrt{D})$ sets. $C(T)$ is similarly stratified.

Every query on a set $S(x)$ in the above discussions is answered by uniting the results from the same queries on the new $S(x)$ and $S'(sr(x))$, where $sr(x)$ is the ancestor of x at the top of x's stratum.

Let the data structure underlying the $S(\cdot)$ sets (e.g., a CST or PST) use $\mathcal{S}(m)$ space, $\mathcal{Q}(m)$ query time, $\mathcal{R}(m) + k$ reporting time, and $\mathcal{U}(m)$ update time or, in the static case, $\mathcal{P}(m)$ preprocessing time. Let D be the depth of the tree being stratified (either T or $C(T)$).

Theorem 4. *A data structure using* $O(\sqrt{D}\mathcal{S}(m))$ *space can be built to support* $edgeQuery(\cdot)$ *in* $O(\mathcal{Q}(m))$ *time,* $edgeExpand(\cdot, \cdot)$ *in* $O(k\mathcal{Q}(m))$ *time, and* $edgeReport(\cdot)$ *in* $O(\mathcal{R}(m) + k)$ *time, where* k *is the number of edges reported. In the dynamic case, insertion and deletion of an edge into* G *take* $O(\sqrt{D}\mathcal{U}(m))$ *time; in the static case, the preprocessing time is* $O(\sqrt{D}\mathcal{P}(m))$.

We can also stratify recursively. Starting with the one-level stratification above, stratify the $\sqrt{D}$ stratum top nodes into $D^{1/4}$ strata of $D^{1/4}$ levels each. Similarly recurse on the nodes within each stratum. We can doubly recurse $\log \log D$ levels until there remain $O(1)$ strata containing $O(1)$ nodes each. Each edge is thus recorded in $O(\log D)$ $S(\cdot)$ and $S'(\cdot)$ sets.

Theorem 5. *A data structure using $O(\mathcal{S}(m) \log D)$ space can be built to support edgeQuery($\cdot$) in $O(\mathcal{Q}(m) \log D)$ time, edgeExpand($\cdot, \cdot$) in $O(k\mathcal{Q}(m) \log D)$ time, and edgeReport($\cdot$) in $O(\mathcal{R}(m) \log D + k)$ time, where k is the number of edges reported. In the dynamic case, insertion and deletion of an edge into G take $O(\mathcal{U}(m) \log D)$ time; in the static case, the preprocessing time is $O(\mathcal{P}(m) \log D)$.*

4.3   Higher Dimensions.

The d-dimensional data structure on nodes of T_1 is a collection of sets $S_d(\cdot)$, such that $S_d(u_1)$ maintains the information as detailed above on hyperedges incident on descendents of $u_1 \in V(T_1)$. Consider such a hyperedge $(x_1, \ldots, x_d)$ $(x_1 \in desc(u_1))$. $S_d(u_1)$ is implemented recursively, as a collection of $S_{d-1}(\cdot)$ sets recording the projections $(x_2, \ldots, x_d)$ of the original hyperedges in $S_d(u_1)$. $S_2(\cdot)$ is the base case, equivalent to the $S(\cdot)$ sets above. There is a separate recursive collection of $S_{i-1}(\cdot)$ sets for each $S_i(\cdot)$ set; no space is allocated for empty sets.

This strategy allows for all except *edgeExpand($\cdot, 1$)* operations. If necessary, we maintain a second set of $S_d(\cdot)$ sets, designating a different tree to be T_1.

We stratify as in Section 4.2. Recall that in the one-level stratification, each original operation engendered two new operations (on the $S(\cdot)$ and $S'(\cdot)$ sets). Denote by $\mathcal{S}_d(m)$, $\mathcal{Q}_d(m)$, $\mathcal{R}_d(m) + k$, $\mathcal{U}_d(m)$, and $\mathcal{P}_d(m)$ the space and query, reporting, update, and preprocessing time bounds, rsp., for the d-dimensional tree-cross product operations. Let D be the depth of the tree (T or $C(T)$). Without stratification, we derive that $\mathcal{S}_d(m) = D\mathcal{S}_{d-1}(m)$, and $\mathcal{Q}_d(m) = \mathcal{Q}_{d-1}(m)$. With one-level stratification, we derive that $\mathcal{S}_d(m) = \sqrt{D}\mathcal{S}_{d-1}(m)$, and $\mathcal{Q}_d(m) = 2\mathcal{Q}_{d-1}(m)$. With recursive stratification, we derive that $\mathcal{S}_d(m) = \log D\mathcal{S}_{d-1}(m)$, and $\mathcal{Q}_d(m) = \log D\mathcal{Q}_{d-1}(m)$. With all methods, the derivation for $\mathcal{R}_d(m)$ follows that for $\mathcal{Q}_d(m)$, and those for $\mathcal{U}_d(m)$ and $\mathcal{P}_d(m)$ follow that for $\mathcal{S}_d(m)$.

Table 1 details some of the resulting bounds, using compressed trees (hence $D = O(\log n)$) and either Overmars' static three-sided range query structure [23] for the static case or Willard's PST [29] for the dynamic case.

These results strictly improve upon what we could derive using classical range searching on the original grid. Consider the two-dimensional case, for example. Overmars' static structure [23] would match only our non-stratified, static space and query time bounds, but his preprocessing time is significantly higher; to reduce the latter would degrade the query time by a $\sqrt{\log n} / \log \log n$ factor. Applying Edelsbrunner's technique [10] to Willard's PST [29] would match only our non-stratified, dynamic bound. Stratification improves all these bounds. We also provide a dynamic solution that achieves logarithmic query **and** update times in almost-linear space. No classical range searching result provides the same bounds. Chazelle [8] provides linear space bounds, but the query and update times are $O(\log^2 n)$, and reporting imposes a non-constant penalty on k.

Table 1. Deterministic, worst-case bounds for d-dimensional tree cross-product operations. $n = |V(G)|$, $m = |E(G)|$. For all methods, the *edgeReport*$(\cdot,\cdot)$ time is $k + f(n)$, and the *edgeExpand*$(\cdot,\cdot)$ time is $k \cdot f(n)$, where k is the number of edges reported, and $f(n)$ is the corresponding *edgeQuery*$(\cdot,\cdot)$ time.

Method	Space	*edgeQuery*$(\cdot,\cdot)$ time	
Static			Preproc. time
No stratification	$O(m\log^{d-1} n)$	$O(\log\log n)$	$O(m\log^{d-1} n)$
One-level strat.	$O(m\log^{\frac{d-1}{2}} n)$	$O(2^{d-1}\log\log n)$	$O(m\log^{\frac{d-1}{2}} n)$
Recursive strat.	$O(m(\log\log n)^{d-1})$	$O((\log\log n)^d)$	$O(m(\log\log n)^{d-1})$
Dynamic			Update Time
No stratification	$O(m\log^{d-1} n)$	$O(\log n/\log\log n)$	$O(\log^d n/\log\log n)$
One-level strat.	$O(m\log^{\frac{d-1}{2}} n)$	$O(2^{d-1}\log n/\log\log n)$	$O(\log^{\frac{d+1}{2}} n/\log\log n)$
Recursive strat.	$O(m(\log\log n)^{d-1})$	$O(\log n(\log\log n)^{d-2})$	$O(\log n(\log\log n)^{d-2})$

5 Applications

5.1 Hierarchical Graph Views. Given a rooted tree T and a graph G, such that the vertices of G correspond to the leaves of T, we say that $U \subseteq V(T)$ *partitions* G if the set $\{leaves(v) : v \in U\}$ partitions $V(G)$. A *view of* G is any $U \subseteq V(T)$ that partitions G. We extend the definitions from Section 2. For any $u, v \in V(T)$ such that neither u nor v is an ancestor of the other, define $E(u,v) = \{\{x,y\} \in E(G) : x \in leaves(u) \wedge y \in leaves(v)\}$. If each edge $e \in E(G)$ has a weight, $w(e) \in \mathcal{G}$, from an abelian group, $(\mathcal{G}, \odot)$, then the *weight* of (u,v) is $w(u,v) = \odot \{w(\{x,y\}) : \{x,y\} \in E(u,v)\}$. For any view U, we define G/U to be the *induced graph* (U, E_U), where $E_U = \{(u,v) \in U \times U : |E(u,v)| > 0\}$.

The *hierarchical graph-view maintenance problem* is to maintain G/U under the following operations on U: *expand*(U,x), where $x \in U$, yields view $U \setminus \{x\} \cup children(x)$; *contract*$(U,x)$, where $children(x) \subseteq U$, yields view $U \setminus children(x) \cup \{x\}$. The problem is motivated by graph visualization applications [7,9].

Let $n = |V(G)|$, $m = |E(G)|$, $p = \min\{depth(T), \log n\}$, and assume without loss of generality that T contains no unary vertices. (Hence $|V(T)| = O(n)$.) Buchsbaum and Westbrook [7] show how, with $O(mp)$ space and $O(mp^2)$ expected preprocessing time, to perform *expand*$(\cdot,\cdot)$ and *contract*$(\cdot,\cdot)$ operations in optimal time: linear in the number of changes to $E(G/U)$. There is an additional $\log n$ factor in the *expand*$(\cdot,\cdot)$ time for weighted graphs. To accommodate updates to G, the space bound becomes $O(mp^2)$, the edge insertion and deletion times are expected $O(p^2 \log n)$; *expand*$(\cdot,\cdot)$ and *contract*$(\cdot,\cdot)$ remain optimal.

By applying tree cross products, we improve the space, update and preprocessing times for unweighted graph-view maintenance. The cost is an increase in *expand*$(\cdot,\cdot)$ time; *contract*$(\cdot,\cdot)$ remains optimal. All of our bounds are deterministic, worst-case; the prior update and preprocessing times [7] are expected.

Set $T_1 = T_2 = T$. Edge $(u,v) \in E(G)$ (ordered by postorder on T) becomes an edge from u in T_1 to v in T_2. An *expand*(U,v) engenders an *edgeExpand*$((u,v),v)$ operation for each $(u,v) \in E(G/U)$ (and symmetrically for $(v,w) \in E(G/U)$).

Induced edges between children of v are found using nearest common ancestors [7]. To implement *contract*(U, v), add an edge to v from each non-child of v adjacent to a child of v in G/U, and remove edges incident on children of v.

Define $\text{Opt}(U, v)$ to be the number of nodes adjacent to children of v in G/U. Denote by U the view before an *expand*$(U, \cdot)$ or *contract*$(U, \cdot)$ and U' the result.

Theorem 6. *On an unweighted graph, with $O(m \log \log n)$ space, we can perform edge insertion and deletion in $O(\log n)$ time, expand(U, v) in $O(\text{Opt}(U', v) \cdot \log n)$ time, and contract(U, v) in $O(\text{Opt}(U, v))$ time. In the static case, with $O(m \log \log n)$ space and preprocessing time, we can perform expand(U, v) in $O(\text{Opt}(U', v) \cdot (\log \log n)^2)$ time and contract(U, v) in $O(\text{Opt}(U, v))$ time.*

Theorem 6 follows from Theorem 5, assuming recursive stratification. One-level stratification improves the *expand*(U, v) times by $\log \log n$ factors but degrades the space, update and preprocessing times by $\sqrt{\log n}/\log \log n$ factors.

5.2   String Matching. Given text $T = x_1 \cdots x_n$ of length n, denote by $T[i, j]$ the substring $x_i \cdots x_j$. The *text indexing with one error problem* is to preprocess T so that, given length-m pattern P, we can compute all locations i in T such that P matches $T[i, i+m-1]$ with exactly one error. Below we assume Hamming distance, i.e., the number of symbols replaced, but the method extends to Levenshtein (edit) distance [19].

This on-line problem differs from *approximate string matching* [14,26], in which both T and P are given off-line. Exact text indexing (finding occurrences of P with no errors), can be solved with $O(n)$ preprocessing time and space, $O(m+k)$ query time, where k is the number of occurrences [21,28], and $O(\log^3 n + s)$ time to insert or delete a length-s substring in T [26].

The work of Ferragina, Muthukrishnan, and de Berg [11] extends to solve text indexing with one error. Given $O(n^{1+\varepsilon})$ preprocessing time and space, queries take $O(m \log \log n + k)$ time. Using the same approach, Amir et al. [2] give a solution with $O(n \log n)$ preprocessing time and space but $O(m\sqrt{\log n}+k)$ query time. Both assume no exact matches occur. We improve these results, achieving $O(m \log \log n + k)$ query time and $O(n\sqrt{\log n})$ space and preprocessing time. Amir et al. [2] extend their solution to the general case with $O(n \log^2 n)$ space and preprocessing time and $O(\log n \log \log n+k)$ query time. We similarly extend our solution, achieving $\log n$ factor improvements in all bounds.

Observe [2,11] that an occurrence of P in T at location i with one error at location $i+j$ implies that $T[i, i+j-1]$ matches $P[1, j]$ and $T[i+j+1, i+m-1]$ matches $P[j+2, m]$. To exploit this, Amir et al. [2] first build suffix trees S_T for T and S_{T^R} for T^R, the reverse string of T, using Weiner's method [28]. Label each leaf in S_T by the starting location of its suffix in T; label each leaf in S_{T^R} by $n - i + 3$, where i is the starting location of the corresponding suffix in T^R. Querying for P is done as follows.

For $j = 1, \ldots, m$ do
1. Find node u, the location of $P[j + 1, m]$ in S_T, if such a node exists.
2. Find node v, the location of $P[1, j - 1]^R$ in S_{T^R}, if such a node exists.
3. If u and v exist, report the intersection of the labels of *leaves*(u) and *leaves*(v).

Steps (1) and (2) can be performed in $O(m)$ time over the progression of the algorithm [2], by implicitly and incrementally continuing the Weiner construction [28] on the suffix trees. By adding edges connecting pairs of identically labeled leaves, Step (3) becomes an *edgeReport(u, v)* operation.

Theorem 7. *Given $O(n\sqrt{\log n})$ preprocessing time and space, we can preprocess a string T of length n, so that, for any string P of length m given on-line, if no exact matches of P occur in T, we can report all occurrences of P in T with one error in $O(m \log \log n + k)$ time, where k is the number of occurrences.*

Each exact match would be reported $|P|$ times. To obviate this problem, we add a third dimension as do Amir et al. [2]. Tree T_3 contains a root and s leaves, each corresponding to one of the $s \leq n$ alphabet symbols. Each edge connecting leaves in S_T and S_{T^R} corresponds to some mismatch position i in T. We extend the (hyper)edge to include leaf $T[i, i]$ in T_3. We extend the *edgeReport(**u**)* semantics to allow the stipulation that any dimension j report elements that are *not* descendents of u_j. (This simply changes the parameters of the queries performed on the $S(\cdot)$ sets.) Step (3) becomes an *edgeReport(u, v, T[i, i])* operation.

Theorem 8. *Given $O(n \log n)$ preprocessing time and space, we can preprocess a string T of length n, so that, for any string P of length m given on-line, we can report all k occurrences of P in T with one error in $O(m \log \log n + k)$ time.*

Grossi and Vitter [15] use a similar strategy to report contiguous ranges in *compressed suffix arrays*, which use only $O(n)$ **bits** to implement all suffix-array functionality on a length-n binary string T. They use two-dimensional, grid range searches that can be equivalently realized by node-intersection queries on suffix trees for T and T^R. As above, tree cross products improve their bounds on the additional suffix-array reporting time, from $O(\log^2 n \log \log n + k)$ to $O(\log n (\log \log n)^3 + k)$, where k is the output size.

5.3 Hammocks. Let $G = (V, E)$ be a directed graph with a designated source node, s, and sink node, t. A node u *dominates* a node v if every path from s to v goes through u. A node v *post-dominates* a node u if every path from u to t goes through v. The *hammock* between two nodes u and v is the set of nodes dominated by u and post-dominated by v. (This modifies the definition due to Kas'janov [18].)

Johnson, Pearson, and Pingali [17] define a canonical, nested hammock structure, which is useful in compiler control-flow analysis, and devise an $O(m)$-time algorithm to discover it. ($n = |V|$, and $m = |E|$.) No previous result, however, allows efficient, on-line queries of the form: return the hammock between two given nodes. Such queries are useful in software system analysis, to detect collections of systems with designated information choke points, e.g., to assess the impact of retiring legacy systems [5].

We can solve such queries as follows. Let T_1 be the dominator tree [1,6] of G, and let T_2 be the dominator tree of the reverse graph of G. The hammock between two nodes u and v in G is the intersection of the set of descendents of u in T_1

with the set of descendents of v in T_2. By adding edges connecting corresponding nodes in T_1 and T_2, this intersection is computed by an *edgeReport*(u, v) query.

Theorem 9. *With* $O(n\sqrt{\log n})$ *(rsp.* $O(n \log \log n)$*) space and preprocessing time, we can compute the hammock between two given nodes in* $O(\log \log n + k)$ *(rsp.,* $O((\log \log n)^2 + k)$*) time, where* k *is the size of the hammock.*

6   Conclusion

Many applications impose balance or low-depth constraints on T, which obviate the sophisticated space-reduction techniques and allow the $S(\cdot)$ sets to be implemented by simple binary search trees, making our tree cross-product framework very practical. Low-degree constraints on T might lead to other simplifications.

How to implement *edgeWeight*$(\cdot, \cdot)$ operations efficiently remains open. It also remains open to unify our graph-view bounds with those of Buchsbaum and Westbrook [7], i.e., to eliminate the penalty that we incur on expand times.

Finally, allowing updates to T remains open.

Acknowledgements. We thank Raffaele Giancarlo and S. Muthukrishnan for helpful discussions and Roberto Grossi for tutelage on compressed suffix arrays.

References

1. S. Alstrup, D. Harel, P. W. Lauridsen, and M. Thorup. Dominators in linear time. *SIAM J. Comp.*, 28(6):2117–32, 1999. 129
2. A. Amir, D. Keselman, G. M. Landau, M. Lewenstein, N. Lewenstein, and M. Rodeh. Indexing and dictionary matching with one error. In *Proc. 6th WADS*, volume 1663 of *LNCS*, pages 181–92. Springer-Verlag, 1999. 121, 122, 128, 129
3. R. A. Baeza-Yates and G. Navarro. A faster algorithm for approximate string matching. In *Proc. 7th CPM*, volume 1075 of *LNCS*, pages 1–23. Springer-Verlag, 1996. 121
4. R. A. Baeza-Yates and G. Navarro. Multiple approximate string matching. In *Proc. 5th WADS*, volume 1272 of *LNCS*, pages 174–84. Springer-Verlag, 1997. 121
5. A. L. Buchsbaum, Y. Chen, H. Huang, E. Koutsofios, J. Mocinego, A. Rogers, M. Jenkowsky, and S. Mancoridis. Enterprise navigator: A system for visualizing and analyzing software infrastructures. Technical Report 99.16.1, AT&T Labs–Research, 1999. Submitted for publication. 121, 129
6. A. L. Buchsbaum, H. Kaplan, A. Rogers, and J. R. Westbrook. A new, simpler linear-time dominators algorithm. *ACM TOPLAS*, 20(6):1265–96, 1998. 129
7. A. L. Buchsbaum and J. R. Westbrook. Maintaining hierarchical graph views. In *Proc. 11th ACM-SIAM SODA*, pages 566–75, 2000. 120, 121, 122, 127, 128, 130
8. B. Chazelle. A functional approach to data structures and its use in multidimensional searching. *SIAM J. Comp.*, 17(3):427–62, 1988. 120, 126
9. C. A. Duncan, M. T. Goodrich, and S. Kobourov. Balanced aspect ratio trees and their use for drawing very large graphs. In *Proc. GD '98*, volume 1547 of *LNCS*, pages 111–24. Springer-Verlag, 1998. 120, 127

10. H. Edelsbrunner. A note on dynamic range searching. *Bull. EATCS*, 15:34–40, 1981. 120, 126

11. P. Ferragina, S. Muthukrishnan, and M. de Berg. Multi-method dispatching: A geometric approach with applications to string matching problems. In *Proc. 31st ACM STOC*, pages 483–91, 1999. 121, 124, 128

12. O. Fries, K. Mehlhorn, S. Näher, and A. Tsakalidis. A $\log\log n$ data structure for three-sided range queries. *IPL*, 25:269–73, 1987. 120, 125

13. H. N. Gabow. Data structures for weighted matching and nearest common ancestors with linking. In *Proc. 1st ACM-SIAM SODA*, pages 434–43, 1990. 124

14. Z. Galil and R. Giancarlo. Data structures and algorithms for approximate string matching. *J. Complexity*, 4:33–78, 1988. 128

15. R. Grossi and J. S. Vitter. Compressed suffix arrays and suffix trees with applications to text indexing and string matching. In *Proc. 32nd ACM STOC*, pages 397–406, 2000. 121, 122, 129

16. D. Harel and R. E. Tarjan. Fast algorithms for finding nearest common ancestors. *SIAM J. Comp.*, 13(2):338–55, 1984. 124

17. R. Johnson, D. Pearson, and K. Pingali. The program structure tree: Computing control regions in linear time. In *Proc. ACM SIGPLAN PLDI '94*, pages 171–85, 1994. 121, 129

18. V. N. Kas'janov. Distinguishing hammocks in a directed graph. *Sov. Math. Dokl.*, 16(2):448–50, 1975. 129

19. V. I. Levenshtein. Binary codes capable of correcting deletions, insertions and reversals. *Sov. Phys. Dok.*, 10:707–10, 1966. 121, 128

20. U. Manber and S. Wu. An algorithm for approximate membership checking with application to password security. *IPL*, 50(4):191–7, 1994. 121

21. E. M. McCreight. A space-economical suffix tree construction algorithm. *J. ACM*, 23(2):262–72, 1976. 128

22. R. Muth and U. Manber. Approximate multiple string search. In *Proc. 7th CPM*, volume 1075 of *LNCS*, pages 75–86. Springer-Verlag, 1996. 121

23. M. H. Overmars. Efficient data structures for range searching on a grid. *J. Alg.*, 9:254–75, 1988. 120, 125, 126

24. P. A. Pevzner and M. S. Waterman. Multiple filtration and approximate pattern matching. *Algorithmica*, 12(1/2):135–54, 1995. 121

25. F. P. Preparata, J. S. Vitter, and M. Yvinec. Output-sensitive generation of the perspective view of isothetic parallelepipeds. *Algorithmica*, 8(4):257–83, 1992. 124

26. S. C. Sahinalp and U. Vishkin. Efficient approximate and dynamic matching of patterns using a labeling paradigm. In *Proc. 37th IEEE FOCS*, pages 320–8, 1996. 128

27. R. E. Tarjan. Applications of path compression on balanced trees. *J. ACM*, 26(4):690–715, 1979. 124

28. P. Weiner. Linear pattern matching algorithms. In *Proc. 14th IEEE Symp. on Switch. and Auto. Thy.*, pages 1–11, 1973. 128, 129

29. D. E. Willard. Examining computational geometry, van Emde Boas trees, and hashing from the perspective of the fusion tree. *SIAM J. Comp.*, 29(3):1030–49, 2000. 120, 125, 126

A $2\frac{1}{10}$-Approximation Algorithm for a Generalization of the Weighted Edge-Dominating Set Problem

Robert Carr[1]*, Toshihiro Fujito[2]**, Goran Konjevod[3]***, and Ojas Parekh[3]†

[1] Sandia National Laboratory, P.O. Box 5800, Albuquerque, NM 87185
[2] Department of Electronics, Nagoya University Furo, Chikusa, Nagoya, 464-8603 Japan
[3] Department of Mathematical Sciences, Carnegie Mellon University, Pittsburgh PA, 15213-3890

Abstract. We study the approximability of the weighted edge-dominating set problem. Although even the unweighted case is *NP*-Complete, in this case a solution of size at most twice the minimum can be efficiently computed due to its close relationship with minimum maximal matching; however, in the weighted case such a nice relationship is not known to exist. In this paper, after showing that weighted edge domination is as hard to approximate as the well studied weighted vertex cover problem, we consider a natural strategy, reducing edge-dominating set to edge cover. Our main result is a simple $2\frac{1}{10}$-approximation algorithm for the weighted edge-dominating set problem, improving the existing ratio, due to a simple reduction to weighted vertex cover, of $2r_{WVC}$, where r_{WVC} is the approximation guarantee of any polynomial-time weighted vertex cover algorithm. The best value of r_{WVC} currently stands at $2 - \frac{\log \log |V|}{2 \log |V|}$. Furthermore we establish that the factor of $2\frac{1}{10}$ is tight in the sense that it coincides with the integrality gap incurred by a natural linear programming relaxation of the problem.

1 Introduction

In an undirected graph $G = (V, E)$, E is a set of *edges*, $\{u, v\}$, where u, v belong to the set of *vertices*, V. An edge e *dominates* all $f \in E$ such that $e \cap f \neq \emptyset$. A set of edges is an *edge-dominating set* (*eds*) if its members collectively dominate all the edges in E. The *edge-dominating set problem* (*EDS*) is then that of finding a minimum-cardinality edge-dominating set, or if edges are weighted by a function $w : E \to \mathbb{Q}_+$, an edge-dominating set of minimum total weight.

* Work supported in part by the United States Department of Energy under contract DE-AC04-94AL85000. (e-mail: bobcarr@cs.sandia.gov)
** (e-mail: fujito@nuee.nagoya-u.ac.jp)
*** Supported in part by an NSF CAREER Grant CCR-9625297. (e-mail: konjevod@andrew.cmu.edu)
† (e-mail: odp@andrew.cmu.edu)

M. Paterson (Ed.): ESA 2000, LNCS 1879, pp. 132–142, 2000.

1.1   Notation

A vertex v *dominates* $u \in V$ if $\{u, v\} \in E$. A v also *covers* all edges incident upon v, or more formally, v covers an edge e if $v \in e$. We overload terminology once again and say that an edge e *covers* a vertex v if $v \in e$. We denote the set of edges that v covers by $\delta(v)$. When we wish to discuss the vertices of an edge set $S \subseteq E$, we define $V(S) = \bigcup_{e \in S} e$. A *matching* is a set of edges M, such that distinct edges e, f in M do not intersect. A maximal matching is one which is not properly contained in any other matching. For $S \subseteq V$, we denote the set $\{e \in E \mid e \cap S = 1\}$ by $\delta(S)$, and we denote the set $\{e \in E \mid e \cap S = 2\}$ by $E(S)$. When given a subset $S \subseteq E$ and a vector $x \in \mathbb{Q}^{|E|}$ whose components correspond to the edges in E, we use $x(S)$, as a shorthand for $\sum_{e \in S} x_e$. Analogously in the case of a function $w^v : V \rightarrow \mathbb{Q}$ or $w : E \rightarrow \mathbb{Q}$ we write $w^v(S) = \sum_{u \in S} w^v(u)$ or $w(S) = \sum_{e \in S} w(e)$, where $S \subseteq V$ or $S \subseteq E$, respectively.

1.2   Related Problems

Yannakakis and Gavril showed that EDS and the minimum maximal matching problem, whose connection to EDS will be presented later, are NP-complete even on graphs which are planar or bipartite of maximum degree 3 [18]. This result was later extended by Horton and Kilakos to planar bipartite, line, total, perfect claw-free, and planar cubic graphs [9]. On the other hand polynomially solvable special cases have been discovered. Chronologically by discovery, efficient exact algorithms for trees [13], claw-free chordal graphs, locally connected claw-free graphs, the line graphs of total graphs, the line graphs of chordal graphs [9], bipartite permutation graphs, cotriangulated graphs [17], and other classes are known.

Although EDS has important applications in areas such as telephone switching networks, very little is known about the weighted version of the problem. In fact, all the polynomial-time solvable cases listed above apply only to the *cardinality* case, although we should note that PTAS's are known for weighted planar [2] and λ-precision unit disk graphs [10]. In particular, while it is a simple matter to compute an edge-dominating set of *size* at most twice the minimum, as any maximal matching will do, such a simple reduction easily fails when arbitrary weights are assigned to edges. In fact the only known approximability result, which follows from a simple reduction to vertex cover, does not seem to have appeared in the literature.

The edge-dominating set problem, especially the weighted version, seems to be the least studied among the other basic specializations of the set cover problem for graphs. The others are called the (weighted) *edge cover* (*EC*), *vertex cover* (*VC*), and *(vertex) dominating set* problems. We seek to obtain a minimum-cardinality (weight) set which covers vertices by edges, edges by vertices, and vertices by vertices respectively. Of these only the weighted edge cover problem is known to be solvable in polynomial time [4,14,16]. Better known and well studied is the dominating set problem. EDS for G is equivalent to the vertex-dominating set problem for the line graph of G. The dominating set problem for general graphs is, unfortunately, equivalent to the set cover problem under an approximation preserving reduction. Although the polynomial-time approximability of set cover is well established and stands at a factor of $\ln |V| + 1$ [11,12,3], it cannot be efficiently approximated better than $\ln |V|$ unless $NP \subseteq DTIME(|V|^{O(\log \log |V|)})$ [5]. The vertex cover problem seems to be the best studied of the bunch and boasts a vast literature. Most known facts and relevant references can be found in the survey by Hochbaum [8]. The best known approximation ratio is $2 - \frac{\log \log |V|}{2 \log |V|}$, and it has been conjectured (see the above survey) that 2 is the best constant approximation factor possible in polynomial time.

In this paper we consider a natural strategy of reducing weighted EDS to the related weighted edge cover problem and establish the approximability of EDS within a factor of $2\frac{1}{10}$. We also obtain the same ratio for the extension in which only a subset of the edges need be dominated. Furthermore the factor of $2\frac{1}{10}$ is tight in the sense that it coincides with the integrality gap incurred by a natural linear programming relaxation of EDS.

2   Approximation Hardness

Yannakakis and Gavril proved the NP-hardness of EDS by reducing VC to it [18]. Although their reduction can be made to preserve approximation quality within some constant factor and thus imply the MAX SNP-hardness of (unweighted) EDS and the non-existence of a polynomial-time approximation scheme (unless $P=NP$) [15,1], it does not preclude the possibility of better approximation of EDS than that of VC. On the other hand, it is quite straightforward to see that the approximation of weighted EDS is as hard as that of weighted VC.

Theorem 1. *Weighted VC can be approximated as well as weighted EDS.*

Proof. Let $G = (V, E)$ be an instance graph for VC with weight function $w^v : V \to \mathbb{Q}_+$. Let s be a new vertex not in V, and construct a new graph $G' = (V \cup \{s\}, E \cup E')$ by attaching s to each vertex of G, that is, $E' = \{\{s, u\} \mid u \in V\}$. Assign a weight function $w' : E \to \mathbb{Q}_+$ to the edges of G' by defining $w'(e) = w^v(u)$ if $e = \{s, u\} \in E'$, and $w'(e) = w^v(u) + w^v(v)$ if $e = \{u, v\} \in E$. By the definition of w', if an edge-dominating set D for G' contains $\{u, v\} \in E$, it can be replaced by the two edges $\{u, s\}, \{v, s\} \in E'$ without increasing the weight of D, so we may assume $D \subseteq E'$. In this case, however, there exists a one-to-one correspondence between vertex covers in G and edge-dominating sets in G', namely $C \overset{\text{def}}{=} V(D) \setminus \{s\}$ in G and D in G', such that $w^v(C) = w'(D)$.

3   Previous Work

3.1   Cardinality EDS: Reduction to Maximal Matching

Obtaining a 2-approximation for the minimum-cardinality edge-dominating set is easy; the following proposition also demonstrates the equivalence of cardinality EDS and minimum-maximal matching.

Proposition 1. *(Harary [7].) There exists a minimum-cardinality edge-dominating set which is a maximal matching.*

Proof. For a set E' of edges, let $adj(E')$ denote the number of (unordered) pairs of adjacent edges in E', that is

$$adj(E') = \frac{1}{2}|\{(e, f) \mid e, f \in E' \text{ and } e \cap f \neq \emptyset\}|.$$

Let $D \subseteq E$ be a minimum-cardinality edge-dominating set. Suppose D is not a matching and let e, $f \in D$ be two adjacent edges, i.e. $e \cap f \neq \emptyset$. Since D is minimal, $D \setminus f$ is not an edge-dominating set. Therefore there exists an edge $g \in E$ adjacent to f, but not to any other member of D. Now the set $D' = D \setminus \{f\} \cup \{g\}$ is another minimum-cardinality edge-dominating set and $adj(D') < adj(D)$. By repeating this exchange procedure on D', we eventually find a minimum edge-dominating set D^* which is a (maximal) matching.

Proposition 2. *Every maximal matching M gives a 2-approximation for the edge-dominating set problem.*

Proof. Let M_1 and M_2 be maximal matchings. The symmetric difference $M_1 \oplus M_2$ consists of disjoint paths and cycles in which edges alternate between those from M_1 and those from M_2. This implies an equal number of edges from M_1 and M_2 in every cycle. By the maximality of M_1 and M_2, every path must contain an edge from each of M_1 and M_2, hence every path contains at most twice as many edges from one as from the other. Letting $k_i = |M_i \cap (M_1 \oplus M_2)|$ for $i = 1, 2$, we now have $k_1 \leq 2k_2$ and $k_2 \leq 2k_1$. Since $|M_i| = |M_1 \cap M_2| + k_i$, it follows that $|M_1| \leq 2|M_2|$ and $|M_2| \leq 2|M_1|$.

3.2   Weighted EDS: Reduction to Vertex Cover

Weighted EDS may be reformulated as finding a set of edges D of minimum weight such that $V(D)$ is a vertex cover of G. This idea leads to a well known $2r_{WVC}$-approximation algorithm, where r_{WVC} is the approximation guarantee of any polynomial-time weighted vertex cover algorithm.

Theorem 2. *(Folklore.) The weighted edge-dominating set problem can be approximated to within a factor of $2r_{WVC}$.*

Proof. Given an instance of weighted EDS, G with weight function $w : E \to \mathbb{Q}_+$, define a vertex-weight function $w^v : V \to \mathbb{Q}_+$ by setting

$$w^v(u) = \min_{e \in \delta(u)} \{w(e)\}$$

for every $u \in V$. Let D^* be a minimum-weight EDS with respect to w, and let C^* be a minimum-weight vertex cover with respect to w^v. Since $V(D^*)$ is a vertex cover for G, $w^v(C^*) \le w^v(V(D^*))$. By the construction of w^v, for each $u \in V(D^*)$

$$w^v(u) \le \min_{e \in \delta(u) \cap D^*} \{w(e)\}.$$

Hence,

$$w^v(C^*) \le w^v(V(D^*)) = \sum_{u \in V(D^*)} w^v(u) \le 2w(D^*). \tag{1}$$

Suppose we use an r_{WVC}-approximation algorithm to obtain a vertex cover C such that $w^v(C) \le r_{WVC} \cdot w^v(C^*)$. We can construct an edge-dominating set D_C from C by selecting a minimum-weight edge in $\delta(u)$ for each $u \in C$. Thus $w(D_C) \le w^v(C)$. Combining this with (1) we have

$$w(D_C) \le r_{WVC} \cdot w^v(C^*) \le 2r_{WVC} \cdot w(D^*),$$

which establishes the theorem.

As mentioned earlier, the smallest value of r_{WVC} currently known for general weighted graphs is $2 - \frac{\log \log |V|}{2 \log |V|}$ [8], yielding an EDS approximation ratio of $4 - \frac{\log \log |V|}{\log |V|}$. Of course, for special classes we can do better. For instance exact polynomial-time algorithms exist for weighted VC on bipartite graphs, yielding a 2-approximation for weighted EDS on bipartite graphs.

4   A $2\frac{1}{10}$-Approximation: Reduction to Edge Cover

4.1   Polyhedra

Given an instance $G = (V, E)$ and a corresponding cost vector $c \in \mathbb{Q}_+^{|E|}$, we may formulate the weighted edge-dominating set problem as an integer program

$$\min \sum_{e \in E} c_e x_e$$

(EDS(G)) subject to:
$$x(\delta(u)) + x(\delta(v)) - x_{uv} \ge 1 \quad \{u, v\} \in E$$
$$x_e \in \{0, 1\} \quad\quad\quad\quad e \in E.$$

The constraints of (EDS(G)) ensure that each edge is covered by at least one edge. Relaxing the 0-1 constraints yields

$$\min \sum_{e \in E} c_e x_e$$

(FEDS(G)) subject to:
$$x(\delta(u)) + x(\delta(v)) - x_{uv} \ge 1 \quad \{u, v\} \in E$$
$$x_e \ge 0 \quad\quad\quad\quad\quad e \in E.$$

We henceforth assume without loss of generality that G has no isolated vertices, since deleting such vertices does not affect an edge-dominating set. In our reduction to edge cover we will also be interested in

$$\min \sum_{e \in E} c_e x_e$$

(FEC(G)) subject to:
$$\begin{aligned} x(\delta(u)) &\geq 1 & u \in V \\ x_e &\geq 0 & e \in E. \end{aligned}$$

It is easy to see that the incidence vector of any edge cover for G satisfies all the constraints in (FEC(G)), hence is feasible for it. However, (FEC) may not have integral optimal solutions in general, to which a unit-weighted triangle attests. The optimal solution for (FEC) has $x_e = 1/2$, for all $e \in E$, for a total weight of $3/2$, while the weight of an integral solution must be at least 2. Thus the inequalities (FEC) are not sufficient to define (EC), the convex hull of the incidence vectors of edge covers. Fortunately, due to a result of Edmonds and Johnson [4], the complete set of linear inequalities describing (EC) is in fact known.

Proposition 3. *(Edmonds and Johnson [4].) The edge cover polytope (EC(G)) can be described by the set of linear inequalities of (FEC(G)) in addition to*

$$x(E(S)) + x(\delta(S)) \geq \frac{|S| + 1}{2} \qquad S \subseteq V, \; |S| \; odd. \qquad (2)$$

4.2   Algorithm

Let x be a feasible solution for (FEDS(G)). Since for each $\{u, v\} \in E$, $x(\delta(u)) + x(\delta(v)) \geq 1 + x_{uv}$, we have $\max\{x(\delta(u)), x(\delta(v))\} \geq \frac{1 + x_{uv}}{2} \geq \frac{1}{2}$. We use this criterion to define a vertex set V_+ as follows. For each edge $\{u, v\} \in E$ we select the endpoint whose fractional degree achieves $\max\{x(\delta(u)), x(\delta(v))\}$ to be in V_+; in the case of a tie, we choose one endpoint arbitrarily. We let $V_- = V \setminus V_+$.

Proposition 4. *V_+ is a vertex cover of G.*

Since an edge cover of a vertex cover is an edge-dominating set, we have reduced the problem at hand to that of finding a good edge cover of the set of vertices V_+. This is not quite the standard edge cover problem, yet a fair amount is known about it. For instance one can reduce this problem to the maximum weight capacitated b-matching problem (see [6, p.259]). In fact a complete polynomial-time separable linear description of the associated polytope is also known [16]. Rather than trouble ourselves with the technicalities that dealing directly with the V_+ edge cover problem imposes, we show how to reduce an instance of this problem to a bona fide instance of weighted edge cover.

We construct a new instance $\bar{G} = (\bar{V}, \bar{E})$ such that there is a one-to-one cost preserving correspondence between V_+ edge covers in G and edge covers of $\bar{G}$. Recall that V_+ and V_- partition V.

Let the vertex set V'_- be a copy of V_-, where $v \in V_-$ corresponds to $v' \in V'_-$. We set $\bar{V} = V \cup V'_-$ and $\bar{E} = E \cup E'$, where E' consists of zero-cost edges, one between each $v \in V_-$ its copy $v' \in V'_-$. Now if $\bar{D}$ is an edge cover of $\bar{G}$, then $\bar{D} \cap E$ must be an edge set of equal cost covering all the vertices in V_+. Conversely if D_+ is an edge set covering all the vertices in V_+, then $D_+ \cup E'$ is an edge cover of $\bar{G}$ of equal cost, since the edges in E' cost nothing.

We are now in a position to describe the algorithm, which may be stated quite simply as

1 Compute an optimal solution x^* for (FED(G)).
2 Compute V_+.
3 Compute and output a minimum-weight set of edges D covering V_+.

The algorithm clearly runs in polynomial time as the most expensive step is solving a compact linear program. Note that steps **2** and **3** may be implemented by the transformation above or by any method the reader fancies; however, the true benefit of the transformation may not be fully apparent until we analyze the approximation guarantee of the algorithm.

4.3   Analysis

As before suppose we are given an instance graph $G = (V, E)$ with no isolated vertices and a nonnegative cost vector c. Let x be some feasible fractional solution for (FEDS(G)). Along the lines of the algorithm, suppose we have computed V_+ and the resulting transformed instance, $\bar{G} = (\bar{V} = V \cup V'_-, \bar{E} = E \cup E')$. Let $\bar{x} = (x, 1^{|E'|}) \in \mathbb{Q}_+^{|\bar{E}|}$; that is, $\bar{x}$ corresponds to the augmentation of the fractional edge-dominating set x by E', a zero-cost set of edges. Note that by construction, $\bar{x}$ is feasible for (FEDS($\bar{G}$)). Similarly we extend c to $\bar{c} = (c, 0^{|E'|}) \in \mathbb{Q}_+^{|\bar{E}|}$. Note that we have $\bar{c} \cdot \bar{x} = c \cdot x$. We may now proceed to show that there is an integral edge cover of $\bar{G}$ which does not cost too much more than our fractional edge-dominating solution, $\bar{x}$.

Theorem 3. *The point $\frac{21}{10}\bar{x}$ is feasible for (EC($\bar{G}$)).*

Proof. Let $\bar{y} = 2\bar{x}$. Suppose u is a vertex in $\bar{V}$. If $u \in V_+$, we have $\bar{x}(\delta(u)) \geq \frac{1}{2}$; otherwise $u \in V_- \cup V'_-$, and we have $\bar{x}_e = 1$ for all $e \in E'$, so in either case

$$\bar{y}(\delta(u)) \geq 1, \tag{3}$$

hence $\bar{y}$ is feasible for (FEC($\bar{G}$)). Yet this is not quite good enough as (FEC($\bar{G}$)) does not have integral extreme points in general, so we extend this by showing that increasing $\bar{y}$ by a $\frac{1}{20}$ fraction places it in (EC($\bar{G}$)). To accomplish this we use the fact that $\bar{x}$ is a fractional edge-dominating set of $\bar{G}$, hence $\bar{y}$ satisfies

$$\bar{y}(\delta(u)) + \bar{y}(\delta(v)) \geq 2 + \bar{y}_{uv}. \tag{4}$$

Armed with this and the constraints of $(\mathrm{FEC}(\bar{G}))$, we proceed to show that $\frac{21}{20}\bar{y}$ also satisfies (2) with respect to $\bar{G}$.

Suppose S is a subset of $\bar{V}$ of odd cardinality; let $s = |S|$. When $s = 1$, the constraints (2) are trivially satisfied by $\bar{y}$, so suppose $s \geq 3$. By combining (3) and (4) we see

$$\bar{y}(\delta(u)) + \bar{y}(\delta(v)) \geq \begin{cases} 2 + \bar{y}_{uv} & \text{if } uv \in \bar{E}, \\ 2 & \text{otherwise.} \end{cases}$$

Summing the appropriate inequality above for each pair $\{u, v\}$ in $S \times S$, where $u \neq v$, we get

$$(s - 1)\bar{y}(\delta(S)) + 2(s - 1)\bar{y}(\bar{E}(S)) = (s - 1) \sum_{u \in S} \bar{y}(\delta(u))$$

$$= \sum_{\{\{u,v\} \in S \times S | u \neq v\}} \bar{y}(\delta(u)) + \bar{y}(\delta(v))$$

$$\geq s(s - 1) + \bar{y}(\bar{E}(S)).$$

Isolating the desired left hand side yields

$$\bar{y}(\delta(S)) + \bar{y}(\bar{E}(S)) \geq \frac{s(s - 1) + (s - 2)\bar{y}(\delta(S))}{2s - 3} \geq \frac{s(s - 1)}{2s - 3}, \text{ for } s \geq 3.$$

Using standard optimization techniques,

$$\max_{s \geq 3, \text{ odd}} \left\{ \frac{\frac{s+1}{2}}{\frac{s(s-1)}{2s-3}} \right\} = \frac{21}{20},$$

which is achieved when $s = 5$.

The theorem implies that $\frac{21}{10}\bar{x}$ is a convex combination of integral edge covers of $\bar{G}$, which by the construction of $\bar{G}$ are also integral edge-dominating sets of $\bar{G}$ containing the edges in E'. Equivalently the theorem implies that $\frac{21}{10}x$ is a convex combination of integral V_+ edge covers in G, which since V_+ is a vertex cover of G, are also integral edge-dominating sets of G. Thus there must be an integral edge-dominating set D of G of cost at most $\frac{21}{10}c \cdot x$. In particular, when x^* is an optimal fractional edge-dominating set of G and z_{EDS} the cost of an optimal integral solution, we find an integral solution of cost at most $\frac{21}{10}c \cdot x^* \leq \frac{21}{10}z_{EDS}$.

Corollary 1. *The point $\frac{21}{10}x$ is feasible for (EDS(G)) when $x \in$ (FEDS(G)).*

Corollary 2. *The algorithm of section 4.2 generates a solution of cost at most $2\frac{1}{10}$ times the optimal.*

Note that when G is bipartite, $\bar{G}$ is bipartite as well. In this case $(\mathrm{FEC}(G))$ forms a totally unimodular constraint set, hence $2\bar{x}$ is feasible for $(\mathrm{EC}(\bar{G}))$.

Proposition 5. *The algorithm of section 4.2 generates a solution of cost at most 2 times the optimal on a bipartite instance graph.*

This is in fact asymptotically tight as Figure 1 demonstrates.

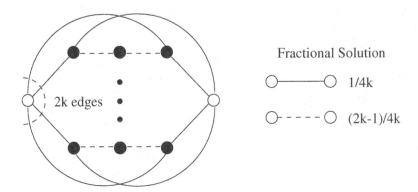

Fig. 1. A fractional extreme point of cost $k + \frac{1}{2}$. The algorithm chooses the darkened vertices as V_+, yielding a solution of cost $2k$, while the optimal integral solution costs $k + 1$.

4.4   An Extension

Suppose we are given an instance in which we are asked to find a minimum-weight edge set which dominates a specified subset $F \subseteq E$. We need only modify our algorithm so that only endpoints of edges in F are considered for V_+. The analysis of the previous section remains essentially the same, and our solution will cover the vertices in V_+ which ensures that the edges in F will be dominated.

4.5   Integrality Gap

Given that weighted EDS is as hard to approximate as weighted VC and that no polynomial-time algorithm with a constant performance guarantee strictly less than 2 is known for the latter, we might indulge in a respite from developing EDS algorithms if the former were shown to be approximable with a factor of 2. Unfortunately, it turns out that as long as our algorithm analysis is based exclusively on the optimal cost of (FEDS) as a lower bound for that of (EDS), we should relinquish such hope. The formulation (FEDS) introduces an *integrality gap*,

$$\max_{G=(V,E),\, c\in\mathbb{Q}_+^{|E|}} \left\{ \frac{\min_{x\in(\text{EDS}(G))} c \cdot x}{\min_{x\in(\text{FEDS}(G))} c \cdot x} \right\}$$

larger than 2. Corollary 1 bounds it above by $2\frac{1}{10}$, and it will be shown below that this is in fact a tight bound.

Consider the complete graph on $5n$ vertices, and let $G_1, \ldots, G_n$ be n vertex disjoint subgraphs, each isomorphic to K_5. Assign to each edge of G_i a weight of 1, and assign to any edge not in any of these subgraphs, some large weight. Let $x_e = 1/7$ if e is an edge of some G_i and $x_e = 0$ otherwise. Then it can be verified that $x(\delta(e)) \geq 1$ for all e, hence x is a feasible solution for $(\text{FEDS}(K_{5n}))$

of cost $\frac{10}{7}n$. On the other hand, any integral solution must cover all but one vertex in the graph. Prohibited to pick an edge outside of some G_i, an integral solution of small cost would choose 3 edges from each of G_i's but one for a total cost of $3n - 1$. Thus the integrality gap of formulation (FEDS) approaches $2\frac{1}{10}$. Although this example establishes the integrality gap of the formulation we employ, our algorithm may still perform provably better. The class of graphs depicted in Figure 1 preclude it from guaranteeing a bound less than 2, even for the unweighted bipartite case, and we offer our gratitude for a proof that it is indeed a 2-approximation.

The integrality gap of (FEDS(G)) is at most 2 when G is bipartite (Proposition 5); in fact it may grow arbitrarily close to 2. Let G be a complete bipartite graph $K_{n,n}$ with unit weights. Then, $x_e = \frac{1}{2n-1}$ for all $e \in E$ is a feasible solution of cost $\frac{n^2}{2n-1}$. Any integral solution must contain k edges since it must cover all of the vertices in at least one vertex class of the bipartition, so the integrality gap must be at least $\frac{n(2n-1)}{n^2} = 2 - \frac{1}{n}$.

References

1. S. Arora, C. Lund, R. Motwani, M. Sudan, and M. Szegedy. Proof verification and hardness of approximation problems. In *Proceedings of the 33rdAnnual IEEE Symposium on Foundations of Computer Science*, pages 14–23, 1992. 134
2. B. Baker. Approximation algorithms for NP-complete problems on planar graphs. *J. ACM*, 41:153–180, 1994. 133
3. V. Chvátal. A greedy heuristic for the set-covering problem. *Math. Oper. Res.*, 4(3):233–235, 1979. 134
4. J. Edmonds and E. Johnson. Matching, a well solved class of integer linear programs. In *Combinatorial Structures and Their Applications*, pages 89–92. Gordon & Breach, New York, 1970. 133, 137
5. U. Feige. A threshold of ln n for approximating set cover. In *Proceedings of the 28thAnnual ACM Symposium on Theory of Computing*, pages 314–318, May 1996. 134
6. M. Grötschel, L. Lovász, and A. Schrijver. *Geometric Algorithms and Combinatorial Optimization*. Springer, 1988. 137
7. F. Harary. *Graph Theory*. Addison-Wesley, Reading, MA, 1969. 135
8. D. S. Hochbaum, editor. *Approximation Algorithms for NP-hard Problems*. PWS Publishing Company, Boston, MA, 1997. 134, 136
9. J. Horton and K. Kilakos. Minimum edge dominating sets. *SIAM J. Discrete Math.*, 6(3):375–387, 1993. 133
10. H. Hunt III, M. Marathe, V. Radhakrishnan, S. Ravi, D. Rosenkrantz, and R. Stearns. A unified approach to approximation schemes for NP- and PSPACE-hard problems for geometric graphs. In *Proc. 2nd Ann. European Symp. on Algorithms*, pages 424–435, 1994. 133
11. D. S. Johnson. Approximation algorithms for combinatorial problems. *J. Comput. System Sci.*, 9:256–278, 1974. 134
12. L. Lovász. On the ratio of optimal integral and fractional covers. *Discrete Math.*, 13:383–390, 1975. 134

13. S. Mitchell and S. Hedetniemi. Edge domination in trees. In *Proc. 8th Southeastern Conf. on Combinatorics, Graph Theory, and Computing*, pages 489–509, 1977.   133
14. K. G. Murty and C. Perin. A 1-matching blossom-type algorithm for edge covering problems. *Networks*, 12:379–391, 1982.   133
15. C. Papadimitriou and M. Yannakakis. Optimization, approximation and complexity classes. *J. Comput. System Sci.*, 43:425–440, 1991.   134
16. W. R. Pulleyblank. Matchings and extensions. In *Handbook of combinatorics*, volume 1, pages 179–232. Elsevier, 1995.   133, 137
17. A. Srinivasan, K. Madhukar, P. Nagavamsi, C. P. Rangan, and M.-S. Chang. Edge domination on bipartite permutation graphs and cotriangulated graphs. *Information Processing Letters*, 56:165–171, 1995.   133
18. M. Yannakakis and F. Gavril. Edge dominating sets in graphs. *SIAM J. Appl. Math.*, 38(3):364–372, 1980.   133, 134

The Minimum Range Assignment Problem on Linear Radio Networks*
(Extended Abstract)

A.E.F. Clementi[1], A. Ferreira[3], P. Penna[1,**], S. Perennes[3], and R. Silvestri[2]

[1] Dipartimento di Matematica, Università di Roma "Tor Vergata",
{clementi,penna}@mat.uniroma2.it
[2] Dipartimento di Matematica Pura e Applicata, Università de L'Aquila,
silver@dsi.uniroma1.it
[3] MASCOTTE Project, I3S/INRIA Sophia Antipolis,
CNRS/INRIA/Université de Nice-Sophia Antipolis,
{ferreira,perennes}@sophia.inria.fr.

Abstract. Given a set S of radio stations located on a line and an integer h $(1 \leq h \leq |S| - 1)$, the MIN ASSIGNMENT problem is to find a range assignment of minimum power consumption provided that any pair of stations can communicate in at most h hops. Previous positive results for this problem were known only when $h = |S| - 1$ (i.e. the unbounded case) or when the stations are equally spaced (i.e. the uniform chain). In particular, Kirousis, Kranakis, Krizanc and Pelc (1997) provided an efficient exact solution for the unbounded case and efficient approximated solutions for the uniform chain, respectively.

This paper presents the first polynomial time, approximation algorithm for the MIN ASSIGNMENT problem. The algorithm guarantees an approximation ratio of 2 and runs in time $O(hn^3)$.

We also prove that, for constant h and for "well spread" instances (a broad generalization of the uniform chain case), we can find a solution in time $O(hn^3)$ whose cost is at most an $(1 + \epsilon(n))$ factor from the optimum, where $\epsilon(n) = o(1)$ and n is the number of stations. This result significantly improves the approximability result by Kirousis *et al* on uniform chains.

Both of our approximation results are obtained by new algorithms that exactly solves two natural variants of the MIN ASSIGNMENT problem that might have independent interest: the *All-To-One* problem (in which every station must reach a fixed one in at most h hops) and the *Base Location* problem (in which the goal is to select a set of *Basis* among the stations and all the other stations must reach one of them in at most $h - 1$ hops).

Finally, we show that for $h = 2$ the MIN ASSIGNMENT problem can be solved in $O(n^3)$-time.

---

* Work partially supported by the RTN Project ARACNE.
** Part of this work has been done while the author was visiting the research center of INRIA Sophia Antipolis.

M. Paterson (Ed.): ESA 2000, LNCS 1879, pp. 143–154, 2000.
© Springer-Verlag Berlin Heidelberg 2000

1   Introduction

At the present time, *Radio Networks* play a vital role in Computer Science and in several aspects of modern life. The proliferation of their applications, such as cellular phones and wireless local area networks, is due to the low cost of infrastructure and flexibility. In general, radio networks are adopted whenever the construction of more traditional networks is impossible or, simply, too expensive.

A *Multi-Hop Packet Radio Network* [10] is a finite set of radio stations located on a geographical region that are able to communicate by transmitting and receiving radio signals. In ad-hoc networks a transmission range is assigned to each station s and any other station t within this range can directly (i.e. by one *hop*) receive messages from s. Communication between two stations that are not within their respective ranges can be achieved by *multi-hop* transmissions. One of the main benefits of ad-hoc networks is the reduction of the power consumption. This can be obtained by suitably varying the transmission ranges.

It is reasonably assumed [10] that the power P_t required by a station t to correctly transmit data to another station s must satisfy the inequality

$$\frac{P_t}{d(t,s)^\beta} > \gamma \tag{1}$$

where $d(t,s)$ is the distance between t and s, $\beta \geq 1$ is the *distance-power gradient*, and $\gamma \geq 1$ is the *transmission-quality* parameter. In an ideal environment (see [10]) $\beta = 2$ but it may vary from 1 to more than 6 depending on the environment conditions of the place the network is located. In the rest of the paper, we fix $\beta = 2$ and $\gamma = 1$, however, our results can be easily extended to any $\beta, \gamma \geq 1$. Given a set $S = \{s_1, \ldots, s_n\}$ of radio stations on the d-dimensional Euclidean space, a *range assignment* for S is a function $r : S \to \mathcal{R}^+$ (where $\mathcal{R}^+$ is the set of non negative real numbers), and the *cost* of r is defined as

$$\mathsf{cost}(r) = \sum_{i=1}^{n} r(s_i)^2.$$

Given an integer h and a set S of stations on the d-dimensional Euclidean space, the MIN d-DIM ASSIGNMENT problem is to find a minimum cost range assignment provided that the assignment ensures the communication between any pair of stations in at most h hops. The MIN 1-DIM ASSIGNMENT problem (i.e. the linear case) will be simply denoted as MIN ASSIGNMENT. In this work we focus on the *linear* case, that is networks that can be modeled as sets of stations located along a line. As pointed out in [9], rather than a simplification, this version of the problem results in a more accurate analysis of the situation arising, for instance, in vehicular technology applications. Indeed, it is common opinion to consider one-dimensional frameworks as the most suitable ones in studying road traffic information systems. Indeed, vehicles follow roads, and messages are to be broadcast along lanes. Typically, the curvature of roads is small in comparison to the transmission range (half a mile up to some few miles).

Motivated by such applications, linear radio networks have been considered in several papers (see for instance [2,5,8,9]).

Tradeoffs between connectivity and power consumption have been obtained in [4,7,11,12].

As for the MIN d-DIM ASSIGNMENT problem, several complexity results have been obtained for the *unbounded case* (i.e. when $h = n - 1$). Under this restriction, MIN 3-DIM ASSIGNMENT is APX-complete (in [7], a polynomial time 2-approximation algorithm is given while in [3], the APX-hardness is proved); the MIN 2-DIM ASSIGNMENT problem is NP-hard [3]; finally, the MIN ASSIGNMENT problem is in P via an $O(n^4)$-time algorithm [7]. On the other hand, few results are known for the general case (i.e. for arbitrary h). In [7], the following bounds on the optimum cost have been proved for a strong restriction of MIN ASSIGNMENT.

Theorem 1 (The Uniform Chain Case [7]). *Let N be a set of n points equally spaced at distance $\delta > 0$ on the same line; let $\mathsf{OPT}_h(N)$ be the cost of an optimal solution for MIN ASSIGNMENT on input h and N. Then, it holds that*

$$- \ \mathsf{OPT}_h(N) = \Theta\left(\delta^2 n^{\frac{2^{h+1}-1}{2^h-1}}\right), \ \textit{for any fixed positive integer } h;$$

$$- \ \mathsf{OPT}_h(N) = \Theta\left(\delta^2 \frac{n^2}{h}\right), \ \textit{for any } h = \Omega(\log n).$$

Furthermore, the two above (implicit) upper bounds can be efficiently constructed.

Although the constructive method of Theorem 1 yields approximated solutions for the uniform chain case, no approximation result is known for more general configurations. Moreover, the approximation ratio guaranteed by Theorem 1, for constant h, *increases with h*. We then observe that for *non constant* values of h such that $h = o(\log n)$ no approximation algorithm *even for the uniform chain* restriction is known.

As for the MIN 2-DIM ASSIGNMENT problem, some upper and lower bounds on the optimal cost function, for constant values of h, have been derived in [4].

We present the first polynomial time approximation algorithm for the MIN ASSIGNMENT problem. The algorithm guarantees an approximation ratio of 2 and runs in time $O(hn^3)$.

Then, we provide a better approximation that works on any family of *well spread* instances and for any constant h; in such instances, the ratio between the maximum and the minimum distance among adjacent stations is bounded by a polylogarithmic function of n (see Sect. 4.2 for a formal definition). More precisely, we show that, for any well spread instance and for any constant h, it is possible to compute in time $O(hn^3)$ a solution whose cost is at most an $(1 + \epsilon(n))$ factor from the optimum, where $\epsilon(n) = o(1)$. Since uniform chains are a (very strong) restriction of well spread instances, our result strongly improves that in Theorem 1 in the case $h = O(1)$. Indeed, the obtained approximation ratio tends to 1 (while, as already observed, the approximation ratio achieved by Theorem 1 is an increasing function of h).

Our approximability results are obtained by exploiting exact solutions for two natural variants of the MIN ASSIGNMENT problem that might be of independent interest:

MIN ALL-TO-ONE ASSIGNMENT Given a set S of stations on the line, a *sink* station $t \in S$, and an integer $h > 0$; find a minimum cost range assignment for S ensuring that any station is able to reach t in at most h hops.

MIN ASSIGNMENT WITH BASIS Given a set S of stations on the line, and an integer $h > 0$; find a minimum cost range assignment for S such that, any station in S is either a *base* (a station is a base if it directly reaches any other station in S) or it reaches a base in at most $h - 1$ hops.

For each of the two above problems, we provide an algorithm, based on dynamic programming, that returns an optimal solution in time $O(hn^3)$.

Finally, we prove that for $h = 2$, the MIN ASSIGNMENT problem can be solved in time $O(n^3)$. This result is obtained by combining the algorithm for the MIN ASSIGNMENT WITH BASIS problem with a simple characterization of the structure of any optimal 2-hops range assignment.

Organization of the paper. In Sect. 1.1 we provide some basic definitions and notation. An efficient solution for the MIN ALL-TO-ONE ASSIGNMENT and the MIN ASSIGNMENT WITH BASIS problem is given in Sect. 2 and in Sect. 3, respectively. The approximability results are contained in Sect. 4. In Sect. 5 we describe an exact algorithm for the case $h = 2$ and in Sect. 6 we discuss some open problems.

1.1   Preliminaries

Let $S = \{s_1, \ldots, s_n\}$ be a set of n consecutive stations located on a line. We denote by $d(i, j)$ the distance between station s_i and s_j. We define $\delta_{\min}(S) = \min\{d(i, i+1) \mid 1 \leq i \leq n - 1\}$, $\delta_{\max}(S) = \max\{d(i, i+1) \mid 1 \leq i \leq n - 1\}$, and $D(S) = d(1, n)$.

Given a range assignment $r : S \to \mathcal{R}^+$, we say that s_i directly (i.e. in one hop) reaches s_j if $r(s_i) \geq d(i, j)$ (in short $i \to_r j$). Additionally, s_i reaches s_j in at most h hops if there exist $h - 1$ stations $s_{i_1}, \ldots, s_{i_{h-1}}$ such that $i \to_r i_1 \to_r i_2, \ldots, \to_r i_{h-1} \to_r j$ (in short $i \to_{r,h} j$). We will omit the subscript r when this will be clear from the context. We will say that r is an h-*assignment* ($1 \leq h \leq n - 1$) if for any pair of stations s_i and s_j, $i \to_{r,h} j$. Notice that h-assignments are exactly the feasible solutions for the instance (h, S) of MIN ASSIGNMENT. The cost of an optimal h-assignment for a given set S of stations is denoted as $\mathsf{OPT}_h(S)$. Given a station s_i we will refer to its *index* as i. Finally, we denote the set of stations $\{s_{i+1}, \ldots, s_{j-1}\}$ by (i, j) and we also use $[i, j]$, $[i, j)$ and $(i, j]$ as a shorthand of $(i - 1, j + 1)$, $(i - 1, j)$ and $(i, j + 1)$, respectively.

2   The MIN ALL-TO-ONE ASSIGNMENT Problem

In this section, we present an efficient method for the MIN ALL-TO-ONE ASSIGNMENT problem which is based on a suitable use of dynamic programming. To this aim, we introduce the following functions.

Definition 1 (All-To-One). *Given a set S of n stations and for any $1 \leq i \leq j \leq n$, we define*

$$\overleftarrow{\mathsf{ALL}}_h(i,j) = \min\{\mathsf{cost}(r) \mid \forall k \in [i,j], \ \ k \!\to_{r,h}\! i\};$$

$$\overrightarrow{\mathsf{ALL}}_h(i,j) = \min\{\mathsf{cost}(r) \mid \forall k \in [i,j], \ \ k \!\to_{r,h}\! j\}.$$

Definition 2 (OR). *Given a set S of n stations and for any $1 \leq i \leq j \leq n$, we define*

$$\mathsf{OR}_h(i,j) = \min\{\mathsf{cost}(r) \mid \forall k \in [i,j], \ \ k \!\to_{r,h}\! i \ \vee \ k \!\to_{r,h}\! j\}.$$

Such functions will also be used in Section 3 in order to solve the MIN ASSIGNMENT WITH BASIS problem.

Lemma 1. *There is an algorithm that, for any set of n stations on the line, for any $1 \leq i \leq j \leq n$, and for any $h \geq 1$, computes $\overleftarrow{\mathsf{ALL}}_h(i,j)$, $\overrightarrow{\mathsf{ALL}}_h(i,j)$, $\mathsf{OR}_h(i,j)$ in time $O(hn^3)$.*

Sketch of Proof. In order to prove the lemma, we need to define two further functions:

$$\overleftarrow{\mathsf{ALL}}_h^*(i,j) = \min\{\mathsf{cost}(r) \mid \forall k \in [i,j), \ \ k \!\to_{r,h}\! i \ \wedge \ j \!\to_r\! i\};$$

$$\overrightarrow{\mathsf{ALL}}_h^*(i,j) = \min\{\mathsf{cost}(r) \mid \forall k \in (i,j], \ \ k \!\to_{r,h}\! j \ \wedge \ i \!\to_r\! j\}.$$

Our next goal is to prove the following recursive equations:

$$\overrightarrow{\mathsf{ALL}}_h(i,j) = \min_{i \leq k < j}\{\overrightarrow{\mathsf{ALL}}_h^*(k,j) + \overrightarrow{\mathsf{ALL}}_{h-1}(i,k)\}; \qquad (2)$$

$$\overleftarrow{\mathsf{ALL}}_h(i,j) = \min_{i < k \leq j}\{\overleftarrow{\mathsf{ALL}}_h^*(i,k) + \overleftarrow{\mathsf{ALL}}_{h-1}(k,j)\}. \qquad (3)$$

In fact, consider the function $\overrightarrow{\mathsf{ALL}}_h(i,j)$ and consider any feasible range assignment r for this function. Let k be the index of the *leftmost* station reaching j in one hop (see Fig. 1). For any station $s \in [i,k)$ it holds that $s \!\to_{r,h}\! j$ but it does not hold that $s \!\to_r\! j$ (by definition of k). It thus easily follows that, for any $s \in [i,k)$, it must be the case that $s \!\to_{r,h-1}\! k$. We also remark that no station in $[i,k]$ uses "bridges" in the interval (k,j). This implies that r, restricted to $[i,k)$, is a feasible range assignment for $\overrightarrow{\mathsf{ALL}}_{h-1}(i,k)$. Furthermore, for any $s \in [k,j]$,

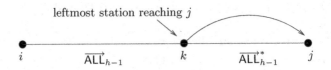

Fig. 1. The proof of Lemma 1.

$s \to_{r,h} j$ without using any "bridge" in $[i, k]$. Hence, r restricted to $[k, j]$ is a feasible assignment for $\overrightarrow{\mathsf{ALL}}_h^*(k, j)$. It thus follows that

$$\mathsf{cost}(r) \geq \overrightarrow{\mathsf{ALL}}_{h-1}(i, k) + \overrightarrow{\mathsf{ALL}}_h^*(k, j)$$

Eq. 3 can be proved by a symmetric argument.

By using similar arguments, it is possible to prove the following other recursive equations

$$\overrightarrow{\mathsf{ALL}}_h^*(i, j) = \min_{i < k \leq j} \{\overrightarrow{\mathsf{ALL}}_h^*(k, j) + \mathsf{OR}_{h-1}(i, k)\} + d(i, j)^2; \tag{4}$$

$$\overleftarrow{\mathsf{ALL}}_h^*(i, j) = \min_{i \leq k < j} \{\overleftarrow{\mathsf{ALL}}_h^*(i, k) + \mathsf{OR}_{h-1}(k, j)\} + d(i, j)^2; \tag{5}$$

$$\mathsf{OR}_h(i, j) = \min_{i \leq k < j} \{\overleftarrow{\mathsf{ALL}}_h(i, k) + \overrightarrow{\mathsf{ALL}}_h(k + 1, j)\}. \tag{6}$$

In what follows, we describe the correct "crossed" recursive computation that will return the outputs of the five functions. The overall computation goes over h *phases*; in the ℓ-th phase, all the functions will be computed for number of hops equal to ℓ.

Phase $\ell = 1$ consists of computing the following values:

$$\forall i < j \ : \overleftarrow{\mathsf{ALL}}_1^*(i, j) = \overleftarrow{\mathsf{ALL}}_1(i, j) = \sum_{k=i+1}^{j} d(i, k)^2$$

$$\forall i < j \ : \overrightarrow{\mathsf{ALL}}_1^*(i, j) = \overrightarrow{\mathsf{ALL}}_1(i, j) = \sum_{k=i}^{j-1} d(k, j)^2$$

$$\mathsf{OR}_1(i, j) = \min_{i \leq k < j} \{\overleftarrow{\mathsf{ALL}}_1(i, k) + \overrightarrow{\mathsf{ALL}}_1(k + 1, j)\}.$$

Notice that, in any Phase $\ell \geq 1$ and for any i, it easily holds that

$$\overleftarrow{\mathsf{ALL}}_\ell^*(i, i) = \overleftarrow{\mathsf{ALL}}_\ell(i, i) = \overrightarrow{\mathsf{ALL}}_\ell^*(i, i) = \overrightarrow{\mathsf{ALL}}_\ell(i, i) = \mathsf{OR}_\ell(i, i) = 0.$$

Now, assume that, at the end of Phase $\ell - 1$, the algorithm has computed the values of the five functions for all possible segments in $[1, n]$ and for number of hops $\ell - 1$. Then, the function $\overleftarrow{\text{ALL}}^*_\ell$ can be computed, by applying Eq. 5, for all the segments in the following order:

$$[1,1], [1,2], \ldots , [1,n], [2,2], [2,3], \ldots , [2,n], \ldots , [n-1,n], [n,n].$$

The opposite order is instead used for computing the values of $\overrightarrow{\text{ALL}}^*_\ell$ by applying Eq. 4. The next two steps (in any order) are the computations of functions $\overrightarrow{\text{ALL}}_\ell$ and $\overleftarrow{\text{ALL}}_\ell$ for any interval in $[1,n]$ by applying Eq. 2 and Eq. 3. The last values computed at Phase ℓ are the $\text{OR}_\ell(i,j)$ for all possible segments (i,j) according to equation 6.

We finally observe that, at every Phase ℓ, we need to compute $O(n^2)$ values, each of them requiring $O(n)$ time.

$\square$

The above lemma easily implies the following theorem.

Theorem 2. *The* MIN ALL-TO-ONE ASSIGNMENT *problem can be solved in time* $O(hn^3)$.

3   The MIN ASSIGNMENT WITH BASIS Problem

In order to provide exact solutions for the MIN ASSIGNMENT WITH BASIS problem, we consider the following definitions.

Definition 3 (base stations). *Let r be a feasible solution for the* MIN ASSIGNMENT *problem on input h and S. A station i is a* base *(in short* B*) if $i \rightarrow_r 1 \wedge i \rightarrow_r n$. Moreover, r is of type* B* *if there is at least one base and any station which is not a base reaches some base in at most $h - 1$ hops. Then,* $\text{BASES}_h(S)$ *denotes the cost of an optimum assignment of type* B*.

Notice that $\text{BASES}_h(S)$ is the optimum for the MIN ASSIGNMENT WITH BASIS problem on input h and S. The main contribution of this section can be stated as follows.

Theorem 3. *For any set S of n stations on the line and for any $1 \leq h \leq n-1$, it is possible to construct an optimum h-assignment of type* B* *for S in time* $O(hn^3)$. *Thus, the* MIN ASSIGNMENT WITH BASIS *problem is in P.*

Sketch of Proof. Let us first consider the indices $i_1^*, \cdots , i_k^*$ of the k bases in the optimal solution (see Fig. 2). It is not hard to prove that, between two consecutive bases, any station must reach one of the two bases in at most $h - 1$ hops. Additionally, the stations in $[1, i_1^*]$ (respectively, $(i_k^*, n]$) must reach in $h-1$ hops the base in i_1^* (respectively, i_k^*).

Thus, given the indices of the bases in an optimal solution, we can use the functions described in Sect. 2 to find the optimal assignment. Notice that if k would be always bounded by a constant then we could try all the possible indices for the bases. However, this is not the case, so a more tricky approach is

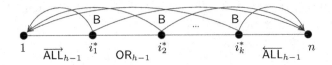

Fig. 2. The structure of the optimum solution for the base location problem.

needed. Basically, we will use the fact that every base "cuts" the instance into two *independent* intervals.

Let us define $\mathsf{BASES}_h(S, i)$ as the cost of the minimum h-assignment of type B^* subject to the rightmost base has index i. Clearly

$$\mathsf{BASES}_h(S) = \min_{1 \le i \le n} \mathsf{BASES}_h(S, i).$$

Let us then see how to compute $\mathsf{BASES}_h(S, i)$ for any i. To this aim we need a function $\mathsf{CHANGE}_h(i, j)$ which, roughly speaking, corresponds to the change of the cost of $\mathsf{BASES}_h(S, i)$ when we set a *new* base $j > i$ (see Fig. 3).

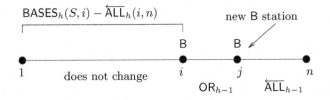

Fig. 3. The proof of Theorem 3.

It is easy to see that the new base j is useless for the stations in $[1, i]$. Indeed, if a station in $[1, i]$ reaches j in at most $h - 1$ hops, then it also reaches i within the same number of hops, thus making j useless. Moreover, the cost of an optimal assignment for $\mathsf{BASES}_h(S, i)$ restricted to $(i, n]$ is $\overleftarrow{\mathsf{ALL}}_{h-1}(i, n)$. The "new cost" due to base j in the interval $(i, n]$ is given by

$$\mathsf{OR}_{h-1}(i, j) + \overleftarrow{\mathsf{ALL}}_{h-1}(j, n) + \max\{d(1, j), d(j, n)\}^2.$$

We can thus define

$$\mathsf{CHANGE}_h(i, j) = \mathsf{OR}_{h-1}(i, j) + \overleftarrow{\mathsf{ALL}}_{h-1}(j, n)$$
$$+ \max\{d(1, j), d(j, n)\}^2 - \overleftarrow{\mathsf{ALL}}_{h-1}(i, n).$$

We can now compute $\mathsf{BASES}_h(S, \cdot)$. Let $\mathsf{BASE}_h^1(S, i)$ be the minimum cost of the range assignment in which there exists *only* one base and its index is i. Then, it holds that

$$\mathsf{BASE}_h^1(S, i) = \overrightarrow{\mathsf{ALL}}_{h-1}(1, i) + \max\{d(1, i), d(i, n)\}^2 + \overleftarrow{\mathsf{ALL}}_{h-1}(i, n). \quad (7)$$

The algorithm starts by computing $\overrightarrow{\text{ALL}}_{h-1}(i,j)$, $\overleftarrow{\text{ALL}}_{h-1}(i,j)$, $\text{OR}_{h-1}(i,j)$ by using the algorithm Lemma 1. Then, $\text{CHANGE}_h(i,j)$ and $\text{BASE}_h^1(S,i)$ can be computed by the above equations, for any $1 \leq i < j \leq n$.

The computation of $\text{BASES}_h(S,j)$ goes over n phases. Phase $j = 1$ corresponds to compute

$$\text{BASES}_h(S,1) = \text{BASE}_h^1(S,1). \tag{8}$$

Assume that at phase $j-1$ $\text{BASES}_h(S,i)$ for any $1 \leq i \leq j-1$ have been already computed. Then, the computation of $\text{BASES}_h(S,j)$ can be carried out according to the following recursive equation:

$$\text{BASES}_h(S,j) = \min_{1 \leq i < j} \left\{ \text{BASE}_h^1(S,j),\ \text{BASES}_h(S,i) + \text{CHANGE}_h(i,j) \right\}.$$

The correctness of the above equation follows from the fact that an optimal assignment for $\text{BASES}_h(S,j)$ either contains only one base and its index is j, or it contains at least one base other than j. In the latter case, as previously observed, the assignment in $[1,i]$ is not used outside. Thus, such an assignment must be optimal for $\text{BASES}_h(S,i)$.

Finally, it is easy to prove that the overall algorithm works in time $O(hn^3)$.

□

4 Approximation Algorithms

4.1 A 2-Approximation Algorithm for MIN ASSIGNMENT

Let us consider a feasible assignment r for the instance $(h, S = \{s_1, \ldots, s_n\})$ of MIN ASSIGNMENT; by definition, it should be clear that r is a feasible assignment also for $\overleftarrow{\text{ALL}}_h(1,n)$ and $\overrightarrow{\text{ALL}}_h(1,n)$. This fact implies the following useful lower bound on the optimum of MIN ASSIGNMENT.

Lemma 2. *For any set of stations $S = \{s_1, \ldots, s_n\}$ and for any $h \geq 1$, it holds that*

$$\text{OPT}_h(S) \geq \max \left\{ \overleftarrow{\text{ALL}}_h(1,n), \overrightarrow{\text{ALL}}_h(1,n) \right\}.$$

By combining the above lemma with the algorithm of Lemma 1, we get the following result (the proof is given in the Appendix).

Theorem 4. *There is an algorithm that, for any set $S = \{s_1, \ldots, s_n\}$ and any $h \geq 1$, computes in time $O(hn^3)$ an h-assignment r_{apx} for S such that $\text{cost}(r_{apx}) \leq 2 \cdot \text{OPT}_h(S)$.*

4.2   Well Spread Instances

This section is devoted to the construction of a better approximation for *well spread instances* of the MIN ASSIGNMENT problem.

Definition 4 (well spread instances). *A family $\mathcal{S}$ of well-spread instances is a family of instances S such that*

$$\delta_{\max}(S) = O(\delta_{\min}(S)\mathrm{poly}(\log n)).$$

The following lemma is proved in the Appendix (by the way, the first item is an easy consequence of Theorem 1).

Lemma 3. *a). For any set S of n stations on the line and for any constant $h \geq 1$, it holds that*

$$\mathsf{OPT}_h(S) = \Omega\left(\delta_{\min}(S)^2 n^{\frac{2^{h+1}-1}{2^h-1}}\right).$$

b). Let $\mathcal{S}$ be a family of well-spread instances. Then, for any instance $S^w \in \mathcal{S}$ of n stations and for any $h \geq 1$,

$$\mathsf{OPT}_h(S^w) = \Omega\left(D(S^w)^2 \frac{n^{\frac{1}{2^h-1}}}{\mathrm{poly}(\log n)}\right).$$

Theorem 5. *There is an algorithm that, for any family $\mathcal{S}$ of well spread instances, for any instance $S^w \in \mathcal{S}$, and for any constant $h > 0$, computes in time $O(hn^3)$ an h-assignment r^{apx} of cost $\mathsf{cost}(r_{apx}) = (1 + \epsilon(n)) \cdot \mathsf{OPT}_h(S^w)$, where $\epsilon(n) = o(1)$.*

Sketch of Proof. We will prove that if $S^w \in \mathcal{S}$ then the optimum $\mathsf{BASES}_h(S^w)$ of the MIN ASSIGNMENT WITH BASIS problem on instance (h, S^w) is such that $\mathsf{BASES}_h(S^w) = (1 + o(1)) \cdot \mathsf{OPT}_h(S^w)$. Let r^{apx} be the range assignment yielded by the algorithm of Theorem 3 in time $O(hn^3)$. Wlog assume that $h \geq 2$. If there is an optimal h-assignment for S^w of type B^* (see Def. 3) then r^{apx} is optimal for S^w. So, assume that all the optimal h-assignments for S^w are not of type B^*. Consider an optimal h-assignment r^{opt}. Let L be the rightmost station reaching station 1 in one hop. It is not hard to see that all the stations that do not use a base to reach both the endpoints must reach L. Define the range assignment r^B in which all the stations different from L keep the range as in r^{opt} while L has now a range sufficient to be a basis.

In this assignment any station reaches some base in at most $h - 1$ hops, so r^B is an h-assignment of type B^*. It turns out that

$$\mathsf{cost}(r^{apx}) \leq \mathsf{cost}(r^B) \leq \mathsf{OPT}_h(S^w) + D(S^w)^2.$$

Since $\mathcal{S}$ is well spread, from Lemma 3(b) we have that, for some constant $c > 0$,

$$\mathsf{OPT}_h(S^w) \geq cD(S^w)^2 \frac{n^{1/(2^h-1)}}{\mathrm{poly}(\log n)}.$$

By combining the last two inequalities, we get

$$\frac{\text{cost}(r^{apx})}{\text{OPT}_h(S^w)} \leq 1 + \frac{D(S^w)^2}{\text{OPT}_h(S^w)} \leq 1 + c\frac{\text{poly}(\log n)}{n^{1/(2^h-1)}}.$$

For constant h and for $n \to \infty$, the above value tends to 1. Hence the theorem follows.
$\square$

5   Min 2-hops Assignment is in P

A suitable application of the algorithms given in Lemma 1 and Theorem 3 will allow us to obtain an algorithm that solves Min 2-hops Assignment in time $O(n^3)$. To this aim, we need to distinguish three types of possible optimal 2-assignments.

Definition 5. *Let r be a 2-assignment. We say that r is of type*

- B^* *if all the stations reach one base in one hop (this is the same as in Def. 3);*
- LR *if there are no base stations and L (R) denotes the rightmost (leftmost) station reaching 1 (n) in one hop, and L is on the left of R;*
- B^*LRB^* *if there is at least a base station and there is no base between L and R, where L (R) denotes the rightmost (leftmost) non-base station reaching 1 (n) in one hop, and L is on the left of R.*

The proof of the following lemma will be given in the full version of the paper.

Lemma 4. *Let S be a set of stations on the line. If r_{opt_2} is an optimal 2-assignment for S, then it is either of type B^*, LR or B^*LRB^*.*

Theorem 6. *The Min 2-hops Assignment problem can be solved in time $O(n^3)$.*

Sketch of Proof. From Lemma 4 the type of an optimal solution for Min 2-Hops Assignment has to be one of the three types listed in Def. 5. As for the first type (i.e. B^*) we apply the algorithm in Theorem 3. As for the second type (i.e. LR), for each possible pair L and R, we split the instance $[1, n]$ in three independent intervals $[1, L]$, (L, R), $[R, n]$. In $[1, L)$ we compute $\overrightarrow{\text{ALL}}_1(1, R)$ in which the stations between L and R are removed. Symmetrically we compute the assignment for $(R, n]$. In (L, R) we compute $\text{AND}_1(L, R)$. The third type (i.e. B^*LRB^*) can be solved by a simple combination of the previous two types.

An optimal 2-assignment can be thus obtained by selecting the minimum cost assignment among all the assignments computed above.
$\square$

6   Open Problems

In 1997, Kirousis *et al* [7] wondered whether efficient solutions for the bounded hops case of the Min Assignment problem can be obtained. This paper provides a first positive answer to this question since, in most practical applications, an easy-to-obtain 2-approximated solution can be reasonably considered good enough. On the other hand, the theoretical question whether MIN ASSIGN-MENT $\in$ P still remains open. However, we believe that exact solutions can be obtained only by algorithms whose time complexity is $\Omega(n^h)$. Finally, it is interesting to observe that a similar "state of the art" holds for another important "Euclidean" problem, i.e., the *Geometric Facility Location Problem* (in short, GFLP) [6]. Indeed, also in this case the Euclidean properties of the support space have been used to derive very good approximation results [1] (much better than those achievable on non-geometric versions of FLP). On the other hand, it is still unknown whether or not GFLP $\in$ P.

References

1. S. Arora, P. Raghavan, and S. Rao. Approximation schemes for the euclidean k-medians and related problem. In *Proc. 30th Annual ACM Symposium on Theory of Computing (STOC)*, pages 106–113, 1998.   154
2. M.A. Bassiouni and C. Fang. Dynamic channel allocation for linear macrocellular topology. *Proc. of ACM Symp. on Applied Computing (SAC)*, pages 382–388, 1998.   145
3. A. Clementi, P. Penna, and R. Silvestri. Hardness results for the power range assignment problem in packet radio networks. *Proc. of RANDOM-APPROX'99, Randomization, Approximation and Combinatorial Optimization*, LNCS(1671):197 – 208, 1999.   145
4. A. Clementi, P. Penna, and R. Silvestri. The power range assignment problem in radio networks on the plane. *17th Annual Symposium on Theoretical Aspects of Computer Science (STACS)*, LNCS(1770):651 – 660, 2000.   145
5. K. Diks, E. Kranakis, D. Krizanc, and A. Pelc. The impact of knowledge on broadcasting time in radio networks. *Proc. 7th European Symp. on Algorithms (ESA)*, LNCS(1643):41–52, 1999.   145
6. D. Hochbaum. Heuristics for the fixed cost median problem. *Math. Programming*, 22:148–162, 1982.   154
7. L. M. Kirousis, E. Kranakis, D. Krizanc, and A. Pelc. Power consumption in packet radio networks. *14th Annual Symposium on Theoretical Aspects of Computer Science (STACS)*, LNCS(1200):363 – 374, 1997.   145, 153
8. E. Kranakis, D. Krizanc, and A. Pelc. Fault-tolerant broadcasting in radio networks. *6th European Symp. on Algorithms (ESA)*, LNCS(1461):283–294, 1998.   145
9. R. Mathar and J. Mattfeldt. Optimal transmission ranges for mobile communication in linear multihop packet radio netwoks. *Wireless Networks*, 2:329–342, 1996.   144, 145
10. K. Pahlavan and A. Levesque. *Wireless Information Networks*. Wiley-Interscince, New York, 1995.   144
11. P. Piret. On the connectivity of radio networks. *IEEE Trans. on Infor. Theory*, 37:1490–1492, 1991.   145
12. S. Ulukus and R.D. Yates. Stochastic power control for cellular radio systems. *IEEE Trans. Commun.*, 46(6):784–798, 1998.   145

Property Testing in Computational Geometry[*]

(Extended Abstract)

Artur Czumaj[1], Christian Sohler[2], and Martin Ziegler[2]

[1] Department of Computer and Information Science, New Jersey Institute of Technology, Newark, NJ 07102-1982, USA. czumaj@cis.njit.edu

[2] Heinz Nixdorf Institute and Department of Mathematics & Computer Science, University of Paderborn, D-33095 Paderborn, Germany. csohler,ziegler@uni-paderborn.de

Abstract. We consider the notion of *property testing* as applied to computational geometry. We aim at developing efficient algorithms which determine whether a given (geometrical) object has a predetermined property Q or is "far" from any object having the property. We show that many basic geometric properties have very efficient testing algorithms, whose running time is significantly smaller than the object description size.

1 Introduction

Property testing is a rapidly emerging new field in computer science with many theoretical as well as practical applications. It is concerned with the computational task of determining whether a given object has a predetermined property or is "far" from any object having the property. It is a relaxation of the standard definition of a decision task, because we allow arbitrary behavior when the object does not have the property, and yet it is close to an object having the property. A notion of property testing was first explicitly formulated in [15] and then extended and further developed in many follow-up works (see, e.g., [1,2,4,7,8,9,10,11,12,16]). Property testing arises naturally in the context of program verification, learning theory, and, in a more theoretical setting, in probabilistically checkable proofs. For example, in the context of program checking, one may first choose to test whether the program's output satisfies certain properties before checking that it is as desired. This approach is a very common practice in software development, where it is (typically) infeasible to require to formally test that a program is correct, but by verifying whether the output satisfies certain properties one can gain a reasonable confidence about the quality of the program's output.

In this work we consider property testing as applied to *geometric objects in Euclidean spaces*. We investigate property testing for some basic and most representative problems/properties in computational geometry. The main goal of this research is to develop algorithms which perform only a very small (sublinear, polylogarithmic, or even a constant) number of operations in order to check with a reasonable confidence the required geometric property.

We study the following family of tasks: *Given oracle access to an unknown geometric object, determine whether the object has a certain predefined property or is "far" from any object having this property.* Distance between objects is measured in terms of

---

[*] Research supported in part by DFG Grant Me872/7-1.

M. Paterson (Ed.): ESA 2000, LNCS 1879, pp. 155–166, 2000.
© Springer-Verlag Berlin Heidelberg 2000

a *relative edit distance*, i.e., an object X (e.g., a set of points in $\mathbb{R}^d$) from a class $\mathcal{C}$ (e.g., a class of all point sets in $\mathbb{R}^d$) is called *ε-far from satisfying a property* Q (over $\mathcal{C}$) if no object Y from $\mathcal{C}$ which differs from X in no more than an ϵ fraction of places (e.g., Y can be constructed from X by adding and removing no more than $\epsilon \cdot |X|$ points) satisfies Q. The notion of *"oracle access"* corresponds to the representation of the objects in $\mathcal{C}$, and it is more problem dependent. And thus, for example, if a geometric object is defined as a collection of points in Euclidean space, then it is reasonable to require that the oracle allows the algorithm to ask only on the coordinate-position of each single input point. If however, a geometric object is, say, a polygon, then the oracle shall typically allow the algorithm to query for the position of each point as well as for each single neighbor of each point in the polygon. In this paper we shall always use the most natural notions of "oracle" for each problem at hand.

An *ε-tester* for a property Q is a (typically randomized) algorithm which always accepts any object satisfying Q and with probability at least $\frac{2}{3}$ rejects any object being ϵ-far from satisfying Q [1]. There are two types of possible complexity measures we focus on: the *query complexity* and the *running time* complexity of an ϵ-tester. The query complexity of a tester is measured only by the number of queries to the oracle, while the running time complexity counts also the time needed by the algorithm to perform other tasks (e.g., to compute a convex hull of some point set).

To exemplify the notion of property testing, let us consider the standard geometrical property Q of a point set P in Euclidean space $\mathbb{R}^d$ of being in *convex position* [2]. In this case, we aim at designing an algorithm which for a given oracle access to P has to decide whether P is in convex position or P is "ϵ-far" from being in convex position. Here, we say a *set P is ε-far from being in convex position*, $0 \leq \epsilon \leq 1$, if for every subset $X \subseteq P$ of size at most $\epsilon \cdot |P|$ the set $P \setminus X$ is not in convex position. Moreover, we assume that the oracle allows to query for the size of P and for the position of each ith point in P.

Another type of property we shall consider in this paper is that of being an *Euclidean minimum spanning tree (EMST)*. That is, for a given set P of points in the plane we have to verify whether a given graph G with the vertex set P is an EMST for P or it is ϵ-far from being an EMST for P. Here we assume the algorithm can query the size of P, the position of each single ith point x in P as well as query the jth neighbor of x in G (with the special symbol indicating non-existence of such a neighbor). We say here that an Euclidean graph G with vertex set P is ϵ-far from being an EMST for P if the minimum spanning tree of P differs from G on at least $\epsilon \cdot |P|$ edges.

1.1 Description of New Results

In this paper we provide efficient ϵ-testers for some basic problems in computational geometry. We investigate problems which we found most representative for our study and which play an important role in the field. The following property testers are studied:

---

[1] One could consider also a *two-sided error* model or consider a confidence parameter δ rather than fixing an error bound $\frac{1}{3}$. These models are not discussed in this paper.

[2] A set of points P in $\mathbb{R}^d$ is in convex position if every point of P is *extreme*, that is, it is a vertex of $\mathrm{CH}(P)$, the convex hull of P.

Problem	Query complexity	Lower bound	Running time
convex position	$\mathcal{O}(n^d/\epsilon)^{1/(d+1)}$	$\Omega(n^d/\epsilon)^{1/(d+1)}$	$\mathcal{O}\left(n^{2/3}(1/\epsilon)^{1/3}\log(n/\epsilon)\right)$ $d=2$ $\mathcal{O}\left(n^{3/4}(1/\epsilon)^{1/4}\log(n/\epsilon)\right)$ $d=3$ $\mathcal{O}\left(n\,\mathrm{polylog}(1/\epsilon)+d\geq 4\right.$ $\left.\left(\frac{n}{\epsilon}\right)^{(\lfloor d/2\rfloor)/(1+\lfloor d/2\rfloor)}\mathrm{polylog}(n)\right)$
disjointness of geometric objects	$\mathcal{O}(\sqrt{n/\epsilon})$	$\Omega(\sqrt{n/\epsilon})$	$T(\mathcal{O}(\sqrt{n/\epsilon}))$
disjointness of polytopes	$\mathcal{O}\left((d/\epsilon)\cdot\log(d/\epsilon)\right)$	$\Omega(1/\epsilon)$	$\mathcal{O}(1/\epsilon)$ deterministic(!)
Delaunay triangulation			
Hamming distance	$\mathcal{O}(n)$	$\Omega(n)$	$\mathcal{O}(n)$
Distance defined in Theorem 5.2	$\mathcal{O}(1/\epsilon)$	$\Omega(1/\epsilon)$	$\mathcal{O}(1/\epsilon)$
EMST	$\mathcal{O}(\sqrt{n/\epsilon}\,\log^2(1/\epsilon))$	-	$\mathcal{O}(\sqrt{n/\epsilon}\,\log^2(1/\epsilon)\log n)$

Table 1. ϵ-testers obtained in the paper. $T(m)$ denote the best known running time of the algorithm deciding given property and we consider lower bounds with respect to the query complexity.

convex position: property that a set of points is in a *convex position*,

disjointness of geometric objects: property that an arrangement of objects is intersection-free,

disjointness of V-polytopes: property that two polytopes are intersection-free,

EMST: property that a given graph is a Euclidean *minimum spanning tree* of a given point set in the plane, and

Delaunay triangulation: property that a triangulation has the Delaunay property.

Table 1 summarizes the bounds of our ϵ-tester developed in the paper in details. As this table shows, all our algorithms are tight or almost tight regarding their query complexity. We also prove a general lower bound which shows that for many problems no *deterministic* ϵ-tester exists with the query complexity $o(n)$ already for $\epsilon = \frac{1}{4}$.

There are two main approaches used in our testers. The first one, which is quite general and is used in all our algorithms, is to sample at random a sufficiently large subset of input basic objects (e.g., input points) and then analyze the sample to obtain information about the whole input. The key issue in this method is to provide a good description of the input's property using a small input's sample. This approach is used in a simple way in our ϵ-testers for disjointness of geometric objects and Delaunay triangulation, where such a description is easy to show. In our ϵ-testers for convex position and disjointness of polytopes this approach is used as well, but it is more complicated to prove the desired property of the whole input out of the analysis of small sample. The second approach (combined with the first one) is used in our tester for EMST: after sampling a portion of the input (which are the vertices of the graph) at random, we enlarge the sample in a systematic way by exploring the paths of the length up to $\mathcal{O}(1/\epsilon)$ starting in the vertices chosen. Only then we can show that the sample chosen can be used to certificate whether the input graph is the EMST of the input points.

We mention also an important feature of all our algorithms, which is that all our testers supply proofs of violation of the property for rejected objects (because of space limitations we do not elaborate on this issue here).

Throughout the paper we always assume that d, the dimension of the Euclidean space under consideration, is a constant. Moreover, we assume the size of the input object to be much bigger than d. To simplify the exposition, we suppose also that all points/arrangements are in general position.

1.2   Motivation and Applications

What does this type of approximation mean. Testing a property is a natural relaxation of deciding that property, and it suggests a certain notion of approximation: the tester accepts an object which either has the property or which is "close" to some other object having the property. Under this notion, in applications where objects close to having the property are almost as good as ones having the property, a tester which is significantly faster than the corresponding decision procedure is a very valuable alternative to the latter. We refer the reader to [7,10,15], where many more detailed applications of this notion of approximation have been discussed (see, especially, [10, Section 1.2.1]).

In the context of computational geometry a related research has been presented in [3] and [14]. For example, Mehlhorn et al. [14] studied *program checkers* for some basic geometric tasks. Their approach is to design simple and efficient algorithms which checks whether an output of a geometric task is correct. It is shown in [14] that some basic geometric tasks can be designed so that program checkers can be obtained in a fairly simple (and efficient) way. This approach has been also successfully implemented as a part of the LEDA system. The main difference between our approach and that in [14] is that we (and, generally, property testing algorithms) do not aim at deciding the property at hand (in this case, whether the task output is correct), but instead we provide a certain notion of *approximation* — we can argue that we do not accept the property (or the algorithm's output) only if it is far from being satisfied. This relaxation enables us to obtain algorithms with running times *significantly* better than those in [14].

Output sensitive algorithms. An important, novel feature of our property testing algorithms is that they can be used to design *output sensitive* algorithms for decision problems. Consider for example, the problem of deciding whether a set P of n points on the plane is in convex position, and if it is not, to report a subset of P which is non-convex. Below we show how our ϵ-tester can be used to obtain an "output sensitive" (randomized) algorithm.

We set first $\epsilon = \frac{1}{2}$ and apply our tester to verify whether P is ϵ-far from being in convex position. If it is so, then we report a subset of P which violates the property of being convex. Otherwise, we decrease ϵ to $\epsilon/2$ and use our tester to verify whether P is ϵ-far from being in convex position. We repeat this procedure until we either reject set P, in which case we report also a subset of P which is non-convex, or we reach $\epsilon \leq 1/n$. In the latter case, we run a deterministic algorithm which asks on positions of all the points (and hence, whose query complexity is linear) and then tests whether P is an extreme point set or not (in the latter case we also return a subset of P which is non-convex). Observe that if P is in convex position, the query complexity of our algorithm (cf. Lemma 3.1) is $\mathcal{O}\left(\sum_{k=1}^{\lceil \log_2 n \rceil} n^{2/3} \cdot 2^{k/3}\right) + \mathcal{O}(n) = \mathcal{O}(n)$, which is clearly optimal. On the other hand, if P is ϱ-far from being in convex position, then with a constant probability our algorithm will reject P for $\epsilon \geq \frac{1}{2}\varrho$. Therefore, the expected query complexity of our algorithm is $\mathcal{O}\left(\sum_{k=1}^{\lceil \log_2(1/\varrho)\rceil} n^{2/3} \cdot 2^{k/3}\right) = \mathcal{O}(n^{2/3}/I^{1/3})$, where I is the number of interior (i.e., non-extreme) points.

We can combine our algorithm with Chan's algorithm [5] to obtain an algorithm that is output-sensitive in h (the number of extreme points) and relative edit distance.

Property testing in computational geometry. Typical applications of property testing are program checking and input filtering. In many applications it is good and often necessary to ensure the correctness of results that are, for example, computed by a possibly unreliable library (e.g., a library that does not use exact precision arithmetic).

Although property testing does not give a 100% guarantee for the correctness, it does provide a good trade-off between running time and safety. A further advantage is the possibility to trade running time against algorithmic simplicity. Since our testers have sublinear running time we do not have to use optimal algorithms when we compute the structures needed while still achieving a sublinear run-time. In this setting especially the query complexity of an algorithm is important.

A particular application of property testing in computational geometry in the context of robustness is *lazy error correction*. It is well known that incrementally computing a complex structure (e.g., a Delaunay triangulation) with fixed precision arithmetic might lead to serious structural defects of the output. Furthermore, early small errors in the structure frequently lead to large structural defects in the final structure. To ensure that the structure is always close to the correct one, we run a property tester after a couple of updates. If the tester rejects, we use some lazy error-correction method (e.g., edge flips with exact arithmetic) to fix the structure and continue. At the end of the algorithm we fix the final structure. We believe this might become an important application of property testing in the context of computational geometry.

2   Disjointness of Generic Geometric Objects

We begin our investigations with a simple tester for the problem of testing whether a collection of objects is disjoint. The main reason to present this algorithm in details is to show the reader the flavor of many testers which run in two phases: (i) sample at random a sufficiently large subset of input objects and then (ii) analyze the sample objects to obtain information about the whole input. This approach works fairly easily for the problem studied in this section; other problems investigated in following sections require more complicated analysis.

The problem of deciding whether an arrangement of objects is intersection-free belongs to the most fundamental problems in computational geometry. A typical example is to verify whether a set of line segments in the plane or a set of hyper-rectangles or polytopes in $\mathbb{R}^d$ is intersection-free. We assume the oracle allows to query for the input objects and we suppose that there is an algorithm A that solves the exact decision problem of testing whether a set of objects is disjoint. Further, we consider the problem only for generic objects[3]. We shall use the following further definitions.

---

[3] That is, we never use any information about the geometric structure of the objects and our solution can be applied to any collection of objects (in any space, not necessarily metric one).

Definition 2.1. *Let $\mathbb{O}$ be a set of any objects in* $\mathbb{R}^d$. *We say $\mathbb{O}$ is* (pairwise) disjoint, *if no two objects in $\mathbb{O}$ intersect; $\mathbb{O}$ is ϵ-far from being pairwise disjoint, if there is no set $T \subseteq \mathbb{O}$ with $|T| \leq \epsilon |\mathbb{O}|$ such that $\mathbb{O} \setminus T$ is disjoint.*

Then the following algorithm is a simple ϵ-tester:

DISJOINTNESS (SET $\mathbb{O}$ consisting of n objects):
    Choose a set $S \subseteq \mathbb{O}$ of size $8\sqrt{n/\epsilon}$ uniformly at random
    Check whether S is disjoint using algorithm A
    if S is disjoint **then** *accept*
    else *reject*

Theorem 2.1. *Algorithm* DISJOINTNESS($\mathbb{O}$) *is an ϵ-tester with the optimal query complexity* $\Theta(\sqrt{n/\epsilon})$. *If the running time of the algorithm A for k objects is $T(k)$, then the running time of* DISJOINTNESS($\mathbb{O}$) *is* $\mathcal{O}(T(8\sqrt{n/\epsilon}))$.

Proof. We prove only that DISJOINTNESS($\mathbb{O}$) is an ϵ-tester and show the required upper bound for the query complexity. (The proof that every ϵ-tester has query complexity of $\Omega(\sqrt{n/\epsilon})$ is omitted.) Clearly, if $\mathbb{O}$ is intersection-free, the algorithm accepts $\mathbb{O}$. So let us suppose that $\mathbb{O}$ is ϵ-far from being intersection-free. In that case we can apply $\frac{1}{2}n\epsilon$ times the following procedure to $\mathbb{O}$: pick a pair of intersecting objects and remove it from $\mathbb{O}$. We can prove that with probability at least $\frac{2}{3}$ at least one of the pairs is chosen to S (because the objects from each pair intersect each other). Since the size of S is $8\sqrt{n/\epsilon}$, the upper bound for the query complexity follows.

3   Convex Position

In this section we consider one of the most classical properties of a set P of points: being in *convex position*. With respect to vertex representations of convex polytopes, it reflects the concept of "minimality" in that no point may be removed without affecting the convex hull of P. Many algorithms (such as the intersection test presented in Section 4) therefore require that their input be in convex position.

Definition 3.1. *A set P of n points in $\mathbb{R}^d$ is in* convex position *iff each point in P is an extreme point of the convex hull* CH(P). *We say P is ϵ-far from being in convex position if no set Q of size ϵn exists s.t. $P \setminus Q$ is in convex position.*

While it is possible to test whether a single point is extreme in $\mathcal{O}(n)$ time [6, Section 3], no such algorithm is known to compute *all* extreme points of a given set P. The fastest algorithm known due to Chan [5] uses data structures for answering many linear programming queries simultaneously and requires time

$$T(n, h) = n \cdot \log^{\mathcal{O}(1)} h + (n\,h)^{\frac{\lfloor d/2 \rfloor}{\lfloor d/2 \rfloor + 1}} \cdot \log^{\mathcal{O}(1)} n \ , \tag{1}$$

where h denotes the output size (number of extreme points). It is also conjectured that the problem of testing whether a set of points P is in convex position is asymptotically as hard as the problem of finding all extreme points of P.

In this section we introduce two ϵ-testers for being in convex position. We begin with the algorithm that has asymptotically optimal query complexity.

CONVEX-A(P):
    Choose a set $S \subseteq P$ of size $36 \cdot \sqrt[d+1]{n^d/\epsilon}$ uniformly at random.
    Compute all h extreme points of S.
    if $h < m$ **then** *reject*
    else *accept*

Lemma 3.1. *Algorithm* CONVEX-A *is an ϵ-tester for the property of being in convex position. Its query complexity is $\mathcal{O}(\sqrt[d+1]{n^d/\epsilon})$ and it can be implemented to run in time $\mathcal{O}\big(T(\sqrt[d+1]{n^d/\epsilon}, \sqrt[d+1]{n^d/\epsilon})\big)$. Furthermore, its query complexity is asymptotically optimal in the sense that every ϵ-tester for convex position has the query complexity $\Omega(\sqrt[d+1]{n^d/\epsilon})$.*

Proof. It is easy to see that Algorithm CONVEX-A accepts sets in convex position. Suppose now that P is ϵ-far from being in convex position. Let $J = \frac{\epsilon n}{d+1}$. We can prove that there exist sets $W_j, U_j \subseteq P$, $j = 1, \ldots, J$ that satisfy the following conditions:

- each set W_j is of size $d + 1$ and all sets W_j are pairwise disjoint,
- for each $u \in U_j$ the set $W_j \cup \{u\}$ is not in convex position,
- each W_j is pairwise disjoint with any U_i, and
- each U_j is of size greater than or equal to $n \cdot (\frac{1-\epsilon}{d+1} - \epsilon)$.

Fix these sets W_j and U_j, $1 \leq j \leq J$. We can prove that with probability at least $\frac{2}{3}$ there is some j such that for some $u \in U_j$ set S contains $W_j \cup \{u\}$. Since $W_j \cup \{u\}$ is not in convex position, it will be detected that S as well as P are not in convex position. Hence, Algorithm CONVEX-A is an ϵ-tester for the property of being in convex position.

The sample complexity follows easily from the bound on the size of S and the running time follows from the definition at the beginning of this section. Because of space limitations we omit here the proof that every ϵ-tester for convex position has the query complexity $\Omega(\sqrt[d+1]{n^d/\epsilon})$.

Now, we describe another algorithm for testing convex position which for $d \geq 4$ achieves better running times if the fastest known implementations are used.

CONVEX-B(P):
    Choose a set $S \subseteq P$ of size $4/\epsilon$ uniformly at random.
    for each $p \in S$ simultaneously
        check whether p is extreme for CH(P)
        if p is not extreme for CH(P) **then exit** and *reject*
    accept

Lemma 3.2. *Algorithm* CONVEX-B *is an ϵ-tester for the property of being in convex position. It can be implemented to run in time $\mathcal{O}\big(T(n, 1/\epsilon)\big)$.*

The following theorem summarizes the running time complexity of algorithms presented in this section. It follows by plugging Chan's bound (1) to Lemmas 3.1–3.2.

Theorem 3.1. *There is an ϵ-tester for convex position with the running time of order*

dimension $d = 2$	dimension $d = 3$	dimensions $d \geq 4$
$n^{2/3}\,(1/\epsilon)^{1/3}\,\log(n/\epsilon)$	$n^{3/4}\,(1/\epsilon)^{1/4}\,\log(n/\epsilon)$	$(n/\epsilon)^{\frac{\lfloor d/2 \rfloor}{\lfloor d/2 \rfloor+1}}\,\log^{\mathcal{O}(1)}(n) + n\,\log^{\mathcal{O}(1)}(1/\epsilon)$

4   Disjointness of V-Polytopes

In this section we consider the property of whether two polytopes $CH(R)$ and $CH(B)$, represented by their vertices (so called V-Polytopes), are disjoint. Denote $n = |P|$ and let $P = R \cup B$.

Definition 4.1. *Two polytopes* $CH(R)$, $CH(B)$ *with finite points sets* $R, B \subseteq \mathbb{R}^d$ *in convex position are ϵ-far from being disjoint, if there is no set* $V \subseteq R \cup B$, $|V| \leq \epsilon \cdot |R \cup B|$ *such that* $CH(R \setminus V)$ *and* $CH(B \setminus V)$ *are disjoint.*

We propose the following simple ϵ-tester for disjointness of V-polytopes.

DISJOINTNESS (R, B):
  Choose a set $S \subseteq P$ of size $\Theta((d/\epsilon) \ln(d/\epsilon))$ uniformly at random
  Test whether $CH(R \cap S)$ and $CH(B \cap S)$ are disjoint
  if $CH(R \cap S)$ and $CH(B \cap S)$ are disjoint **then** *accept*
  else *reject*

Theorem 4.1. *Algorithm* DISJOINTNESS *is ϵ-tester for disjointness of V-polytopes with the query complexity* $\mathcal{O}((d/\epsilon) \ln(d/\epsilon))$.

Proof. Since the query complexity follows immediately from the upper bound for the size of S, we focus only on showing that DISJOINTNESS is a proper ϵ-tester. It is easy to see that if R and B are disjoint then DISJOINTNESS always accepts the input. So let us suppose that the input is ϵ-far from being disjoint. For any hyperplane h we denote by h^- and h^+ two halfspaces induced by h. Two convex polytopes are disjoint iff they can be separated by some hyperplane. Therefore, our goal is to ensure that with probability at least $\frac{2}{3}$ the sample set S contains, for every hyperplane h, a *"witness"* that h does not separate $CH(R)$ from $CH(B)$. It is easy to see that such a witness exists if we could ensure that for every hyperplane h set S contains two points a, b such that either (a, b) or (b, a) belong to one of the following four sets: $(R \cap h^-) \times (R \cap h^+)$, $(R \cap h^-) \times (B \cap h^-)$, $(R \cap h^+) \times (B \cap h^+)$, and $(B \cap h^-) \times (B \cap h^+)$. With our choice for the size of S, we can show that for at least one of these four sets, the both sets in the Cartesian product are of cardinality at least $\epsilon n/2$. Let such two sets be called *representative* for h. To complete the proof of the theorem we only must show that S intersects each representative set for *every* hyperplane h in $\mathbb{R}^d$. And this follows from the result on the randomized construction of ε-nets (with $\epsilon = 2\varepsilon$) [13]. Indeed, from the result due to Haussler and Welzl [13] it follows that the set S is with probability at least $\frac{2}{3}$ an $(\epsilon/2)$-net of each of R and B for the range space $(\mathbb{R}^d, \mathcal{H})$. Therefore, by the definition of ε-nets (see, e.g., [13]), S intersects every representative of each hyperplane h in $\mathbb{R}^d$.

One can also easily improve the query complexity of algorithm DISJOINTNESS in the special case when $d = 2$ and the input polygons are stored in a sorted array.

Theorem 4.2. *For all pairs of polygons in* $\mathbb{R}^2$ *represented by a sorted array there exists a **deterministic** ϵ-tester for disjointness of V-polytopes with the query complexity and the running time of* $\mathcal{O}(1/\epsilon)$.

5 Euclidean Minimum Spanning Tree

In this section we consider the problem to determine if a given input graph G is a Euclidean Minimum Spanning Tree (EMST). The vertices of G are labeled with their position in the plane. We have oracle access to each ith vertex v and the jth neighbor of v (with a special symbol indicating non-existance of such a neighbor). Since the degree of each vertex of the EMST is a constant (it is at most 5), in this model we may think of the graph represented by the adjacency list representation.

The distance measure for the EMST problem is defined as follows:

Definition 5.1. *Given a point set P of n points in general position in the plane and a graph G whose vertices are labeled with the points in P. G is said to be ϵ-far from being the Euclidean minimum spanning tree T, if G has edit distance at least ϵn from T. The edit distance between G and T is the minimum number of edge deletions and insertions to construct T from G.*

In any testing algorithm it is very important to develop methods to reject inputs that are far away from the desired property. It is known that the longest edge in a cycle of the graph does not belong to the minimum spanning tree. We can therefore reject G, if we find a cycle in the complete Euclidean graph whose longest edge belongs to the input graph (we call this a 'bad' cycle). We start with a useful lemma about the EMST (which holds for any MST).

Lemma 5.1. *Let P be a point set in the plane and let $e = (p_1, p_2)$ be an edge of the EMST. Further, let P' be a subset of P and let $p_1, p_2 \in P'$. Then e belongs to the EMST of P'.*

The lemma above implies that we can reject G if we find a bad cycle in the complete Euclidean graph of *a subset* of P. Until the end of this paragraph we assume that G is ϵ-far from the EMST. Under the assumptions that (1) G is connected, (2) its straight-line embedding is crossing-free and (3) there are no crossings with edges of the correct EMST, we can show that there are many bad cycles in the complete Euclidean graph of P. In fact, there are many bad cycles even if G is only close to properties (1)-(3). Also note that these properties are necessary conditions for G to be an EMST. Therefore, our tester for the EMST runs in two phases. It first checks (with a property tester) whether G satisfies properties (1)-(3). If the input graph passes these tests but is ϵ-far from the EMST, it must have many "bad" cycles. We then run a tester that finds such a cycle w.h.p. and we are done.

We begin with testing whether an input graph is connected and its straight-line embedding is crossing-free. Additionally, we reject graphs with many vertices having degree larger than 5 (we omit this issue in the extended abstract).

Definition 5.2. *Let P be a set of n points in general position in the plane and let $G = (V, E)$ be a graph whose vertices are labeled with the points in P, $|V| = n$. Let S be the set of subsets E' of E s.t. $G' = (V, E \setminus E')$ is crossing-free. We say G is ϵ-far from being a straight-line, crossing-free, connected embedding, if $\min_{E' \in S}\{|E'| + \# \text{ connected components in } G'\} \geq \epsilon n$.*

For connectivity in graphs represented by the adjacency list there is a test developed in [11] which runs in $\mathcal{O}(\frac{1}{\epsilon})$ time. We combine this algorithm with the tester DISJOINTNESS($\mathbb{O}$) developed in Section 2 to obtain the following result.

Lemma 5.2. *The property of being a straight-line, crossing-free, connected embedding can be tested in $\mathcal{O}(\sqrt{\frac{n}{\epsilon}} \log n)$ time.*

We continue our studies with property (3). We need some further notation:

Definition 5.3. *Let P be a point set in the plane and G be a graph whose vertices are labeled with the points in P. The EMST-completion $C(G)$ of G is the straight-line embedding of G together with all segments of the EMST of P.*

We can view $C(G)$ either as a set of segments or as a labeled graph. In the following we will use these both interpretations. From now on we call an edge in $C(G)$ *red*, if it does not belong to the input graph and *blue* otherwise. Since G and the EMST are crossing-free, there can only be red blue intersections in $C(G)$. The following lemma shows that in order to detect a red-blue intersection, it is sufficient to find the blue segment and one endpoint of the red segment.

Lemma 5.3. *Let AB be a blue and CD be a red segment and let them intersect. Then AB is not in the EMST of any set containing either $\{A, B, C\}$ or $\{A, B, D\}$.*

One can show that if $C(G)$ is ϵ-far from (red-blue) intersection-free, then it is sufficient to sample a set of $\mathcal{O}(\sqrt{\frac{n}{\epsilon}})$ points to reject G. Therefore, our algorithm samples a random set S of $\mathcal{O}(\sqrt{\frac{n}{\epsilon}})$ points from P and adds the neighbors in the input graph of each point to S. Then it computes the subgraph G' of G induced by S. This can be easily done in $\mathcal{O}(|S|)$ time, if for each point in S its degree in G is constant. On the other hand, if we detect a vertex with degree larger than 5 we can immediately reject the input. After that we compute the EMST of S and then $C(G')$ in $\mathcal{O}(|S| \log |S|)$ time. Using a sweep-line algorithm we can check whether $C(G')$ is intersection-free in $\mathcal{O}(|S| \log |S|)$ time. We reject the input, if an intersection has been found. Repeating this procedure a constant number of times we achieve the desired $\frac{2}{3}$ probability.

If the input graph was not rejected so far, our goal is to find short cycles in $C(G)$ with at most two red edges. We call a cycle *bad*, if it has length at most $\frac{32}{\epsilon}$ and if it contains at most 2 red edges. (The number of short cycles with less than two red edges might be small and cycles with more than 2 red edges are more difficult to detect.)

Lemma 5.4. *Let $C(G)$ be the EMST-completion of G, let G be not $\frac{\epsilon}{100}$-far from a crossing-free, connected graph and let $C(G)$ have at most $\frac{\epsilon}{100}$ red-blue intersections. Then there are at least $\frac{\epsilon n}{12}$ bad cycles in $C(G)$.*

We find bad cycles by sampling a set of random directed edges and then we walk in both directions along the boundary of the face incident to each edge (we can think of G as a planar map).

Each of the at most $\frac{\epsilon n}{50}$ edges that can be removed to obtain an intersection-free embedding might destroy two bad cycles (we cannot walk along the boundary to close the face). Therefore, $\frac{\epsilon n}{12} - \frac{\epsilon n}{25} \geq \frac{\epsilon n}{25}$ bad cycles remain.

We distinguish between three types of bad cycles. Type i cycles, $0 \leq i \leq 2$, contain i red edges. Type 0 and type 1 cycles are easy to detect. It suffices to sample one of the boundary edges and our walk procedure will find the complete face (recall that the red edges are defined implicitly).

We classify the type 2 cycles according to the length of their longer blue chain into sets C_i. Each C_i contains all cycles whose longer chain has length l, $2^i \leq l < 2^{i+1}$. Each C_i is partitioned into subclasses $C_{i,j}$. $C_{i,j}$ contains all cycles whose shorter chain has length k, $2^j \leq k < 2^{j+1}$. For each $C_{i,j}$ we sample two different sets S_1 and S_2, one to detect the longer chain and one for the shorter one. Set S_1 has size $\frac{1}{l}\sqrt{\frac{n}{\epsilon}}$ and set S_2 has size $\frac{1}{k}\sqrt{\frac{n}{\epsilon}}$. The probability that a type 2 cycle in class $C_{i,j}$ is detected is $\mathcal{O}(\frac{kl \cdot |S_1| \cdot |S_2|}{n^2}) = \mathcal{O}(\frac{1}{\epsilon n})$. Therefore, the probability that any bad cycle is found is $(1 - \mathcal{O}(\frac{1}{\epsilon n}))^{\mathcal{O}(\epsilon n)} = O(1)$. Again amplification yields the desired bound of $\frac{2}{3}$. The query complexity of the algorithm is $\mathcal{O}(\sqrt{\frac{n}{\epsilon}} \log^2(\frac{1}{\epsilon}))$ and its running time is $\mathcal{O}(\sqrt{\frac{n}{\epsilon}} \log^2(\frac{1}{\epsilon}) \log n)$.

Theorem 5.1. *There is a randomized ϵ-tester for the EMST with $\mathcal{O}(\sqrt{\frac{n}{\epsilon}} \log^2(\frac{1}{\epsilon}))$ query complexity and running time $\mathcal{O}(\sqrt{\frac{n}{\epsilon}} \log(\frac{1}{\epsilon})^2 \log n)$.*

6 Delaunay Triangulation

We summarize our results for the Delaunay triangulation in the following theorems.

Theorem 6.1. *There is no sublinear property $\frac{1}{4}$-tester for Delaunay triangulations using Hamming distance like measure (edge deletion and insertion).*

Theorem 6.2. *If we define a triangulation to be ϵ-far from being Delaunay if there are at least ϵn edges that can be flipped to improve the minimal angle locally, then there exists an ϵ-tester with the running time of $\mathcal{O}(1/\epsilon)$.*

7 Deterministic Lower Bounds

We present a simple technique to prove lower bounds for deterministic property test algorithms and apply it to the problems described in the previous sections. Our model for the deterministic algorithms is as follows. The input is given as an array of input items. The array has size n and we may access any item by its index in constant time.

Let A be a deterministic tester for property Q. Let I be an instance of size n with property Q. By definition A must accept I. Let $T(n)$ be the running time of A. Clearly, A can access at most $T(n)$ items of I. We color the accessed items red and all other items blue. Since A is deterministic, changing the blue items does not affect the outcome of the algorithm. Thus, if we can construct an instance that is ϵ-far from P by changing only blue items regardless how the red items are chosen by the algorithm then we could obtain a lower bound for the problem at hand. We can apply this approach to obtain lower bounds which are summarized in the following theorem.

Theorem 7.1. *There is no deterministic property $\frac{1}{4}$-tester with $o(n)$ query complexity for the following problems (using the distance measure defined in this paper and the relative edit distance for sorting): Sorting, Disjointness of Objects, Convex Position, Disjointness of Polytopes, and EMST.*

8   Acknowledgments

The second author would like to thank Stefan Funke for the helpful discussion about lazy error correction.

References

1. N. Alon, E. Fischer, M. Krivelevich, and M. Szegedy. Efficient testing of large graphs. In *Proc. 40th IEEE FOCS*, pp. 656–666, 1999. 155
2. N. Alon, M. Krivelevich, I. Newman, and M. Szegedy. Regular languages are testable with a constant number of queries. In *Proc. 40th IEEE FOCS*, pp. 645–655, 1999. 155
3. H. Alt, R. Fleischer, M. Kaufmann, K. Mehlhorn, S. Näher, S. Schirra, and C. Uhrig. *Algorithmica*, 8(5/6):365-389, 1992. 158
4. M. Blum, M. Luby, and R. Rubinfeld. Self-testing/correcting with applications to numerical problems. *Journal of Computer and System Sciences*, 47(3):549–595, 1993. 155
5. T. M. Chan. Output-sensitive results on convex hulls, extreme points, and related problems. *Discrete & Computational Geometry*, 16:369–387, 1996. 159, 160
6. M. Dyer and N. Megiddo. Linear programming in low dimensions. In J. E. Goodman and J. O'Rourke, eds., *Handbook of Discrete and Computational Geometry*, ch. 38, pp. 699–710, CRC Press, Boca Raton, FL, 1997. 160
7. F. Ergün, S. Kannan, S. Ravi Kumar, R. Rubinfeld, and M. Viswanathan. Spot-checkers. In *Proc. 30th ACM STOC*, pp. 259–268, 1998. 155, 158
8. F. Ergün, S. Ravi Kumar, and R. Rubinfeld. Approximate checking of polynomials and functional equations. In *Proc. 37th IEEE FOCS*, pp. 592–601, 1996. 155
9. O. Goldreich, S. Goldwasser, E. Lehman, and D. Ron. Testing monotonicity. In *Proc. 39th IEEE FOCS*, pp. 426–435, 1998. 155
10. O. Goldreich, S. Goldwasser, and D. Ron. Property testing and its connection to learning and approximation. *Journal of the ACM*, 45(4):653–750, 1998. 155, 158
11. O. Goldreich and D. Ron. Property testing in bounded degree graphs. In *Proc. 29th ACM STOC*, pp. 406–415, 1997. 155, 164
12. O. Goldreich and D. Ron. A sublinear bipartiteness tester for bounded degree graphs. *Combinatorica*, 19(3):335–373, 1999. 155
13. D. Haussler and E. Welzl. Epsilon-nets and simplex range queries. *Discrete & Computational Geometry*, 2:127–151, 1987. 162
14. K. Mehlhorn, S. Näher, M. Seel, R. Seidel, T. Schilz, S. Schirra, and C. Uhrig. Checking geometric programs or verification of geometric structures. *Computational Geometry: Theory and Applications*, 12:85–103, 1999. 158
15. R. Rubinfeld and M. Sudan. Robust characterization of polynomials with applications to program testing. *SIAM Journal on Computing*, 25(2):252–271, 1996. 155, 158
16. R. Rubinfeld. Robust functional equations and their applications to program testing. In *Proc. 35th IEEE FOCS*, pp. 288–299, 1994. 155

On R-Trees with Low Stabbing Number

Mark de Berg[1], Joachim Gudmundsson[2], Mikael Hammar[2], and Mark
Overmars[1]

[1] Department of Computer Science, Utrecht University,
PO Box 80.089, 3508 TB Utrecht, the Netherlands.
[2] Department of Computer Science, Lund University,
Box 118, 221 00 Lund, Sweden.

Abstract. The R-tree is a well-known bounding-volume hierarchy that
is suitable for storing geometric data on secondary memory. Unfortunately, no good analysis of its query time exists. We describe a new algorithm to construct an R-tree for a set of planar objects that has provably
good query complexity for point location queries and range queries with
ranges of small width. For certain important special cases, our bounds
are optimal. We also show how to update the structure dynamically, and
we generalize our results to higher-dimensional spaces.

1 Introduction

Researchers in computational geometry have developed data structures for many
types of queries on geometric data: point-location structures, range-searching
structures, nearest-neighbor searching structures, and so on. The asymptotic
worst-case behavior of these data structures is usually quite good—or at least
close to the theoretical lower bounds. In practice, however, other kinds of data
structures are often used. One reason is that in many applications storage is
a very critical issue: $\Theta(n \log n)$ storage and even linear storage with a large
constant factor can already be too much. Another reason is that the structures
developed in computational geometry are usually dedicated to a very specific
setting: a structure for searching with rectangular ranges in a set of line segments
will not work for searching with rectangular ranges in a set of curve segments, or
for searching with circular ranges in a set of line segments. In a typical application
one needs to perform several different types of queries, and it is desirable to have
a data structure that supports all (or at least many) of them.

An example of a versatile structure that is used in many applications is
the bounding-volume hierarchy. This is a tree structure, whose leaves store the
geometric data objects and whose internal nodes store a bounding box (or some
other bounding volume) for the objects in the subtree rooted at that node. A
bounding-volume hierarchy uses linear space and it can store any type of objects.
It can perform range queries with any type of range (which means it can also do
point location, since this is simply a range query with a point range).

The R-tree, which was proposed by Guttmann [7], is a bounding-volume hierarchy that is suitable for storing data on secondary storage. It can be considered

M. Paterson (Ed.): ESA 2000, LNCS 1879, pp. 167–178, 2000.
© Springer-Verlag Berlin Heidelberg 2000

a geometric version of a B-tree: all leaves are at the same depth, and all internal nodes (except for the root) have degree between t and $2t$, for a fixed parameter t which we call the *minimum degree* of the R-tree.[1] The root has a degree between 2 and $2t$. An internal node stores a bounding box for each of its subtrees; these bounding boxes are used to decide whether or not to visit a subtree when querying with a query range. The depth of an R-tree storing n objects in its leaves is $\Theta(\log n/\log t)$. The idea, like for B-trees, is to choose t as large as possible in order to minimize the depth of the tree, while making sure that each internal node still fits into one page of external memory. The R-tree is one of the most widely used geometric data structure in Geographic Information Systems—see for example the survey articles by Nievergelt and Widmayer [8] or by Six and Widmayer [10].

The key to the efficiency of an R-tree is how the underlying objects are grouped together in subtrees. Intuitively, for each subtree we would like the objects in its leaves to be clustered, so that their bounding box does not have too much empty space or overlap too many other bounding boxes. A number of heuristics has been proposed to achieve this [2,3,5,6,7,9]. To our knowledge no construction algorithm has been described resulting in a structure with provably efficient worst-case performance. The only analytic result that we know of is by Faloutsos et al. [4]. Their setting is rather limited, however: they consider a 1-dimensional version of the R-tree, and assume that the input intervals have only one or two different sizes and that they are distributed uniformly. For this case they bound the number of nodes visited when answering a point-location query. They consider two heuristics to build the R-tree, and obtain bounds that are roughly $\Theta(\log n/\log t)$. Another result is by Becker et al. [1], who gives an optimal solution to a problem arising for some of the heuristics used to update an R-tree dynamically. The goal of our paper is to describe an algorithm for constructing R-trees whose worst-case query performance is good. We show this for point-location queries and for range queries with ranges of small width. Next we discuss our results in more detail.

Let $\mathcal{S}$ be a set of n objects for which we wish to construct an R-tree. A range query on $\mathcal{S}$ asks for all objects in $\mathcal{S}$ intersecting a query range Q. A point-location query is a range query where the query range is a point. Such queries are performed by traversing the tree starting at the root, visiting only subtrees whose bounding box is intersected by Q. The efficiency of the query procedure is determined by the number of nodes visited, since this number equals the number of disc accesses.

We define the *stabbing number* of a set of rectangles in the plane as the maximum number of rectangles stabbed by (that is, containing) any query point. For example, a set of disjoint rectangles has stabbing number equal to one. The worst-case number of nodes of the R-tree visited when answering a point-location query corresponds to the stabbing number of the set of bounding boxes stored in the tree. The stabbing number of $\mathcal{R}_{\mathcal{S}}$, the set of bounding boxes of the objects in

---

[1] The original definition allows between t and s rectangles for some given s with $s \geqslant 2t$, but for concreteness we assume $s = 2t$.

S, may already be n—take a set of n diagonal line segments that are very close together. Hence, we cannot achieve a sublinear bound on the number of visited nodes for general scenes. Therefore we will express our bounds in terms of σ, the stabbing number of $\mathcal{R}_S$. A second parameter that we will use in our analysis is ρ, the *x-scale factor* (or *scale factor* for short) of S. This is the ratio of the largest x-extent to the smallest x-extent of the objects in S. (The x-extent of an object is the length of its projection onto the x-axis.) The scale factor has also been used by Zhou and Suri [11] for the analysis of a bounding-box heuristic, giving bounds on the number of intersections among the bounding boxes as compared to the number of intersections among the original objects.

We will prove that our construction algorithm produces an R-tree such that any point-location visits $O((\sigma + \lceil \log \rho \rceil) \log n / \log t)$ nodes. When σ and ρ are constant, which we expect to be true in many applications, this is optimal. (In fact, our result is slightly more general than this—see the remark below Theorem 1.) We can get rid of the dependency of ρ at the expense of an extra $O(\log n)$ factor, leading to an $O(\sigma \log^2 n / \log t)$ bound on the number of visited nodes. We also analyze the number of nodes visited by a range query. Here we obtain a bound of $O((\sigma + \lceil \log \rho \rceil + w + k) \log n / \log t)$, where w is the ratio of the x-extent of the query range to the smallest x-extent of any object in S and k is the number of reported objects.

Finally, we generalize our results to higher dimensions, and show how to update the R-tree dynamically.

2   The Construction

Let S be a set of n disjoint objects in the plane, and let $\mathcal{R} = \mathcal{R}_S$ be the set of bounding boxes of these objects. Let σ be the stabbing number of $\mathcal{R}$, that is, the maximum number of rectangles in $\mathcal{R}$ containing any query point. For convenience of presentation we shall sometimes pretend that $\mathcal{R}$ is the set for which we want to construct an R-tree. Of course, the R-tree for $\mathcal{R}$ is exactly the R-tree for S that we are looking for. Let ρ denote the scale factor of $\mathcal{R}$ as defined above (which is equal to the scale factor of S).

Before we proceed, let's give a more precise definition of the R-tree and of the terminology and notation that we will use. An R-tree for S is a tree $\mathcal{T}$ with the following properties.

- Each leaf node of $\mathcal{T}$ (except when it is also the root) contains between t and $2t$ rectangles from $\mathcal{R}$. With each rectangle, a pointer to the corresponding object in S is stored.
- All leaves of $\mathcal{T}$ are at the same level.
- Each internal node ν of $\mathcal{T}$ stores for each of its subtrees the bounding box of all the rectangles stored in the leaves of that subtree.
  The bounding box of all bounding boxes stored at ν is denoted by $b(\nu)$. In other words, $b(\nu)$ is the bounding box of $\mathcal{R}(\nu)$, the set of rectangles stored in the subtree rooted at ν. We say that $\mathcal{R}(\nu)$ is the *defining set* of $b(\nu)$. Notice that $b(\nu)$ is not stored at ν, but that it will be stored at the parent of ν.

– The root node of T has between 2 and $2t$ children, unless it is also a leaf. In the latter case it can contain between 1 and $2t$ rectangles from $\mathcal{R}$, with pointers to the corresponding objects in $\mathcal{S}$.

Sets with scale factor at most two. When the scale factor of $\mathcal{R}$ is two or less, we can proceed as follows. Assume without loss of generality that the smallest x-extent of any rectangle in $\mathcal{R}$ is equal to one. We partition the plane into vertical strips of unit width. We associate each rectangle in $\mathcal{R}$ with the strip containing its left edge, where strips are closed to the left and open to the right. A strip that has no rectangles associated with it is called *empty*, otherwise it is *non-empty*. Let $s_1, \ldots, s_k$ be the sequence of strips starting at the leftmost non-empty strip and ending at the rightmost non-empty strip. Notice that the sequence can contain empty strips—see Figure 1. Denote the set of rectangles associated to s_i by $\mathcal{R}(s_i)$, and let $n_i := |\mathcal{R}(s_i)|$. We number the rectangles in $\mathcal{R}$ in a bottom-to-top fashion, based on the strips: the rectangles associated to the leftmost strip are numbered $r_1, \ldots, r_{n_1}$, the rectangles associated to the second leftmost non-empty strip are numbered $r_{n_1+1}, r_{n_1+2}, \ldots$, and so on. We call the resulting ordering on the rectangles the *strip order*. Figure 1 illustrates it. The

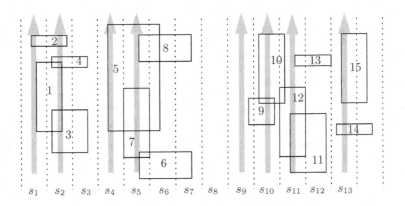

Fig. 1. The strip order.

following observation will be crucial; it follows trivially from the fact that the x-extents of the rectangles are between one and two.

Observation 1 *Any rectangle intersecting a given strip s_i must be assigned to s_i, to s_{i-1}, or to s_{i-2}.*

The bounds (on the number of nodes visited by a query) we shall prove later apply to any R-tree that respects the strip order, that is, any R-tree for which the left-to-right order of the rectangles in the leaves corresponds to the strip order. For concreteness we will describe a simple bottom-up procedure to construct such an R-tree. An alternative way to construct the R-tree is by inserting the rectangles one at a time, as described in Section 4. Since the latter method already give good behavior in terms of number of disc accesses, namely $\Theta(n \log n / \log t)$, we do not analyze the number of disc accesses for the bottom-up method.

The bottom-up construction works as follows. The bottom level of the R-tree consists of leaf nodes whose defining set have between t and $2t$ rectangles. This is achieved by simply letting the first (leftmost) leaf contain the first t rectangles in the strip order (that is, $\{r_1, \ldots, r_t\}$), the second leaf the next t rectangles, and so on. This continues until the number of remaining rectangles is at most $2t$, which are then put into the last (rightmost) leaf.

The level above the leaf level is built on the rectangles $b(\nu)$ for the leaf nodes ν, in the same way as the level before: the rectangles are put into groups of size t with the last group containing at most $2t$ rectangles. (Recall that the notation $b(\nu)$ is used to denote the bounding box of all rectangles in the subtree rooted at ν. Hence, for a leaf ν, we have that $b(\nu)$ denotes the bounding box of all rectangles stored in ν.) The ordering on the rectangles used to do the grouping is the left-to-right ordering of the leaves corresponding to the rectangles.

The remaining levels of the R-tree are constructed in the same way, always using the bounding boxes of the subtrees on the previous level. The process ends when the number of rectangles we are dealing with falls below $2t$. We then finish the R-tree by putting all rectangles into a single root node.

The method of constructing R-trees by first ordering the rectangles along a 1-dimensional curve has also been used by other authors [3,9]. It has been observed that the main drawback of this method is that it disregards the sizes of the rectangles. Therefore we developed a new method, presented below, to deal with rectangles that differ a lot in size. Our analysis of the query complexity given in the next section is new as well.

The general case. So far we assumed that ρ, the scale factor of the set $\mathcal{R}$ of rectangles, is at most two. The algorithm we developed can also be used for larger ρ, but the dependency of the query complexity on ρ will be linear. We now describe a method that reduces the dependency to logarithmic.

The idea is to partition $\mathcal{R}$ into $m := \lceil \log \rho \rceil$ subsets, each with scale factor at most two. Let $\mathcal{R}_1, \ldots, \mathcal{R}_m$ be these subsets. For each $\mathcal{R}_i$ we can construct an R-tree with the algorithm described earlier. Since the depths of these R-trees may be different, however, we cannot simply merge them by constructing a tree on top of these R-trees. Another problem arising with this approach is that we may get internal nodes with too few children.

We therefore define a new ordering on the rectangles, as follows: rectangles are ordered by the index number of the set $\mathcal{R}_i$ they are in, and rectangles with the same index number are ordered using the strip order, as above. In other words, to obtain the sorted sequence of rectangles, we concatenate the sorted sequences for $\mathcal{R}_1, \ldots, \mathcal{R}_m$ in that order. We call the new order the *index-strip order*.

Now that we have a well-defined order on the rectangles, we can construct the R-tree as before (either using the bottom-up procedure, or the insertion algorithm described later).

Unbounded scale factors. When the scale factor gets really large, the method above gives rise to many subsets $\mathcal{R}_i$ and the resulting query complexity will not be very good (see below). We can overcome this problem with a simple trick: we replace the x-coordinate of the vertical edges of the rectangles by their rank. This way the x-'coordinates' that we are dealing with are integers between 1 and n, so the scale factor is bounded by n. We then apply our algorithm to these normalized rectangles. Conversion of the resulting R-tree to an R-tree for the original rectangles is trivial: simply replace the x-'coordinates' of the edges of the bounding boxes by the original coordinates. The latter step does not influence the stabbing number. In the analysis given next, we can thus replace ρ by n if that gives a better result.

3   Analysis of the Query Complexity

Point-location queries. Suppose we perform a point-location query in the R-tree $\mathcal{T}$ with a point q. The number of nodes visited by the query procedure equals the number of bounding boxes stored in $\mathcal{T}$ stabbed by q.

Let ℓ_i denote the left bounding line of the strip s_i. We say that a bounding box b *straddles* ℓ_i if the defining set of b contains rectangles assigned to strips to the left of ℓ_i as well as rectangles assigned to strips to the right of ℓ_i. The following basic property of the construction will be important.

Lemma 1. *Let $\mathcal{T}$ be an R-tree constructed using the index-strip order for a set of rectangles with scale factor ρ. Let $\mathcal{B}_j(l)$ be the collection of bounding boxes of all nodes at a given level l in $\mathcal{T}$ with the property that the defining set of the bounding box has only rectangles from $\mathcal{R}_j$. For each line ℓ bounding a strip, the number of bounding boxes in $\mathcal{B}_j(l)$ straddling ℓ is at most one.*

Proof. By construction, the defining sets of the bounding boxes stored at level l form a disjoint partition of $\mathcal{R}$. Moreover, the left-to-right order of (the defining sets of) the nodes is consistent with the index-strip order. Consider all defining sets containing only rectangles from $\mathcal{R}_j$. Since we use the strip order within a set, there is at most one such defining set that has both a rectangle whose left edge is to the left of ℓ and a rectangle whose left edge is to the right of ℓ.   □

We can now prove a bound on the complexity of a point-location query.

Theorem 1. *Let S be a set of n objects in the plane such that the set of bounding boxes of S has stabbing number σ and scale factor ρ. For a given t, we can construct an R-tree of minimum degree t for S such that the number of nodes visited when answering a point-location query is $O((\sigma + \lceil \log \rho \rceil) \log n / \log t)$.*

Proof. Let l be a fixed level in the R-tree $\mathcal{T}$. Define $m := \lceil \log \rho \rceil$. We will show that the stabbing number of $\mathcal{B}(l)$, the set of bounding boxes of the nodes at level l, is at most $3\sigma + 13m/2$.

Let q be a query point, and let s_i be the strip containing q. We consider three categories of bounding boxes in $\mathcal{B}(l)$ stabbed by q.

- *category (i): bounding boxes whose defining subset has rectangles from more than one of the subsets $\mathcal{R}_j$.*

Because in the index-strip ordering rectangles with the same index are consecutive, there can be at most $m/2$ such bounding boxes.

- *category (ii): bounding boxes not in category (i) that straddle ℓ_i, ℓ_{i-1}, or ℓ_{i-2}.*

By Lemma 1 there are at most m such bounding boxes per bounding line (at most one for each subset $\mathcal{R}_j$),
leading to at most $3m$ such bounding boxes in total.
- *category (iii): bounding boxes not in category (i) whose defining set contains only rectangles assigned to s_i, or only rectangles assigned to s_{i-1}, or only rectangles assigned to s_{i-2}.*

Consider the bounding boxes whose defining set has only rectangles assigned to s_i. Such bounding boxes may have a defining set containing both a rectangle with bottom edge below q and one with bottom edge above q, as shown in Figure 2(a). Because of the ordering scheme within a strip, there are at most

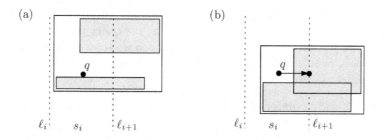

Fig. 2. Illustration for the proof of Theorem 1.

m such bounding boxes (at most one per subset $\mathcal{R}_j$). Otherwise, the defining set of the bounding box has a rectangle $[x : x'] \times [y : y']$ with $q_y \in [y : y']$, as in Figure 2(b). Because the x-extent of each rectangle is at least the width of s_i, this means that such a rectangle must be stabbed by the orthogonal projection of q onto ℓ_{i+1}. There can be no more than σ such rectangles and, consequently, no more than σ such bounding boxes.
This shows that there are at most $\sigma + m$ bounding boxes stabbed by q whose defining set has only rectangles assigned to s_i. A similar argument works for bounding boxes stabbed by q whose defining set has only rectangles assigned to s_{i-1}, or to s_{i-2}. The only difference is that we now need to consider the projection of q onto ℓ_i, and onto ℓ_{i-1} respectively.

Adding up the bounds for each of the cases, we get a total bound of $3\sigma + 13m/2$. Multiplying by the number of levels gives the desired bound.   □

By applying the normalization described in the previous section, we can replace the factor ρ by n.

Corollary 1. *Let $\mathcal{S}$ be a set of n objects in the plane such that the set of bounding boxes of $\mathcal{S}$ has stabbing number σ and scale factor ρ. For a given t, we can construct an R-tree of minimum degree t for $\mathcal{S}$ such that the number of nodes visited when answering a point-location query is $O((\sigma + \log n) \log n / \log t)$.*

Remark. The only way in which the scale factor ρ plays a role in the proof of Theorem 1, is that it ensures that we can partition $\mathcal{R}$ into a logarithmic number of subsets with scale factor at most two. In general, our method gives a bound of $O(\sigma m \log n / \log t)$ for sets of rectangles that can be partitioned into m such subsets, even when $\log \rho$ is larger than m. For instance, if $\mathcal{R}$ contains three classes of rectangles—the large rectangles, the intermediate ones, and the small ones—each with scale factor at most two, then our method will work well even when the large rectangles are much larger than the small ones. Such a behavior may well occur for practical inputs.

Range-searching queries. Now suppose we want to perform a range query with an axis-parallel rectangular range Q. Let w denote the ratio of the x-extent of Q to the smallest x-extent of any object in $\mathcal{S}$. We call w the *width* of the range. Furthermore, let k denote the number of objects reported by the range query. We first analyze the number of nodes visited by the query procedure in terms of w, k and the parameters introduced earlier. Then we show that in general (that is, for ranges that can be unbounded in both x- and y-direction) one cannot obtain similar (logarithmic) bounds.

Theorem 2. *The number of nodes visited when answering a range query with an axis-parallel rectangular range of width w is $O((\sigma + \lceil \log \rho \rceil + w + k) \log n / \log t)$.*

Proof. Define $m := \lceil \log \rho \rceil$. Let l be a fixed level in the R-tree $\mathcal{T}$, and let $\mathcal{B}(l)$ be the set of bounding boxes of the nodes at level l. We start by showing that the number of bounding boxes in $\mathcal{B}(l)$ intersecting the query range Q is $O(\sigma + m + w + k)$.

We consider five categories of bounding boxes in $\mathcal{B}(l)$ intersecting Q.

- *category (i): bounding boxes whose defining set has rectangles in more than one subset $\mathcal{R}_j$.*

  There are at most $m/2$ such bounding boxes.

- *category (ii): bounding boxes not in category (i) containing a corner of Q.*

  From the proof of Theorem 1, it follows that there are at most $(3\sigma + 6m)$ such bounding boxes per corner.

- *category (iii): bounding boxes not in category (i) straddling one of the lines $\ell_{i-2}, \ldots, \ell_j$.*

Fix a subset $\mathcal{R}_k$. Let $s_i, \ldots, s_j$ be the strips defined for $\mathcal{R}_k$ that are intersected by Q. Note that $j - i \leqslant w_k + 1$, where w_k is the ratio of the width of Q and the width of the strips defined for $\mathcal{R}_k$. By Lemma 1 there are at most $j - i + 3 \leqslant w_k + 3$ such bounding boxes for a subset $\mathcal{R}_k$. Since $w_k \leqslant w/2^k$, the total number of all bounding boxes of category (iii) is $\sum_{1 \leqslant k \leqslant m} w_k + 3 = O(w + m)$.

- *category (iv): bounding boxes not intersecting the top or bottom edge of Q and not in categories (i)–(iii).*

Such bounding boxes are either fully contained in Q or they intersect the left or right edge of Q. In the former case we can charge the intersection to one (in fact, many) objects intersecting Q. In the latter case this is possible as well: Consider for example a bounding box b intersecting the left edge, e, of Q. Its defining set must have a rectangle whose right edge is to the right of e and a rectangle whose left edge is to the left of e. Since b does not straddle any strip boundary, and the x-extent of any object is at least the strip width, this implies that b must contain an object intersecting Q. Hence, the total number of bounding boxes of category (iv) is $O(k)$.

- *category (v): bounding boxes intersecting the top or bottom edge of Q and not in one of the previous categories.*

There are two cases. One is where the defining set of such a bounding box b has one rectangle whose bottom edge is below the bottom edge of Q and one rectangle whose bottom edge is above the top edge of Q. For each $\mathcal{R}_k$, this can happen only once for each of the strips defined for $\mathcal{R}_k$ and intersected by Q. Hence, there are at most $w_k + 3$ such bounding boxes for $\mathcal{R}_k$, where w_k is defined as in case (iii), giving $O(w + m)$ such bounding boxes in total. In the other case the defining set of b must contain a rectangle whose top or bottom edge is contained fully in Q. This means that the object contained in this rectangle intersects Q. The total number of bounding boxes of category (v) is therefore $O(w + m + k)$.

Adding up the bounds for each of the cases, we get a total bound of $O(\sigma + m + w + k)$ for the number of nodes visited on a fixed level l. Over all levels we thus get a bound of $O((\sigma + m + w + k) \log n / \log t)$. $\qquad\square$

Can we improve on this result? In particular, one would hope that it is possible to get rid of the dependence on w. Unfortunately, the next theorem shows that in this case one cannot get bounds close to the ones we just obtained. The proof is given in the full version of the paper.

Theorem 3. *For any n, there is a set $\mathcal{S}$ of n disjoint unit squares such that for any R-tree with minimum degree t on $\mathcal{S}$, there is a rectangular query range for which the query procedure will visit $\Omega(\sqrt{n}/\sqrt{t})$ nodes even though the range does not intersect any of the squares.*

4   Dynamization

For the dynamic version, we assume that all coordinates are integers. Hence, the smallest x-extent that can ever occur is equal to one. To define the index-strip ordering we used in the previous section more formally, we define the following functions for a rectangle r:

$$index\text{-}nbr(r) := \lceil \log(x\text{-}extent(r)) \rceil$$
$$strip\text{-}width(r) := 2^{index\text{-}nbr(r)-1}$$
$$strip\text{-}nbr(r) := \lceil (x\text{-coordinate of left edge of } r)/strip\text{-}width(r) \rceil$$
$$y\text{-}nbr(r) := y\text{-coordinate of the bottom edge of } r$$

We now define the following representation for r:

$$rep(r) := (index\text{-}nbr(r), strip\text{-}nbr(r), y\text{-}nbr(r)).$$

If we are working in the real RAM model we cannot use ceil/floor functions. However, we can easily compute $index\text{-}nbr(r)$ for a given r in $O(\log \rho)$ time, where ρ is the scale factor. Similarly, we can compute $strip\text{-}nbr(r)$ in $O(\log x_{\max})$ time, where $x_{\max}$ is the maximum x-coordinate that ever occurs.

Observation 2 *The ordering on the rectangles in $R = \{r_1, \ldots, r_n\}$ induced by a lexicographical ordering on the representations $rep(r_i)$ is equal to the index-strip ordering as used in Section 2.*

This observation implies that if we augment the R-tree with some additional information, we can (almost—see below) use standard B-tree algorithms for insertions and deletions. The extra information is needed to be able to locate the position of a new rectangle in the leaf-level of the R-tree. Recall that a bounding box b stored in an internal node v is the bounding box for the set of rectangles stored in the subtree of some child w_b of v. The extra information we need to store with b is the representation $rep(r_b)$, where r_b is the minimum (according to the index-strip order) rectangle stored in the subtree of w_b. Even though we now have a well-defined order on the rectangles, and we can search on this order in the B-tree, there is still one complication: B-trees store numbers both in the leaves and in the internal nodes. That is, internal nodes store not just splitting values to guide the search, but they store the actual numbers. This is not the case in an R-tree: The extra information we store is $rep(r_b)$ and not the actual rectangle r_b, which is still stored at the leaf level. Hence, the update algorithms need to be changed slightly. In the full paper we discuss how to do this. We obtain the following theorem.

Theorem 4. *The number of disc accesses for updates in the R-tree is $O(\frac{\log n}{\log t})$.*

Remark. The extra information will force us to choose the minimum degree smaller, roughly by a factor of two. This implies that the depth will increase by a factor of roughly $(1 + \log t)/\log t$.

5 Higher-Dimensional R-Trees

The approach for the planar case extends easily into higher dimensions. For instance, suppose we have a set S of n objects in 3-dimensional space. As before, we let σ denote the stabbing number of the set R of bounding boxes of the objects in S. We let ρ_x and ρ_y denote the x-scale factor and the y-scale factor of R, respectively.

First assume that $\rho_x \leqslant 2$ and $\rho_y \leqslant 2$. We partition space into columns by planes orthogonal to the x-axis and planes orthogonal to the y-axis. The spacing of the planes equals the minimum x-extent and y-extent, respectively, of the objects. We number the columns in a bottom-to-top fashion. We assign each box in R to the column containing its front left edge (that is, the vertical edge with smallest x- and y-coordinate). We then number the boxes in R according to the ordering of the columns, where within each column we order the boxes based on the z-coordinate of their bottom facet. The latter ordering is done bottom-up. Given this new definition of the strip ordering, the construction proceeds in exactly the same way as in the planar case. Also the construction for sets where the scale factors are more than two is similar: we partition R into $\lceil \log \rho_x \rceil \cdot \lceil \log \rho_y \rceil$ subsets with scale factors at most two, we apply the standard construction algorithm, and merge whenever necessary. The analysis of the number of nodes visited when answering a point-location query is very similar to the planar case. When ρ_x and/or ρ_y are very large, we can again use normalization to improve the bounds. We get the following result.

Theorem 5. *Let S be a set of n objects in 3-space such that the set of bounding boxes of S has stabbing number σ. Let ρ_x and ρ_y denote the x-scale factor and the y-scale factor of R, respectively. For a given t, we can construct in $O(n \log n)$ time an R-tree of minimum degree t for S such that the number of nodes visited when answering a point-location query is $O(\sigma \lceil \log \rho_x \rceil \cdot \lceil \log \rho_y \rceil \log n / \log t)$. Alternatively, we can obtain an R-tree where $O(\sigma \log^3 n / \log t)$ nodes are visited in a point-location query.*

The results also generalize to dimensions higher than three. For each extra dimension, the number of visited nodes gets multiplied by the logarithm of the scale factor in that dimension (or, alternatively, by $\log n$).

We cannot obtain bounds for range searching that are similar to the planar case. The reason is that even when $\sigma = 1$ it can happen that a range with small width intersects many of the boxes in R without intersecting any of the corresponding objects in S.

6 Concluding Remarks

We have given an algorithm to construct R-trees for sets of n objects in the plane and in higher dimensional spaces. We analyzed the number of nodes visited when answering a point-location query in terms of n, and σ (the stabbing number of the initial bounding boxes), and ρ (the scale factor). When σ and ρ are constant, our results are optimal.

Our results might be improved in several ways. First of all, it would be interesting to reduce the dependency on σ in our bounds. Ideally, we would like to replace σ by $\lceil \sigma/t \rceil$. Another question is whether it is possible to improve the $O(\log^2 n/\log t)$ bound that we get for constant σ to $O(\log n/\log t)$.

Finally, it would be nice to find another way to deal with scale factors larger than two. Our method of partitioning the set into subsets with scale factor two or less works fine in theory, but it is questionable whether it works well in practice.

Acknowledgment

We thank Frank van der Stappen for stimulating discussions during the early stages of this research.

References

1. B. Becker, P. Franciosa, S. Gschwind, T. Ohler, G. Thiemt, and P. Widmayer. Enclosing many boxes by an optimal pair of boxes. In *Proc. 9th Annual Symposium on Theoretical Aspects of Computer Science (STACS)*, LNCS 577, pages 475–486, 1992. 168
2. N. Beckmann, H.-P. Kriegel, R. Schneider, and B. Seeger. The R*-tree: an efficient and robust access method for points and rectangles. In *Proc. ACM-SIGMOD International Conference on Management of Data*, pages 322–331, 1990. 168
3. C. Faloutos and I. Kamel. *Packed R-trees using fractals*. Report CS-TR-3009, University of Maryland, College Park, 1992. 168, 171
4. C. Faloutos, T. Sellis, and N. Roussopoulos. Analysis of object oriented spatial access methods. In *Proc. ACM-SIGMOD International Conference on Management of Data*, pages 426–439, 1987. 168
5. D.M. Gavrila. R-tree index optimization. In *Proc. 6th Int. Symp. on Spatial Data Handling*, pages 771–791, 1994. 168
6. D. Greene. An implementation and performance analysis of spatial data access methods. In *Proc. 5th International Conference on Data Engineering*, pages 606–615, 1989. 168
7. A. Guttmann. R-trees: a dynamic indexing structure for spatial searching. In *Proc. ACM-SIGMOD International Conference on Management of Data*, pages 47–57, 1984. 167, 168
8. J. Nievergelt and P. Widmayer. Spatial data structures: concepts and design choices. In M. van Kreveld, J. Nievergelt, T. Roos, and P. Widmayer (eds.), *Algorithmic Foundations of Geographic Information Systems*, LNCS 1340, pages 153–198, 1997. 168
9. N. Roussopoulos and D. Leifer. Direct spatial search on pictorial databases with packed R-trees. In *Proc. ACM-SIGMOD International Conference on Management of Data*, 1985. 168, 171
10. H.-W. Six and P. Widmayer. Spatial access structures for geometric databases. In B. Monien and T. Ottmann (eds.), *Data Structures and Efficient Algorithms*, LNCS 594, pages 214–232, 1992. 168
11. Y. Zhou and S. Suri. Analysis of a bounding box heuristic for object intersection. In *Proc. 10th Annual Symposium on Discrete Algorithms (SODA)*, 1999. To appear in *Journal of the ACM*. 169

K-D Trees Are Better when Cut on the Longest Side

Matthew Dickerson[1], Christian A. Duncan[2], and Michael T. Goodrich[3*]

[1] Dept. of Math & Comp. Sci., Middlebury College, Middlebury, VT 05753
`dickerso@middlebury.edu`
[2] Algorithms & Complexity Group, Max-Planck-Inst. für Informatik, D-66123
Saarbrücken, Germany `duncan@mpi-sb.mpg.de`
[3] Dept. of Comp. Sci., Johns Hopkins Univ., Baltimore, MD 21218
`goodrich@jhu.edu`

Abstract. We show that a popular variant of the well known k-d tree data structure satisfies an important packing lemma. This variant is a binary spatial partitioning tree T defined on a set of n points in $\mathbb{R}^d$, for fixed $d \geq 1$, using the simple rule of splitting each node's hyper-rectangular region with a hyperplane that cuts the longest side. An interesting consequence of the packing lemma is that standard algorithms for performing approximate nearest-neighbor searching or range searching queries visit at most $O(\log^{d-1} n)$ nodes of such a tree T in the worst case. Traditionally, many variants of k-d trees have been empirically shown to exhibit polylogarithmic performance, and under certain restrictions in the data distribution some theoretical expected case results have been proven. This result, however, is the first one proving a worst-case polylogarithmic time bound for approximate geometric queries using the simple k-d tree data structure.

1   Introduction

The humorist Erma Bombeck is quoted as once praying,

> "Lord, if you can't make me thin, then make my friends look fat."

In computational geometry, however, we desire objects that are fat, not thin. That is, we desire objects that have a bounded aspect ratio, and this desire is particularly true in the context of spatial partitioning data structures. However it turns out occasional skinny objects are acceptable as long as there are not too many of them and their neighbors are not skinny. Indeed, a recent paper on this topic by Maneewongvatana and Mount [14] can be viewed as turning the Bombeck quote around by stating,

> "It's okay to be skinny, if your friends are fat."

---

* This research partially supported by NSF grant CCR-9732300 and ARO grant DAAH04-96-1-0013.

M. Paterson (Ed.): ESA 2000, LNCS 1879, pp. 179–190, 2000.

In the same way, we are interested in studying spatial partitioning data structures that allow some regions to be skinny so long as the regions around them tend to be fat. We formalize this into a framework we call the *quasi-balanced aspect ratio* tree framework, or *quasi-BAR* tree, for short. We show that trees that fall into this class satisfy an important packing lemma that implies they can be used efficiently for approximate nearest-neighbor and range searching queries. The major result of this paper is to show that a popular variant of the well-known k-d tree data structure falls into this framework. This fact implies that several standard k-d tree searching algorithms actually run in polylogarithmic time when applied to this k-d tree variant.

1.1   Prior Related Work

Bentley [3] introduced the k-d tree data structure for storing a set S of n points in $\mathbb{R}^d$, for fixed constant $d \geq 1$.[1] The idea behind this structure is quite simple. We recursively assume that we have a bounding box, B, that contains all the points in S. We choose one of the d coordinate directions and find the median point in S with respect to this direction. We then cut the box B by a hyperplane perpendicular to this direction so as to go through this median point, and we recurse on the two regions and point sets that this cutting determines. Thus, by repeating this operation until the number of points inside the bounding box is less than some constant, we define a binary spatial partitioning (BSP) tree structure that has $O(\log n)$ depth. Variants of the k-d tree structure are distinguished by the heuristic applied to determine the cut directions. In the original paper by Bentley [3], the heuristic was to simply alternate between the d possible directions in a round-robin fashion.

Friedman, Bentley, and Finkel [12] study an alternate definition, where the cut is defined perpendicular to the direction with *maximum spread*, that is, the direction where the difference between coordinate values in that direction in S is largest. They show that for data distributions with bounded density, k-d trees defined using this heuristic can achieve $O(\log n)$ expected query times for approximate nearest-neighbor and range searching. Silva-Filho [17] studies the choosing of a cutting hyperplane based on probabilistic considerations. Sproull [18] considers several other heuristics for k-d trees in practice. One standard simplification is to use the following splitting rule:

The Longest Side Rule: Choose a splitting hyperplane perpendicular to the longest side of the bounding box B.

This heuristic is often used in practice, since it is so simple to implement and tends to mimic the behavior of the maximum-spread heuristic.

Bentley [4] reports on experimental results for k-d trees defined using the maximum-spread heuristic, showing that for a variety of input distributions this variation performs remarkably well for nearest-neighbor and other searches. We are not familiar with any previous work that reports a non-trivial worst-case

---

[1] We assume the dimension d is fixed throughout this paper.

upper bound for using a k-d tree for nearest-neighbor searching, however, or even approximate nearest-neighbor searching. Still, for range searching queries, where one wishes to count (or report) the points inside a given axis-parallel hyper-rectangle, Lee and Wong [13] show that the standard k-d tree structure can be used to achieve a worst-case query time of $\Theta(n^{(d-1)/d})$ (plus output size in the reporting case). Silva-Filho [16] shows that this bound even holds in the average case for range searching in the standard k-d tree structure.

Deviating from the strict k-d tree approach, there have been several data structures developed for efficiently performing approximate nearest-neighbor searching and range search queries. For example, fair-split trees [5], defined by Callahan and Kosaraju, achieve logarithmic query time for approximate nearest-neighbor searching, although the trees they define do not necessarily have $O(\log n)$ depth. The balanced box-decomposition (BBD) trees of Arya et al. [2,1], on the other hand, have $O(\log n)$ depth and have regions with good aspect ratio, and achieve logarithmic-time performance for approximate nearest-neighbor and range searching. The BBD tree deviates from the k-d tree approach by introducing holes into the middle of regions. The balanced-aspect ratio (BAR) trees of Duncan, Goodrich, and Kobourov [9,10] achieve similar bounds to those of Arya et al., but do so using simple hyperplane cuts at each internal node. However, their BSP trees are not strictly k-d trees as the cuts need not be axis-aligned and do not always equally subdivide the set of points in a region.

Previous work most similar to ours is a recent paper by Maneewongvatana and Mount [14], which provides a packing lemma analogous to one we derive. Their lemma is defined for a specific k-d tree splitting rule, called the *sliding-midpoint split*, that allows the resulting k-d tree to have $\Theta(n)$ depth in the worst case. Thus, it does not seem possible to use this tree to achieve polylogarithmic query times in the worst case for approximate nearest-neighbor and range searching queries. In addition, the proof technique used by Maneewongvatana and Mount is quite different from the one of this paper.

1.2   Our Results

In this paper, we present a general framework for proving that families of BSP trees have good worst-case performance for approximate nearest-neighbor and range searching queries. The framework is based on a relaxed form of the BAR tree of Duncan, Goodrich, and Kobourov [9,10], which we call the *quasi-BAR tree*. The most important component of our approach is that BSP trees falling into our framework satisfy an important packing lemma, which implies efficient performance of several approximate query operations. Since these structures are straight-forward partitioning trees, we only need linear space to store them as well.

In Section 3, we show that a k-d tree defined using the *longest-side* splitting rule falls into our quasi-BAR tree framework. Thus, longest-side k-d trees consequently achieve polylogarithmic worst-case performance for approximating geometric queries. Although these bounds are certainly not better than the logarithmic worst-case bounds of fair-split trees [5], BBD trees [2,1], and BAR trees [9,10], they are nonetheless intriguing for a number of reasons. For example, many empirical results (e.g., see [4,8,12]) show that k-d trees defined with the maximum-spread and longest-side splitting rules perform well in practice, but no worst-case theoretical evidence has been given to support these observations. Our analysis shows that k-d trees defined by the longest-side splitting rule require that we visit at most $O(\log^d n)$ nodes in the worst case to answer any approximate nearest-neighbor and range searching queries on a set of n points in $\mathbb{R}^d$, for any fixed constant $d \geq 1$. We also show that such k-d trees perform exact orthogonal range searches in the plane quite efficiently. We discuss the main ideas behind these results in the sections that follow.

2   A Framework for BSP Trees with Packing Lemmas

Prior work indicates that the performances of BBD [2,1] and BAR [9,10] trees are rooted mainly in the bounds from important packing lemmas. This property is due to the manner in which standard BSP searching algorithms proceed down a search tree to answer a given query. In this section we show how to extend these previous specific packing lemmas into a framework for establishing packing lemmas for other types of BSP trees. Thus, for the sake of completeness, let us review some BSP search algorithms.

2.1   A Standard BSP Nearest-Neighbor Searching Algorithm

Let S be a set of n points in $\mathbb{R}^d$, and let T be a BSP tree defined on S. We use $\delta(p, q)$ to denote the distance between a point p and a point q in $\mathbb{R}^d$. We often refer to a node u in a given tree with an associated region R_u. In these cases, for convenience we use $\delta(u, q) = \delta(R_u, q)$, where $\delta(R_u, q)$ is the distance from q to the region R_u.

Definition 2.1. *For a set S of points in $\mathbb{R}^d$, a query point $q \in \mathbb{R}^d$, and $\epsilon > 0$, a point $p \in S$ is a $(1 + \epsilon)$-nearest neighbor of q if $\delta(p, q) \leq (1 + \epsilon)\delta(p^*, q)$, where p^* is the true nearest neighbor to q.*

In other words, such a p is within a constant error factor of the true nearest neighbor. Given a query point p and an error parameter $\epsilon > 0$ we can use the following algorithm, which is similar to an algorithm of Arya et al. [2], to find an approximate nearest-neighbor to p in S.

We initialize a priority queue Q with the root node of T. Let p be the current nearest neighbor identified during our search, initially some point at ∞. At every stage, extract from Q the node u that is the nearest to q. If $(1+\epsilon)\delta(u, q) \geq \delta(p, q)$, we exit and return p as the $(1 + \epsilon)$-approximate nearest neighbor. The following operations are repeated until the next node is extracted from Q. If u is not a leaf, let u_1 and u_2 be u's children. Without loss of generality, let u_1 be the node nearer to q, i.e. $\delta(u_1, q) \leq \delta(u_2, q)$. We insert u_2 onto the queue and continue with u_1, bypassing the extraction process. If u is a leaf node, we let $S' = S \cap u$ be the (constant-size) set of data points in u. For all $p' \in S'$, if $\delta(p', q) < \delta(p, q)$, we let $p \leftarrow p'$. We continue by extracting the next nearest node from Q.

Definition 2.2. *For a set S of points in $\mathbb{R}^d$, a query point $q \in \mathbb{R}^d$, and $\epsilon > 0$, a point $p \in S$ is a $(1 - \epsilon)$-farthest neighbor of q if $\delta(p, q) \leq \delta(p^*, q) - \epsilon D$, where p^* is the true farthest neighbor to q and D is the diameter of the point set.*

Definition 2.3. *For a set S of points in $\mathbb{R}^d$, a query region Q with diameter O_Q, and $\epsilon > 0$, an ϵ-approximate range query returns (or counts) a set S' such that $S \cap Q \subseteq S' \subseteq S$ and for every point $p \in S'$, $\delta(p, Q) \leq \epsilon O_Q$.*

In the full version, we also review the standard BSP algorithms for finding approximate farthest neighbors and approximate range searching.

2.2   Quasi-BAR Trees

A foundational construct in our quasi-BAR tree framework is the need to bound non-trivially the number of regions *piercing* a set we call a region annulus.

Definition 2.4. *For any region R, we define a region annulus with radius r $A_{R,r}$ to be the set of all real points $p \in \mathbb{R}^d$ such that $p \notin R$ and $\delta(p, R) < r$. A region R' pierces $A_{R,r}$ if and only if there exists two real points $q_1, q_2 \in R'$ such that $q_1 \in R$ and $q_2 \notin R \cup A_{R,r}$.*

Basically, a region annulus is the set of points outside but near the border of R. If R were a spherical region with radius r', this would be the traditional notion of an annulus with radii r' and $r' + r$, respectively. A region R' pierces this annulus if it lies partially inside R and partially farther than r away.

Definition 2.5. *Given any region annulus A and BSP tree T, let $\mathcal{P}_T(A)$ denote the largest set of disjoint nodes in T whose associated regions pierce the region annulus A. A class of binary space partitioning trees is a $\rho(n)$-quasi-BAR tree if, for any tree T in the class constructed on a set S of n points in $\mathbb{R}^d$ and any region annulus A, $|\mathcal{P}_T(A)| \leq \rho(n)V_A/r^d$, where V_A and r are the volume and associated radius (intuitively, the "width") of A, respectively (see Definition 2.4).*

In other words, the number of nodes with regions intersecting an annulus A in a quasi-BAR tree defined on a set of n points is bounded by a function of n times the "relative thickness" of A. The advantage of the quasi-BAR tree definition is that it allows us to prove the following theorems.

Theorem 2.6. *Suppose we are given a $\rho(n)$-quasi-BAR tree T with depth $D_T = \Omega(\log n)$ constructed on a set S of n points in $\mathbb{R}^d$. For any query point q, the standard search algorithms find respectively a $(1+\epsilon)$-nearest and a $(1-\epsilon)$-farthest neighbor to q in $O(\epsilon^{1-d}\rho(n)D_T)$ time.*

Theorem 2.7. *Suppose we are given a $\rho(n)$-quasi-BAR tree T with depth D_T constructed on a set S of n points in $\mathbb{R}^d$. For any convex query region Q, one can perform a counting (or reporting) ϵ-approximate range searching query in T in $O(\epsilon^{1-d}\rho(n)D_T)$ time (plus output size in the reporting case). For any general non-convex query region Q, the time required is $O(\epsilon^{-d}\rho(n)D_T)$ (plus output size).*

The proofs for these theorems follow directly from proofs in Arya et al. [2] and Duncan et al. [10]. The main concept is in proving that all visited "leaf" nodes pierce a region annulus A such that $V_A/r^d = O(\epsilon^{1-d})$. We leave the proof for the full version of the paper.

So long as a class of trees satisfies a packing lemma, we can allow for skinny regions and still achieve good worst-case performance in several approximate geometric queries. The important feature in such cases is to minimize the packing function $\rho(n)$. The difficulty is in developing a data structure that actually guarantees a non-trivial packing function. Surprisingly, such a structure already exists and is a commonly used k-d tree variant.

3   Longest-Side K-d Trees

"... just don't have too many other skinny friends."

We begin our proof of the existence of a non-trivial quasi-BAR tree by reviewing the definition of the longest-side k-d tree. We show that these regions, which may be skinny, do not have many skinny regions nearby of comparable size.

Definition 3.1. *The* longest-side k-d tree *is the tree constructed by recursively dividing the point set associated with each node u in half by cutting perpendicular to the longest axis-orthogonal side of R_u.*

The establishment of a non-trivial packing function $\rho(n)$ for the longest-side k-d trees depends upon the following *Hypercube Stabbing Lemma*, whose proof technique may potentially be of use in analyzing other BSP trees.

Lemma 3.2 (Hypercube Stabbing Lemma). *Suppose we are given a longest-side k-d tree T constructed on a set S of n points in $\mathbb{R}^d$. Let L be the largest set of disjoint nodes in T such that every region in L intersects at least two opposing sides of a hypercube H. Then $|L|$ is $O(\log^{d-1} n)$.*

Proof. Let h be the side length of H. Let us assume, wlog, that the hypercube H is fully contained in the region R associated with the root of T. To generate the largest set L of disjoint nodes in T intersecting at least two opposing sides of H, we first observe that if a node intersects a side of H then all of its ancestors must also intersect this side. Therefore, L consists of all nodes in T none of whose children intersect two opposing sides of H. We now show that $|L|$ is $O(2^d \log^{d-1} n)$.

For this analysis, let us classify a node u in T and its associated region R_u by the number of sides of H that R_u intersects. In particular, let the class $\mathcal{C}(i, j)$ denote the set of regions in T such that each region R intersects i pair of opposing sides of H and j *other* sides (that is, without their "partners" in the same direction). Notice that $i + j \leq d$. More importantly, for any region in L belonging to class $\mathcal{C}(i, j)$ we know $i \geq 1$ (by the definition of L). Also, note that the root of T is in the class $\mathcal{C}(d, 0)$.

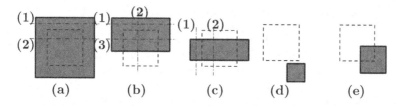

Fig. 1. The various possibilities for the class $\mathcal{C}$ in the plane. The hypercube H is shown in outline while a region in T is shown shaded. We label the various cuts by the classes of child regions they produce. Examples of classes (a) $\mathcal{C}(2,0)$, (b) $\mathcal{C}(1,1)$, and (c) $\mathcal{C}(1,0)$. Regions (d) and (e) cannot be in L because they do not intersect two opposing sides

Let R, associated with some internal node v in T, be a region belonging to class $\mathcal{C}(i,j)$. Since we are only concerned with regions intersecting at least one pair of opposing sides of H, we look at those cases where $i \geq 1$. For the cut c that divides R into two subregions R_1 and R_2, consider the possible cases of classes to which the two children belong.

1. The cut c may be entirely outside of H. In this case, one of v's child regions, say R_2, lies entirely outside of H. Hence, $R_2 \in \mathcal{C}(0,0)$ and $R_1 \in \mathcal{C}(i,j)$.
2. The cut c intersects the inside of H and is perpendicular to a dimension in which the region R spans H on both sides. In this case, both R_1 and R_2 intersect H but with $i-1$ opposing pairs and $j+1$ other sides. Therefore, $R_1, R_2 \in \mathcal{C}(i-1, j+1)$.
3. The cut c intersects the inside of H and is perpendicular to a dimension in which the region R intersects H on just one side. In this case, both R_1 and R_2 intersect H. Moreover, both R_1 and R_2 have the same number of dimensions in which they span H on both sides. One of these regions, however, has one fewer dimension in which it spans H on a single side. That is, without loss of generality, $R_1 \in \mathcal{C}(i,j)$ and $R_2 \in \mathcal{C}(i, j-1)$.

We argue that these are all the possible cases. Case 1 includes all cuts that do not intersect H. If the cut intersects the inside of H, then it is perpendicular to some dimension, k. Along the dimension k, the region R must intersect 2 sides of H, 1 side of H, or 0 sides of H. The first two of these possibilities are covered by Cases 2 and 3 above. The last possibility cannot occur, however, because of the longest-side splitting rule. In particular, if a region R intersects two opposing sides of H, then the longest side of R is at least h, but the side of R in dimension k must have length less than h.

We can now proceed to our evaluation by utilizing a recurrence relation. Let u be a node in T whose associated region R is in $\mathcal{C}(i,j)$. We define $c(u,i,j)$ to be the largest number of disjoint regions associated with descendents of u which intersect at least one pair of opposing sides of H. In other words, every region belongs to a class $\mathcal{C}(i',j')$ for some $i' \geq 1$. Let u_1 and u_2 be u's two children.

The recurrence follows directly from the three cases above:

$$c(u, i, j) = \max \begin{cases} c(u_1, i, j) \\ c(u_1, i-1, j+1) + c(u_2, i-1, j+1) \\ c(u_1, i, j) + c(u_2, i, j-1) \end{cases}$$

Our base cases occur either when u is a leaf l_u (u_1 and u_2 do not exist) or when $i = 0$. In the former $c(l_u, i, j) = 1$. In the latter $c(u, 0, j) = 0$ by definition. We use induction to show that $c(u, i, j) \leq 2^i (\log |u|)^{i+j-1}$, where $|u|$ is the number of points in the subtree rooted at u. For convenience, we maintain that every leaf node has at least two points in it. This simply implies that $\log |u| \geq 1$. We also assume that $|u|$ is an even number, therefore, $|u_1| = |u_2| = |u|/2$. The base cases follow from brute force. Our analysis relies on two fundamental inequalities for $a \geq 1$:

$$a^b \geq (a-1)^b \qquad \text{For } b \geq 0$$
$$a^b \geq (a-1)^b + (a-1)^{b-1} \text{ For } b \geq 1$$

We show that the inductive case holds for any of the three possible recurrences. If u is not a leaf node, $\log |u| \geq 1$, and since $i \geq 1$, we know that $i + j - 1 \geq 0$.

Case 1. Let $c(u, i, j) = c(u_1, i, j)$. Using our inductive hypothesis, we have

$$\begin{aligned} c(u, i, j) &= c(u_1, i, j) \\ &\leq 2^i (\log |u_1|)^{i+j-1} \\ &= 2^i (\log |u| - 1)^{i+j-1} \\ &\leq 2^i (\log |u|)^{i+j-1}. \end{aligned}$$

Case 2. Let $c(u, i, j) = c(u_1, i-1, j+1) + c(u_2, i-1, j+1)$. By induction,

$$\begin{aligned} c(u, i, j) &= c(u_1, i-1, j+1) + c(u_2, i-1, j+1) \\ &\leq 2^{i-1} (\log |u_1|)^{i+j-1} + 2^{i-1} (\log |u_2|)^{i+j-1} \\ &= 2^i (\log |u| - 1)^{i+j-1} \\ &\leq 2^i (\log |u|)^{i+j-1}. \end{aligned}$$

Case 3. Let $c(u, i, j) = c(u_1, i, j) + c(u_2, i, j-1)$. In this case, recall that the cut lies in a direction in which the region R_u intersects H on only one side. This implies that $j \geq 1$ and thus that $i + j - 1 \geq 1$. Again by induction,

$$\begin{aligned} c(u, i, j) &= c(u_1, i, j) + c(u_2, i, j-1) \\ &\leq 2^i (\log |u_1|)^{i+j-1} + 2^i (\log |u_2|)^{i+j-2} \\ &= 2^i (\log |u| - 1)^{i+j-1} + 2^i (\log |u| - 1)^{i+j-2} \\ &\leq 2^i (\log |u|)^{i+j-1}. \end{aligned}$$

The solution to the recurrence relation is now proven. We only need to recall that the region associated with the root of the tree T spans H on both sides in every dimension. Thus, $|L| \leq c(T, d, 0) \leq 2^d (\log n)^{d-1}$. $\square$

This Hypercube Stabbing Lemma might at first seem to be unrelated to the definition of a quasi-BAR tree, which depends heavily on the notion of a region annulus, but this is not the case, as we show in the following theorem.

Theorem 3.3. *Suppose we are given a longest-side k-d tree T constructed on a set S of n points in $\mathbb{R}^d$. Then the packing function $\rho(n)$ of T for a region annulus A is $O(\log^{d-1} n)$. That is, the class of longest-side k-d trees is an $O(\log^{d-1} n)$-quasi-BAR tree.*

Proof. Let L be a set of disjoint regions from T piercing $A = A_{Q,r}$. Let $R \in L$ be any such region piercing A. Notice that $O_R \geq r/2$; therefore, we know that the longest side of the bounding box of R is certainly greater than $r/\sqrt{d}$. Let $\mathcal{H}$ be the smallest set of disjoint hypercubes with side length $r/(2\sqrt{d})$ that completely cover A. Notice that $|\mathcal{H}|$ is $O(V_A/r^d)$. Now, any region R that pierces A must intersect two opposing sides of at least one hypercube in $\mathcal{H}$ (see Figure 2). For any hypercube $H \in \mathcal{H}$, let $L' \subseteq L$ be the subset of regions in L that intersect two opposing sides of H. From the Hypercube Stabbing Lemma (3.2), we know that $|L'| = O(\log^{d-1} n)$. Therefore, we know that

$$|\mathcal{P}_T(A)| = |L| = |L'||\mathcal{H}| \leq c \log^{d-1} n V_A/r^d,$$

for some constant $c > 0$. From Definition 2.5, the class of longest-side k-d trees is therefore an $O(\log^{d-1} n)$-quasi-BAR tree.   □

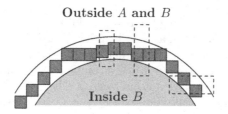

Fig. 2. An annulus A with associated region B. Any region in L, shown here with dashed lines, must cross two opposing sides of one of the hypercube boxes inside A

4   Applications

The theorems from previous sections immediately imply additional corollaries.

Corollary 4.1. *Suppose we are given a longest-side k-d tree T constructed on a set S of n points in $\mathbb{R}^d$. For any query point q, the standard algorithms for finding respectively a $(1 + \epsilon)$-nearest or a $(1 - \epsilon)$-farthest neighbor to q run in $O(\epsilon^{1-d} \log^d n)$ time in the worst case.*

Corollary 4.2. *Suppose we are given a longest-side k-d tree T constructed on a point set S in $\mathbb{R}^d$. For any convex query region, the standard algorithm for counting (or reporting) in an ϵ-approximate range searching query runs in $O(\epsilon^{1-d} \log^d n)$ worst-case time (plus output size in the reporting case). For any general query region, the worst-case time required is $O(\epsilon^{-d} \log^d n)$ (plus output size).*

4.1   Exact Range-Searching in the Plane

In $\mathbb{R}^2$, there is an additional interesting result that can also be proven for the longest-side k-d tree if we enhance each node's region to maintain the bounding box of the point set in the region, rather than just the region itself. We define the bounding box of a region to be the smallest axis-orthogonal box containing all points in the region. This assumption induces only a constant factor overhead and is also common in practice because of the time and space saved. Using this enhancement, we can actually show that any orthogonal range query in $\mathbb{R}^2$ can be answered *exactly* with a running time dependent only on the set size and the aspect ratio of the query. Thus, if the query region is fat, that is, has a low aspect ratio, then the running time becomes polylogarithmic.

Theorem 4.3. *Suppose we are given a bounding-box longest-side k-d tree T constructed on a set S of n points in $\mathbb{R}^2$. Let Q be any orthogonal query region with dimensions w_x and w_y and wlog let $w_x \geq w_y$. The standard range searching algorithm reports an exact orthogonal range query in $O((w_x/w_y) \log^2 n + k \log n)$ time, where k is number of points reported.*

Proof. Let us begin by breaking the region Q into $\alpha = \lceil w_x/w_y \rceil$ squares, labeled Q_i. Notice that each square has side length w_y (see Figure 3a). Let L be the set of all visited nodes which are neither trivially accepted nor rejected. Recall that a node is trivially accepted if it lies completely within the query region Q and trivially rejected if it does not intersect Q. The nodes in L are then exactly those nodes which do partially intersect Q. Let L' be the subset of L such that for any node $u \in L$ with child nodes u_l and u_r, $u \in L'$ if and only if $u_l, u_r \notin L$. Let $L'' \subseteq L'$ be the set of nodes in L' which intersect two opposing sides of Q. Furthermore, let $L_i \subseteq L''$ be the set of nodes in L'' which intersect two opposing sides of Q_i. From the Hypercube Stabbing Lemma (3.2), we know that $|L_i|$ is $O(\log n)$. Thus, $|L''|$ is $O(\alpha \log n)$.

Let us now look at the set $K = L' \setminus L''$. This is a disjoint set of nodes which intersect Q but do not intersect two opposing sides. First, notice that there can be at most 4 nodes in K which intersect two sides of Q, namely the four regions at the corners (see Figure 3b). We now use the bounding box enhancement. If the bounding box of a region R intersects only one side of Q, then there must exist a point p in R which lies inside of Q. Therefore, we know that $|K| \leq k + 4$.

If we combine our results, we see that

$$|L| \leq |L'| \log n = (|L''| + |K|) \log n \in O(\alpha \log^2 n + k \log n).$$

Consequently, the running time is $O(|L| + k) = O(\alpha \log^2 n + k \log n)$.   □

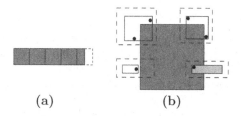

(a)   (b)

Fig. 3. (a) Breaking a region into a chain of squares. (b) A particular square showing potential region intersections *not* intersecting at least two opposing sides. The original regions prior to the bounding box are shown dashed. Notice only one type can intersect the square without being at a corner, and this intersection must contain a point of S

This result, although surprising, is still not better than the range searching data structure of Chazelle [6], which performs exact queries in the plane in $O(\log n)$ time and $O(n)$ space regardless of the aspect ratio of the query. It is simply interesting to note that such exact queries are achievable using a widely-used k-d tree data structure.

5   Conclusion

In this paper, we have developed a general framework for BSP trees that satisfy a packing lemma that qualifies them to be quasi-BAR trees. We also described some important applications that can be solved using this type of tree. In particular, we showed that the well-known longest-side k-d tree falls into this framework and is, therefore, capable of polylogarithmic time approximation queries. Although certainly not a theoretical improvement on the BBD tree structure [2,1] or the BAR tree structure [9,10], this result helps explain why longest-side k-d trees perform well in practice.

Several researchers [4,15,19] have studied the dynamic behavior of k-d trees, where items can be inserted and/or removed. A natural open question, then, is whether one can show if a natural dynamic variant of the k-d tree fits into our quasi-BAR tree framework and exhibits worst-case polylogarithmic update times and worst-case polylogarithmic query times for approximate nearest-neighbor and range searching. Another natural open question is whether any other variations of the k-d tree, such as maximum-spread, are $\rho(n)$-quasi-BAR trees for some polylogarithmic $\rho(n)$ function.

Acknowledgement

We would like to thank David Mount for several helpful conversations relating to the topics of this paper.

References

1. S. Arya and D. M. Mount. Approximate range searching. In *Proc. 11th Annu. ACM Sympos. Comput. Geom.*, pages 172–181, 1995. 181, 182, 189

2. S. Arya, D. M. Mount, N. S. Netanyahu, R. Silverman, and A. Wu. An optimal algorithm for approximate nearest neighbor searching. In *Proc. 5th ACM-SIAM Sympos. Discrete Algorithms*, pages 573–582, 1994. 181, 182, 184, 189

3. J. L. Bentley. Multidimensional binary search trees used for associative searching. *Commun. ACM*, 18(9):509–517, 1975. 180

4. J. L. Bentley. *K*-d trees for semidynamic point sets. In *Proc. 6th Annu. ACM Sympos. Comput. Geom.*, pages 187–197, 1990. 180, 182, 189

5. P. B. Callahan and S. R. Kosaraju. A decomposition of multidimensional point sets with applications to *k*-nearest-neighbors and *n*-body potential fields. *J. ACM*, 42:67–90, 1995. 181

6. B. Chazelle. A functional approach to data structures and its use in multidimensional searching. *SIAM J. Comput.*, 17:427–462, 1988. 189

7. J. R. Driscoll, H. N. Gabow, R. Shrairaman, and R. E. Tarjan. Relaxed heaps: An alternative to Fibonacci heaps with applications to parallel computation. *Commun. ACM*, 31:1343–1354, 1988.

8. C. A. Duncan. *Balanced Aspect Ratio Trees.* Ph.D. thesis, Dept. of Computer Science, Johns Hopkins Univ., 1999. 182

9. C. A. Duncan, M. T. Goodrich, and S. G. Kobourov. Balanced aspect ratio trees and their use for drawing very large graphs. In *Proc. Graph Drawing '98, LNCS 1547*, pages 111–124. Springer-Verlag, 1998. 181, 182, 189

10. C. A. Duncan, M. T. Goodrich, and S. G. Kobourov. Balanced aspect ratio trees: Combining the advantages of *k*-d trees and octrees. In *Proc. 10th ACM-SIAM Symp. on Discrete Algorithms (SODA)*, pages 300–309, 1999. 181, 182, 184, 189

11. M. Fredman and R. E. Tarjan. Fibonacci heaps and their uses in improved network optimization problems. *J. ACM*, 34:596–615, 1987.

12. J. H. Friedman, J. L. Bentley, and R. A. Finkel. An algorithm for finding best matches in logarithmic expected time. *ACM Trans. Math. Softw.*, 3:209–226, 1977. 180, 182

13. D. T. Lee and C. K. Wong. Worst-case analysis for region and partial region searches in multidimensional binary search trees and balanced quad trees. *Acta Inform.*, 9:23–29, 1977. 181

14. S. Maneewongvatana and D. M. Mount. It's okay to be skinny, if your friends are fat. In *4th Annual CGC Workshop on Computational Geometry*, 1999. 179, 181

15. M. H. Overmars and J. van Leeuwen. Dynamic multi-dimensional data structures based on quad- and *k*-d trees. *Acta Inform.*, 17:267–285, 1982. 189

16. Y. V. Silva Filho. Average case analysis of region search in balanced *k*-d trees. *Inform. Process. Lett.*, 8:219–223, 1979. 181

17. Y. V. Silva Filho. Optimal choice of discriminators in a balanced *k*-d binary search tree. *Inform. Process. Lett.*, 13:67–70, 1981. 180

18. R. F. Sproull. Refinements to nearest-neighbor searching. *Algorithmica*, 6:579–589, 1991. 180

19. M. J. van Kreveld and M. H. Overmars. Divided *k*-d trees. *Algorithmica*, 6:840–858, 1991. 189

On Multicriteria Online Problems[*]

Michele Flammini[1] and Gaia Nicosia[2]

[1] Dipartimento di Matematica Pura ed Applicata, University of L'Aquila,
Via Vetoio loc. Coppito, I-67100 L'Aquila, Italy.
flammini@univaq.it
[2] Dipartimento di Informatica ed Automazione, University of Roma Tre,
Via della Vasca Navale 79, I-00146 Roma, Italy.
nicosia@dia.uniroma3.it

Abstract. In this paper we consider multicriteria formulations of classical online problems where an algorithm must simultaneously perform well with respect to two different cost measures. The performance of the algorithm is compared with that of an adversary that serves the sequence of requests selecting one of the possible optimal offline strategies according to a given selection function. We consider a parametric family of functions based on their monotonicity properties which covers all the possible selections. Then, we provide a universal multicriteria algorithm that can be applied to different online problems. For the multicriteria k-server formulation, for each function class, such an algorithm achieves competitive ratios that are only an $O(\log W)$ multiplicative factor away from the corresponding lower bounds that we determine for the class, where W is the maximum edge weight. We then show how to extend our results to other multicriteria online problems sharing similar properties.

1 Introduction

In this paper we consider instances of the k-server problem where each edge of the graph G is assigned a time and a length. k servers must move among the nodes in order to fulfill the various requests and to minimize both the total traveling time and length.

The above problem belongs to the class of the so called *multicriteria* or *multiobjective* problems, where the same solution has to be evaluated with respect to different cost measures. Such problems have been widely investigated in recent years with special attention to the determination of good approximate solutions [13,14,16,20,22]. For instance, one can ask for the determination of a spanning tree of a graph whose global cost is low with respect to two different weightings of the edges [17,19], or such that the diameter is low with respect to first weighting and has low global cost with respect to the second one [18].

In this paper we extend the multicriteria setting to online problems where the algorithm must satisfy a series of requests arriving along the time without

---

[*] Work partially supported by the Italian "Progetto Cofinanziato: Resource Allocation in Computer Networks".

M. Paterson (Ed.): ESA 2000, LNCS 1879, pp. 191–201, 2000.
© Springer-Verlag Berlin Heidelberg 2000

knowledge of the future requests (see [4] for a survey). Assuming the existence of two cost measures for serving requests, the algorithm must compare favorably for both measures with respect to any algorithm knowing all the sequence in advance. This concept is not completely new in the area. For instance, in [9] the authors propose algorithms for routing and bandwidth allocation which simultaneously approximate fair and max throughput solutions.

A multicriteria algorithm can fight against two different types of adversaries. The first one is more powerful than the algorithm, as it can serve any sequence of requests differently for the two cost measures. This translates in finding, among all possible algorithms that are c_1-competitive with respect to the first cost measure, one that has the lowest competitive ratio c_2 with respect to the second one. Basically, it requires the use of standard competitive techniques with the additional constraint that, while minimizing the competitive ratio of one cost function, the ratio for the other measure must be kept below a certain limit.

In this paper we consider a more realistic type of adversary, called *fair*, that has the same constraints of the algorithm, that is, it is compelled to serve the sequence of requests in the same way for both the two cost measures. In fact, this gives a more realistic estimate of how the algorithm is paying (in terms of competitive ratio) its missing knowledge of the future and seems to correspond to a more appropriate view of the multicriteria setting. In this case, since any offline strategy for a sequence of requests σ is characterized by a pair of costs, there is not a unique optimal strategy for the adversary, but a set $\mathcal{F}(\sigma)$ of non-dominated or incomparable minimal strategies. Thus, we assume the existence of a selection function f that associates to each σ a solution in $\mathcal{F}(\sigma)$ the performance of the algorithm is compared with.

We consider a complete classification of the selection functions based on their monotonicity properties. A function is called *monotone* if both its two optimal offline costs do not decrease as the sequence of requests evolves, while is called Δ-monotone for a parameter $\Delta \leq 1$ if the decrease is bounded by a multiplicative factor equal to Δ. We then give a universal algorithm for the multicriteria k-server problem with both competitive ratios $O(\frac{k \log W}{\Delta})$ versus any Δ-monotone function. Moreover, we show a corresponding lower bound of $\Omega(\frac{k}{\Delta})$ stating that our algorithm in every case looses only an $O(\log W)$ multiplicative factor. Such results are then extended to other multicriteria online problems dealing with homogeneous optimization criteria and sharing similar properties.

The paper is organized as follows. The next section is introductory and gives all the necessary background and definitions. In Section 3 we present a simplified version of our algorithm that works for strictly monotone selection functions and in Section 4 we extend it to the class of the Δ-monotone functions for every $\Delta \leq 1$. In Section 5 we generalize all the results to other online problems and finally, in Section 6, we give some conclusive remarks and discuss some open problems.

2   Preliminaries

An algorithm A for an online problem is faced with a sequence $\sigma = \langle r_1, \ldots, r_m \rangle$ of requests and has to satisfy each incoming request r_i without any knowledge of the successive ones. The way A serves r_i causes a certain cost $cost(A, r_i)$ and affects the future costs.

The performance of an online algorithm is usually compared with the one of an optimal offline algorithm that knows the sequence σ in advance.

Definition 1. [11,21] *A is said c-competitive if, for all possible sequences σ, $cost(A, \sigma) = \sum_i cost(A, r_i) \leq c \cdot opt(\sigma) + \alpha$, where $opt(\sigma)$ is the cost paid to serve σ by an optimal offline algorithm that knows all the sequence σ in advance and α is a suitable constant independent from the length m of σ.*

This worst-case analysis is usually performed by assuming that the sequence of requests is supplied by an adversary that at each step provides the request in such a way to maximize the ratio between the overall cost paid by A and its own (optimal offline) cost. Hence, Definition 1 can be rephrased by saying that $cost(A, \sigma) \leq c \cdot opt(\sigma) + \alpha$, where σ is any sequence generated by the adversary.

One of the problems considered in this paper is the k-server problem (see [6,8,15]) where, given a weighted graph $G = (V, E)$ with strictly positive edge weights and k mobile servers in G, a request generated at a particular node $u \in V$ involves the movement of one server from its current position $v \in V$ to u, incurring a cost equal to the distance traveled in G. The cost $cost(A, \sigma)$ paid by A over σ is the total distance traveled by all the servers.

In the bicriteria formulation of an online problem, it is assumed that the cost of serving each request r_i is evaluated simultaneously with respect to two different cost functions $cost_1$ and $cost_2$. Hence, the execution of an online algorithm A over σ is characterized by a pair of costs $(cost_1(A, \sigma), cost_2(A, \sigma))$. For instance, one could consider the cost of moving k servers in a graph $G = (V, E)$ with respect to two different weight measures associated to the edges of G.

A solution for σ with costs $(cost_{1B}(\sigma), cost_{2B}(\sigma))$ provided by an offline algorithm B is a *dominating solution* if and only if $cost_{1C}(\sigma) \geq cost_{1B}(\sigma)$ and $cost_{2C}(\sigma) \geq cost_{2B}(\sigma)$ for any other algorithm C.

Unfortunately, in most cases a dominating solution does not exist since the multiple objectives are conflicting and no solution can dominate all the others. Hence, in general there exist different nondominated minimal (or optimal) offline strategies.

Definition 2. *Given any sequence of requests σ, a solution for σ with costs $(cost_{1B}(\sigma), cost_{2B}(\sigma))$ provided by an offline algorithm B is a nondominated solution if and only if no algorithm C exists such that $cost_{1C}(\sigma) < cost_{1B}(\sigma)$ and $cost_{2C}(\sigma) \leq cost_{2B}(\sigma)$, or $cost_{1C}(\sigma) \leq cost_{1B}(\sigma)$ and $cost_{2C}(\sigma) < cost_{2B}(\sigma)$.*

The set $\mathcal{F}(\sigma)$ of nondominated solutions associated to σ thus forms the frontier of the optimal offline strategies, sometimes called *Pareto frontier* or *efficient frontier.* In order to evaluate the performance of an online algorithm A, we assume the existence of a selection criteria f known by A that associates to each σ a nondominated strategy in $\mathcal{F}(\sigma)$ with costs $(opt_1(f, \sigma), opt_2(f, \sigma))$.

Definition 3. *Given a selection criteria f, an algorithm A is said (c_1, c_2)-competitive versus f if for all possible sequences σ, $cost_1(A, \sigma) \leq c_1 \cdot opt_1(f, \sigma) + \alpha_1$ and $cost_2(A, \sigma) \leq c_2 \cdot opt_2(f, \sigma) + \alpha_2$, where $(opt_1(f, \sigma), opt_2(f, \sigma))$ is the pair of costs paid by the optimal offline strategy in $\mathcal{F}(\sigma)$ selected by f and α_1, α_2 are suitable constants.*

Again, the analysis can be performed by assuming that the sequence of requests is supplied by an adversary that knowing f generates σ in order to maximize the two competitive ratios. Consequently, in Definition 3 it is possible to say that σ is any sequence generated by the adversary.

A reasonable assumption over the selection function f is that both $opt_1(f, \sigma)$ and $opt_2(f, \sigma)$ do not decrease as σ evolves or their decrease can be suitably bounded.

Definition 4. *A selection function f is monotone if and only if, for any sequence of requests σ and for every prefix σ' of σ, $opt_1(f, \sigma) + \alpha_1 \geq opt_1(f, \sigma')$ and $opt_2(f, \sigma) + \alpha_2 \geq opt_2(f, \sigma')$, where α_1 and α_2 are suitable constants.*

Definition 5. *Given any $\Delta \leq 1$, a selection function f is Δ-monotone if and only if, for any sequence of requests σ and for every prefix σ' of σ, $opt_1(f, \sigma) + \alpha_1 \geq \Delta \cdot opt_1(f, \sigma')$ and $opt_2(f, \sigma) + \alpha_2 \geq \Delta \cdot opt_2(f, \sigma')$, where α_1 and α_2 are suitable constants.*

The constants α_1 and α_2 in Definitions 4 and 5 may depend on the problem characteristics, but not on the sequence σ.

Example 1. Consider the bicriteria 1-server problem. Every selection function f is $\frac{1}{W}$-monotone and any f such that for every prefix σ' of σ the strategy $f(\sigma) \in \mathcal{F}(\sigma)$ is an extension of $f(\sigma') \in \mathcal{F}(\sigma')$ is strictly monotone.

Let f_1, f_2 and f_3 be the following selection functions:

1. f_1 chooses a nondominated solution in $\mathcal{F}(\sigma)$ that minimizes $opt_1(\sigma)$;
2. given a real number β, f_2 chooses a nondominated solution in $\mathcal{F}(\sigma)$ such that $opt_1(\sigma) \approx \beta opt_2(\sigma)$, i.e., minimizing $|opt_1(\sigma) - \beta opt_2(\sigma)|$;
3. f_3 chooses a nondominated solution in $\mathcal{F}(\sigma)$ such that $opt_1(\sigma)$ does not exceed a given budget $b \cdot m$ proportional to the length m of σ and such that $opt_2(\sigma)$ is minimum.

It is possible to show that f_1 and f_2 are monotone, while f_3 is not Δ-monotone for $\Delta > \frac{1}{W}$. This very poor monotonicity behavior of f_3 is due to the fact that not only $f_3(\sigma) \in \mathcal{F}(\sigma)$ is not an extension of $f_3(\sigma') \in \mathcal{F}(\sigma')$, but it can be extremely different. In this case the algorithm is paying both the fact that it does not know the future and that it is considering $f_3(\sigma')$ and not the restriction of $f_3(\sigma)$ on the prefix σ' of σ.

As we will see in the sequel, for $O(\frac{1}{W})$-monotone functions the performance of any algorithm in general can be so poor, that a very simple one gives asymptotically optimal competitive ratios.

3   An Universal Algorithm for Monotone Selections

In this section we present a universal algorithm for the bicriteria k-server problem versus monotone selection functions. As we will see in the sequel, our results can be extended to all Δ-monotone functions with $\Delta \leq 1$ and to other bicriteria formulations.

In the bicriteria k-server we are given a graph $G = (V, E)$ where each edge $e \in E$ has a pair of weights $(w_1(e), w_2(e))$. Without loss of generality we can assume that all the edge weights are greater or equal to 1, as otherwise it is possible to scale up all the weights $w_1(e)$ and $w_2(e)$ obtaining an equivalent instance for which any (c_1, c_2)-competitive algorithm translates directly into an (c_1, c_2)-competitive algorithm for the initial one, and vice versa. Moreover, let W be the maximum edge weight with respect to both the two weight measures. For the sake of brevity in the following, when clear from the context, we will denote $opt_1(f, \sigma)$ and $opt_2(f, \sigma)$ simply as $opt_1(\sigma)$ and $opt_2(\sigma)$, respectively.

The basic idea downstanding our algorithm is due to the following fact.

Fact 1 *Given any sequence σ, a selection function f and a positive real λ such that $opt_1(\sigma) = \lambda \cdot opt_2(\sigma)$, let $U(A)$ be an algorithm consisting in moving the servers as a c-competitive algorithm A for the instance of the (single-criterion) k-server problem with the graph G_λ having edge weights $w(e) = w_1(e) + \lambda w_2(e)$ for each $e \in E$. Then, $cost_1(U, \sigma) \leq 2c \cdot opt_1(\sigma) + \alpha_1$ and $cost_2(U, \sigma) \leq 2c \cdot opt_2(\sigma) + \alpha_2$, where α_1 and α_2 are suitable constants.*

Proof. Let $opt(\lambda, \sigma)$ be the optimal offline cost to serve σ on G_λ. Clearly, $opt(\lambda, \sigma) \leq opt_1(\sigma) + \lambda opt_2(\sigma)$. The cost incurred by A on G_λ for σ can be written as

$$cost(A, \sigma) = cost_1(U, \sigma) + \lambda cost_2(U, \sigma) \leq c \cdot opt(\lambda, \sigma) + \alpha$$

$$\leq c(opt_1(\sigma) + \lambda opt_2(\sigma)) + \alpha,$$

for a suitable constant α.

Since by hypothesis $opt_1(\sigma) = \lambda opt_2(\sigma)$, $cost_1(U, \sigma) \leq 2c \cdot opt_1(\sigma) + \alpha$ and $cost_2(U, \sigma) \leq 2c \cdot opt_2(\sigma) + \frac{\alpha}{\lambda}$. The thesis follows by observing that $1/W \leq \lambda \leq W$ and thus $\frac{\alpha}{\lambda}$ is bounded by a constant. $\square$

Clearly λ cannot be determined in advance and therefore the above algorithm $U(A)$ is not online. Our method to get an online version of $U(A)$ is based on the observation that working with an approximation of the real λ gives competitive ratios close to those in Fact 1 and that λ can be learned online during the processing of the sequence σ. Our algorithm at each step maintains the invariant that the used λ is a 2-approximation of the correct value of the currently processed prefix σ' of σ, that is $\frac{1}{2}\frac{opt_1(\sigma')}{opt_2(\sigma')} \leq \lambda \leq 2\frac{opt_1(\sigma')}{opt_2(\sigma')}$. More precisely $U(A)$ performs the steps described in Figure 1.

By construction, the sequence σ can be partitioned in phases during which λ is not modified by the algorithm. Let λ_i be the value of λ used during the i-th

---

- Let $\lambda = 1$;
- upon the arrival of a new request r
  - let σ' be the current sequence of requests (r included);
  - if $\lambda < \frac{1}{2}\frac{opt_1(\sigma')}{opt_2(\sigma')}$ or $\lambda > 2\frac{opt_1(\sigma')}{opt_2(\sigma')}$ then let $\lambda = \frac{opt_1(\sigma')}{opt_2(\sigma')}$;
  - serve r moving the servers in G as algorithm A in G_λ.

Fig. 1. The algorithm $U(A)$.

phase. We call a 1-phase (resp. 2-phase) a phase after which λ doubles (resp. halves) its value, i.e., $\lambda_{i+1} = 2\lambda_i$ (resp. $\lambda_{i+1} = \lambda_i/2$). Moreover, let $\sigma[i-1,i]$ be the subsequence of σ corresponding to the requests of the i-th phase and $\sigma[i] = \sigma[0,i]$ the prefix of σ constituted by all the requests till the i-th phase included. Since f is monotone, if the i-th phase is a 1-phase (resp. 2-phase) opt_1 (resp. opt_2) during the phase has at least to double its value, that is $opt_1(\sigma[i]) \geq 2opt_1(\sigma[i-1])$ (resp. $opt_2(\sigma[i]) \geq 2opt_2(\sigma[i-1])$). Notice that the strategy selected by f in $\mathcal{F}(\sigma[i])$ in general is not an extension of the one in $\mathcal{F}(\sigma[i-1])$.

The following lemma is an easy consequence of the fact that the real value of λ is always comprised between $1/W$ and W.

Lemma 1. *Let h_1 and h_2 be respectively the number of 1- and 2-phases of σ. Then, $|h_1 - h_2| \leq \log W$.*

The costs $cost_1(U, \sigma[i-1,i])$ and $cost_2(U, \sigma[i-1,i])$ paid by the algorithm $U(A)$ during the i-th phase can be suitably bounded.

Lemma 2. *During the i-th phase, $cost_1(U, \sigma[i-1,i]) \leq 3c \cdot opt_1(\sigma[i]) + \alpha_1$ and $cost_2(U, \sigma[i-1,i]) \leq 3c \cdot opt_2(\sigma[i]) + \alpha_2$, where α_1 and α_2 are suitable constants.*

Proof. Similarly as in Fact 1, the cost incurred by A on G_{λ_i} for $\sigma[i-1,i]$ can be written as

$$cost(A, \sigma[i-1,i]) = cost_1(U, \sigma[i-1,i]) + \lambda_i cost_2(U, \sigma[i-1,i]) \leq$$

$$c \cdot opt(\lambda_i, \sigma[i-1,i]) + \alpha \leq c \cdot opt(\lambda_i, \sigma[i]) + \alpha \leq$$

$$c(opt_1(\sigma[i]) + \lambda_i opt_2(\sigma[i])) + \alpha.$$

Since $\frac{1}{2}\frac{opt_1(\sigma[i])}{opt_2(\sigma[i])} \leq \lambda_i \leq 2\frac{opt_1(\sigma[i])}{opt_2(\sigma[i])}$, $cost_1(U, \sigma[i-1,i]) \leq 3c \cdot opt_1(\sigma[i]) + \alpha$ and $cost_2(U, \sigma[i-1,i]) \leq 3c \cdot opt_2(\sigma[i]) + \alpha/\lambda_i$. $\square$

Summing up the costs paid by the algorithm over all the phases, we get the following theorem.

Theorem 2. *The algorithm $U(A)$ defined above is $((3\log W + 13)c, (3\log W + 13)c)$-competitive for the bicriteria k-server problem versus any monotone selection function f.*

Proof. Let h_1 and h_2 be the number of 1-phases and 2-phases in the final sequence σ. So σ has $h_1 + h_2$ phases plus a last possibly incomplete phase.

The cost paid by the algorithm on the first cost measure is

$$cost_1(U, \sigma) = \sum_{i=1}^{h_1+h_2+1} cost_1(U, \sigma[i-1, i]) \leq \sum_{i=1}^{h_1+h_2+1} (3c \cdot opt_1(\sigma[i]) + \alpha_1).$$

At each 1-phase i opt_1 at least doubles its cost, i.e. $opt_1(\sigma[i+1]) \geq 2opt_1(\sigma[i])$. Therefore, informally speaking, if σ were composed only of 1-phases then the above summation would follow a geometric behavior and $cost_1(U, \sigma)$ would be roughly twice the last term $(3c \cdot opt_1(\sigma) + \alpha_1)$. Unfortunately, such a doubling of opt_1 does not hold for the 2-phases, and the worst case for $cost_1$ occurs when the last phases are all 2-phases. By Lemma 1, $|h_1 - h_2| \leq \log W$ and there are at most $\log W$ final 2-phases plus the incomplete last one. Before the last $\log W$ 2-phases the worst case is completed alternating 1- and 2-phases. Therefore,

$$cost_1(U, \sigma) \leq \sum_{i=1}^{h_1+h_2+1} (3c \cdot opt_1(\sigma[i]) + \alpha_1) \leq$$

$$(\log W + 4)(3c \cdot opt_1(\sigma)) + \sum_{i=1}^{h_1+h_2+1} \alpha_1.$$

Since $h_1 \leq \log opt_1(\sigma)$ and $h_2 \leq h_1 + \log W \leq \log opt_1(\sigma) + \log W$,

$$cost_1(U, \sigma) \leq \left((\log W + 4) + \alpha_1 \frac{2(\log opt_1(\sigma)) + (\log W) + 1}{3c \cdot opt_1(\sigma)} \right) 3c \cdot opt_1(\sigma) <$$

$$(3 \log W + 13)c \cdot opt_1(\sigma),$$

for $opt_1(\sigma)$ suitably large.

A symmetric argument holds for $cost_2(U, \sigma)$, thus proving the theorem.   □

The analysis in the previous theorem is asymptotically tight in the sense that there exist examples for which the both two competitive ratios achieved by $U(A)$ are at least order of $c \log W$. Moreover, it can be shown that the use of better approximations of λ might only decrease the competitive ratios of a small multiplicative constant and cannot get rid of the $\log W$ factor.

As a corollary of the above theorem and of the result in [12] with $c = 2k - 1$, there exists a $((6 \log W + 26)k, (6 \log W + 26)k)$-competitive algorithm for the bicriteria k-server problem. Since a trivial lower bound of k for both ratios comes directly from the single-criterion k-server by considering identical edge weights $w_1(e) = w_2(e)$ for each $e \in E$, our algorithm looses only an order of $\log W$ multiplicative factor.

4   Extension to Δ-Monotone Selections

In this section we briefly sketch the extension of our algorithm $U(A)$ to the class of the Δ-monotone selection functions for any fixed $\Delta \leq 1$. More details will appear in the full version of the paper.

Let us define for each sequence σ, $maxopt_1(\sigma) = \max_{\sigma' \ prefix \ of \ \sigma} opt_1(\sigma')$ and $maxopt_2(\sigma) = \max_{\sigma' \ prefix \ of \ \sigma} opt_2(\sigma')$.

Clearly the $maxopt_1$ and $maxopt_2$ functions are monotone in the sense that both cannot decrease as the sequence of requests grows. Substituting in $U(A)$ opt_1 and opt_2 with $maxopt_1$ and $maxopt_2$, we obtain a new algorithm with the same competitive ratios with respect to $maxopt_1$ and $maxopt_2$. Since by definition of Δ-monotone function for every σ it results $opt_1(\sigma) + \alpha_1 \geq \Delta maxopt_1(\sigma)$ and $opt_2(\sigma) + \alpha_2 \geq \Delta maxopt_2(\sigma)$ for two suitable constants α_1 and α_2, the following theorem can be easily proven.

Theorem 3. $U(A)$ is $\left(\frac{(3 \log W + 13)c}{\Delta}, \frac{(3 \log W + 13)c}{\Delta} \right)$-competitive for the bicriteria k-server problem versus any Δ-monotone selection function f, $\Delta \leq 1$.

Again, using the result of [12], there exists a $\left(\frac{(26 + 6 \log W)k}{\Delta}, \frac{(26 + 6 \log W)k}{\Delta} \right)$-competitive algorithm. Moreover, also in this case our algorithm is only an order of $\log W$ multiplicative factor away from the lower bound.

Theorem 4. For each fixed Δ such that $\frac{1}{W} \leq \Delta \leq \frac{1}{2}$, there exist instances of the bicriteria k-server problem with Δ-monotone selection functions for which no algorithm can be (c_1, c_2)-competitive with $c_1 + c_2 \leq \frac{k}{\Delta}$.

Before concluding the section, let us remark that for $o(\frac{\log W}{W})$-monotone functions there exists a much simpler algorithm reaching asymptotically better competitive ratios. It consists simply in performing the $(2k - 1)$-competitive algorithm of [12] on the single-criterion instance obtained by the linear combination of the two criteria with $\lambda = 1$. In fact, by considerations analogous to those in Fact 1, exploiting the property that for every σ the ratios $\frac{opt_1(\sigma)}{opt_2(\sigma)}$ and $\frac{opt_2(\sigma)}{opt_1(\sigma)}$ are bounded by W, it is possible to show that the obtained algorithm is $((2k - 1)(W + 1), (2k - 1)(W + 1))$-competitive versus every possible selection function f.

5   Generalizations

The results shown in the previous sections can be extended to many other bicriteria formulations of classical online problems. In particular, if we redefine W as the maximum ratio over each possible sequence σ between $opt_1(\sigma)$ and $opt_2(\sigma)$ and vice versa, the same claims hold under the following conditions:

1. the optimization criteria must be *homogeneous*, that is, the minimization of a same function under two different cost measures, so that their linear combination into a corresponding single-criterion problem is well defined;

2. any solution for the combined single-criterion combination must directly translate into a feasible solution for both the two cost measures of the bicriteria formulation;

3. the ratio W must be bounded and independent of the length m of the sequence σ.

An example of problem satisfying such properties is Metrical Tasks Systems, where each request consists in the execution of one of n tasks $t_1, t_2, \ldots, t_n$. Each t_i can be processed in a state s_j belonging to a finite set S incurring a cost equal to $t_i(j)$. Given a metric function defining a distance $d_{j,h}$ for each pair of states s_j and s_h in S, the objective of the algorithm is to determine the state in which to execute each task in order to minimize the overall cost of the schedule, i.e., the sum of the costs due to the transitions between the states and the processing of the tasks (see [5]).

In the bicriteria version of the problem two different costs $t_i^1(j)$ and $t_i^2(j)$ are associated to each task t_i, and two different distances $d_{j,h}^1$ and $d_{j,h}^2$ to each pair of states s_j and s_h in S. As it can be easily checked, all the above conditions hold with $W = \max\left(\max_{j,h,i}\left(\frac{d_{j,h}^1+t_i^1(h)}{d_{j,h}^2+t_i^2(h)}, \frac{d_{j,h}^2+t_i^2(h)}{d_{j,h}^1+t_i^1(h)}\right)\right)$. Here our results are still valid if there are cases such that both $d_{j,h}^1 + t_i^1(h) = 0$ and $d_{j,h}^2 + t_i^2(h) = 0$. This situation can be taken into account either excluding such ratios in the definition of W or considering them simply equal to 1.

When suitable subsets of (single-criterion) instances of Metrical Task Systems are considered for which better competitive ratios can be achieved (like instances that model the k-server problem), our method can be applied if such subsets are closed under linear combination, that is when the combination according to any possible λ of any two instances in a subset still belongs to the subset.

Another example of problem satisfying the properties above is the bicriteria version of Multiprocessor Scheduling, where jobs arrive one by one and are to be assigned to one of p processors in order to minimize the maximum completion time (see for instance [10,1]). Associated with each job i are two, possibly different, processing times $t_{1,i}$ and $t_{2,i}$. Again, all the above conditions are satisfied with $W = \max(\frac{t_{1,max}}{t_{2,min}}, \frac{t_{2,max}}{t_{1,min}})$.

Multiprocessor scheduling with the additional constraint that each job has to meet a given deadline does not satisfy the above properties. In fact, a job executed before its deadline in the single-criterion combined version does not necessarily meet the original two deadlines in the bicriteria formulation. Therefore, the single-criterion solution might not be feasible for the bicriteria, thus violating condition 2.

More details and applications will appear in the full version of the paper.

6   Conclusion

We have provided a universal online algorithm for the multicriteria k-server problem and other problems sharing similar properties.

We have considered a *fair* adversary which seems to give a more realistic estimation of the algorithm performances and to our opinion corresponds to a more appropriate view of the multicriteria setting. However, one could compare the online algorithm also with an unfair adversary that can serve the sequence differently for the different cost measures. As already mentioned in the introduction, this basically requires standard online techniques with additional competitive constraints.

Another natural open question concerns the elimination of the order of $\log W$ multiplicative gap between the shown lower and upper bounds on the competitive ratios. Moreover, it would be nice to extend our results to any number of objectives.

Another interesting issue to be considered concerns algorithms with good performances versus more specific selections. To this aim observe that, for $k > 1$ servers, the functions f_1 and f_2 of Example 1 like f_3 have a very poor monotonicity behavior. However, algorithms with good competitive ratios for these selections can be determined by a slight modification of $U(A)$.

More details will appear in the full version of the paper.

References

1. Albers, S.: Better bounds for online scheduling. SIAM Journal on Computing **29**(2) (1999) 459 - 473. 199
2. Awerbuch, B., Azar, Y., Plotkin, S.: Throughput-competitive online routing. Proceedings of 34th Symposium on Foundations of Computer Science (1993) 32-40.
3. Awerbuch, B., Azar, Y.: Local optimization of global objectives: competitive distributed deadlock resolution and resource allocation. Proceedings of the 35th Symposium Foundations of Computer Science (1994) 240-249.
4. Borodin, A., El-Yaniv, R.: Online Computation and Competitive Analysis. Cambridge University Press (1998). 192
5. Borodin, A., Linial, N., Saks, M.: An Optimal Online Algorithm for Metrical Task Systems. Proceedings of the 19nd Annual ACM Symposium on the Theory of Computing (1987) 373-382. 199
6. Chrobak, M., Larmore, L. L.: The server problem and online games. On-Line Algorithms, Proceedings of a DIMACS Workshop, DIMACS Series in Discrete Mathematics and Computer Science **7** (1991) 11-64, American Mathematical Society, New York. 193
7. Fiat, A., Karp, R. M., Luby, M., McGeoch, L. A., Sleator, D. D., Young, N. E.: On competitive paging algorithms. Journal of Algorithms. **12** (1991) 685-699 .
8. Fiat, A., Rabani, Y., Ravid, Y.: Competitive k-server algorithms. Proceedings of the 31th Symposium Foundations of Computer Science (1990) 454-463. 193
9. Goel, A., Meyerson, A., Plotkin, S.: Combining Fairness with Throughput: Online Routing with Multiple Objectives. Proceedings of the 32nd Annual ACM Symposium on the Theory of Computing (2000). 192
10. Graham, R.: Bounds on Certain Multiprocessor Timing Anomalies. SIAM Journal on Applied Mathematics **17** (1969) 416–429. 199
11. Karlin,A. R., Manasse, M. S., Rudolph, L., Sleator, D. D.: Competitive snoopy caching. Agorithmica **3** (1988) 79-119. 193

12. Koutsoupias, E., Papadimitriou, C.: On the k-server conjecture. Journal of the ACM **42** (1995) 971-983. 197, 198

13. Krumke, S. O., Noltermeier, H., Wirth, H.-C., Marathe, M. V., Ravi, R., Ravi, S. S., Sundaram, R.: Improving spanning trees by upgrading nodes. Theoretical Computer Science **221** (1999) 139-155. 191

14. Lin, J. H., Vitter, J. S.: ϵ-approximations with minimum packing constraint violation. Proceedings of the 24th Annual ACM Symposium on the Theory of Computing (1992) 771-782. 191

15. Manasse, M. S., McGeoch, L. A., Sleator, D.D.: Competitive algorithms for server problems. Journal of Algorithms **11** (1990) 208-230. 193

16. Marathe, M. V., Ravi R.,Sundaram, R.: Service constrained network design problems. Proceedings of the Scandinavian Workshop on Algorithmic Theory '96, LNCS Vol. 1097 (1996) 28-40. 191

17. Marathe, M. V., Ravi, R., Sundaram, R., Ravi, S. S., Rosenkrantz, D. J., Hunt, H. B.: Bicriteria Network Design Problems. Proceedings of the 22th ICALP, LNCS Vol. 944 (1995) 487-498 . 191

18. Ravi, R.: Rapid Rumor Ramification: Approximating the minimum broadcast time. Proceedings of the 25th Annual IEEE Foundations of Computer Science (1994) 202-213. 191

19. Ravi, R., Goemans, M. X.: The constrained Minimum Spanning Tree Problem. Proceedings of the Scandinavian Workshop on Algorithmic Theory '96, LNCS Vol. 1097 (1996) 66-75 . 191

20. Ravi, R., Marathe, M. V., Ravi, S. S., Sundaram, R., Rosenkrantz, D. J., Hunt, H. B.: Many Birds with one stone: Multi-objective approximation algorithms. Proc. of the 25th Annual ACM Symposium on Theory of Computing (1993) 438-447. 191

21. Sleator, D. D., Tarjan, R. E.: Amortized efficiency of list update and paging rules. Communications of the ACM **28**(2) (1985) 202-208. 193

22. Warburton, A.: Approximation of Pareto optima in multiple-objective, shortest path problems. Operations Research **35** (1987) 70-79. 191

Online Scheduling Revisited

Rudolf Fleischer[1] and Michaela Wahl[2]

[1] University of Waterloo, Department of Computer Science
email: rudolf@uwaterloo.ca
[2] Max-Planck-Institut für Informatik, Saarbrücken
email: altherr@mpi-sb.mpg.de

Abstract. We present a new online algorithm, MR, for non-preemptive scheduling of jobs with known processing times on m identical machines which beats the best previous algorithm for $m \geq 64$. For $m \to \infty$ its competitive ratio approaches $1 + \sqrt{\frac{1+\ln 2}{2}} < 1.9201$.

1 Introduction

Scheduling problems are of great practical interest. However, even some of the simplest variants of the problem are not fully understood. In this paper, we study the classical problem of scheduling jobs online on m identical machines without preemption, i.e., the jobs arrive one at a time with known processing times and must immediately be scheduled on one of the machines, without knowledge of what jobs will come afterwards, or how many jobs are still to come. The goal is to achieve a small *makespan* which is the total processing time of all jobs scheduled on the most loaded machine. Since the jobs must be scheduled online we cannot expect to achieve the minimum makespan whose computation would require a priori knowledge of all jobs (even then computing the minimum makespan is difficult, i.e., NP-hard [12]). The quality of an online algorithm is therefore measured by how close it comes to that optimum [4,10]. It is said to be *c-competitive* if its makespan is at most c times the optimal makespan for all possible job sequences.

Graham [14] showed some 30 years ago that the List algorithm which always puts the next job on the least loaded machine is exactly $(2 - \frac{1}{m})$-competitive. Only much later were better algorithms designed. A series of papers improved the upper bound on the competitive ratio of the online scheduling problem first to $(2 - \frac{1}{m} - \epsilon_m)$ [11,5], then to 1.986 for $m \geq 70$ [2], then to 1.945 [16], and finally to 1.923 [1]. On the other hand, the lower bound for the problem was raised in similarly small steps: From 1.707 for $m \geq 4$ [9], to 1.837 for $m \geq 3454$ [3], to 1.852 for $m \geq 80$ [1], and finally to 1.85358 for $m \geq 80$ [13].

Better bounds are known for a few special cases. For $m = 2$ and $m = 3$, the lower bound is $(2 - \frac{1}{m})$ [9], i.e., List is optimal. And for $m = 4$, a 1.733-competitive algorithm is known [7]. The best lower bound for randomized algorithms is $\frac{e}{e-1} \approx 1.58$ for large m [6,18], and at least for $m \leq 5$ randomized algorithms can beat the best deterministic lower bound [17]. For scheduling with

M. Paterson (Ed.): ESA 2000, LNCS 1879, pp. 202–210, 2000.

preemption the competitive ratio is exactly $\frac{e}{e-1}$ for $m \to \infty$ [8]. For more results on scheduling see the recent survey chapters in [4,10,15].

In this paper we propose another small improvement on the upper bound of the competitiveness of the online scheduling problem, decreasing it to $1 + \sqrt{\frac{1+\ln 2}{2}} < 1.9201$ for $m \to \infty$. For $m \geq 64$ we beat the best previous bound of 1.923 [1]. Our new algorithm, called MR (the authors' initials), tries to schedule jobs either on the least loaded machine as long as its load is relatively low, or on a certain machine with medium-high load if it can safely do this. This is explained in Section 2. In Sections 3 and 4 we establish the competitiveness of MR. The proof is quite straightforward and relatively simple (compared to proofs of previous algorithms). In particular, we give an intuitive explanation for our choice of the crucial parameters of MR optimizing its performance. And contrary to the most recent previous papers [1,16] we even derive a closed formula for c.

So far, all proofs for the competitiveness of algorithms beating Graham's LIST algorithm make use of three elementary lower bounds on the optimal makespan (see (C1)-(C3) in Section 3). We conjecture that better upper bounds are possible if more lower bounds are added to this list (for example one involving the size of the $(2m + 1)$th largest job), but currently we do not know how to do this.

2　The Algorithm

We assume that at each time step $t = 1, 2, 3, \ldots$ a new job J^t with processing time p^t (also called the *size* of the job) arrives which must be scheduled online on one of m identical machines. The *current schedule* at time t is an assignment of the first t jobs on the machines. The *load* of a machine at a given time is the sum of the sizes of the jobs assigned to it at that time. We assume that the machines are always ordered by decreasing load, i.e., M_1^t is the machine with the highest load and M_m^t is the machine with the lowest load, after J^t was scheduled. The load of M_j^t at time t will be denoted by l_j^t, $1 \leq j \leq m$. Thus, l_1^t is always the makespan after the first t jobs have been scheduled. At time $t = 0$, all machines have load 0. For $j = 1, \ldots, m$ let D_j^t be the average load of machines $M_j^t, \ldots, M_m^t$ at time t, i.e., $D_j^t = \frac{1}{m-j+1} \cdot \sum_{k=j}^{m} l_k^t$. Let $D^t = D_1^t$ be the average load of all machines at time t (which is independent of the current schedule).

Our new algorithm MR is parameterized by the competitive ratio c we hope to achieve. We show in the next section that MR works well if we choose $c \geq 1 + \sqrt{\frac{1+\ln 2}{2}}$. Let $i = \left\lceil \frac{5c-2c^2-1}{c} \cdot m \right\rceil - 1$ be a 'magic' number which will also be explained in the next section, and let $k = 2i - m$. We call MR's schedule *flat* at time t if $l_k^{t-1} < \frac{2(c-1)}{2c-3} \cdot D_{i+1}^{t-1}$, otherwise *steep*. We say J^t is *scheduled flatly* (*steeply*) if MR's schedule is flat (steep) at time t.

As the worst case example for Graham's List algorithm shows [14] one should try to avoid situations were all machines have approximately the same load. So we try to make flat schedules steeper again by scheduling new jobs on some medium-loaded machine, if possible. All previous algorithms [11,2,16,1] are based on this basic idea, however they differ in their definitions of flatness and medium-loadedness.

Algorithm MR

Let p^t be the size of the next job J^t to be scheduled.
If the current schedule is steep or if $p^t + l_i^{t-1} > c \cdot D^t$ then schedule J^t on the smallest machine M_m^{t-1}, else schedule it on M_i^{t-1}.

Note that l_i^{t-1} is the load of the ith machine prior to scheduling J^t, and D^t is the average load of all machines after scheduling J^t.

Theorem 1 ((Main theorem)). *The competitive ratio of* MR *approaches* $1 + \sqrt{\frac{1 + \ln 2}{2}} < 1.9201$ *for* $m \to \infty$. $\qquad\square$

3   Proof of the Main Theorem

Let $c = 1 + \sqrt{\frac{1 + \ln 2}{2}} \approx 1.9201$. Then $i \approx 0.639m$, $k \approx 0.278m$, and $\frac{2(c-1)}{2c-3} \approx 2.19$. The following proof is done under the assumption that $m \to \infty$ but it can easily be adjusted to finite m, however at the cost of a slightly higher value of c (Lemmas 8 and 9 are the critical ones). Note that we can therefore ignore the rounding in the definition of i.

Consider an arbitrary sequence of jobs to be scheduled online. Let t^f be the number of jobs in this sequence. We may assume w.l.o.g. that MR's makespan is defined by the machine receiving the last job J^{t^f}. Let $a = p^{t^f}$ be the size of J^{t^f} and let b be the load of the machine which receives J^{t^f} (not including p^{t^f}). Then MR's final makespan is $a + b$. Let $D = D^{t^f}$ be the final average load of all machines. Furthermore, let P_{m+1} be the size of the $(m+1)$th largest job in the sequence; if $t^f \leq m$ then $P_{m+1} = 0$.

We know the following three lower bounds for the optimal makespan (the first two were introduced by Graham [14], the third seems to be first used by Galambos and Woeginger [11]): D, a, and $2P_{m+1}$. Thus, the following three inequalities must hold for MR to fail to be c-competitive.

$$
\begin{array}{lll}
(C1) & a + b > c \cdot D & \Leftrightarrow \quad a > c \cdot D - b \\
(C2) & a + b > c \cdot a & \Leftrightarrow \quad a < \frac{b}{c-1} \\
(C3) & a + b > c \cdot 2P_{m+1} &
\end{array}
$$

We assume from now that (C1) and (C2) hold. We will show that then (C3) cannot hold, thus proving the Main Theorem. Since $a + b < \frac{c}{c-1} \cdot b$ by (C2) it suffices to show that $P_{m+1} \geq d$ where

$$d = \frac{b}{2(c-1)}.$$

Note that $b - d = \frac{2c-3}{2(c-1)} \cdot b$, and the schedule is flat at time t if $l_k^{t-1} < \frac{b}{b-d} \cdot D_{i+1}^{t-1}$. We call jobs of size at least d *big*. We prove $P_{m+1} \geq d$ by showing that the last job is big and that each machine receives at least one big job before the last job arrives.

We call a machine *full* if its current load is at least b, *half-full* if its current load is at least $b - d$ but less than b, and *half-empty* if its current load is less than $b - d$. Note that MR does not know whether a machine is full or not before the end of the sequence. A *filling job* is a job scheduled on a non-full machine which is full afterwards, i.e., the job *fills* the machine. Slightly abusing this term, we also call the last job a filling job. In the following we are only interested in filling jobs.

(C1) implies that MR must schedule the last job J^{t^f} on the smallest machine $M_m^{t^f-1}$. In particular, all machines have load at least b before J^{t^f} is scheduled. Therefore, there are exactly $m + 1$ filling jobs, and we prove that each of them is big. Note that any filling job scheduled on a half-empty machine is big. (C1) and (C2) together imply

$$b > (c-1) \cdot D$$

justifying the name 'half-full' for machines with load at least $b - d > (c - \frac{3}{2}) \cdot D$.

Lemma 2. *The last job is scheduled flatly.*

Proof. Assume the schedule is steep at time t^f. Then $l_k^{t^f-1} \geq \frac{2(c-1)}{2c-3} \cdot b$ and therefore

$$D \geq \frac{k \cdot \frac{2(c-1)}{2c-3} \cdot b + (m-k) \cdot b}{m}$$
$$> \frac{m + \frac{k}{2c-3}}{m} \cdot (c-1) \cdot D$$
$$\approx 1.22 \cdot D,$$

a contradiction. □

For $j = 1, \ldots, m+1$ let t_j be the time the jth filling job arrives. And for $j = 1, \ldots, m$ let $D_j = D_j^{t_j-1}$ be the average load of machines $M_j^{t_j-1}, \ldots, M_m^{t_j-1}$ (which are exactly the non-full machines at that time) just prior to filling the jth machine. Clearly, $D_{i+1} \leq \cdots \leq D_m < D$.

From Lemma 2 we conclude that there is a minimal index $s \leq m$ such that $J^{t_s+1}, J^{t_s+2}, \ldots, J^{t_m+1}$ are scheduled flatly. If $s < i$ then we choose $s = i$.

Lemma 3. $J^{t_s+1}, J^{t_s+2}, \ldots, J^{t_m+1}$ *are big.*

Proof. Consider one of these jobs at time t. Since machine M_i^{t-1} is already full, J^t can only be filling if it is scheduled on M_m^{t-1}. If this machine is half-empty then J^t must be big to fill it. If it is half-full then $D^t = D$ if $t = t^f$, or

$$D^t \geq \frac{i \cdot l_i^{t-1} + b + (m - i - 1) \cdot (b - d)}{m}$$

if $t < t^f$ because the first i machines have load at least l_i^{t-1}, one machine just becomes full, and all other machines are half-full; if $t = t^f$ then all machines have load at least b. But MR will only schedule J^t on the smallest machine if

$$\begin{aligned}
p^t &> c \cdot D^t - l_i^{t-1} \\
&\geq c \cdot \frac{(i+1) \cdot b + (m - i - 1) \cdot (b - d)}{m} - b \\
&= c \cdot b - \frac{m - i - 1}{m} \cdot c \cdot d - b \\
&= 2(c - 1)^2 \cdot d + (4c - 2c^2 - 1) \cdot d \\
&= d .
\end{aligned}$$

The second inequality holds because $\frac{ci}{m} - 1 \approx 0.22 > 0$ and $l_i^{t-1} \geq b$.   □

Note that in the proof of the lemma in the worst case $l_i^{t-1} = b$ and then our choice of $i = \frac{5c - 2c^2 - 1}{c} \cdot m - 1$ is the smallest possible to prove the claim.

Lemma 4. $D_{i+1} < b - d$. *And if* $s > i$ *then* $D_s < b - d$.   □

We prove the lemma in the next section and continue with the proof of the Main Theorem.

Lemma 5. $J^{t_1}, J^{t_2}, \ldots, J^{t_s}$ *are big.*

Proof. We show that all these jobs are scheduled on a half-empty machine (and hence are big). Let $u = \max\{i + 1, s\}$.

If $s > i$ then $J^{t_{i+1}}, J^{t_{i+2}}, \ldots, J^{t_s}$ are filling jobs and must therefore be scheduled on the smallest machine. But that machine is half-empty just before time t_s by Lemma 4, so it is half-empty at any time before t_s.

For $j = k + 1, \ldots, i$, the schedule is steep at time t_j because

$$l_k^{t_j - 1} \geq b = \frac{2(c - 1)}{2c - 3} \cdot (b - d) \overset{(L4)}{>} \frac{2(c - 1)}{2c - 3} \cdot D_{i+1} \geq \frac{2(c - 1)}{2c - 3} \cdot D_{i+1}^{t_j - 1} .$$

Hence $J^{t_{k+1}}, J^{t_{k+2}}, \ldots, J^{t_i}$ are also scheduled on the smallest machine.

This also implies $l_i^{t_k} \leq l_{i+1}^{t_{k+1}} \leq \cdots \leq l_{i+(m-i)}^{t_{k+(m-i)}} = l_m^{t_i}$. For $j = 1, \ldots, k$, J^{t_j} is either scheduled on $M_i^{t_j - 1}$ or on $M_m^{t_j - 1}$. But

$$l_m^{t_j - 1} \leq l_i^{t_j - 1} \leq l_i^{t_j} \leq l_i^{t_k} \leq l_m^{t_i} \leq D_{i+1} \overset{(L4)}{<} b - d$$

and hence $J^{t_1}, J^{t_2}, \ldots, J^{t_k}$ are scheduled on a half-empty machine.   □

This concludes the proof of the Main Theorem.

4   Proof of Lemma 4

Let $\alpha = \frac{b}{D}$. We first consider the case $s = i$.

We assume $D_{i+1} \geq b - d$ and show that this would imply that all the jobs $J^{t_{i+1}}, \ldots, J^{t_m}$ together already have a higher load than is possible. Therefore the assumption must be wrong. For $j = 1, \ldots, m - i$ let

$$\omega_j = \alpha \cdot \left(\frac{c-1}{c} - \frac{m-i-1}{2(c-1)m} \right) \cdot \left(1 + \frac{c}{m}\right)^{j-1} + \frac{\alpha}{c} .$$

We derive the contradiction by first showing that $\omega_j \cdot D$ is a lower bound for $D^{t_{i+j}}$, i.e., the average load after scheduling $J^{t_{i+j}}$, and then showing that $\omega_{m-i} > 1$.

Since $s = i$, $J^{t_{i+1}}, \ldots, J^{t_m}$ are scheduled flatly but are not scheduled on machine M_i (they are filling jobs). The following observation follows directly from the definition of MR.

Observation 6. *Let b' be the current load of M_i, let p be the size of the next job J, and let $\omega \cdot D$ be the average load before scheduling J for some $\omega \leq 1$. If the current schedule is flat and J cannot be scheduled on M_i then $p > c\omega D - b'$.* $\square$

We even have the stronger bound $p > c \cdot (\omega D + \frac{p}{m}) - b'$ which would improve the lower bound on p by a factor of $\frac{1}{1 - \frac{c}{m}}$, but under the assumption $m \to \infty$ this factor equals 1.

Lemma 7. $D^{t_{i+j}} \geq \omega_j \cdot D$ for $j = 1, \ldots, m - i$.

Proof. We prove the lemma by induction on j. Let $\omega'_j = \frac{D^{t_{i+j}}}{D}$, for $j = 1, \ldots, m - i$. The proof of Lemma 3 shows that $J^{t_{i+1}}$ cannot have size smaller than d, and the smallest size is only possible if the first i machines are just full prior to scheduling $J^{t_{i+1}}$, i.e., $l_1^{t_{i+1}-1} = \ldots = l_i^{t_{i+1}-1} = b$. After scheduling $J^{t_{i+1}}$ we have $i + 1$ machines of size at least b and $m - i - 1$ machines of average size at least $b - d$. Therefore

$$\omega'_1 \geq \frac{(i+1)\cdot\alpha+(m-i-1)\cdot\frac{2c-3}{2(c-1)}\cdot\alpha}{m}$$
$$= \alpha \cdot \frac{(2c-3)m+i+1}{2(c-1)m}$$
$$= \alpha \cdot \left(\frac{c-1}{c} - \frac{m-i-1}{2(c-1)m} \right) \cdot 1 + \frac{\alpha}{c}$$
$$= \omega_1 .$$

For $j > 1$, if we assume that $l_i^{t_{i+j}-1} = b = \alpha D$ then Observation 6 implies
$$\omega'_j \geq \omega_{j-1} + \frac{p^{t_{i+j}}}{mD}$$
$$\geq \omega_{j-1} + \frac{c\omega_{j-1}-\alpha}{m}$$
$$= \omega_{j-1} \cdot \left(1 + \frac{c}{m}\right) - \frac{\alpha}{m}$$
$$= \left[\alpha \cdot \left(\frac{c-1}{c} - \frac{m-i-1}{2(c-1)m} \right) \cdot \left(1 + \frac{c}{m}\right)^{j-2} + \frac{\alpha}{c} \right] \cdot \left(1 + \frac{c}{m}\right) - \frac{\alpha}{m}$$
$$= \alpha \cdot \left(\frac{c-1}{c} - \frac{m-i-1}{2(c-1)m} \right) \cdot \left(1 + \frac{c}{m}\right)^{j-1} + \frac{\alpha}{c}$$
$$= \omega_j .$$

But is the assumption $l_i^{t_{i+j}-1} = b$ justified? Observation 6 seems to imply that the lower bound on the size of the next job is smaller if the load on machine M_i is higher. However, if we want to decrease the lower bound on the next job by some $\epsilon > 0$ then we must first increase the load of M_i to at least $b + \epsilon$ (at some time prior to time $t_{i+j} - 1$), so the net effect of that action does not lead to a smaller average load than the lower bound obtained with our assumption. □

Lemma 8. $\omega_{m-i} > 1$.

Proof. $\alpha > c - 1$ implies

$$
\begin{aligned}
\omega_{m-i} &= \alpha \cdot \left(\tfrac{c-1}{c} - \tfrac{m-i-1}{2(c-1)m} \right) \cdot \left(1 + \tfrac{c}{m} \right)^{m-i-1} + \tfrac{\alpha}{c} \\
&= \alpha \cdot \left(\tfrac{c-1}{c} - \tfrac{1 - \frac{5c-2c^2-1}{c}}{2(c-1)} \right) \cdot \left(1 + \tfrac{c}{m} \right)^{\left(1 - \frac{5c-2c^2-1}{c}\right) \cdot m} + \tfrac{\alpha}{c} \\
&> \tfrac{\left(2(c-1)^2 - (2c^2+1-4c)\right) \cdot \left(1 + \frac{c}{m}\right)^{\frac{2c^2+1-4c}{c} \cdot m} + 2(c-1)}{2c} \\
&= 1 + \tfrac{e^{2c^2+1-4c} - 2}{2c} .
\end{aligned}
$$

For the last equality we used our assumption $m \to \infty$ which implies $(1 + \tfrac{c}{m})^m = e^c$. Therefore, $\omega_{m-i} > 1$ if $c \geq 1 + \sqrt{\tfrac{1+\ln 2}{2}}$.   □

This concludes the proof of the case $s = i$ for $m \to \infty$. For given finite m we can easily compute c such that the lemma holds. For example, if $m = 64$ then we can choose $i = 40$, $k = 16$, and $c = 1.9229 < 1.923$ which is the best previous bound [1] (Lemma 9 below is also true with these parameters).

Some tedious analysis shows that if $c = 1 + \sqrt{\tfrac{1+\ln 2}{2}}$ then the proof of Lemma 8 works *only* for $i = \tfrac{5c-2c^2-1}{c} \cdot m - 1$. If c is chosen bigger then there is some interval around this value of i which works fine, and if c is chosen smaller then no value of i works.

We now consider the case $s > i$. We assume $D_s \geq b - d$ and derive the same contradiction as in the previous case.

Lemma 9. $D^{t_s} \geq \omega_{s-i} \cdot D$.

Proof. Prior to scheduling J^{t_s} we have

$$
D_{i+1}^{t_s-1} \geq \frac{(s-1-i) \cdot b + (m-s+1) \cdot (b-d)}{m-i} .
$$

Since J^{t_s} is scheduled steeply we conclude

$$
\begin{aligned}
\frac{D^{t_s}}{D} &\geq \frac{k \cdot l_k^{t_s-1} + (s-k) \cdot b + (m-s) \cdot (b-d)}{mD} \\
&\geq \frac{k \cdot \frac{2(c-1)}{2c-3} \cdot \frac{(s-1-i) \cdot b + (m-s+1) \cdot (b-d)}{m-i} + (m-k) \cdot b - (m-s) \cdot d}{mD} \\
&= \tfrac{\alpha}{m} \cdot \left(k \cdot \tfrac{2(c-1)}{2c-3} \cdot \left(1 - \tfrac{m-s+1}{2(c-1)(m-i)} \right) + m - k - \tfrac{m-s}{2(c-1)} \right) .
\end{aligned}
$$

We call this term E_s. Note that E_s is linearly growing in s. On the other hand, ω_{s-i} is exponentially growing in s. Observing that $E_{i+1} = \omega_1 = \omega_{i+1-i}$ and $E_m \geq 1.33 \cdot \alpha > 1.09 \cdot \alpha \geq w_{m-i}$ for $m \to \infty$ concludes the proof. $\qquad\square$

Now we can proceed as in the proof of Lemma 7 proving $D^{t_s+j} \geq \omega_{s-i+j} \cdot D$ for $j = 0, \ldots, m - s$. Note that the steepness condition only affects the load of the first k machines which never become the ith machine in the worst case scenario (with minimal load increases) of the proof of Lemma 7. Therefore the assumption about the size of the ith machine in that proof is still justified. This concludes the proof of Lemma 4.

5   Acknowledgements

We thank Erik Demaine for suggesting the term 'half-empty'. We also thank an unknown referee whose detailed comments were very valuable. `Maple` was also of great help.

References

1. S. Albers. Better bounds for online scheduling. In *Proceedings of the 29th ACM Symposium on the Theory of Computation (STOC'97)*, pages 130–139, 1997. To appear in *SIAM Journal on Computing*. 202, 203, 208
2. Y. Bartal, A. Fiat, H. Karloff, and R. Vohra. New algorithms for an ancient scheduling problem. *Journal of Computer and System Sciences*, 51:359–366, 1995. A preliminary version was published in *Proceedings of the 24th ACM Symposium on the Theory of Computation (STOC'92)*, pages 51–58, 1992. 202, 203
3. Y. Bartal, H. Karloff, and Y. Rabani. A better lower bound for on-line scheduling. *Information Processing Letters*, 50:113–116, 1994. 202
4. A. Borodin and R. El-Yaniv. *Online Computation and Competitive Analysis*. Cambridge University Press, Cambridge, England, 1998. 202, 203
5. R. Chandrasekaran, B. Chen, G. Galambos, P. R. Narayanan, A. van Vliet, and G. J. Woeginger. A note on 'An on-line scheduling heuristic with better worst case ratio than Graham's list scheduling'. *SIAM Journal on Computing*, 26(3):870–872, 1997. 202
6. B. Chen, A. van Vliet, and G. J. Woeginger. A lower bound for randomized on-line scheduling algorithms. *Information Processing Letters*, 51:219–222, 1994. 202
7. B. Chen, A. van Vliet, and G. J. Woeginger. New lower and upper bounds for on-line scheduling. *Operations Research Letters*, 16:221–230, 1994. 202
8. B. Chen, A. van Vliet, and G. J. Woeginger. An optimal algorithm for preemptive on-line scheduling. *Operations Research Letters*, 18:300–306, 1994. A preliminary version was published in *Proceedings of the 2nd European Symposium on Algorithms (ESA'94)*. Springer Lecture Notes in Computer Science 855, pages 300–306, 1994. 203
9. U. Faigle, W. Kern, and G. Turán. On the performance of on-line algorithms for partition problems. *Acta Cybernetica*, 9:107–119, 1989/90. 202
10. A. Fiat and G. Woeginger, editors. *Online Algorithms — The State of the Art*. Springer Lecture Notes in Computer Science 1442. Springer-Verlag, Heidelberg, 1998. 202, 203

11. G. Galambos and G. J. Woeginger. An on-line scheduling heuristic with better worst case ratio than Graham's list scheduling. *SIAM Journal on Computing*, 22(2):349–355, 1993. 202, 203, 204

12. M. R. Garey and D. S. Johnson. *Computers and Intractability — A Guide to the Theory of NP-Completeness*. W. H. Freeman and Company, New York, 1979. 202

13. T. Gormley, N. Reingold, E. Torng, and J. Westbrook. Generating adversaries for request-answer games. In *Proceedings of the 11th ACM-SIAM Symposium on Discrete Algorithms (SODA'00)*, pages 564–565, 2000. 202

14. R. L. Graham. Bounds for certain multiprocessing anomalies. *Bell System Technical Journal*, 45:1563–1581, 1966. 202, 203, 204

15. D. S. Hochbaum, editor. *Approximation Algorithms for NP-Hard Problems*. PWS Publishing Company, Boston, MA, 1996. 203

16. D. R. Karger, S. J. Phillips, and E. Torng. A better algorithm for an ancient scheduling problem. *Journal of Algorithms*, 20(2):400–430, 1996. A preliminary version was published in *Proceedings of the 5th ACM-SIAM Symposium on Discrete Algorithms (SODA'94)*, pages 132–140, 1994. 202, 203

17. S. Seiden. Randomized algorithms for that ancient scheduling problem. In *Proceedings of the 5th Workshop on Algorithms and Data Structures (WADS'97)*. Springer Lecture Notes in Computer Science 1272, pages 210–223, 1997. 202

18. J. Sgall. A lower bound for randomized on-line multiprocessor scheduling. *Information Processing Letters*, 63:51–55, 1997. 202

I/O-Efficient Well-Separated Pair Decomposition and Its Applications[*]

(Extended Abstract)

Sathish Govindarajan[1], Tamás Lukovszki[2], Anil Maheshwari[3], and Norbert Zeh[3]

[1] Duke University, gsat@cs.duke.edu
[2] University of Paderborn, Heinz-Nixdorf-Institut, tamas@hni.upb.de
[3] Carleton University, {maheshwa,nzeh}@scs.carleton.ca

Abstract. We present an external memory algorithm to compute a well-separated pair decomposition (WSPD) of a given point set P in $\mathbb{R}^d$ in $O(sort(N))$ I/Os using $O(N/B)$ blocks of external memory, where N is the number of points in P, and $sort(N)$ denotes the I/O complexity of sorting N items. (Throughout this paper we assume that the dimension d is fixed). We also show how to dynamically maintain the WSPD in $O(\log_B N)$ I/O's per insert or delete operation using $O(N/B)$ blocks of external memory. As applications of the WSPD, we show how to compute a linear size t-spanner for P within the same I/O and space bounds and how to solve the K-nearest neighbor and K-closest pair problems in $O(sort(KN))$ and $O(sort(N+K))$ I/Os using $O(KN/B)$ and $O((N+K)/B)$ blocks of external memory, respectively. Using the dynamic WSPD, we show how to dynamically maintain the closest pair of P in $O(\log_B N)$ I/O's per insert or delete operation using $O(N/B)$ blocks of external memory.

1 Introduction

Many geometric applications require computations involving the set of all distinct pairs of points (and their distances) from a set P of N points in d-dimensional Euclidean space (e.g. nearest neighbor for each point). Voronoi diagrams and multi-dimensional divide and conquer are the traditional techniques used for solving several distance based geometric problems, especially in two and three dimensions. But in d dimensions, the worst case size of Voronoi diagrams can be $\Omega(N^{\lfloor d/2 \rfloor})$, and divide and conquer will require an exponent in the polylog factor depending on the dimension. Callahan and Kosaraju [6] introduced the WSPD data structure to cope with higher dimensional geometric problems. It consists of a binary tree T whose leaves are the points in P, with internal nodes representing the subsets of points within the subtree, and a list of "well-separated" pairs of subsets of P, each of which is a node in T. Intuitively a pair $\{A, B\}$ is well separated, if the distance between A and B is significantly greater than the distance between any two points within A or B. It turns out that for many problems it is sufficient to perform only a constant number of operations on pair $\{A, B\}$ instead of performing $|A||B|$ separate operations on the corresponding pairs of points. Moreover for fixed d, a WSPD of $O(N)$ pairs of subsets can be constructed. This has resulted in fast sequential, parallel,

[*] Research supported by NSERC, NCE GEOIDE, and by DFG-SFB376.

M. Paterson (Ed.): ESA 2000, LNCS 1879, pp. 221–232, 2000.
© Springer-Verlag Berlin Heidelberg 2000

and dynamic algorithms for a number of problems on point-sets. Here we extend these results to external memory.

Previous Work   In the Parallel Disk Model (PDM) [16], there is an external memory consisting of D disks attached to a machine with internal memory size M. Each of the D disks is divided into blocks of B consecutive data items. Up to D blocks, at most one per disk, can be transferred between internal and external memory in a single I/O-operation. The complexity of an algorithm in the PDM model is the number of I/O operations it performs. Relevant research work in the external memory (EM) setting can be found in the survey of Vitter [15]. In the PDM model it has been shown that sorting an array of size N takes $sort(N) = \Theta((N/DB)\log_{(M/B)}(N/B))$ I/Os [16,15]. Scanning an array of size N takes $scan(N) = \Theta(N/DB)$ I/Os.

For the geometric problems discussed in this paper, the best resources will be [10] for WSPD and its applications and [14] for results on proximity problems. We omit the discussion on the state of the art, importance and significance of these problems here due to the lack of space, and refer the reader to these references.

New Results   In this paper we present external memory algorithms for the following problems for a set P consisting of N points in d-dimensional Euclidean space: In Sections 3 and 4, we present an algorithm to compute the WSPD data structure for P; it requires $O(sort(N))$ I/Os using $O(N/B)$ blocks of external memory. In Section 5, we present an algorithm to dynamically maintain the WSPD in $O(\log_B N)$ I/O's per insert/delete operation using $O(N/B)$ blocks of external memory. In Section 6.1, we present an algorithm to compute a linear size spanner for P; it requires $O(sort(N))$ I/Os using $O(N/B)$ blocks of external memory. Choosing spanner edges carefully, we can guarantee a diameter of $2\log N$ for the constructed spanner. We also show that $\Omega(\min\{N, sort(N)\})$ I/Os are required to compute any t-spanner of a given point set, for any $t > 1$. In Section 6.2, we present an algorithm to compute K-nearest neighbors, i.e. for each point $a \in P$, compute the K-nearest points to a in $P - \{a\}$; this takes $O(sort(KN))$ I/Os using $O(KN/B)$ blocks of external memory. In Section 6.3, we present an algorithm to compute K-closest pairs, i.e. report the K smallest interpoint distances in P; this takes $O(sort(N+K))$ I/Os using $O((N+K)/B)$ blocks of external memory. In Section 6.4, we present an algorithm to dynamically maintain the closest pair of P in $O(\log_B N)$ I/O's per insert or delete operation using $O(N/B)$ blocks of external memory.

In [5] an $O(\log_B N)$ algorithm for the dynamic closest pair problem in higher dimensions is given. Their approach uses the Topology B-tree data structure. For the remaining problems, our results are the only efficient external memory algorithms known in higher dimensions. For the K-nearest neighbor and K-closest pair problems, optimal algorithms were presented in [12] for the case where $d = 2$ and $K = 1$. In [2] it is shown that computing the closest pair of a point set requires $\Omega(sort(N))$ I/Os, which implies the same lower bound for the general problems we consider in this paper. I/O-efficient construction of fault-tolerant spanners and bounded degree spanners for point sets in the plane and of spanners for polygonal obstacles in the plane has been discussed in [13].

2 Preliminaries

For a given point set P, the *bounding rectangle* $R(P)$ is the smallest rectangle containing all points in P, where a rectangle is the Cartesian product $[x_1, x_1'] \times [x_2, x_2'] \times \cdots \times [x_d, x_d']$ of a set of closed intervals. The *length* of R in dimension i is $l_i(R) = x_i' - x_i$. The maximum and minimum lengths of R are $l_{\max}(R) = \max\{l_i(R) : 1 \le i \le d\}$ and $l_{\min}(R) = \min\{l_i(R) : 1 \le i \le d\}$. When all lengths of R are equal, R is a cube; we denote it's side length by $l(R) = l_{\max}(R) = l_{\min}(R)$. Let $i_{\max}(R)$ be the dimension such that $l_{i_{\max}}(R) = l_{\max}(R)$. For a point set P, let $l_i(P) = l_i(R(P))$, $l_{\max}(P) = l_{\max}(R(P))$, $l_{\min}(P) = l_{\min}(R(P))$, and $i_{\max}(P) = i_{\max}(R(P))$. Let $d(x,y)$ denote the Euclidean distance between points x and y. For point sets X and Y, let $d(X,Y) = \min\{d(x,y) : x \in X \wedge y \in Y\}$. Given a *separation constant* s, we say that two point sets A and B are well-separated if $R(A)$ and $R(B)$ can be enclosed in two d-balls of radius r such that the distance between the two balls is at least sr. We define the *interaction product* $A \otimes B$ of two point sets A and B as $A \otimes B = \{\{a,b\} : a \in A \wedge b \in B \wedge a \ne b\}$. A *well-separated realization* of $A \otimes B$ is a set $\{\{A_1, B_1\}, \ldots, \{A_k, B_k\}\}$ with the following properties [9]:

(R1) $A_i \subseteq A$ and $B_i \subseteq B$, for $1 \le i \le k$,
(R2) $A_i \cap B_i = \emptyset$, for $1 \le i \le k$,
(R3) $(A_i \otimes B_i) \cap (A_j \otimes B_j) = \emptyset$, for $1 \le i < j \le k$,
(R4) $A \otimes B = \bigcup_{i=1}^{k} A_i \otimes B_i$.
(R5) A_i and B_i are well-separated, for $1 \le i \le k$.

Intuitively, this means that for every pair $\{a,b\}$ of points $a \in A$ and $b \in B$, there is a unique pair $\{A_i, B_i\}$ such that $a \in A_i$ and $b \in B_i$ and that for any pair $\{A_i, B_i\}$ the distance between the points in one of the sets is small compared to the distance between the sets. We can use a binary tree T to define a partition of a point set P into subsets. In particular, there is a leaf in T for each point in P. An internal node represents the set of points associated with its descendant leaves. We refer to a node representing a subset $A \subseteq P$ as node A. A leaf representing point $a \in P$ is referred to as leaf a. The parent of a node A in T is denoted by $p(A)$. We say that a realization of $A \otimes B$ *uses* a tree T if all sets A_i and B_i in the realization are nodes in T. A *well-separated pair decomposition* of a point set P is a binary tree T associated with P and a well-separated realization of $P \otimes P$ that uses T. A *split* of a point set P is a partition of P into two non-empty point sets lying on either side of a hyperplane perpendicular to one of the coordinate axes, and not intersecting any points in P. A *split tree* T of P is a binary tree whose nodes are associated with subsets of P, defined as follows: If $P = \{x\}$, T contains a single node x. Otherwise, we use a split to partition P into two subsets P_1 and P_2; T consists of two split trees for point sets P_1 and P_2 whose roots are the children of the root node P of T. For a node A in T, we define the *outer rectangle* $\hat{R}(A)$ as follows: For the root P, let $\hat{R}(P)$ be an open d-cube centered at the center of $R(P)$ with $l_i(\hat{R}(P)) = l_{\max}(P)$. For all other nodes A, the hyperplane used for the split of $p(A)$ divides $\hat{R}(p(A))$ into two open rectangles. Let $\hat{R}(A)$ be the one that contains A. A *fair split* of A is a split of A where the hyperplane splitting A is at distance at least $l_{\max}(A)/3$ from each of the two boundaries of $\hat{R}(A)$ parallel to it. A split tree formed using only fair splits is called a *fair split tree*. A *partial fair split tree* is defined in the same way; but the leaves may represent point sets instead of single points.

3   Constructing a WSPD

In this section we assume that we are given a fair split tree $T = (V,E)$ for a point set P. Also, let every interior node A of T be labeled with its bounding rectangle $R(A)$. Every leaf is labeled with the point it represents. Given this input, we show how to find a well-separated pair decomposition of P I/O-efficiently. In the output, every pair $\{A_i, B_i\}$ in the WSPD will be represented by the IDs of the two nodes A_i and B_i in T.

Our approach is based on [9], which we sketch next. Given a postorder traversal of T, denote the postorder number of a node A in T by $\eta(A)$. Define an ordering $\prec$ on the nodes of T as $A \prec B$ if and only if either $l_{\max}(A) < l_{\max}(B)$ or $l_{\max}(A) = l_{\max}(B)$ and $\eta(A) < \eta(B)$. Given the sequence of nodes in T sorted by $\succ$, define $v(A)$ to be the position of node A in this sequence.

Denote the internal memory algorithm of [9] to construct a well-separated realization of $P \otimes P$ by COMPUTEWSR. It takes the tree T as an input. For every internal node X of T, let A and B be the two children of X. Then COMPUTEWSR calls a procedure FINDPAIRS with argument (A,B). Given a pair (A',B') as an input, procedure FINDPAIRS does the following: First it ensures that $B' \prec A'$, by swapping A' and B' if necessary. If A' and B' are well-separated, the pair $\{A',B'\}$ is added to the well-separated realization. Otherwise, let A'_1 and A'_2 be the children of A' in T. The procedure invokes itself recursively with pairs (A'_1,B') and (A'_2,B') as arguments.

The recursive invocation pattern of procedure FINDPAIRS can be described using the concept of *computation trees*. The root of such a tree T' is a pair $\{A,B\}$ such that A and B are the children of an internal node X of T. A node $\{A',B'\}$ in T' is a leaf if A' and B' are well-separated. Otherwise, let $B' \prec A'$ and A'_1 and A'_2 be the two children of A'; then node $\{A',B'\}$ has two children $\{A'_1,B'\}$ and $\{A'_2,B'\}$ in T'. Tree T' represents the recursive invocations of procedure FINDPAIRS made to compute a realization of $A \otimes B$. Every leaf of T' corresponds to a pair $\{A_i,B_i\}$ in the well-separated realization of $P \otimes P$. In [9] it is shown that there are only $O(N)$ pairs in this realization, which implies that the total size of all computation trees is $O(N)$.

We simulate COMPUTEWSR in external memory by applying *time-forward processing* [11] to the forest of computation trees. The difficulty is that this forest is not known in advance, and we have to generate it on the fly while processing it.

We use the split tree T to guide the processing of F. Assume that we are given the forest $F = (V_F, E_F)$ of computation trees and the fair split tree $T = (V,E)$. We sort the nodes of T by the $\succ$-relation and define a mapping $\mu : V_F \to V$ as $\mu(\{A,B\}) = A$, assuming w.l.o.g. that $B \prec A$. This mapping naturally defines an edge set $\mu(E_F)$ between the vertices in V as $\mu(E_F) = \{(\mu(v),\mu(w)) : (v,w) \in E_F\}$. Note that $\mu(E_F)$ may be a multiset. Given this mapping μ we process F by applying time-forward processing to the multigraph $\mu(F)$. In particular, we process the nodes of $\mu(F)$ sorted by $\succ$, which guarantees that we process the source vertex of every edge before the target vertex. Indeed, for any edge (A,A') in $\mu(E_F)$, either $\{A,A'\}$ is a vertex in F with $A' \prec A$ or A' is a child of A, which also implies that $A' \prec A$.

As F is *not* given beforehand, we have to show how to generate the edges of $\mu(F)$ while processing $\mu(F)$. For every node $A \in V$ with children A_1 and A_2, we store a label $\lambda(A)$ consisting of $v(A_1)$, $v(A_2)$, $R(A_1)$, and $R(A_2)$. Given an edge $(\{A,B\},\{A',B\}) \in E_F$, where A' is a child of A in T, the information $\phi(\mu(\{A,B\}))$ sent along edge

$(\mu(\{A,B\}),\mu(\{A',B\}))$ consists of $v(A')$, $v(B)$, $R(A')$, and $R(B)$. Now consider a node $A \in V$ receiving some tuple describing pair $\{A,B\}$ as input. We have to show how to generate the two children $\{A_1,B\}$ and $\{A_2,B\}$ of $\{A,B\}$ in F (if any) as well as the two edges $(A,\mu(\{A_1,B\}))$ and $(A,\mu(\{A_2,B\}))$ from this input and the label $\lambda(A)$. From the received information, decide whether A and B are well-separated. If they are, add $\{A,B\}$ to the well-separated realization, and proceed to the next input received by A. Otherwise, as the input has been sent to A, $B \prec A$. Hence, $\{A_1,B\}$ and $\{A_2,B\}$ are indeed the two children of $\{A,B\}$ in F. Generate the description of these two pairs from the received information about B and the information about A_1 and A_2 stored in $\lambda(A)$. Compare $v(A_1)$ and $v(A_2)$ with $v(B)$ to compute $\mu(\{A_1,B\})$ and $\mu(\{A_2,B\})$. Queue $\{A_1,B\}$ with priority $v(\mu(\{A_1,B\}))$ and $\{A_2,B\}$ with priority $v(\mu(\{A_2,B\}))$.

To bound the I/O-complexity of this procedure, observe that we queue a constant amount of information per edge in F. As there are $O(N)$ edges in F, applying known results [11,1] about time-forward processing, we obtain the following lemma.

Lemma 1. *Given a set P of N points in $\mathbb{R}^d$, a fair split tree T of P with $O(N)$ nodes and a separation constant $s > 0$, a well-separated realization of P can be computed in $O(\text{sort}(N))$ I/Os using $O(N/B)$ blocks of external memory. Together with T this gives a well-separated pair decomposition of P.*

4   Constructing a Fair Split Tree

Our algorithm to construct a fair split tree T for a given point set P follows the framework of the optimal PRAM algorithm in [4].[1] The idea is to construct T recursively. First we construct a partial fair split tree T' whose leaves have size $O(N^\alpha)$ for some constant $1 - \frac{1}{6d} \leq \alpha < 1$. Then we recursively build split trees for the leaves, proceeding with an optimal internal memory algorithm for every leaf whose size is at most M. We will show how to implement one such recursive step in $O(\text{sort}(N))$ I/Os using $O(N/B)$ blocks of external memory, which leads to the following result.

Lemma 2. *Given a set P of N points in $\mathbb{R}^d$, a fair split tree for P can be computed in $O(\text{sort}(N))$ I/Os using $O(N/B)$ blocks of external memory.*

From Lemmas 1 and 2 we obtain the following result.

Theorem 1. *Given a set P of N points in $\mathbb{R}^d$ and a separation constant $s > 0$, a well-separated pair decomposition for P can be computed in $O(\text{sort}(N))$ I/Os using $O(N/B)$ blocks of external memory.*

To construct a partial fair split tree T', we first construct a compressed version T_c as in [4]. First each dimension is partitioned into slabs containing N^α points each. Every rectangle R associated with a node in T_c satisfies the following three invariants: (1) In each dimension at least one side of R lies on a slab boundary; (2) If R' is the largest rectangle contained in R all of whose sides fall onto slab boundaries, then either $l_i(R') = l_i(R)$ or $l_i(R') \leq \frac{2}{3}l_i(R)$; (3) $l_{\min}(R) \geq \frac{1}{3}l_{\max}(R)$. Now if we want to split rectangle R associated with an internal node v into smaller rectangles, we split R along its longest

---

[1] We do not simulate the PRAM algorithm as this would lead to a much higher I/O complexity.

side according to the following three cases: (1) $l_{\max}(R) = l_{i_{\max}}(R')$: We find the slab boundary that comes closest to splitting R into equal-sized pieces. If this slab boundary is at distance at least $\frac{1}{3}l_{\max}(R)$ from either side of R, we split R along this slab boundary. Otherwise, we split R into two equal-sized pieces. This case produces the two resulting rectangles R_1 and R_2. (2) $l_{\max}(R') \geq \frac{8}{81}l_{\max}(R)$: If $l_{i_{\max}}(R') \geq \frac{1}{3}l_{\max}(R)$, we split R along the side of R' that is not shared with R. Otherwise, we find the unique integer j such that $y = \frac{2}{3}\left(\frac{4}{3}\right)^j l_{\max}(R')$ lies in the interval $\left(\frac{1}{2}, \frac{2}{3}\right] l_{\max}(R)$ and split along a hyperplane that is at distance y from the slab boundary shared by R' and R. This case produces the rectangle containing R'. The other rectangle is being ignored for the time being; we reattach it later. (3) $l_{\max}(R') < \frac{8}{81}l_{\max}(R)$: In this case, R and R' share a unique corner. We construct a d-cube sharing the same corner with R and with side length $\frac{3}{2}l_{\max}(R')$. This case produces the cube C. Later we have to construct a sequence of fair splits cutting R down to C.

The construction of T_c takes $O(sort(N))$ I/Os using standard external-memory graph techniques. The reattachment of the missing Case 2 leaves takes sorting and scanning. Next we describe how to expand compressed edges corresponding to a Case 3 split.

Such a Case 3 edge corresponds to a rectangle R that has been "shrunk" to a d-cube C with the following properties: (P1) C and R share exactly one corner, and C is contained in R. (P2) Any rectangle contained in R and not intersecting C contains at most $O(N^\alpha)$ points. (P3) $l(C) < \frac{8}{81}l_{\max}(R)$ and $l_{\min}(R) \geq \frac{1}{3}l_{\max}(R)$.

Properties P1 and P3 guarantee that there exists a sequence of fair splits that produces C from R. Property P2 guarantees that for every split in this sequence, the rectangle not containing C can be made a leaf of T'. We show how to construct these sequences of fair splits for all Case 3 edges in $O(sort(N))$ I/Os. We also have to assign the points in P to the containing leaves of T', as this information is needed for the next level of recursion in the fair split tree construction algorithm.

Every point of P is contained in some leaf rectangle or in some region $R \setminus C$, where (R, C) is a compressed edge produced by Case 3. We compute an assignment of the points in P to these regions similarly to the corresponding step in the optimal PRAM-algorithm in [4]. This takes a constant number of sorting and scanning steps.

To replace each edge corresponding to Case 3 by a sequence of fair splits, we simulate one phase of the sequential algorithm of [9]. We sort the points in $R \setminus C$ in each dimension, producing a sorted list L_i^R of points for each dimension i. Consider the current rectangle R which we want to split in dimension $i_{\max}$, producing rectangles R_1 and R_2, where R_1 contains the cube C. If $l_{\max}(R) > 3l(C)$, split R in half. Otherwise, choose a hyperplane containing a side of C for the split. We use list $L_{i_{\max}}^R$ to decide whether there are any points in R_2. If so, we perform a split producing two new nodes R_1 and R_2 of T'' and assigning the points in R_2 to the leaf R_2. Otherwise, shrink R to R_1, not producing any new nodes. Shrinking R does not cost any I/Os as the coordinates of R are maintained in internal memory. This approach ensures that a linear number of splits are performed in the number of points in $R \setminus C$. Thus, in total only $O(N)$ splits for all edges (R, C) to be expanded are performed. We have to show how to maintain the lists L_i^R representing the set of points in the current rectangle R.

Assume that we split R in dimension i and that this produces two new rectangles R_1 and R_2. R_1 contains the cube C and R_2 does not. We scan L_i^R from the tail until we find the first point that is in R_1. All points after this position in L_i^R are not in R_1 and are therefore removed and put into the point list of leaf R_2. As R_1 is going to act as rectangle R for the next split, we need to delete all points in R_2 from all lists L_j^R. The above scan takes care of deleting these points from L_i^R. However, we cannot afford to delete these points from all the other lists L_j^R, $j \neq i$, because this would cause many random accesses or would require a scan over all these lists per split. Instead we delete points lazily. When splitting the rectangle R_1 in dimension j, which produces rectangles R_3 and R_4, where R_3 contains C, we scan L_j^R from the tail until we find the first point in R_3. For each point visited by this scan we test whether it is in R_4. If it is, it is appended to the point set of leaf R_4. Otherwise, as it is not in R_3 and not in R_4, it must be contained in a leaf produced by a previous split and can therefore be discarded. In total, we scan every list at most once, for a total of $O(scan(N))$ I/Os.

Up to this point, the construction algorithm has in fact computed a "super-tree" T'' of T', as some leaves produced by the algorithm may be empty. Even though some leaves of T'' are empty, the size of T'' is $O(N)$ [4]. Thus, given the assignment of point sets to the leaves of T'' as computed above, it takes $O(sort(N))$ I/Os using standard external memory graph techniques to remove empty leaves and all internal nodes such that all descendant leaves of at least one of its two children are empty. The result is the desired partial fair split tree T'.

5   Dynamic Well-Separated Pair Decomposition

Dynamic Fair-Split Tree   We present an I/O-efficient algorithm for dynamically maintaining the fair-split tree using $O(\log_B N)$ I/Os per insert or delete operation. We follow the approach of [8].

The main idea is to maintain a rectangle $\tilde{R}(A)$, such that $R(A) \subseteq \tilde{R}(A)$, for each node $A \in T$ that is not modified during updates. A *fair cut* of any rectangle $\tilde{R}$ is defined as partitioning the rectangle using an axis-parallel hyperplane that is at distance at least $\frac{1}{3} l_{\max}(\tilde{R})$ from its nearest parallel sides. Note that a fair cut of $\tilde{R}(A)$ satisfies the fair split condition on $R(A)$, since $l_{\max}(R(A)) \leq l_{\max}(\tilde{R}(A))$ and $R(A) \subseteq \tilde{R}(A)$ at any moment in time. Let $R \rightarrow R'$ indicate that R' can be constructed from R by a sequence of fair cuts.

We maintain a binary tree T under insertions and deletions in which each node satisfies the following invariants:

(I1) For all internal nodes A with children A_1 and A_2, there exists a *fair cut* that partitions $\tilde{R}(A)$ into rectangles R_1 and R_2 such that $R_1 \rightarrow \tilde{R}(A_1)$ and $R_2 \rightarrow \tilde{R}(A_2)$.
(I2) For all leaves a, $\tilde{R}(a) = a$.

It is easy to see that T is a fair-split tree. Invariant I1 ensures the existence of a fair cut, which satisfies the fair split property.

Insertion: When we want to insert a point a, we find the deepest node A in T such that $a \in \tilde{R}(A)$. Let R_1 and A_1 have the same meaning as in Invariant I1 and assume that $a \in R_1$. We insert a new node B, which replaces A_1 as a child of A. We make A_1 and a children of B. We show how to construct $\tilde{R}(B)$ satisfying Invariant I1.

Consider Invariant I1 for node A. Since $R_1 \rightarrow \tilde{R}(A_1)$, there exists a sequence of fair cuts that construct $\tilde{R}(A_1)$ from R_1. Clearly the last rectangle R in the sequence of cuts that satisfies $a \in R$ can be assigned as $\tilde{R}(B)$. However, this does not give us an efficient way to find $\tilde{R}(B)$, since the sequence of cuts may be long. In [8] it is shown that given the deepest node A, $\tilde{R}(B)$ can be computed in constant time.

We can compute A using $O(\log_B N)$ I/Os by posing it as a deepest-intersect query on the Topology B-tree [5].

Deletion: To delete a point a, delete the leaf a and compress the parent node $p(a)$. Note that this preserves the invariants.

Theorem 2. *The fair-split tree of a point set P can be maintained using $O(N/B)$ disk blocks and $O(\log_B N)$ I/Os per insert or delete operation.*

Maintaining the Pairs Dynamically We follow the approach of [8]. We use the following characterization of well-separated pairs. An ordered pair of nodes $(A,B), A,B \in T$ is well-separated if and only if it satisfies the following conditions:

1. The point sets A and B are well-separated,
2. Nodes $p(A)$ and B are not well-separated, and
3. $l_{\max}(R(B)) < l_{\max}(R(p(A))) \leq l_{\max}(R(p(B)))$.

It is easy to construct examples where the insertion of a new point can add a linear number of well-separated pairs. Thus, we cannot do a trivial update of the pairs upon insertion or deletion. The main idea of the approach is to anticipate all but a constant number of pairs that would be added when a new point is inserted. So, when a point is inserted or deleted, we need to update only a constant number of pairs. In [8] it is shown that this invariant can be maintained only if the point distribution of P is uniform. If it is not uniform, we add dummy points to make it uniform. In [8] it is shown that it is necessary to add only $O(1)$ dummy points per well-separated pair.

When a point p is inserted, we compute a c-approximation of the distance between p and its nearest neighbor in P, for some constant $c = f(s)$. In [5] it is shown how to compute the approximate nearest neighbor in $O(\log_B N)$ I/Os using a Topology B-tree.

Let d_p be the approximate distance computed above. It can be shown that the interactions between p and points $\{q \in P \mid d(q,p) > d_p\}$ are already covered in the current WSPD. We need to compute well-separated pairs that account for interactions between p and points within distance d_p from p. The number of such pairs needed is $O(1)$. We can compute these pairs as follows:

Extract all nodes $A \in T$ whose bounding rectangles $R(A)$ overlap the cube C centered at p with length $l(C) = d_p$ and satisfy the conditions of a WSPD defined earlier in this subsection. This can be done by filtering C through a Topology-B-tree that corresponds to the fair-split tree. Since the number of pairs extracted is $O(1)$, we can perform this extraction using $O(\log_B N)$ I/O's.

We need to add dummy points to make sure that the region covering the newly computed pairs (A,B) is uniform. Note that for the existing pairs, we have already added the required dummy points. A straightforward way is to add a mesh of suitably spaced points covering the region around $R(A)$ and $R(B)$. We insert the dummy points and compute the pairs corresponding to them as described above. The total number of dummy points added is $O(1)$ and the number of pairs added to the WSPD is $O(1)$. Thus, the entire process requires $O(\log_B N)$ I/Os. The deletion of a point p can be taken care of using a similar procedure.

Theorem 3. *Let P be a set of N points in $\mathbb{R}^d$. The well-separated pair decomposition of P with respect to a fair-split tree T can be maintained under insertions and deletions using $O(\log_B N)$ I/Os per update and $O(N/B)$ disk blocks.*

6   Applications of the WSPD

6.1   *t*-Spanners

Given a point set P, let $G = (P,E)$ be a graph whose edges have weights corresponding to the Euclidean distance between their two endpoints. G is called a t-spanner for P if for any two points p and q in P, the shortest path from p to q in G is at most t times longer than the Euclidean distance between p and q. In [9] it is shown that the following graph $G = (P,E)$ is a t-spanner of linear size for the point set P: For every node $A \in T$ choose a representative $q(A) \in A$. For every pair $\{A_i, B_i\}$ in the given WSPD of P, add an edge $\{q(A_i), q(B_i)\}$ to E. In [3] it is shown that one can compute a spanner of diameter at most $2\log N$ by choosing representatives carefully. In particular, we partition the edges in T into heavy and light edges. Given a node A with children A_1 and A_2, edge (A,A_1) is heavy and edge (A,A_2) is light if $|A_1| \geq |A_2|$. Otherwise, edge (A,A_2) is heavy and edge (A,A_1) is light. Every node of T is contained in a unique chain of heavy edges starting at a leaf. For node A, let $q(A)$ be the leaf whose chain contains A.

This assignment of representatives corresponds to sending representatives up the tree; at every node A receiving representative $q(A_1)$ and $q(A_2)$ from its children, we choose $q(A)$ as the one sent along the heavy edge. The decision which edge is heavy can be made by sending sizes $|A_1|$ and $|A_2|$ along with the representative $q(A_1)$ and $q(A_2)$. This can be done using time-forward processing.

Theorem 4. *Given a set P of N points in $\mathbb{R}^d$, it takes $O(sort(N))$ I/Os and $O(N/B)$ blocks of external memory to compute a t-spanner G of linear size and diameter at most $2\log N$ for P.*

The next theorem states that the I/O-complexity of the spanner construction is practically optimal.

Theorem 5. *It takes $\Omega(\min\{N, sort(N)\})$ I/Os to compute a t-spanner, $t > 1$, for a point set P of N points in $\mathbb{R}^d$.*

Proof. We prove the lemma for $d = 1$. The proof generalizes to higher dimensions by placing all points on a straight line in $\mathbb{R}^d$. Given a list X of items $x_1, \ldots, x_N$ and a permutation $\sigma : [1,N] \to [1,N]$, we are finally interested in computing the sequence $x_{\sigma(1)}, \ldots, x_{\sigma(N)}$. Let σ be given as a sequence $S = \langle \sigma(1), \ldots, \sigma(N) \rangle$. We map data item x_i to the point $i + \frac{1}{t}$ and $\sigma(i)$ to the point $\sigma(i)$. Any t-spanner, $t > 1$, of the resulting point set must contain edges $(\sigma(i), x_{\sigma(i)})$. We first remove all edges that are not of this type and then reverse the permutation performed on elements $\sigma(i)$ by the spanner construction algorithm to arrange the remaining edges in the order $(\sigma(1), x_{\sigma(1)}), \ldots, (\sigma(N), x_{\sigma(N)})$. This reversal can be done by recording the I/Os performed by the spanner construction algorithm and playing this I/O-sequence backward, exchanging the meaning of read and write operations.

The construction of the point set takes $O(scan(N))$ I/Os. The reversal of the permutation takes $O(T_S(N))$ I/Os, where $T_S(N)$ is the I/O-complexity of the spanner algorithm. In total, we can compute any permutation in $O(T_S(N))$ I/Os, so that $T_S(N) = \Omega(perm(N)) = \Omega(\min\{N, sort(N)\})$.   □

6.2 K-Nearest Neighbors

In this section we show how to compute the K-nearest neighbors for every point $p \in P$. The construction follows the sequential algorithm in [9] for this problem.

Lemma 3 ([9]). *Let $\{A, B\}$ be a pair in a well-separated realization of $P \otimes P$ with separation $s > 2$. If there is a point $b \in B$ that is a K-nearest neighbor of a point $a \in A$, then $|A| \leq K$.*

Given a set B, let O_B be the center of $R(B)$. Then we divide the space around B into a constant number of cones with apex O_B such that for any two points a and a' in the same cone, $\angle a O_B a' < \frac{s}{s+1}$.

Lemma 4 ([9]). *Let a point set X and a cone c with apex O_B be given, such that for any two points a and a' in c, $\angle a O_B a' < \frac{s}{s+1}$. Let $X_c = (x_1, \ldots, x_l)$ be the set of points in X that lie in c, sorted by distance from O_B. For $i > K$, no point in B can be a K-nearest neighbor of x_i.*

Based on Lemma 3 the algorithm in [9] first extracts all pairs $\{A_i, B_i\}$ with $|A_i| \leq K$. This can be done in $O(sort(N) + scan(KN))$ I/Os using time-forward processing and a reverse preorder traversal of T. For a node B in T, let $\{A'_1, B\}, \ldots, \{A'_q, B\}$ be the set of pairs such that $|A'_i| \leq K$. Note that the sets A'_i are pairwise disjoint. We store the set $f(B) = \bigcup_{i=1}^{q} A'_i$ as a candidate set at node B in T. The construction of sets $f(B)$ takes sorting and scanning and thus $O(sort(KN))$ I/Os. Then the set of candidates is narrowed down by a top-down procedure in T, which can be realized using time-forward processing. At the root node B of T, the algorithm partitions the space around O_B into cones as described above and uses K-selection in each cone to find the K points in $f(B)$ that fall into this cone and are closest to O_B. By Lemma 4, all other points in $f(B)$ cannot have a K-nearest neighbor in B. Call the resulting set of points $N(B)$. As $N(B)$ contains at most K points per cone and there are $O(1)$ cones, $|N(B)| = O(K)$. The set $N(B)$ is passed on to the children B_1 and B_2 of B. The points possibly having a closest neighbor in B_i are in the set $f(B_i) \cup N(B)$. Thus, the algorithm applies the same K-selection algorithm to this set to construct $N(B_i)$ and pass it on to B_i's children. As we send only $O(K)$ points along every edge of T, this takes $O(sort(KN))$ I/Os.

Finally every leaf b stores a set $N(b)$ of size $O(K)$ such that b is a potential K-nearest neighbor for the points in $N(b)$. The total size of all these sets is $O(KN)$. Now we build sets $N'(a)$ for all points $a \in P$ such that $b \in N'(a)$ if and only if $a \in N(b)$ and apply

K-selection to each set $N'(a)$ separately to find the K-nearest neighbors of point a. This takes sorting of KN points and a standard K-selection algorithm, hence, $O(sort(KN))$ I/Os.

Theorem 6. *Given a set P of N points in $\mathbb{R}^d$, it takes $O(sort(KN))$ I/Os and $O(KN/B)$ blocks of external memory to find the K-nearest neighbors for all points in P.*

6.3 K-Closest Pairs

Given a well-separated realization of $P \otimes P$, let the pairs $\{A_i, B_i\}$ be sorted by increasing distances $d(R(A_i), R(B_i))$. For any such pair, $|A_i \times B_i| = |A_i||B_i|$. To find the K closest pairs, we first determine the smallest i such that $\sum_{j=1}^{i} |A_j||B_j| \geq K$ and retrieve all pairs $\{A, B\}$ such that $d(R(A), R(B)) \leq (1 + 4/s)r$, where $r = d(R(A_i), R(B_i))$. We extract the set of pairs $\{a, b\}$ such that $a \in A$ and $b \in B$ for some pair $\{A, B\}$ that we retrieved.

In [7] it is shown that the K closest pairs are among the extracted point pairs and that the total number of extracted point pairs is $O(N + K)$. Given this set of point pairs, we apply K-selection again to find the K closest ones. Given all point pairs, this takes $O(scan(N + K))$ I/Os. We have to show how to extract all candidate pairs $\{a, b\}$.

Sorting the pairs $\{A_i, B_i\}$ by their distances $d(R(A_i), R(B_i))$ between the bounding rectangles takes $O(sort(N))$ I/Os. Then it takes a single scan to extract all pairs with $d(R(A), R(B)) \leq (1 + r/s)r$. We extract the points in sets A_i and B_i in the same way as in the previous section and sort these at most $O(N + K)$ points to guarantee that the points in the same pair $\{A_i, B_i\}$ are stored consecutively. We can now construct the set of all candidate point pairs by scanning the resulting point list. We obtain the following result.

Theorem 7. *Given a set P of N points in $\mathbb{R}^d$, it takes $O(sort(N + K))$ I/Os and $O((N + K)/B)$ blocks of external memory to compute the K closest pairs in P.*

6.4 Dynamic Closest Pair

We show how to maintain the closest pair of points in P under insertion and deletion. We make the following simple observation:

Lemma 5. *Let P be a point set of n points in $\mathbb{R}^d$. Let (a, b), $a, b \in P$ be the closest pair in P. Then, the pair $\{\{a\}, \{b\}\}$ belongs to the well-separated pair decomposition of P.*

Proof. Let (A, B) be the pair in the well-separated decomposition such that $a \in A, b \in B$ and b is a nearest neighbor of a. Applying Lemma 3, we have $|A| = 1$. Similarly, a is a nearest neighbor of b. By Lemma 3, we have $|B| = 1$. $\square$

Denote the well-separated pairs of the form $\{\{a\}, \{b\}\}$ as singleton pairs. We maintain the singleton pairs in a B-tree based on the distance between them. The closest pair is the singleton pair with minimum distance.

During an insert or delete operation, update the well-separated pairs and insert or delete the singleton pairs in the B-tree, respectively. Since only $O(1)$ well-separated pairs change during an insert or delete operation, we obtain the following theorem.

Theorem 8. *Let P be a point set of n points in $\mathbb{R}^d$. The closest pair in P can be maintained using $O(\log_B N)$ I/Os per insert or delete operation.*

Acknowledgments

We would like to thank Pankaj Agarwal, Pat Morin, and Michiel Smid for helpful discussions.

References

1. Lars Arge. The buffer tree: A new technique for optimal I/O-algorithms. In *Proceedings of the Workshop on Algorithms and Data Structures*, volume 955 of *Lecture Notes in Computer Science*, pages 334–345, 1995. 225
2. Lars Arge and P. B. Miltersen. On showing lower bounds for external-memory computational geometry problems. In J. Abello and J. S. Vitter, editors, *External Memory Algorithms and Visualization*. AMS, 1999. 222
3. S. Arya, D. M. Mount, and M. Smid. Randomized and deterministic algorithms for geometric spanners of small diameter. In *35th IEEE Symposium on Foundations of Computer Science (FOCS'94)*, pages 703–712, 1994. 229
4. P. B. Callahan. Optimal parallel all-nearest-neighbors using the well-separated pair decomposition. In *Proc. FOCS'93*, pages 332–341, 1993. 225, 226, 227
5. P. B. Callahan, M. Goodrich, and K. Ramaiyer. Topology B-trees and their applications. In *Proc. WADS'95*, LNCS 955, pages 381–392, 1995. 222, 228
6. P. B. Callahan and S. R. Kosaraju. A decomposition of multi-dimensional point sets with applications to k-nearest-neighbors and n-body potential fields. In *Proc. STOC'92*, pages 546–556, 1992. 221
7. P. B. Callahan and S. R. Kosaraju. Faster algorithms for some geometric graph problems in higher dimensions. In *Proc. SODA'93*, pages 291–300, 1993. 231
8. P. B. Callahan and S. R. Kosaraju. Algorithms for dynamic closest pair and n-body potential fields. In *Proc. SODA'95*, pages 263–272, 1995. 227, 228
9. P. B. Callahan and S. R. Kosaraju. A decomposition of multidimensional point sets with applications to k-nearest neighbors and n-body potential fields. *Journal of the ACM*, 42:67–90, 1995. 223, 224, 226, 229, 230
10. Paul B. Callahan. *Dealing with Higher Dimensions: The Well-Separated Pair Decomposition and Its Applications*. PhD thesis, Johns Hopkins University, Baltimore, Maryland, 1995. 222
11. Yi-Jen Chiang, Michael T. Goodrich, Edward F. Grove, Roberto Tamassia, Darren Erik Vengroff, and Jeffrey Scott Vitter. External-memory graph algorithms. In *Proceedings of the 6th Annual ACM-SIAM Symposium on Discrete Algorithms*, January 1995. 224, 225
12. Michael T. Goodrich, Jyh-Jong Tsay, Darren Erik Vengroff, and Jeffrey Scott Vitter. External-memory computational geometry. In *Proceedings of the 34th Annual IEEE Symposium on Foundations of Computer Science*, November 1993. 222
13. T. Lukovszki, A. Maheshwari, and N. Zeh. I/O-efficient spanner construction in the plane. http://www.scs.carleton.ca/~nzeh/Pub/spanners00.ps.gz, 2000. 222
14. Michiel Smid. Closest-point problems in computational geometry. In J.-R. Sack and J. Urrutia, editors, *Handbook of Computational Geometry*, pages 877–936. North-Holland, 2000. 222
15. J. S. Vitter. External memory algorithms. In *Proceedings of the 17th Annual ACM Symposium on Principles of Database Systems*, June 1998. 222
16. J.S. Vitter and E.A.M. Shriver. Algorithms for parallel memory I: Two-level memories. *Algorithmica*, 12(2–3):110–147, 1994. 222

Higher Order Delaunay Triangulations[*]

Joachim Gudmundsson[1], Mikael Hammar[1], and Marc van Kreveld[2]

[1] Dept. of Computer Science, Lund University, Box 118, 221 00 Lund, Sweden.
{Joachim.Gudmundsson,Mikael.Hammar}@cs.lth.se
[2] Dept. of Computer Science, Utrecht University, P.O.Box 80.089, 3508 TB Utrecht,
the Netherlands. marc@cs.uu.nl

Abstract. For a set P of points in the plane, we introduce a class of triangulations that is an extension of the Delaunay triangulation. Instead of requiring that for each triangle the circle through its vertices contains no points of P inside, we require that at most k points are inside the circle. Since there are many different *higher order Delaunay triangulations* for a point set, other useful criteria for triangulations can be incorporated without sacrificing the well-shapedness too much. Applications include realistic terrain modelling, and mesh generation.

1 Introduction

One of the most well-known and useful structures studied in computational geometry is the Delaunay triangulation [7,11,20]. It has applications in spatial interpolation between points with measurements, because it defines a piecewise linear interpolation function. The Delaunay triangulation also has applications in mesh generation for finite element methods. In both cases, the usefulness of the Delaunay triangulation as opposed to other triangulations is the fact that the triangles are well-shaped. It is well-known that the Delaunay triangulation of a set P of points maximizes the smallest angle, over all triangulations of P.

One specific use of the Delaunay triangulation for interpolation is to model elevation in Geographic Information Systems. The so-called *Triangulated Irregular Network*, or *TIN*, is one of the most common ways to model elevation. Elevation is used for hydrological and geomorphological studies, for site planning, for visibility impact studies, for natural hazard modelling, and more. Because a TIN is a piecewise linear, continuous function which is generally not differentiable at the edges, these edges play a special role. In elevation modelling, one usually tries to make the edges of the TIN coincide with the ridges and valleys of the terrain. Then the rivers that can be predicted from the elevation model are a subset of the edges of the TIN. When one obtains a TIN using the Delaunay triangulation of a set of points, the ridges and valleys in the actual terrain will not always be as they appear in the TIN. The so-called 'artificial dam' in valleys is a well-known artifact in elevation models, Fig. 1. It appears when a Delaunay edge crosses a valley from the one hillside to the other hillside, creating a local minimum in

---

[*] This research is partially supported by the ESPRIT IV LTR Project No. 21957 (CGAL)

M. Paterson (Ed.): ESA 2000, LNCS 1879, pp. 232–243, 2000.

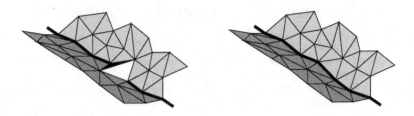

Fig. 1. Artificial dam that interrupts a valley line (left), and a correct version obtained after one flip (right).

the terrain model slightly higher up in the valley. It is known that in real terrains such local minima are quite rare [13]. These artifacts need to be repaired, if the TIN is to be used for realistic terrain modelling [21], in particular for hydrological purposes [18,19,22]. If the valley and ridge lines are known, these can be incorporated by using the constrained Delaunay triangulation [6,9,17]. The cause of problems like the one mentioned above may be that the Delaunay triangulation is a structure defined for a planar set of points, and doesn't take into account the third dimension. One would like to define a triangulation that is both well-shaped and has some other properties as well, like avoiding artificial dams. This lead us to define *higher order Delaunay (HOD) triangulations,* a class of triangulations for any point set P that allows some flexibility in which triangles are actually used. The Delaunay triangulation of P has the property that for each triangle, the circle through its vertices has no points of P inside. A k-order Delaunay (k-OD) triangulation has the relaxed property that at most k points are inside the circle. The idea is then to develop algorithms that compute some HOD triangulation that optimizes some other criterion as well. Criteria one can think of are minimizing the number of local minima, and minimizing the number of local extrema. The former criterion deals with the artificial dam problem, and the latter criterion also deals with interrupted ridge lines. For finite element method applications, criteria like minimizing the maximum angle, area triangle, and degree of any vertex may be of use [3,4,5].

In Section 2 we define HOD triangulations and show some easy properties. In Section 3 we give an algorithm to compute which edges can be included in a k-OD triangulation. The algorithm runs in $O(nk^2 + n \log^3 n)$ expected time. In Section 4 we consider 1-OD triangulations, and prove more specific, useful results in this case. In Section 5 we give the applications. We show that for 1-OD triangulations, most of the criteria we study can be optimized in $O(n \log n)$ time. We also give some approximation algorithms for general k-OD triangulations.

The proofs omitted in this extended abstract can be found in the full version.

2   Definitions and Preliminaries

We first define higher order Delaunay edges, higher order Delaunay triangles, and higher order Delaunay triangulations. Given two vertices u and v we will

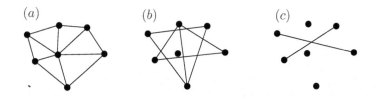

Fig. 2. (a) The 0-OD edges (b) extended with the new 1-OD edges, and (c) the new 2-OD edges.

denote by $\overline{uv}$ the edge between u and v and by $\vec{uv}$ the directed line segment from u to v. Furthermore, the unique circle through three vertices u, v and w is denoted $C(u, v, w)$, and the triangle defined by u, v and w is denoted $\triangle uvw$. We will throughout this article assume that P is non-degenerate, that is, no three points of P lie on a line and no four points of P are co-circular.

Definition 1. *Let P be a set of points in the plane. For $u, v, w \in P$:*

- *An edge $\overline{uv}$ is a k-order Delaunay edge (or k-OD edge) if there exists a circle through u and v that has at most k points of P inside, Fig 2;*
- *A triangle $\triangle uvw$ is a k-order Delaunay triangle (or k-OD triangle) if the circle through u, v, and w has at most k points of P inside.*
- *A triangulation of P is a k-order Delaunay triangulation (or k-OD triangulation) of P if every triangle of the triangulation is a k-OD triangle.*

For $k = 0$, the definitions above match the usual Delaunay edge and triangle definitions.

Lemma 1. *Let P be a set of points in the plane.*
(i) *Every edge of a k-OD triangle is a k-OD edge.*
(ii) *Every edge of a k-OD triangulation is a k-OD edge.*
(iii) *Every k-OD edge with $k > 0$ that is not a 0-OD edge intersects a Delaunay edge.*

Note that the converse of Lemma 1(i) is not true. Not every triangle consisting of three k-OD edges is a k-OD triangle. Figure 3a shows an example where three 1-OD edges form a 3-OD triangle. A natural question to ask is whether any k-OD edge or any k-OD triangle can be part of some k-OD triangulation. Put differently, can k-OD edges exist that cannot be used in any k-OD triangulation? Indeed, such edges (and triangles) exist. In the next section we'll give a method to test for any k-OD edge if it can be extended to a k-OD triangulation. Figure 3b shows an example where $\triangle uvx$ is a 1-OD triangle that cannot be included in a 1-OD triangulation since $\triangle uvy$ is not a 1-OD triangle. To distinguish between 'useful' and 'non-useful' k-OD edges we use the following definition.

Definition 2. *Let P be a set of points in the plane. A k-OD edge $\overline{uv}$ with $u, v \in P$ is useful if there exists a k-OD triangulation that includes $\overline{uv}$. A k-OD*

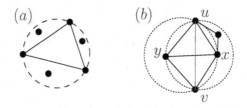

Fig. 3. (a) Not every triangle with three 1-OD edges is a 1-OD triangle. (b) Not every 1-OD triangle ($\triangle uvx$) can be included in a 1-OD triangulation.

triangle $\triangle uvw$ with $u, v, w \in P$ is valid *if it doesn't contain any point of P inside and its three edges are useful.*

There is a close connection between k-OD edges and higher order Voronoi diagrams.

Lemma 2. *Let P be a set of n points in the plane, let $k \leq n/2 - 2$, and let $u, v \in P$. The edge $\overline{uv}$ is a k-OD edge if and only if there are two adjacent faces, F_1 and F_2, in the order-$(k+1)$ Voronoi diagram such that u is in the set of points that induces F_1 and v is in the set of points that induces F_2.*

Since the worst case complexity of order-k Voronoi diagrams is $O(k(n - k))$ [16], it follows that $O(n + nk)$ pairs of points can give rise to a k-OD edge. These pairs can be computed in $O(nk \log n + n \log^3 n)$ expected time [1].

3   Higher Order Delaunay Triangulations for General k

In this section we show some properties of k-OD edges. We also give an efficient way to compute all useful k-OD edges of a point set P. In Section 5 we will use these results to compute k-OD triangulations that take into account some other criterion.

There are always Delaunay edges that must be present in a k-OD triangulation, for example the convex hull of the point set. The *k-OD triangulation skeleton* is defined as the set of Delaunay edges that does not intersect any useful k-OD edges. A first step of completing the k-OD skeleton would be to decide which k-OD edges are possible to insert into a k-OD triangulation. The main result in this section is Lemma 4, which allow us to perform a simple and fast test to check if a k-OD edge is useful or not. The result also implies heuristics for computing a k-OD triangulation; this given in Section 5.2. We next study the properties of an arbitrary k-OD edge $\overline{uv}$. The following observation and its corollary follows from [7], Lemma 9.4.

Observation 1 *For any k-OD edge $\overline{uv}$ and any Delaunay edge $\overline{sp}$ that intersects $\overline{uv}$, the circle $C(u, v, s)$ contains p.*

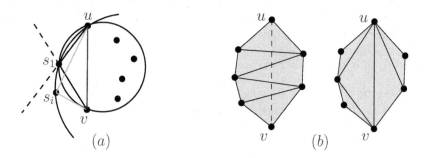

Fig. 4. (a) If $\overline{uv}$ is useful then $\triangle us_1v$ is a useful k-OD triangle. (b) The hull of the k-OD edge $\overline{uv}$ (shaded) and the completion of $\overline{uv}$.

Corollary 1. *Consider a k-OD edge $\overline{uv}$. Any circle through u and v that does not contain any vertices to the left (right) of $\vec{vu}$ contains all vertices to the right (left) of $\vec{vu}$ that are incident to Delaunay edges that intersect $\overline{uv}$.*

3.1   Testing a k-OD Edge for Usefulness

To decide if a k-OD edge, $\overline{uv}$ can be included in any k-OD triangulation one has to check if the point set can be k-OD triangulated with the edge $\overline{uv}$. We will show that it suffices to check only two circles through the points u, v, and one other point, and count how many points they contain. For simplicity, we will assume, without loss of generality, that $\overline{uv}$ is vertical and that u is above v.

Lemma 3. *Let $\overline{uv}$ be a k-OD edge and let s_1 be the point to the left (right) of $\vec{vu}$, such that the circle $C(u, s_1, v)$ contains no points to the left (right) of $\vec{vu}$. If $\triangle us_1v$ is not a k-OD triangle then $\overline{uv}$ is not useful.*

We would like to go a step further than the result of Lemma 3, namely, prove that if the 'first' triangle to the left side of $\vec{uv}$, $\triangle us_1v$, is valid, and the symmetrically defined triangle on the right side of $\vec{uv}$ is valid, then $\overline{uv}$ is useful. We show this result constructively, by giving a method that gives a k-OD triangulation that includes $\overline{uv}$. It appears that we only have to compute a triangulation of a small region near $\overline{uv}$, called the hull of $\overline{uv}$. The hull is defined as follows.

Definition 3. *The* hull *of a k-OD edge $\overline{uv}$ is the closure of the union of all Delaunay triangles whose interior intersects $\overline{uv}$, Fig. 4b. We say that the hull of a k-OD edge $\overline{uv}$ is* complete *if it is triangulated. The additional edges needed to triangulate the hull are called the* completion *of $\overline{uv}$, Fig. 4b.*

The following algorithm computes a triangulation of the hull. Let $\overline{uv}$ be a k-OD edge. Let p_1 be the point to the right of $\vec{vu}$ such that the part of $C(u, v, p_1)$ to the right of $\vec{vu}$ is empty. Note that p_1 must be a vertex on the boundary of the hull of $\overline{uv}$. Add the two edges $\overline{up_1}$ and $\overline{vp_1}$ to the graph. Continue like this recursively for the two edges $\overline{up_1}$ and $\overline{vp_1}$ until the hull of $\overline{uv}$ to the right of $\vec{vu}$ is completely triangulated. The same procedure is then performed on the left side of $\vec{vu}$. The obtained triangulation is called the *greedy triangulation* of the hull of $\overline{uv}$, see Figure 5. The next corollary shows that the hull is a simple polygon consisting of at most $2k + 2$ vertices.

Corollary 2. *The Delaunay edges intersecting one useful k-OD edge $\overline{uv}$ are connected to at most k vertices on each side of the k-OD edge.*

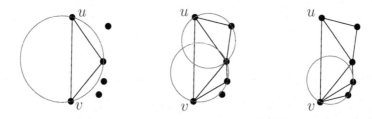

Fig. 5. Greedy triangulation to the right of $\vec{vu}$.

Lemma 4. *Let $\overline{uv}$ be a k-OD edge, let s be the point to the left of $\vec{vu}$, such that the circle $C(u, s, v)$ contains no points to the left of $\vec{vu}$. Let s' be defined similarly but to the right of $\vec{vu}$. The edge $\overline{uv}$ is a useful k-OD edge if and only if $\triangle uvs$ and $\triangle uvs'$ are valid.*

After preprocessing, we can test efficiently whether an edge $\overline{uv}$ is a useful k-OD edge. First, locate u and v in the Delaunay triangulation, and traverse it from u to v along $\overline{uv}$ from Delaunay triangle to Delaunay triangle. Collect all intersected Delaunay edges. If there are more than $2k - 1$ of them, stop and report that $\overline{uv}$ is not a k-OD edge. Otherwise, determine the endpoints of intersected Delaunay edges to the left of $\vec{vu}$ and to the right of $\vec{vu}$. If there are more than k on one side, stop with the same result. Next, determine s and s' as in the lemma just given. Now we must test how many points lie in the circles $C(u, s, v)$ and $C(u, s', v)$ to determine usefulness. To this end, we use a data structure for storing a set of points that can report all the ones that lie in a given query circle. An $O(n \log n)$ size structure exists that answers such queries in $O(\log n + m)$ time, when m points are reported [2]. In our application we abort reporting as soon as the number of answers exceeds k, so the query time becomes $O(\log n + k)$.

In Section 2 we showed that all k-OD edges can be determined in $O(nk \log n + n \log^3 n)$ expected time using an algorithm for higher order Voronoi diagrams [1]. There are $O(n + nk)$ edges to be tested, so it takes $O(nk^2 + n \log^3 n)$ expected time overall to determine all useful k-OD edges.

4   First Order Delaunay Triangulations

We examine the special structure of any 1-OD edge $\overline{uv}$ further. We already observed in Corollary 2 that if $\overline{uv}$ is a useful 1-OD edge, and not a 0-OD edge, then it intersects exactly one Delaunay edge. We again assume without loss of generality that $\overline{uv}$ is vertical and that u is above v.

Lemma 5. *Every useful 1-OD edge intersects at most one useful 1-OD edge.*

Proof. From Corollary 2 we already know that any Delaunay edge at most intersect one 1-OD edge. It remains to prove the lemma for any non-Delaunay, useful 1-OD edge $\overline{uv}$. Let $\overline{sy}$ be the Delaunay edge $\overline{uv}$ intersects. If there exists a useful 1-OD edge that intersects the Delaunay edge $\overline{su}$ it must be connected to y, according to Corollary 2. Denote this edge $\overline{xy}$. Since $\overline{xy}$ is a 1-OD edge x must be connected to u and s by Delaunay edges. Consider the circle $C(u, v, y)$. This circle contains only s by Observation 1 and from the fact that $\overline{uv}$ is useful. The circle $C(u, y, x)$ can be obtained by expanding $C(u, v, y)$ until it hits x while releasing v. Since we let go of v, both v and s is contained in $C(u, y, x)$. Thus $\triangle xuy$ cannot be a 1-OD triangle and hence $\overline{xy}$ cannot be a useful 1-OD edge, which contradicts our assumption. By symmetry this holds for any edge intersecting $\overline{uy}$, $\overline{sv}$, or $\overline{vy}$. □

The lemma just given shows that if $\overline{uv}$ is a useful 1-OD edge that is not Delaunay, then the four Delaunay edges $\overline{us}$, $\overline{uy}$, $\overline{sv}$, and $\overline{vy}$ must be in every 1-OD triangulation. Given a triangulation $\mathcal{T}$ and two edges e_1 and e_2 in $\mathcal{T}$, we say that e_1 and e_2 are independent if they are not incident to the same triangle in $\mathcal{T}$. From Corollary 2 and Lemma 5 we obtain the main result of this section.

Corollary 3. *Every 1-OD triangulation can be obtained from a Delaunay triangulation by flips of independent Delaunay edges.*

It is easy to see that—given the Delaunay triangulation—all 1-OD edges can be determined in linear time. In $O(n \log n)$ time, we can find out which ones are useful.

5   Applications: Triangulations with Additional Criteria

Recall from the introduction that Delaunay triangulations are often used in terrain modelling, because they give well-shaped triangles. However, artifacts like artificial dams may arise. Since the Delaunay triangulation is completely specified by the input points (in non-degenerate cases), there is no flexibility to incorporate other criteria into the triangulation, which is why higher order Delaunay triangulations were introduced. In this section we show how to avoid artificial dams in higher order Delaunay triangulations, and deal with a number of other criteria as well. Many of these criteria can be optimized for 1-OD triangulations, which is what we will show first. Then we give some approximation algorithms to incorporate such criteria in k-OD triangulations.

5.1   Applications for 1-OD Triangulations

Minimizing the Number of Local Minima To minimize the number of local minima is straight-forward if the domain is the class of 1-OD triangulations.

Lemma 6. *Removing a local minimum by adding 1-OD triangles never prevents any other local minimum from being removed.*

Theorem 1. *An optimal 1-OD triangulation with respect to minimizing the number of local minima can be obtained by flips of independent Delaunay edges in $O(n \log n)$ time.*

Minimizing the Number of Local Extrema The number of local extrema— minima and maxima—can also be efficiently minimized over all 1-OD triangulations. In the previous subsection we could choose the edge in any quadrilateral that connects to the lowest of the four points. But if we want to minimize local minima and maxima we get conflicts: it can be that the one edge of a convex quadrilateral gives an additional local minimum and the other edge gives a local maximum. Consider the subdivision S consisting of all edges that must be in any 1-OD triangulation, so S contains triangular and convex quadrilateral faces only. Consider the set of points that either have no lower neighbors or no higher neighbors; they are extremal in S. Consider the graph $G = (M, A)$, where M is the set of nodes representing the local extrema, and two nodes m, m' are connected if they represent points on the same quadrilateral face and the one triangulating edge makes that m is not a local extremum and the other triangulating edge makes that m' is not a local extremum.

Lemma 7. *Any quadrilateral face of S defines at most one arc in G, and this arc connects a local minimum to a local maximum.*

From the lemma it follows that G is bipartite, because every arc connects a local minimum to a local maximum. G may contain many isolated nodes; these can be ignored. For any node incident to only one arc, we can choose to make the point representing that node to be non-extremal, without giving up optimality (minimum number of local extrema). If there are no nodes connected to only one other node, all nodes appear in cycles. Since the graph is bipartite, every cycle has even length. Take any cycle (of even length). Now all nodes in the cycle can be made non-extremal: we assign one quadrilateral (represented by the arc) to one incident extremum of S and choose the triangulating edge to make it non-extremal. We can repeat to treat nodes with only one incident arc, and even cycles, until no more extrema can be removed by triangulating edges. Then we complete the triangulation of S in any manner. This greedy, incremental method completes the subdivision S to a 1-OD triangulation that minimizes the number of local minima and maxima. The algorithm can be implemented to run in linear time when S is given.

Theorem 2. *An optimal 1-OD triangulation with respect to minimizing the number of local extrema can be determined in $O(n \log n)$ time.*

Other Criteria In visualization applications it is sometimes important to construct planar drawings with small degree and large angles between the edges. Thus, a natural optimization criteria for a 1-OD triangulation would be to minimize the maximum degree, since the Delaunay triangulation already maximizes the minimum angle. Besides visualization applications [8], constructing drawing with high angular resolution is important in the design of optical communication networks [10]. The problem of minimizing the maximum degree have been studied in several papers [12,14,15]. We know of no polynomial-time algorithm that gives an optimal solution to this optimization problem. The more general problem to complete the triangulation of a biconnected planar graph while minimizing the maximum degree is known to be NP-Hard. There are efficient approximation algorithms though [15]. Based on it, we can show:

Theorem 3. *There is an $O(n \log n)$ time algorithm to triangulate the 1-OD triangulation skeleton S such that the degree of the triangulation is at most $\lceil \frac{3}{2} \Delta(S) \rceil + 13$, where $\Delta(S)$ is the maximum degree of S.*

As was pointed out in the introduction criteria like minimizing the maximum angle and minimizing the maximum area triangle may be of use for finite element method applications. These criteria (together with a number of other criteria not mentioned above) are trivial to optimize if the edges in the domain are useful 1-OD edges.

Theorem 4. *For a Delaunay triangulation of a set P of n points in the plane, an optimal 1-OD triangulation can be obtained by flips of independent Delaunay edges in $O(n \log n)$ time for each one of the following criteria: (i) minimizing the maximal area triangle, (ii) minimizing the maximal angle, (iii) maximizing the minimum radius of a circumcircle, (iv) maximizing the minimum radius of an enclosing circle, (v) minimizing the sum of inscribed circle radii, (vi) minimizing the number of obtuse angles, and (vii) minimizing the total edge length.*

5.2   Applications for k-OD Triangulations

It appears to be difficult to obtain general, optimization results for all of the criteria listed before, given a value of $k \geq 2$. When k is so large that every pair of points gives a useful edge (like $k = n - 3$), then certain criteria can be optimized. For example, when minimizing the number of local minima, we can choose an edge from every point to the global minimum (in non-degenerate cases), so that there is only one local minimum. For minimizing the maximum angle and some other criteria, various results are known [3].

To develop approximation algorithms for k-OD triangulations, we need to determine how many hulls a single hull can intersect. To this end, we first prove an upper bound on the maximum number of useful k-OD edges that intersect a given Delaunay edge. Figure 6 shows that $\Omega(n)$ 2-OD edges can intersect a given Delaunay edge. But these 2-OD edges cannot all be useful. The next lemma shows that the maximum number of useful k-OD edges intersecting a given Delaunay edge doesn't depend on n, but only on k.

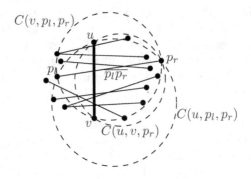

Fig. 6. Many 2-OD edges can intersect a Delaunay edge $\overline{uv}$.

Lemma 8. *There are $O(k^2)$ useful k-OD edges that intersect any Delaunay edge.*

Lemma 9. *Let $\overline{uv}$ be a useful k-OD edge and let H be its hull. The number of hulls of useful k-OD edges that intersect the interior of H is $O(k^3)$.*

The hull intersection graph G is defined as follows. There is a node for the hull of every useful k-OD edge. Two nodes are connected by an arc if their hulls intersect, that is, there exists a point that is interior to both hulls. The choice of one useful k-OD edge e possibly prohibits the choice of any other useful k-OD edge whose hull intersects the hull of e. In any case, if we choose an independent set of nodes in graph G, we get a set of hulls of useful k-OD edges that can be used together in a k-OD triangulation.

Suppose that all useful k-OD edges and their hulls have been computed, and the hull intersection graph G as well. Choose any useful k-OD edge e that removes a local minimum. Mark the node for the hull of e in G, and also the adjacent nodes. Choose the greedy triangulation of the hull of e. The choice of e prevents at most $O(k^3)$ other useful k-OD edges to be chosen. Therefore, at most $O(k^3)$ points can be prevented from not being a local minimum. Repeat to choose a useful k-OD edge that avoids another local minimum, provided its node in G is unmarked, until no such choice exists.

The same approach can be used for local extrema.

Theorem 5. *Let m be the smallest number of local minima (or extrema) in any k-OD triangulation of a set P of points. There is a polynomial time algorithm that computes a k-OD triangulation of P with at most $O(m \cdot k^3)$ local minima (or extrema).*

Next we consider minimizing the number of obtuse angles in k-OD triangulations in the case that k is constant. Consider the Delaunay triangulation. For any obtuse angle at a point p, it can be avoided if there is a useful k-OD edge that splits it into two non-obtuse angles. It can also be avoided if there are two useful k-OD edges that split it into three non-obtuse angles. We need not explicitly look at more than two edges, because there will always be two that already work. The completion to a k-OD triangulation may have more useful k-OD edges incident to p.

It is easy to test if a single useful k-OD edge can avoid an obtuse angle at a point p. To test whether two useful k-OD edges are needed, we can test any pair e, e' at point p. When we remove the Delaunay edges intersecting e or e', we get a region—the union of the hulls of e and e'—that must be triangulated to a k-OD triangulation. Since k is constant, we can try every possible triangulation of the region. This leads to:

Theorem 6. *Let k be a constant, and let m be the smallest number of obtuse angles in any k-OD triangulation of a set P of points. There is a polynomial time algorithm that computes a k-OD triangulation of P with at most $O(m \cdot k^3)$ obtuse angles.*

6   Conclusions and Directions for Further Research

This paper introduced a class of triangulations that generalizes the Delaunay triangulation: the empty-circle property for the triangles in the Delaunay triangulation is replaced by requiring that the circle of each triangle contains at most some given number k of points. Such a triangulation is called a k-OD triangulation. For any point set, there may be several different k-OD triangulations. Therefore, one can study the optimization of some geometric criterion over all possible k-OD triangulations of a given point set. Such optimizations have applications in terrain modelling for GIS and in mesh generation for finite element methods.

The class of 1-OD triangulations was studied in more detail, and because of its special properties, most of the criteria could be optimized efficiently. For minimizing the maximum degree we obtained an approximation result. In the general case we gave some initial results. Obviously, optimization or better approximation results are a topic of future research. Another issue we would like to address is experimental: how much benefit can one obtain for the criterion of optimization, with respect to the Delaunay triangulation?

References

1. P. K. Agarwal, M. de Berg, J. Matoušek, and O. Schwarzkopf. Constructing levels in arrangements and higher order Voronoi diagrams. *SIAM J. Comput.*, 27:654–667, 1998.   235, 237
2. A. Aggarwal, M. Hansen, and T. Leighton. Solving query-retrieval problems by compacting Voronoi diagrams. In *Proc. 22nd Annu. ACM Symp. Theory Comput.*, pages 331–340, 1990.   237
3. M. Bern. Triangulations. In Jacob E. Goodman and Joseph O'Rourke, editors, *Handbook of Discrete and Computational Geometry*, chapter 22, pages 413–428. CRC Press LLC, Boca Raton, FL, 1997.   233, 240

4. M. Bern, D. Dobkin, and D. Eppstein. Triangulating polygons without large angles. *Internat. J. Comput. Geom. Appl.*, 5:171–192, 1995. 233

5. M. Bern and D. Eppstein. Mesh generation and optimal triangulation. In D.-Z. Du and F. K. Hwang, editors, *Computing in Euclidean Geometry*, volume 4 of *Lecture Notes Series on Computing*, pages 47–123. World Scientific, Singapore, 2nd edition, 1995. 233

6. L. P. Chew. Constrained Delaunay triangulations. *Algorithmica*, 4:97–108, 1989. 233

7. M. de Berg, M. van Kreveld, M. Overmars, and O. Schwarzkopf. *Computational Geometry: Algorithms and Applications*. Springer-Verlag, Berlin, 1997. 232, 235

8. G. Di Battista, P. Eades, R. Tamassia, and I. G. Tollis. Algorithms for drawing graphs: an annotated bibliography. *Comput. Geom. Theory Appl.*, 4:235–282, 1994. 240

9. B. Falcidieno and C. Pienovi. Natural surface approximation by constrained stochastic interpolation. *Comput. Aided Design*, 22(3):167–172, 1990. 233

10. M. Formann, T. Hagerup, J. Haralambides, M. Kaufmann, F. T. Leighton, A. Simvonis, E. Welzl, and G. Woeginger. Drawing graphs in the plane with high resolution. *SIAM J. Comput.*, 22:1035–1052, 1993. 240

11. S. Fortune. Voronoi diagrams and Delaunay triangulations. In Jacob E. Goodman and Joseph O'Rourke, editors, *Handbook of Discrete and Computational Geometry*, chapter 20, pages 377–388. CRC Press LLC, Boca Raton, FL, 1997. 232

12. A. Garg and R. Tamassia. Planar drawings and angular resolution: Algorithms and bounds. In *Proc. 2nd Annu. European Symp. Algorithms*, volume 855 of *LNCS*, pages 12–23. Springer-Verlag, 1994. 240

13. M.F. Hutchinson. Calculation of hydrologically sound digital elevation models. In *Proc. 3th Int. Symp. on Spatial Data Handling*, pages 117–133, 1988. 233

14. K. Jansen. One strike against the min-max degree triangulation problem. *Comput. Geom. Theory Appl.*, 3:107–120, 1993. 240

15. G. Kant and H. L. Bodlaender. Triangulating planar graphs while minimizing the maximum degree. *Information and Computation*, 135:1–14, 1997. 240

16. D. T. Lee. On *k*-nearest neighbor Voronoi diagrams in the plane. *IEEE Trans. Comput.*, C-31:478–487, 1982. 235

17. J.J. Little and P. Shi. Structural lines, TINs, and DEMs. In T.K. Poiker and N. Chrisman, editors, *Proc. 8th Int. Symp. on Spatial Data Handling*, pages 627–636, 1998. 233

18. D.R. Maidment. GIS and hydrologic modeling. In M.F. Goodchild, B.O. Parks, and L.T. Steyaert, editors, *Environmental modeling with GIS*, pages 147–167. Oxford University Press, New York, 1993. 233

19. D.M. Mark. Automated detection of drainage networks from digital elevation models. *Cartographica*, 21:168–178, 1984. 233

20. A. Okabe, B. Boots, and K. Sugihara. *Spatial Tessellations: Concepts and Applications of Voronoi Diagrams*. John Wiley & Sons, Chichester, UK, 1992. 232

21. B. Schneider. Geomorphologically sound reconstruction of digital terrain surfaces from contours. In T.K. Poiker and N. Chrisman, editors, *Proc. 8th Int. Symp. on Spatial Data Handling*, pages 657–667, 1998. 233

22. D.M. Theobald and M.F. Goodchild. Artifacts of TIN-based surface flow modelling. In *Proc. GIS/LIS*, pages 955–964, 1990. 233

On Representations of Algebraic-Geometric Codes for List Decoding

Venkatesan Guruswami[1] and Madhu Sudan[2][*]

[1] Laboratory for Computer Science, Massachusetts Institute of Technology,
Cambridge, MA 02139. `venkat@theory.lcs.mit.edu`
[2] Laboratory for Computer Science, Massachusetts Institute of Technology,
Cambridge, MA 02139. `madhu@mit.edu`

Abstract. We show that all algebraic-geometric codes possess a succinct representation that allows for the list decoding algorithms of [15,7] to run in polynomial time. We do this by presenting a root-finding algorithm for univariate polynomials over function fields when their coefficients lie in finite-dimensional linear spaces, and proving that there is a polynomial size representation given which the root finding algorithm runs in polynomial time.

1 Introduction

The *list decoding* problem for an error-correcting code is that of reconstructing a list of all codewords within a specified Hamming distance from a received word. List decoding, which was introduced independently by Elias and Wozencraft [4,18], offers a potential for recovery from errors beyond the traditional "error-correction radius" (i.e., half the minimum distance) of a code.

There has recently been a spurt of activity in the design of *efficient* list decoding algorithms for recovery from very high noise for several families of algebraic error-correcting codes [17,15,7,8]. In particular, Shokrollahi and Wasserman [15] gave a list decoding algorithm for algebraic-geometric codes. Subsequently the authors [7] presented improvements to their algorithm, and increased further the bound on the number of errors that could be efficiently recovered from. Neither result is proven, however, to work unconditionally in polynomial time for all algebraic-geometric codes. Rather, the results are conditioned upon the existence of an efficient root-finding algorithm over the underlying algebraic function field. In turn, efficiency of known root-finding algorithms are dependent upon the size of the representation of certain basis elements of the underlying function field. Finally, no sufficiently succinct representations were known for the basis elements in the standard representation.

In this note, we examine and resolve this issue. We show that there exists a (non-standard) representation of the "basis" elements of a function field that is succinct and at the same time allows for efficient root-finding. As a consequence

---

[*] Supported in part by an MIT-NEC Research Initiation Award, a Sloan Foundation Fellowship and NSF Career Award CCR-9875511.

M. Paterson (Ed.): ESA 2000, LNCS 1879, pp. 244–255, 2000.

we show that every algebraic-geometric code has a succinct (polynomial-sized) representation such that the algorithms of [15,7] run in polynomial time.

Our result shares an interesting role in the broader issue of "representation of codes" versus "complexity of decoding algorithms". It is well-known that the representation of an error-correcting code can have crucial impact on its decoding complexity. Some representations of a code allow for efficient decoding, while other representations of the same code may not yield an equally efficient algorithm for decoding. In fact this non-trivial feature laid the foundation for a public-key cryptosystem proposed by McEliece [11]. On the other hand some error-correcting codes are inherently hard to decode. This was established by Bruck and Naor [3] who proved that there are codes which are hard to decode even with unlimited preprocessing time and hence independent of representation. Our result shows that algebraic-geometric codes (henceforth AG-codes) do not belong to this class, i.e., they admit a polynomial size representation given which they can be list decoded in polynomial time up to a large radius. This representation may itself be hard to find; we do not address the complexity of finding it.

Our result may also be viewed as a parallel to results of Pellikaan [12] and Kotter [9] who proved, using the elegant idea of *error-correcting pairs*, that a succinct representation of AG-codes exists for the purpose of unambiguous decoding (up to essentially half the minimum distance). In this work we show that a similar result holds for list decoding as well.

We now give further details of the exact question we address. Existing list decoding algorithms for AG-codes [15,7] run in polynomial time by assuming that operations over the underlying function field can be done efficiently, and in this work we would like to prove that the algorithm runs in polynomial time given a certain polynomial size representation of the code, without *any further assumptions*. Current list decoding algorithms for AG-codes (see [15,7]) involve a two-step process. The first step is an "interpolation" step where one finds a (univariate) polynomial over the function field K that "fits" the received word. In the second step all roots of this polynomial in a B certain linear subspace of K are determined using a root-finding algorithm and these include all the candidate codewords within the list decoding radius.

The first of these steps works by exploiting linearity and solves the interpolation problem by finding a non-zero solution to a homogeneous linear system (the system is guaranteed to have a non-zero solution by a simple rank argument). The second step is more tricky, and Gao and Shokrollahi [5] recently gave a (randomized) polynomial time algorithm for root finding of univariate polynomials over function fields of nonsingular plane curves. Their algorithm would actually work over any function field modulo a specific representation of the function, in time polynomial in the size of their representation. However, the representation is not known to be succinct for all AG-codes. For a specific class of AG-codes, namely Hermitian Codes, [13] shows how to implement both steps, and in particular the root-finding step, efficiently. Wu and Siegel [19] present a different root-finding algorithm but which still assumes efficient operations over the underlying function field. Root-finding algorithms in [1] also assume efficient operations over the function field, and do not address if this can *always* be achieved through a succinct representation of the function field. In this work, we present a root-finding algorithm for general function fields which in fact uses the same approach as the one in [5]; our main contribution is a proof that the representation of the function field required by this algorithm is in fact terse, i.e., has size bounded by a polynomial in the relevant parameters. This in turn implies that there is a polynomial size representation of AG-codes given which they can be efficiently list decoded up to the radius bound achieved by [7].

Organization: We describe the basic notions about algebraic function fields that we will need in Section 2. In Section 3, we give a high level "algebraic" description of the list decoding algorithm of [7], and the root-finding algorithm that we will use, and then address the representation issues that these algorithms motivate. Section 4 describes the root finding algorithm formally for the representation we choose in Section 3. Finally, in Section 5, we give a formal description of the list decoding algorithm of [7] and state our main theorem that AG-codes have a succinct representation given which they can be efficiently list decoded.

2   Algebraic-Geometric Codes: Preliminaries

We now abstract the main notions associated with the theory of algebraic function fields that will be important to us. The interested reader may find further details in [16,6]. In what follows we assume familiarity with the basic notions of field extensions.

An extension field K of a field k, denoted K/k, is an *algebraic function field* over k if the following conditions are satisfied: (i) There is an element $x \in K$ that is transcendental over k such that K is a finite extension of $k(x)$, and (ii) k is algebraically closed in K (i.e., the only elements in K that are algebraic over k are those in k).

For our applications to AG-codes, we will be interested in the case when k is a finite field, i.e., $k = \mathbb{F}_q$ for some prime power q. Thus $K = \mathbb{F}_q(X)[y_1, y_2, \ldots, y_m]$ where each y_i satisfies some polynomial equation over $\mathbb{F}_q(X)[y_1, \ldots, y_{i-1}]$. For the rest of this section, K will denote the function field in question.

Places, Valuations, and Degrees: A function field $K/\mathbb{F}_q$ has a set of *places* $\mathbb{P}_K$ and the associated set of *valuations*, given by a valuation map $v : \mathbb{P}_K \times K \to \mathbb{Z} \cup \{\infty\}$. The exact definition of these notions can be found, for instance, in [16]; we only abstract some properties relevant to us below.

Intuitively the places correspond to "points" on the algebraic curve associated with the function field K, and the valuation map tells us how many poles or zeroes a function in K has at a specific place in $\mathbb{P}_K$. It has the property that for any $f \in K$, there are only finitely many places $P \in \mathbb{P}_K$ such that $v(P, f) \neq 0$. As is normal practice, for each $P \in \mathbb{P}_K$, we denote by $v_P : K \to \mathbb{Z} \cup \{\infty\}$, the map $v_P(\cdot) = v(P, \cdot)$ which tells how many zeroes or poles a given function has at P (with the convention $v_P(0) = \infty$). The valuation v_P at any place satisfies the following properties:

(a) $v_P(a) = \infty$ iff $a = 0$ and $v_P(a) = 0$ for all $a \in \mathbb{F}_q \setminus \{0\}$.
(b) $v_P(ab) = v_P(a) + v_P(b)$ for all $a, b \in K \setminus \{0\}$.
(c) $v_P(a + b) \geq \min\{v_P(a), v_P(b)\}$.

Associated with every place is a *degree* abstracted via the map $\deg : \mathbb{P}_K \to \mathbb{Z}^+$. The degree, $\deg(P)$, of any place P is a positive integer and intuitively means the following: when we pick a function $f \in K$ which has no poles at P and "evaluate" it at P, we get a value in the field $\mathbb{F}_{q^{\deg(P)}}$. Thus places of degree one correspond to *rational points* on the curve.

Evaluations of functions at places: We can abstract the notion of *evaluation* of elements of the function field at the places by a map $\mathsf{eval} : K \times \mathbb{P}_K \to \bar{\mathbb{F}}_q \cup \{\infty\}$ (here $\bar{\mathbb{F}}_q$ is the algebraic closure of $\mathbb{F}_q$). This map has the following properties:

(i) $\mathsf{eval}(f, P) = \infty$ iff $v_P(f) < 0$, and $\mathsf{eval}(f, P) = 0$ iff $v_P(f) > 0$ for every $P \in \mathbb{P}_K$ and $f \in K$.
(ii) If $f \in K$, $P \in \mathbb{P}_K$ and $v_P(f) \geq 0$, then $\mathsf{eval}(f, P) \in \mathbb{F}_{q^{\deg(P)}}$.
(iii) The map eval respects field operations; i.e., if $v_P(f_1) \geq 0$ and $v_P(f_2) \geq 0$, then $\mathsf{eval}(f_1 + f_2, P) = \mathsf{eval}(f_1, P) + \mathsf{eval}(f_2, P)$, and $\mathsf{eval}(f_1 * f_2, P) = \mathsf{eval}(f_1, P) * \mathsf{eval}(f_2, P)$.

Divisors: The *divisor group* $\mathcal{D}_K$ of the function field K is a free abelian group on $\mathbb{P}_K$. An element D of $\mathcal{D}_K$ is thus represented by the formal sum $\sum_{P \in \mathbb{P}_K} a_P P$ where $a_P = 0$ for all but finitely many P; we say $D \succeq 0$ if $a_P \geq 0$ for all $P \in \mathbb{P}_K$. The support of a divisor D, denoted $\mathrm{supp}(D)$, is the (finite) set $\{P \in \mathbb{P}_K : a_P \neq 0\}$. The degree map extends naturally to a homomorphism $\deg : \mathcal{D}_K \to \mathbb{Z}$ as $\deg(\sum_P a_P P) = \sum_P a_P \deg(P)$.

For every $f \in K \setminus \{0\}$, there is an associated divisor, called the principal divisor and denoted (f), which is defined by $(f) = \sum_P v_P(f) P$. The following result is well known, and just states that every non-zero function in the function field has an equal number of zeroes and poles.

Theorem 1. *For any function field $K/\mathbb{F}_q$ and any $f \in K \setminus \{0\}$, $\deg((f)) = 0$.*

For every divisor $D \in \mathcal{D}_K$, one can define the linear space of functions $\mathcal{L}(D)$ as

$$\mathcal{L}(D) = \{f \in K : (f) + D \succeq 0\}.$$

For example for a divisor $D = aQ - bP$ where $P, Q \in \mathbb{P}_K$ and $a, b > 0$, $\mathcal{L}(D)$ is the space of all functions that have at least b zeroes at P and at most a poles at Q. It is known that for any divisor $D \succeq 0$, $\mathcal{L}(D)$ is a finite-dimensional vector space over $\mathbb{F}_q$ and $\dim(\mathcal{L}(D)) \leq 1 + \deg(D)$ (see [16] for a proof). A lower bound on $\dim(\mathcal{L}(D))$ is given by the celebrated Riemann-Roch theorem for function fields:

Theorem 2. [Riemann-Roch]: *Let $K/\mathbb{F}_q$ be any function field. There is a non-negative integer g, called the* genus *of $K/\mathbb{F}_q$, such that*

(a) For any divisor $D \in \mathcal{D}_K$, $\dim(\mathcal{L}(D)) \geq \deg(D) - g + 1$.

(b) There is an integer c, depending only on $K/\mathbb{F}_q$, such that $\dim(\mathcal{L}(D)) = \deg(D) - g + 1$ whenever $\deg(D) \geq c$. (In fact $c \leq 2g - 1$.)

Algebraic-geometric codes: We are now ready to define the notion of an AG-code (also known as *geometric Goppa code*). Let $K/\mathbb{F}_q$ be an algebraic function field of genus g, let $Q, P_1, P_2, \ldots, P_n$ be *distinct* places of degree one in $\mathbb{P}_K$, and let $G = P_1 + P_2 + \cdots + P_n$ and $D = \alpha Q$ be divisors of $K/\mathbb{F}_q$ (note that $\operatorname{supp}(G) \cap \operatorname{supp}(D) = \emptyset$).

The algebraic-geometric code $\mathcal{C}_\mathcal{L}(G, D) = \mathcal{C}_\mathcal{L}(G, \alpha, Q)$ is defined by[1]

$$\mathcal{C}_\mathcal{L}(G, \alpha, Q) := \{(\operatorname{eval}(f, P_1), \ldots, \operatorname{eval}(f, P_n)) : f \in \mathcal{L}(\alpha Q)\} \subseteq \mathbb{F}_q^n .$$

(Note that $\operatorname{eval}(f, P_i) \in \mathbb{F}_q$ since $v_{P_i}(f) \geq 0$ and $\deg(P_i) = 1$.) The following Proposition follows from the Riemann-Roch theorem and quantifies the parameters of these codes.

Proposition 3 *Suppose that $\alpha < n$. Then $\mathcal{C}_\mathcal{L}(G, \alpha, Q)$ is an $[n, k, d]_q$ code with $k = \dim(\mathcal{L}(\alpha Q)) \geq \alpha - g + 1$ and $d \geq n - \alpha$ (thus $k + d \geq n + 1 - g$).[2]*

3   Representation Issues

We now give a high level description of the list decoding algorithm of [7] and the root-finding algorithm over function fields that it uses. These will motivate the issues relating to representation of the function field that we address in this section.

The list decoding algorithm for an AG-code $\mathcal{C}_\mathcal{L}(G, \alpha, Q)$ over a function field $K/\mathbb{F}_q$ with $G = P_1 + P_2 + \cdots + P_n$ where $n = \deg(G)$, on input a received word $\boldsymbol{y} = (y_1, \ldots, y_n) \in \mathbb{F}_q^n$, works, at a high level, as follows:

Procedure LIST-DECODE$_{\mathcal{C}_\mathcal{L}(G, \alpha, Q)}(\boldsymbol{y}, e)$

Input: $\boldsymbol{y} \in \mathbb{F}_q^n$; error bound e.

Output: All elements $h \in \mathcal{L}(\alpha Q)$ such that $\operatorname{eval}(h, P_i) \neq y_i$ for at most e values of $i \in \{1, \ldots, n\}$.

1. "Fit" the pairs (y_i, P_i) by a "suitable" polynomial $H \in K[T]$. Specifically pick parameters ℓ, r, s appropriately (the exact choice will be specified in the full description of Section 5 and is not necessary for understanding the algorithm) and find a *non-zero* $H = \sum_{j=0}^s a_j T^j$ with $a_j \in \mathcal{L}((\ell - j\alpha)Q)$, $0 \leq j \leq s$, such that for $1 \leq i \leq n$, $H(h) \in \mathcal{L}(-rP_i)$ whenever $\operatorname{eval}(h, P_i) = y_i$.

---

[1] It is clear that the defined space is a linear space.

[2] The notation $[n, k, d]_q$ code stands, as usual, for a code over $\mathbb{F}_q$ of blocklength n, rate k and minimum distance d.

2. Find all roots $h \in \mathcal{L}(\alpha Q)$ of H and for each of them check if $\mathsf{eval}(h, P_i) \neq y_i$ for at most e values of i, and if so, output h.

Rationale behind the approach: Note that if $\mathsf{eval}(h, P_i) = y_i$ for at least $n - e$ values of i, then $H(h)$ has at least $r(n - e)$ zeroes (among the points $P_1, P_2, \ldots, P_n$ counting multiplicities). Also, by the choice of H, if $h \in \mathcal{L}(\alpha Q)$, we have $H(h) \in \mathcal{L}(\ell Q)$, and thus $H(h)$ has at most ℓ poles. Thus if $r(n - e) > \ell$, then $H(h) = 0$, or in other words h is a root of $H \in K[T]$. The parameters r, ℓ are chosen so that the algorithm is able to correct a "large" number of errors e.

The root-finding step referred to above can be implemented by the following algebraic procedure which finds all roots (in K) of a polynomial over K whose coefficients are restricted to lie in a linear space $\mathcal{L}(D)$ for some divisor $D \in \mathcal{D}_K$.

Procedure ROOT-FIND$_{(K,D)}(H)$

Input: A degree m polynomial $H = \sum_{i=0}^{m} a_i T^i \in K[T]$ where each $a_i \in \mathcal{L}(D)$.

Output: All roots of H that lie in K.

1. Find a divisor D' that depends on D such that all roots of H lie in $\mathcal{L}(D')$.
2. "Reduce" H modulo a place $R \in \mathbb{P}_K$ of large enough degree, say r (i.e., compute $b_i = \mathsf{eval}(a_i, R)$ for $0 \leq i \leq m$ and consider the polynomial $P = \sum_{i=0}^{m} b_i Y^i \in \mathbb{F}_{q^r}[Y]$).
3. Compute the roots, say $\alpha_1, \ldots, \alpha_t$, of P that lie in $\mathbb{F}_{q^r}$ using a root-finding algorithm for finite fields.
4. For each α_j, $1 \leq j \leq t$, "find" $\beta_i \in \mathcal{L}(D')$ such that $\mathsf{eval}(\beta_i, R) = \alpha_i$.

The above algebraic procedures raise several questions on how to represent the various objects associated with a function field, so as to be able to perform the associated computations efficiently. Since K is infinite, we will not try to represent *all* elements of K. Rather, we will always focus attention only on elements of K from the finite-dimensional space $\mathcal{L}(D)$. For any $D \succeq 0$, we can represent elements of $\mathcal{L}(D)$ as vectors in $\mathbb{F}_q^{\dim(\mathcal{L}(D))}$ which represent their coordinates with respect to some fixed basis of $\mathcal{L}(D)$ over $\mathbb{F}_q$. Since $\dim(\mathcal{L}(D)) \leq \deg(D) + 1$, this representation will be small provided $\deg(D)$ is not too large. Indeed, Step 1 of the decoding algorithm above is handled in [7] using such ideas.

The more tricky issue is in the root-finding step where we need to "evaluate" an $f \in \mathcal{L}(D)$ at some place $R \in \mathbb{P}_K$ of interest. To aid this we "represent" a place R by the values $\mathsf{eval}(\phi_i, R)$, for $1 \leq i \leq p$ where $p = \dim(\mathcal{L}(D))$ and $\phi_1, \ldots, \phi_p$ is a basis of $\mathcal{L}(D)$ over $\mathbb{F}_q$. Together with the representation of any element of $\mathcal{L}(D)$ as a linear combination of the ϕ_i's this clearly enables us to evaluate any element of $\mathcal{L}(D)$ at R. Since each $\mathsf{eval}(\phi_i, R) \in \mathbb{F}_{q^r}$ where $r = \deg(R)$, one can "represent" R, for purposes of evaluation by members of $\mathcal{L}(D)$, as an element of $\mathbb{F}_{q^r}^p$. (We assume that some standard representation of elements of $\mathbb{F}_{q^r}$.) It should be somewhat clear that given these representations, the algebraic procedures discussed at the beginning of this section can in fact be turned into efficient algorithms. The next couple of sections prove formally that this is indeed the case.

4   Efficient Root-Finding

In this section we consider the problem of finding the roots of a univariate polynomial $H \in K[T]$. We are interested in this task since it forms the critical component in list decoding algorithms for AG-codes. Considering the application to list decoding in mind, we only need to solve special instances of the univariate root-finding problem. Namely, we assume the input polynomial $H \in K[T]$ of degree m has all its coefficients in a linear subspace $\mathcal{L}(D)$ for some divisor $D \succeq 0$ of K, and the coefficients are input in the form discussed in the previous section. (For applications to AG-codes, the divisor D is actually a *one-point* divisor, i.e., is of the form ℓQ for some place Q of degree one, and moreover we will only be interested in roots that lie in $\mathcal{L}(\alpha Q)$ for some $\alpha \leq \ell$. We, however, present a root-finding algorithm that works for any divisor, as this could be of independent interest.) In addition to this "uniform" input H, the algorithm also uses a non-uniform input, namely a place R of large degree, which only depends on D and does not depend on the degree of the input polynomial.

Before we specify the algorithm formally, we recall the high level description outlined in Section 3. The root-finding algorithm, on input a polynomial $H = \sum_{i=0}^{m} a_i T^i \in K[T]$ where each $a_i \in \mathcal{L}(D)$, first reduces H modulo R by evaluating each a_i at R. This gives a polynomial P of degree m over $\mathbb{F}_{q^r}$ (r is the degree of R). The algorithm then finds the roots in $\mathbb{F}_{q^r}$ of P, and then "lifts" these roots to give corresponding elements in K which are roots of H.

Algorithm ROOT-FIND$_{(K,D)}(H)$

Non-uniform input: (This depends only on D and is independent of the actual input) A place $R \in \mathbb{P}_K$ such that $\deg(R) = r > \deg(D)|\mathrm{supp}(D)|$ (such a place necessarily exists; see Lemma 1). The place R is represented as an s-tuple $(\zeta_1^R, \ldots, \zeta_s^R)$ over $\mathbb{F}_{q^r}$ comprising of evaluations of the basis functions $\mathcal{B}' = \{\phi_1, \phi_2, \ldots, \phi_p, \phi_{p+1}, \ldots, \phi_s\}$ of $\mathcal{L}(D')$ at R. Here $D' = \sum_{P \in \mathrm{supp}(D)} \frac{\deg(D)}{\deg(P)} P$, $s = \dim(\mathcal{L}(D'))$, and the basis $\mathcal{B}'$ *extends* a basis $\mathcal{B} = \{\phi_1, \ldots, \phi_p\}$ of $\mathcal{L}(D)$.

Input: A polynomial $H = a_0 + a_1 T + \cdots + a_m T^m$ of degree m in $K[T]$ where each $a_i \in \mathcal{L}(D)$ for some divisor D of K. Each a_i is presented as a p-tuple $(a_{i1}, \ldots, a_{ip})$ over $\mathbb{F}_q$ where $p = \dim(\mathcal{L}(D))$, $a_{ij} \in \mathbb{F}_q$ (this is to interpreted as $a_i = \sum_{j=1}^{p} a_{ij}\phi_j$ where $\mathcal{B} = \{\phi_1, \phi_2, \ldots, \phi_p\}$ is a basis of $\mathcal{L}(D)$ over $\mathbb{F}_q$).

Output: All the roots of H that lie in K. (The roots can have poles only at places in $\mathrm{supp}(D)$ and moreover cannot have too many poles at these places; in fact they all lie in $\mathcal{L}(D')$; see Lemma 2.)

Step 1: Reduce H modulo R to obtain a polynomial $h \in \mathbb{F}_{q^r}[T]$: i.e., for $0 \leq i \leq m$, compute

$$b_i = \mathsf{eval}(a_i, R) = \sum_{j=1}^{p} a_{ij}\zeta_j^R \in \mathbb{F}_{q^r},$$

and set $h[T] = b_0 + b_1 T + \cdots + b_m T^m$.

Step 2: Find all the (distinct) roots $\alpha_1, \alpha_2, \ldots, \alpha_t$ of h that lie in $\mathbb{F}_{q^r}$ using a standard root finding algorithm for finite fields. (This can be accomplished in deterministic poly(q, r) time by an algorithm due to Berlekamp [2].)

Step 3 (Recovering the original roots): For each $\alpha_i \in \mathbb{F}_{q^r}$ such that $h(\alpha_i) = 0$ "find" the *unique* $\beta_i \in \mathcal{L}(D')$ such that $\mathsf{eval}(\beta_i, R) = \alpha_i$, in terms of its coefficients $c_{ij} \in \mathbb{F}_q$ with respect to the basis $\mathcal{B}'$ of $\mathcal{L}(D')$. (Recall that $D' = \sum_{P \in \mathrm{supp}(D)} \frac{\deg(D)}{\deg(P)} P$ and thus $\deg(D') = \deg(D)|\mathrm{supp}(D)|$, so Lemma 3 proves that such a β_i, if any, is unique.) For each i, the c_{ij}'s can be found by solving $\sum_{j=1}^{s} c_{ij} \zeta_j^R = \alpha_i$ which can be viewed as a linear system of equations over $\mathbb{F}_q$ (by fixing some representation of elements of $\mathbb{F}_{q^r}$ over $\mathbb{F}_q$).

Step 4: Output the list of roots $\{\beta_1, \ldots, \beta_t\}$.

It is clear, given the Lemmas referred to in the algorithm, that the algorithm correctly finds all roots of H in K. Moreover, it clearly runs in polynomial time given the non-uniform input. We thus get our main theorem of this section:

Theorem 4. *There is an efficient root-finding algorithm that, for any function field K and any divisor $D \succeq 0$, given an "advice" that depends only on D and is of size polynomial in $\deg(D)$, finds, in $\mathrm{poly}(m, \deg(D))$ time, the roots of any degree m polynomial in $K[T]$ whose coefficients all lie in $\mathcal{L}(D)$.*

Lemma 1. *For any function field K, there exists a place of degree m in $\mathbb{P}_K$ for every large enough integer m.*

Proof: By the Hasse-Weil inequality, the number N_m of places of degree m in $\mathbb{P}_K$ satisfies $|N_m - q^m - 1| \le 2g q^{m/2}$ where g is the genus of the function field K. Hence if $m \ge m_0$ where m_0 is the smallest integer that satisfies $\frac{q^{m_0} - 1}{2q^{m_0/2}} > g$, then $N_m \ge 1$. $\qquad\square$

Lemma 2. *If all coefficients of a non-zero polynomial $H \in K[T]$ lie in $\mathcal{L}(D)$ for some divisor $D \succeq 0$, then for all roots $\alpha \in K$ of H, $\alpha \in \mathcal{L}(D')$ where $D' = \sum_{P \in \mathrm{supp}(D)} \frac{\deg(D)}{\deg(P)} P$. (Note that $\deg(D') = \deg(D)|\mathrm{supp}(D)|$.)*

Proof: We first prove that if α has a pole at some place $P \in \mathbb{P}_K$, then $P \in \mathrm{supp}(D)$. Indeed, suppose $H(\alpha) = 0$ and α has a pole at some $P \in \mathbb{P}_K \setminus \mathrm{supp}(D)$. Since none of the coefficients of H have a pole at P, it follows that $H(\alpha)$ has a pole at P (in fact $v_P(H(\alpha)) \le -m$ where m is the degree of H), and hence certainly $H(\alpha) \ne 0$, a contradiction.

We next prove that for $P \in \mathrm{supp}(D)$, $v_P(\alpha) \ge -\frac{\deg(D)}{\deg(P)}$. Indeed, let $H[T] = a_m T^m + \ldots + a_1 T + a_0$ where each $a_j \in \mathcal{L}(D)$, and let $D = \sum_{R \in \mathrm{supp}(D)} e_R R$ where each $e_R > 0$. Clearly $v_P(a_j \alpha^j) \ge -e_P + j v_P(\alpha)$ since $v_P(a_j) \ge -e_P$. If $v_P(\alpha) \ge 0$ we are done, so assume $v_P(\alpha) < 0$. Hence

$$v_P(H(\alpha) - a_m \alpha^m) = v_P\left(\sum_{j=0}^{m-1} a_j \alpha^j\right) \ge -e_P + (m-1) v_P(\alpha) . \qquad (1)$$

We now upper bound $v_P(a_m)$. Since $\deg((a_m)) = 0$ and $a_m \in \mathcal{L}(D)$, we have

$$\sum_{R \in \mathrm{supp}(D)} v_R(a_m) \deg(R) \leq 0,$$

and this together with $v_R(a_m) \geq -e_R$ for every R gives

$$v_P(a_m) \deg(P) \leq \sum_{R \in \mathrm{supp}(D) \backslash P} e_R \deg(R) = \deg(D) - e_P \deg(P) . \qquad (2)$$

Thus we have

$$v_P(a_m \alpha^m) \leq \frac{\deg(D)}{\deg(P)} - e_P + m v_P(\alpha) . \qquad (3)$$

Since $H(\alpha) = 0$, we must have $v_P(a_m \alpha^m) = v_P(H(\alpha) - a_m \alpha^m)$. Using Equations (1) and (3) this gives $-e_P + (m-1) v_P(\alpha) \leq \frac{\deg(D)}{\deg(P)} - e_P + m v_P(\alpha)$ which gives $v_P(\alpha) \geq -\frac{\deg(D)}{\deg(P)}$, as desired. $\qquad \square$

Lemma 3. *If $f_1, f_2 \in \mathcal{L}(A)$ for some divisor $A \succeq 0$ and $\mathsf{eval}(f_1, R) = \mathsf{eval}(f_2, R)$ for some place R with $\deg(R) > \deg(A)$, then $f_1 = f_2$.*

Proof: Suppose not, so that $f_1 - f_2 \neq 0$. Then, by Theorem 1, $\deg((f_1 - f_2)) = 0$. But $f_1 - f_2 \in \mathcal{L}(A)$ and $v_R(f_1 - f_2) \geq 1$, so that $\deg((f_1 - f_2)) \geq \deg(R) - \deg(A) > 0$, a contradiction. Hence $f_1 = f_2$. $\qquad \square$

5   An Explicit List Decoding Algorithm

We now present an algorithm for list decoding AG-codes that runs in polynomial time given a (non-standard) representation of the code. The underlying algorithm is the same as the one in [7]; we refer the reader to [7] for details on the development of the algorithm, the specific choice of parameters and the correctness of the algorithm. We reproduce the algorithm here mainly to bring out the fact that the algorithm runs in polynomial time given the representation it assumes. The reader unfamiliar with [7] may find it useful to ignore the specific choice of parameters and just observe the parallel to the skeleton algorithm mentioned in Section 3.

Let $\mathcal{C} = \mathcal{C}_{\mathcal{L}}(G, \alpha, Q)$ be an AG-code where $G = P_1 + P_2 + \cdots + P_n$ and $Q \notin \mathrm{supp}(G)$. Let $\alpha < n$ and let g be the genus of K. Assume $\alpha > 2g - 2$ so that by the Riemann-Roch theorem $\dim(\mathcal{L}(\alpha Q)) = \alpha - g + 1$. Thus the rate k of $\mathcal{C}$ is $\alpha - g + 1$ and its designed distance is $d = n - \alpha$.

Algorithm $LIST\text{-}DECODE_{\mathcal{C}_{\mathcal{L}}(G, \alpha, Q)}(\boldsymbol{y}, e)$

<u>Parameters:</u> The algorithm computes and uses parameters r, ℓ (based on n, e, g).

<u>Representation of $\mathcal{C}_{\mathcal{L}}(G, \alpha, Q)$:</u> Fix a basis $\{\phi_{j_1} : 1 \leq j_1 \leq \ell - g + 1\}$ of $\mathcal{L}(\ell Q)$ such that $v_Q(\phi_{j_1}) \geq -(j_1 + g - 1)$ (i.e., ϕ_{j_1} has at most $j_1 + g - 1$ poles at Q). As shown in [7], for each i, $1 \leq i \leq n$, there exists a basis $\{\psi_{j_3, P_i} : 1 \leq j_3 \leq \ell - g + 1\}$ of $\mathcal{L}(\ell Q)$ such that $v_{P_i}(\psi_{j_3, P_i}) \geq j_3 - 1$. The explicit information which the decoding algorithm needs are the following:

(a) The values $\mathrm{eval}(\phi_{j_1}, P_i) \in \mathbb{F}_q$ for $1 \leq i \leq n$ and $1 \leq j_1 \leq \ell - g + 1$.
(b) The set $\{\alpha_{P_i, j_1, j_3} \in \mathbb{F}_q : 1 \leq i \leq n, 1 \leq j_1, j_3 \leq \ell - g + 1\}$ such that for every i and every j_1, we have $\phi_{j_1} = \sum_{j_3} \alpha_{P_i, j_1, j_3} \psi_{j_3, P_i}$ (as elements in K).
(c) A place $R \in \mathbb{P}_K$ with $\deg(R) = s > \ell$ represented through the evaluations of ϕ_{j_1} for $1 \leq j_1 \leq \ell - g + 1$ at R (these lie in $\mathbb{F}_{q^s}$).

Input: $\boldsymbol{y} = (y_1, y_2, \ldots, y_n) \in \mathbb{F}_q^n$; error bound e.
Output: A list of all codewords C in $\mathcal{C}_{\mathcal{L}}(G, \alpha, Q)$ such that $C_i \neq y_i$ for at most e values of i.

Step 0: Computer parameters r, ℓ such that

$$r(n - e) > \ell \text{ and } \frac{(\ell - g)(\ell - g + 1)}{2\alpha} > n \binom{r + 1}{2}.$$

In particular set

$$r \stackrel{\text{def}}{=} 1 + \left\lfloor \frac{(2g + \alpha)n - 2ge + \sqrt{((2g + \alpha)n - 2ge)^2 - 4(g^2 - 1)((n - e)^2 - \alpha n)}}{2((n - e)^2 - \alpha n)} \right\rfloor,$$

$$\ell \stackrel{\text{def}}{=} r(n - e) - 1$$

Also set $s = \frac{\ell - g}{\alpha}$.

Step 1: Find $H \in \mathcal{L}(\ell Q)[T]$ of the form $H[T] = \sum_{j_2 = 0}^{s} \sum_{j_1 = 1}^{\ell - g + 1 - \alpha j_2} h_{j_1 j_2} \phi_{j_1} T^{j_2}$, i.e., find values of the coefficients $\{h_{j_1, j_2} \in \mathbb{F}_q\}$ such that the following conditions hold:

1. At least one h_{j_1, j_2} is non-zero.
2. For every $i \in [n]$, $\forall j_3, j_4, j_3 \geq 1, j_4 \geq 0$ such that $j_3 + j_4 \leq r$,

$$h_{j_3, j_4}^{(i)} \stackrel{\text{def}}{=} \sum_{j_2 = j_4}^{s} \sum_{j_1 = 1}^{\ell - g + 1 - \alpha j_2} \binom{j_2}{j_4} y_i^{j_2 - j_4} \cdot h_{j_1, j_2} \alpha_{x_i, j_1, j_3} = 0.$$

/* This ensures that $\mathrm{eval}(h, P_i) = y_i$ implies $H(h) \in \mathcal{L}(-rP_i)$ for each i. */

Step 2: Using the root-finding algorithm described in Section 4 together with the place R which is supplied to the algorithm, "find" all roots $h \in \mathcal{L}(\alpha Q) \subseteq \mathcal{L}(\ell Q)$ of the polynomial $H \in K[T]$. For each such h, check if $\mathrm{eval}(h, P_i) = y_i$ for at least $(n - e)$ values of i, and if so, include h in output list. (Since h is "found" by finding its coefficients with respect to the basis functions ϕ_{j_1} and the algorithm is given the values $\mathrm{eval}(\phi_{j_1}, P_i)$ for $1 \leq i \leq n$ and $1 \leq j_1 \leq \ell - g + 1$, $\mathrm{eval}(h, P_i)$ can be computed efficiently.)

Step 3: Output the list of "codewords" h found in Step 2.

Since Step 1 just involves solving a homogeneous linear system of equations and Step 2 involves root-finding for which we gave an efficient algorithm in Section 4, it is clear that the above algorithm runs in polynomial time given the representation of the code it takes as input. We can thus state our main result:

Theorem 5 (Main theorem). *For every AG-code $C_L(G, \alpha, Q)$, there is a representation of the code of size polynomial in n (here $n = \deg(G)$), given which one can list decode from up to $n - \sqrt{n(n-d)}$ errors in polynomial time where $d = n - \alpha$ is the designed distance of the code.*

6   Conclusions

We have shown that AG-codes admit a representation given which the list decoding algorithm of [7] runs in polynomial time. It would be interesting to examine whether, for specific AG-codes which beat the Gilbert-Varshamov bound, say the codes based on the Garcia-Stichtenoth tower of function fields [6], this representation can also be *found* in polynomial time.

References

1. D. AUGOT AND L. PECQUET. A Hensel lifting to replace factorization in list decoding of algebraic-geometric and Reed-Solomon codes. Manuscript, April 2000. 246

2. E. R. BERLEKAMP. Factoring polynomials over large finite fields. *Mathematics of Computations*, 24 (1970), pp. 713-735. 251

3. J. BRUCK AND M. NAOR. The hardness of decoding linear codes with preprocessing. *IEEE Trans. on Information Theory*, Vol. 36, No. 2, March 1990. 245

4. P. ELIAS. List decoding for noisy channels. *Wescon Convention Record*, Part 2, Institute of Radio Engineers (now IEEE), pp. 94-104, 1957. 244

5. S. GAO AND M. A. SHOKROLLAHI. Computing roots of polynomials over function fields of curves. In *Proceedings of the Annapolis Conference on Number Theory, Coding Theory, and Cryptography*, 1999. 245, 246

6. A. GARCIA AND H. STICHTENOTH. Algebraic function fields over finite fields with many rational places. *IEEE Trans. on Info. Theory*, 41 (1995), pp. 1548-1563. 246, 254

7. V. GURUSWAMI AND M. SUDAN. Improved decoding of Reed-Solomon and Algebraic-geometric codes. *IEEE Trans. on Information Theory*, 45 (1999), pp. 1757-1767. Preliminary version appeared in *Proc. of FOCS'98*. 244, 245, 246, 248, 249, 252, 254

8. V. GURUSWAMI AND M. SUDAN. List decoding algorithms for certain concatenated codes. *Proc. of STOC 2000*, to appear. 244

9. R. KOTTER. A unified description of an error locating procedure for linear codes. *Proc. of Algebraic and Combinatorial Coding Theory*, 1992. 245

10. R. MATSUMOTO. On the second step in the Guruswami-Sudan list decoding algorithm for AG-codes. Technical Report of IEICE, pp. 65-70, 1999.

11. R. J. McELIECE. A public-key cryptosystem based on algebraic coding theory. DSN Progress Report 42-44, Jet Propulsion Laboratory. 245

12. R. PELLIKAAN. On decoding linear codes by error correcting pairs. *Eindhoven Institute of Technology*, preprint, 1988. 245

13. R. R. NIELSEN AND T. HOHOLDT. Decoding Hermitian codes with Sudan's algorithm. In *Proceedings of AAECC-13, LNCS 1719, Springer-Verlag*, 1999, pp. 260-270. 245

14. R. R. NIELSEN AND T. HOHOLDT. Decoding Reed-Solomon codes beyond half the minimum distance. In *Coding Theory, Cryptograhy and Related areas*, (eds. Buchmann, Hoeholdt, Stichtenoth and H. tapia-Recillas) pp. 221-236, Springer 1999.

15. M. A. SHOKROLLAHI AND H. WASSERMAN. List decoding of algebraic-geometric codes. *IEEE Trans. on Information Theory*, Vol. 45, No. 2, March 1999, pp. 432-437. 244, 245

16. H. STICHTENOTH. *Algebraic Function Fields and Codes*. Springer-Verlag, Berlin, 1993. 246, 247

17. M. SUDAN. Decoding of Reed-Solomon codes beyond the error-correction bound. *Journal of Complexity*, 13(1):180-193, March 1997. 244

18. J. M. WOZENCRAFT. List Decoding. *Quarterly Progress Report*, Research Laboratory of Electronics, MIT, Vol. 48 (1958), pp. 90-95. 244

19. XIN-WEN WU AND P. H. SIEGEL. Efficient list decoding of algebraic geometric codes beyond the error correction bound. In *Proc. of International Symposium on Information Theory*, June 2000, to appear. 245

Minimizing a Convex Cost Closure Set

Dorit S. Hochbaum[*1] and Maurice Queyranne[2]

[1] Department of Industrial Engineering and Operations Research
and Walter A. Haas School of Business,
University of California, Berkeley
hochbaum@ieor.berkeley.edu
[2] University of British Columbia
Vancouver, British Columbia V6T1W5, Canada

Abstract. We study here a convex optimization problem with variables subject to a given partial order. This problem is a generalization of the known maximum (or minimum) closure problem and the isotonic regression problem. The algorithm we devise has complexity $O(mn \log \frac{n^2}{m} + n \log U)$, for U the largest interval of values associated with a variable. For the quadratic problem and for the closure problem the complexity of our algorithm is strongly polynomial, $O(mn \log \frac{n^2}{m})$. For the isotonic regression problem the complexity is $O(n \log U)$.

1 Introduction

An application that occurs frequently when observing objects and then attempting to fit the observations with preset ranking order requirement with minimum deviation is formalized as minimization over partial order constraints. The formal definition of the problem is given for a partial order graph $G = (V, A)$ and a convex function associated with each $j \in V$. The convex cost closure problem is then,

$$
\begin{array}{ll}
\text{(CCC)} \quad \text{Min} & \sum_{j \in V} f_j(x_j) \\
\text{subject to} & x_i - x_j \geq 0 \quad \forall (i, j) \in A \\
& \ell_j \leq x_j \leq u_j \text{ integer} \quad j \in V.
\end{array}
$$

This problem generalizes the isotonic regression problem in which the partial order is linear – the arcs of A are all of the form $(i, i+1)$. Another well known problem that (CCC) generalizes is the minimum closure problem:

$$
\begin{array}{ll}
\text{(Minimum} \quad \text{Min} & \sum_{j \in V} w_j \cdot x_j \\
\text{Closure)} \quad \text{subject to} & x_i - x_j \geq 0 \quad \forall (i, j) \in A \\
& 0 \leq x_j \leq 1 \text{ integer} \quad j \in V.
\end{array}
$$

Picard established in 1976 [Pic76] that the closure problem is equivalent to a minimum cut problem on an associated graph to which we add a source and a sink. This construction is described in Section 3.

---

* Research supported in part by NSF award No. DMI-9713482

M. Paterson (Ed.): ESA 2000, LNCS 1879, pp. 256–267, 2000.
© Springer-Verlag Berlin Heidelberg 2000

The challenge of convex optimization problem is that searching for a minimum of such function involves an unavoidable factor such as $\log U$ in the running time, for $U = \max_i\{u_i - \ell_i\}$. Although one can replace other parameters that depend on the variability of the functions, the running time cannot be made strongly polynomial using the arithmetic complexity model (see [Hoc94] for details on this result). The algorithm presented here differs from previous algorithms in that the search for the minima of the convex functions is disjoint from the rest of the algorithm and conducted as a post-processing step. The main body of the algorithm identifies disjoint intervals that are guaranteed to contain the optimal values of each variable and satisfy the partial order constraints. The run time of the algorithm is the fastest known for the problem, and it either matches or improves the complexity of algorithms for special cases of the problem.

We assume the unit cost complexity model: The structure of the convex functions is not restricted, so it is necessary to define the complexity model used. We assume the existence of an oracle returning function values for every argument input in $O(1)$. We will only be interested with arguments that are integers, or lie on a grid of ϵ granularity when we consider the continuous problem. Any arithmetic operation or comparison involving functions values is assumed to executed in unit time. We further assume that the sum of two convex functions can be determined in $O(1)$. Derivatives, or rather subgradients, are also required. For the integer problems considered we let

$$f_j'(x) = f_j(x+1) - f_j(x).$$

Note that the feasibility of the problem can be verified in linear time. The feasibility verification algorithm is given in the full version of this paper.

2   Related Literature and Applications

We details several classes of applications for the convex closure problem in the expanded version of this paper. These problems include the *selection of discrete contingent projects*; see Picard and Queyranne [PQ82] for a more detailed discussion. Another problem which is a special case of (CCC) was addressed by Maxwell and Muckstadt, [MM83]. They consider *nested power-of-two policies* in a *multi-stage production/inventory* problem. Statistical problems of *partially ordered estimation*, Veinott [Vei81], Barlow et al., [BBB72], can also be modelled as problem (CCC). The isotonic regression problem was initially considered in the context of linearly ordered estimation problems.

The convex closure problem has been previously addressed by Picard and Queyranne [PQ85]. They proposed an algorithm solving the problem with running time of $O(n(mn \log \frac{n^2}{m} + n \log U))$. Ahuja, Hochbaum and Orlin [AHO99], addressed a generalization of (CCC) – a convex cost dual of minimum cost network flow. Their algorithm for this convex cost dual of minimum cost flow has running time of $O(mn \log \frac{n^2}{m} \log(nU))$.

The method of Hochbaum and Naor, [HN94] solves integer problems on monotone inequalities in two variables at most per inequality. Monotone inequality is of the form $ax - by \leq c$, where the coefficients of x and y are of opposite signs. Obviously, the constraints of (CCC) are monotone inequalities. The algorithm of Hochbaum and Naor runs in pseduopolynomial time $O(\sum_i (u_i - \ell_i) mU \log \frac{n^2 U}{m})$. It is possible to combine that algorithm with a scaling approach that is implemented for (CCC) in time $O(mn \log U \log \frac{n^2 U}{m})$.

The algorithm described here solves (CCC) in time $O(mn \log \frac{n^2}{m} + n \log U)$. It should be noted that the first factor is reflects the run time required to solve the minimum closure problem, and the second factor is required to find an integer minimum of n convex functions. It is likely that if a faster algorithm for the minimum closure problem is discovered then the run time of the algorithm can be respectively improved. For the second term the factor of $\log U$ cannot be avoided as any algorithm solving a constrained nonlinear and nonquadratic optimization problem may not run in strongly polynomial time, [Hoc94]. When the functions f_i are quadratic convex, the algorithm runs in time $O(mn \log \frac{n^2}{m} + n \log n)$, which is strongly polynomial . For the isotonic regression problem the running time improves to $O(n \log n + n \log U)$, and thus the complexity of our algorithm is $O(n \log(\max\{n, U\}))$.

There are efficient algorithms known for solving the special cases of (CCC). Any maximum flow algorithm can be used to solve the minimum (or maximum) closure problem. The most efficient algorithm known to date solving the maximum flow and thus the minimum cut and the closure problems has complexity of $O(mn \log \frac{n^2}{m})$ due to Goldberg and Tarjan, [GT88].

The isotonic regression problem is an instance of (CCC) defined on a linear order. Ahuja and Orlin report on a $O(n \log U)$ time algorithm for this problem (personal communication). The problem has been studied extensively in the statistical study of observations. The book of Barlow et. al. [BBB72] provides an excellent review of applications and algorithms for this problem as well as the (CCC) problem.

3   Solving the Minimum Closure Problem

A set of nodes S in a directed graph $G = (V, A)$ is said to be *closed*, if all predecessor nodes of S are also included in S, i.e. if $j \in S$ and $(i, j) \in A$ then $i \in S$. Equivalently, S is said to be closed if there are no incoming arcs into S.

We review here the reduction of Picard [Pic76]) demonstrating that the minimum closure problem is solved using a minimum cut procedure. We first define an s, t-graph associated with the minimum closure problem: The graph has a node j associated with each variable x_j. A source and sink nodes, s and t are now added to the graph. If the weight of the variable w_j is positive then node j has an arc from the source into it with capacity w_j. If the node has weight w_j which is negative, then there is an arc from j to t with capacity $-w_j$. Let V^+ be the set of nodes with positive weights, and V^- the set of nodes with negative weights.

Each inequality $x_i \geq x_j$ is associated with an arc (i,j) of infinite capacity. Consider any finite s,t-cut in the graph that partitions the set of nodes to two subsets $\{s\} \cup S$ and $\{t\} \cup \bar{S}$. It is easy to see that $\bar{S}$ is a closed set if there are no infinite capacity arcs from S to $\bar{S}$.

We denote by (A,B) the collection of arcs with tails at A and heads at B. The corresponding sum of capacities of these arcs is denoted by $C(A,B)$, $C(A,B) = \sum_{i \in A, j \in B} c_{ij}$ where c_{ij} is the capacity of arc (i,j). Let $w(A) = \sum_{j \in A} w_j$.

Given a finite cut $(\{s\} \cup S, \bar{S} \cup \{t\})$.

$$\min_{\bar{S} \subseteq V}[C(\{s\} \cup S, \bar{S} \cup \{t\})] = \min_{\bar{S} \subseteq V} \sum_{j \in \bar{S} \cap V^+} w_j + \sum_{j \in S \cap V^-}(-w_j)$$
$$= \min_{\bar{S} \subseteq V} \sum_{j \in \bar{S} \cap V^+} w_j - (\sum_{i \in V^-} w_i - \sum_{i \in \bar{S} \cap V^-} w_i)$$
$$= \min_{\bar{S} \subseteq V} \sum_{j \in \bar{S}} w_j - w(V^-).$$

In the last expression the term $w(V^-)$ is a constant. Thus the closed set $\bar{S}$ of minimum weight is also the sink set of a minimum cut and, vice versa – the sink set of a minimum cut (which has to be finite) also minimizes the weight of the closure.

4   A Threshold Theorem

The threshold theorem is the corner stone of our algorithm. The theorem is an extension of a corresponding result of Picard and Queyranne, [PQ85].

Let α be a scalar. Let w_i be the subgradient of f_i at α, $w_i = f_i'(\alpha) = f_i(\alpha+1) - f_i(\alpha)$. We set the subgradient value of $f_i'(x)$ to be equal to M at values of $x > u_i$, and to $-M$ for values $x < \ell_i$, for M a suitably large value.

Consider now the minimum closure problem with variable weights equal to w_i. The theorem establishes that the variables x_i in the problem (CCC) satisfy that all variables corresponding to elements i in the minimum weight closed set S^* are at least α for any optimal solution $\mathbf{x}$, $x_i > \alpha$. And all the variables corresponding to elements j in the complement of S^* satisfy that $x_j \leq \alpha$.

In case there are several optimal minimum closed sets we term a *minimal* minimum closed set that minimum closed set that does not contain other minimum closed sets. Similarly we define a *maximal* minimum closed set, as a minimum closed set that is not contained in another minimum closed set.

Theorem 1. *Let $w_i = f_i'(\alpha)$ be the weight assigned to node i, $i = 1, \ldots, n$ in a minimum closure problem defined on the partial order graph $G = (V, A)$. Let S^* be the minimal minimum weight closed set in this graph. Then an optimal solution $\mathbf{x}^*$ to the convex cost closure problem satisfies, $x_i^* > \alpha$ if $i \in S^*$ and $x_i^* \leq \alpha$ if $i \in \bar{S}^*$.*

Proof: Let S^* be a minimal minimum weight closed set, and suppose it violates the theorem. Then there is a subset $S^o \subseteq S^*$ such that at an optimal solution $\mathbf{x}'$, $x_j' \leq \alpha$ for all $j \in S^o$.

Since at the optimum $x_j' > \alpha$ for $j \in S^* \setminus S^o$, then the set $S^* \setminus S^o$ must be closed, as it has no predecessors (larger values) in S^o. But this set is not a minimal minimum closed set as S^* is minimal. Thus the weight of nodes in S^o – the total sum of subgradients – is negative, $\sum_{j \in S^o} f'_j(\alpha) < 0$. Furthermore increasing the values of all x'_j in this set to α does not violate feasibility, since the values of their predecessors in $S^* \setminus S^o$ are all $> \alpha$. Thus replacing x'_j for $j \in S^o$ by α is feasible and strictly reduces the weight of the closure compared to an optimal solution. This contradicts the assumption that $\mathbf{x}'$ is optimal.

An analogous contradiction is reached if we assume that an optimal solution has in the set $\bar{S}^*$ a variable with value $> \alpha$.

$\square$

As a result of the theorem, we can decompose the set of nodes into subsets that imply a narrowing of the interval in which the optimal value of the respective variable is to be found.

5   The Algorithm

One possible approach for solving the problem is to identify for each variable an interval where it optimal value lies so that for any selection of value in those intervals the resulting assignment is feasible. This can be done by recursively applying the minimum closure procedure to each subset generated in the decomposition resulting from the threshold theorem. A direct application of this method is relatively expensive in terms of complexity. The key to our approach is to utilize *parametric minimum cut* to generate all the decompositions. This can be done, as is shown here, by adapting the method of Gallo, Grigoriadis and Tarjan [GGT89] that works in the same running time as a single maximum flow or minimum closure procedure.

5.1   The Parametric Graph G_λ

We create a graph with parametric capacities, $G_\lambda = (V, A)$. Each node $j \in V$ has an arc from s going into the node with capacity $\max\{0, f'_j(\lambda)\}$, and an arc from j to the sink t with capacity $-\min\{0, f'_j(\lambda)\}$. The capacities of the arcs adjacent to source in this graph are monotone nondecreasing as a function of λ, and the arcs adjacent to sink are all with capacities that are monotone nonincreasing as a function of λ. Note that each node is connected with a positive capacity arc, either to source, or to sink, but not to both. Denote the source set of a minimum cut in the graph G_λ by S_λ.

Let ℓ be the lowest lower bound on any of the variables and u the largest upper bound. Consider varying the value of λ in the interval $[\ell, u]$. As the value of λ increases, the sink set becomes smaller and contained in the previous sink sets corresponding to smaller values of λ. Our goal is to identify for each variable the largest value of λ, $\underline{\lambda}$, so that the variable is still in the minimum closed set, and the smallest value of λ, $\overline{\lambda}$, so the variable is not in the closed set. Once this interval is established, we conclude that the value of x_j in an optimal solution is the value attaining minimum of

$$\min_{x \in [\underline{\lambda}, \overline{\lambda}]} f_j(x).$$

For larger values of λ additional nodes join the source set of the cut: When the value of λ is below a certain threshold then the minimum cut is attained for the source set containing only node s. As λ grows the value of the cut changes as a function of λ which is the sum of the capacities of the nodes adjacent to source that are in the sink set, and nodes adjacent to sink that are in the source set. That function has a breakpoint where a node switches from the sink set to the source set – called *node shifting breakpoint*. The value of $\underline{\lambda}$ for node x_j is the value of the breakpoint when node j joins the source set of the minimum cut. The value of $\overline{\lambda}$ is the next node shifting breakpoint.

Since the source set of a minimum cut can only grow with λ, then

$$S_k \subseteq S_{k+1}, \quad \text{for all } k.$$

5.2 Identifying an Integer Node-Shifting Breakpoint

We consider the convex functions $f_j(x)$ to be piecewise linear segments connecting the values of $f_j(k)$ on integer points $\ell_j \leq k \leq u_j$. For such functions the derivatives at the integer points are not well defined, and indeed could be any subgradient of the function at the respective integer point. We will consider the derivative $f_j'(x)$ to be a step function with the value in the interval $(k-1, k]$ equal to $f_j(k) - f_j(k-1)$.

We denote a maximal minimum cut source set in G_λ by $S_\lambda^{\max}$ and a minimal minimum cut source set, by $S_\lambda^{\min}$.

The source set of a minimum cut of G_λ remains invariant for $\lambda \in (k-1, k]$. Thus, in order to verify that λ is a node-shifting breakpoint, it suffices to compare $S_\lambda^{\max}$ with $S_{\lambda+\epsilon}^{\min}$ for $\epsilon > 0$ small enough. In our case we only consider integer values of λ and $\epsilon = 1$ is a small enough value : If $S_\lambda^{\max} \subset S_{\lambda+1}^{\min}$, then λ is a node shifting breakpoint.

The existence of a breakpoint in an interval (λ_1, λ_2) is confirmed if and only if $S_{\lambda_1}^{\max} \subset S_{\lambda_2}^{\min}$.

Parametric Analysis We assume henceforth that our minimum cut algorithm delivers as output both $S_\lambda^{\min}$ and $S_\lambda^{\max}$. We further assume that the procedure **min-cut**(G_λ) returns both the minimal and maximal source sets of minimum cuts (if different), $S_\lambda^{\min}$, $S_\lambda^{\max}$ and $S_{\lambda+1}^{\min}$.

For a given interval (λ_1, λ_2) where arc capacities do not contain a breakpoint b_j where a node switches from being connected to source to being connected to sink, we can find all node shifting breakpoints by using the procedure **parametric**.

Procedure parametric $(\lambda_1, \lambda_2, S_{\lambda_1}^{\max}, S_{\lambda_2}^{\min})$
Contract: $s \leftarrow s \cup S_{\lambda_1}^{\max}$, $t \leftarrow t \cup \bar{S}_{\lambda_2}^{\min}$. If $V = \{s, t\}$, or, if $\lambda_2 - \lambda_1 \leq 1$, halt "no breakpoints".
Else, let $\lambda^* = \lfloor \frac{\lambda_1 + \lambda_2}{2} \rfloor$.

Call **min-cut**(G_{λ^*}) for the output $S_{\lambda^*}^{\min}$, $S_{\lambda^*}^{\max}$
If λ^* is a breakpoint, output λ^*. Else continue,
    Call **parametric** $(\lambda_1, \lambda^*, S_{\lambda_1}^{\max}, S_{\lambda^*}^{\min})$
    Call **parametric** $(\lambda^*, \lambda_2, S_{\lambda^*}^{\max}, S_{\lambda_2}^{\min})$
end

The choice of λ^* as the median in the interval (λ_1, λ_2) is unimportant. Any integer value interior to the interval will work correctly. To verify whether λ^* is a breakpoint, it is equivalent to one of the conditions being satisfied:
(i) $S_{\lambda_1}^{\max} \subset S_{\lambda^*}^{\min}$ and $\lambda^* - \lambda_1 \leq 1$, or
(ii) $S_{\lambda^*}^{\max} \subset S_{\lambda_2}^{\min}$ and $\lambda_2 - \lambda^* \leq 1$.

The analysis of the complexity of the procedure follows arguments used in [GGT89]. The details are omitted in this extended abstract. The running time $T(m, n)$ satisfies the recursion:

$$T(m, n) = T(m_1, n_1) + T(m_2, n_2) + 2Q \log \frac{n_1^2}{m_1} m_1 n_1.$$

The solution is $T(m, n) = O(mn \log \frac{n^2}{m})$.

5.3   The Algorithm

Let ℓ be the lowest lower bound on any of the variables and u the largest upper bound. Let $U = u - \ell$.

procedure convex closure $(G, f_j, j = 1, \ldots, n)$
___

Step 1: Call **parametric** $(\ell, u, \emptyset, V)$.
    Let the output be a set of up to n breakpoints $\lambda_1, \lambda_2, \ldots, \lambda_q$.

Step 2: Find the minimum of each convex function $f_j(x)$, min_j, for $j = 1, \ldots, n$.
    $f_j(min_j) = \min_{x = \ell_j, \ldots, u_j} f_j(x)$.

Step 3: Output the optimal solution $\mathbf{x}^*$ where,

$$x_j^* = \begin{cases} \lambda_{j_i - 1} + 1 & \text{if } min_j \leq \lambda_{j_i - 1} \\ \lambda_{j_i} & \text{if } min_j \geq \lambda_{j_i} \\ min_j & \text{if } \lambda_{j_i - 1} \leq min_j \leq \lambda_{j_i}. \end{cases}$$

At the end of step 1 the optimal value of x_i falls in the interval $(\lambda_{j_i - 1}, \lambda_{j_i}]$.

In step 2, identifying the minimum min_j such that $f_j(min_j) = \min_{x = \ell_j, \ldots, u_j} f_j(x)$ amounts to finding the value in the sorted array of derivative values $f_j'(k)$ the first index k so that $f_j'(k)$ is nonnegative. This can be accomplished using binary search in $O(\log U)$ time per function and $O(n \log U)$ total complexity.

The parametric search generating all breakpoints is implemented in time of $O(mn \log \frac{n^2}{m})$. The second step of the algorithm requires finding all the integer minima of the convex functions for a total complexity of $O(n \log U)$. Step 3 is accomplished in $O(n)$ time. The total run time of the algorithm is thus, $O(mn \log \frac{n^2}{m} + n \log U)$.

6   Special Cases

6.1   The Quadratic Convex Closure Problem

Nonlinear and nonquadratic optimization problems with linear constraints were proved impossible to solve in strongly polynomial time in a complexity model of the arithmetic operations, comparisons, and the rounding operation, [Hoc94]. That negative result however is not applicable to the quadratic case, thus it may be possible to solve constrained quadratic optimization problems in strongly polynomial time.

In the quadratic case our algorithm is implemented to run in strongly polynomial time. This is easily achieved since finding the minima in the second phase of the algorithm amounts to solving a linear equation in one variable. So the run time required to finding all the minima is $O(n)$ and thus the overall run time of the algorithm is dominated by the complexity of the minimum cut,

$$O(mn \log \frac{n^2}{m}).$$

6.2   Isotonic Regression

The isotonic regression problem is a special case of (CCC) where the order is linear and the corresponding graph $G = (V, A)$ is a path from node n to node 1. In other words, the inequalities associated are of the type,

$$x_i \leq x_{i+1} \text{ for all } i = 1, \ldots, n.$$

There are only n possible cuts in such graph, each with a source set S_i of the form, $S_i = \{1, \ldots, i\}$. Each cut is thus $(\{1, \ldots, i\}, \{i+1, \ldots, n\})$. The minimum cut for such graphs is trivially identified in $O(n)$ time, by comparing the capacities of the n possible cuts. The capacity of cut $(S_i, \bar{S}_i)$ is computed in $O(1)$ by subtracting from the capacity of $(S_{i-1}, \bar{S}_{i-1})$ the weight of node i, w_i. Indeed, if the weight w_i is positive then it contributes w_i to the capacity of the cut $(S_{i-1}, \bar{S}_{i-1})$, but not to the capacity of the cut $(S_i, \bar{S}_i)$. If $w_i < 0$ then node i contributes $-w_i$ to the capacity of $(S_i, \bar{S}_i)$, but 0 to the capacity of $(S_{i-1}, \bar{S}_{i-1})$.

Consider the closure graph in which each node has a weight $f'_j(x)$ associated with it for a given value of x. Minimizing the value of the cut is equivalent to minimizing the sum of weights of the sink set, (see Section 3). Alternatively, the cut is minimized when the weight of the corresponding source set is maximized thus seeking an index i to maximize $F_i(x) = \sum_{j=1}^{i} f'_j(x)$. We thus conclude,

Lemma 1. *If* $\sum_{j=1}^{i} f'_j(x) = \max_{k=1,\ldots,n} \sum_{j=1}^{k} f'_j(x)$ *then the minimum cut in the graph* G_x *is* $(S_i, \bar{S}_i)$.

Consider the partial sum functions $F_1(x), F_2(x), \ldots, F_n(x)$ where $F_i(x) = \sum_{j=1}^{i} f'_j(x)$. Recall that the functions $f'_j(x)$ are monotone nondecreasing in x. Denote the roots of the partial sum functions by b_i. So $F_i(b_i) = 0$. If the function is negative in the interval $[\ell, u]$ then we let $b_i = u + 1$. If the function is positive throughout the interval then we let $b_i = \ell - 1$. Let $b_{i_1} = \min_i b_i$, then for $x \leq b_{i_1}$ the optimal minimum cut is $(\emptyset, V)$. For this cut, the maximum weight source set is empty, since all the partial sums of weights are nonpositive. The value of $\lambda_1 = b_{i_1}$ is thus a breakpoint beyond which, for $x > b_{i_1}$, the source set of the minimum cut is $\{1, \ldots, i_1\}$.

As the value of x increases sufficiently so that $\sum_{j=i_1+1}^{i_2} f_j'(x) = F_{i_2}(x) - F_{i_1}(x) \geq 0$, the nodes $\{i_1, \ldots, i_2\}$ join the source set of the minimum cut. In other words, the second breakpoint is the smallest value λ_2 so that there is an index $i_2 > i_1$ such that $F_{i_2}(\lambda_2) - F_{i_1}(\lambda_2) = 0$. The general procedure is as follows:

---

procedure Isotonic regression breakpoints

$i_0 = 0$, $\lambda_0 = \ell - 1$, $k = 1$

while $i_{k-1} < n$, do

Find smallest integer value of λ_k such that for $i_k > i_{k-1}$, $F_{i_k}(\lambda_k) - F_{i_{k-1}}(\lambda_k) \geq 0$.

$k \leftarrow k + 1$

repeat

Output $\lambda_1, \ldots, \lambda_k$.

end

A naive implementation of this algorithm has n iterations with each iteration involving the finding of the roots of $O(n)$ functions. The total complexity is $O(n^2 \log U)$. We can do better with the implementation of the parametric search algorithm that requires the solution of the minimum cut problem for a specific parameter value in $O(n)$. However, each time the procedure calls for the minimum cut, the weights of the nodes must be updated for the new parameter value. This update requires $O(n)$ operation. The additional work of finding the roots of the n functions adds up to a total complexity of $O(n^2 + n \log U)$.

To achieve a better running time we investigate further the properties of the partial sum functions $F_i(x)$.

Lemma 2. *For $i < j$ and the functions F_i and F_j, if for some value of the argument λ, $F_i(\lambda) < F_j(\lambda)$ then $F_i(x) < F_j(x)$ for any $x > \lambda$.*

Proof: $F_j(x) - F_i(x)$ is a sum of monotone nondecreasing functions $\sum_{k=i+1}^{j} f_k'(x)$. Thus the difference $F_j(x) - F_i(x) > F_j(\lambda) - F_i(\lambda) > 0$ and can only increase as the value of x grows. Thus the two functions do not intersect for any value of $x > \lambda$. □

An immediate corollary of the lemma is that any pair of functions F_i, F_j can intersect at most once.

Consider the upper envelope of the functions F_i represented as an array of functions and breakpoints $(\ell, 0, b_{i_1}, F_{i_1}, b_{i_2}, F_{i_2}, \ldots, b_{i_n}, F_{i_n}, u)$. The functions on the envelope have the property that for all j,

$$F_{i_k}(x) \geq F_j(x) \quad x \in [b_{i_k-1}, b_{i_k}].$$

From the lemma it follows that the upper envelope of the partial sums functions has at most n breakpoints where the function on the envelope changes. The first breakpoint is b_{i_1}. The next breakpoint occurs for a value of x when some function $F_{i_2}(x) = F_{i_1}(x)$. It is easy to see from procedure Isotonic regression breakpoints that the list of breakpoints of this envelope is precisely the list of the breakpoints that determines the sequence of cuts.

The following *sweep* algorithm may be used to find the upper or upper envelope of a set of functions: Partition arbitrarily the set of functions into two equal size sets $\mathcal{F}_1$, $\mathcal{F}_2$. Compute recursively the upper envelopes of $\mathcal{F}_1$, $\mathcal{F}_2$. Let E_1, E_2 denote the two resulting upper envelopes. *Sweep* the two upper envelopes E_1, E_2 from left to right, and compute the upper envelope of the two upper envelopes. For a detailed description of the above algorithm the reader is referred to [SA95] (page 134–136) or to [BKOS97].

It remains to show how to implement the sweep algorithm for our particular set of functions. Instead of partitioning arbitrarily the set of functions we choose the partition of $\mathcal{F}_1 = \{1, \ldots, n\}$ to $\{1, \ldots, \lfloor \frac{n}{2} \rfloor\}$ and $\mathcal{F}_2 = \{\lfloor \frac{n}{2} + 1, \ldots, n \rfloor\}$. That is, one set contains the lower half set of indices and the other the upper half set of indices.

Consider the first breakpoint in E_1 and in E_2 (recall that at that point the partial sum values are still 0). If the first breakpoint of E_1 is larger then the first breakpoint of E_2 than the first portion of E_1 is below the first portion of E_2. From the lemma no pair of the functions from the two sets intersect, and the entire envelope E_1 lies below the envelope E_2. Thus the merged envelope is E_2.

If, on the other hand, the first breakpoint of E_1 is smaller than the first breakpoint of E_2 then there could be a point where a functions from $\mathcal{F}_2$ crosses a function from $\mathcal{F}_1$. We consider the array of breakpoints of the envelope E_1 for the last breakpoint where it is still above E_2. Similarly, we search the array of breakpoints of the envelope E_2 for the last breakpoint where it is still below E_1. Since there are $O(n)$ breakpoints per envelope, the search for that breakpoint is done by binary search in $O(\log n)$ steps. The intersection point is then to be determined between the function between this breakpoint and the next one on each envelope. Finding the intersection of $F_i(x)$ and $F_j(x)$ takes at most $O(\log U)$ steps.

Thus the merger of two envelopes of functions is executed in $O(\log n + \log U)$. Since there are at most n mergers in the procedure, the total running time is $O(n \log n + n \log U)$.

Once all the upper envelope has been identified we have the implied source sets of the associated cuts:

$$\{1, \ldots, i_1\}, \ \{1, \ldots, i_2\}, \ldots, \{1, \ldots, i_q\}.$$

If $i \in \{i_{k-1}, \ldots, i_k\}$ then $x_i^* \in [b_{i_{k-1}}, b_{i_k}]$. It remains to apply the equivalent of step 3 in procedure convex closure in order to determine an optimal solution:

$$x_i^* = \begin{cases} b_{i_{k-1}} + 1 & \text{if } min_i \le b_{i_{k-1}} \\ b_{i_k} & \text{if } min_i \ge b_{i_k} \\ min_i & \text{if } b_{i_{k-1}} \le min_i \le b_{i_k}. \end{cases}$$

Thus the total complexity of the algorithm for the isotonic regression problem is $O(n(\log U + \log n))$. In the quadratic case this leads to a complexity of $O(n \log n)$.

7   The Continuous Convex Closure Problem

When solving the problem in continuous variables, one has to determine how to output the solution. Let an ϵ-accurate solution (introduced in [HS90]), $\mathbf{x}^{(\epsilon)}$, be specified as an integer multiple of ϵ, i.e. it lies on the so-called ϵ-grid. The solution is such that there is an optimal vector $\mathbf{x}^*$ so that, $\|\mathbf{x}^{(\epsilon)} - \mathbf{x}^*\|_\infty < \epsilon$.

The continuous problem is then solved using the same algorithm as used for the integer case. The only modification required is in the finding of ϵ-accurate roots of the functions. This can be done in $O(\log(U/\epsilon))$ per function using binary search. In the parametric analysis procedure the choice of λ^* is such that it is any interior point in the interval (λ_1, λ_2) that lies on the ϵ-grid. The complexity of the algorithm is thus the complexity of finding the roots of the n functions plus the complexity of a minimum cut, $O(mn \log \frac{n^2}{m} + n \log(U/\epsilon))$.

Acknowledgement

The first author wishes to extend thanks to Sariel Har-Peled for his input on the state-of-the art of manipulating lower and upper envelopes of functions and pointing out references [BKOS97] and [SA95].

References

[AHO99]   R. K. Ahuja, D. S. Hochbaum and J. B. Orlin. Solving the convex cost integer dual network flow problem. Proceedings of IPCO'99, G. Cornuejols, R.E. Burkard and G.J. Woeginger (Eds.), Lecture Notes in Computer Science 1610, 31-34, 1999. 257

[BBB72]   R. E. Barlow, D. J. Bartholomew, J. M. Bremer and H. D. Brunk. Statistical Inference Under Order Restrictions. Wiley, New York, 1972. 257, 258

[BKOS97]   M. de Berg and M. van Kreveld and M. Overmars and O. Schwarzkopf. *Computational Geometry: Algorithms and Applications.* Springer-Verlag, Berlin, 1997. 265, 266

[CLR89]   T. H. Cormen, C. E. Leiserson and R. L. Rivest. *Introduction to Algorithms,* MIT Press,Cambridge Mass., 1989.

[GGT89]   G. Gallo, M. D. Grigoriadis and R. E. Tarjan. A fast parametric maximum flow algorithm and applications. *SIAM Journal of Computing,* Vol. 18, No. 1 1989, 30–55. 260, 262

[GT88]   A. V. Goldberg and R. E. Tarjan. A New Approach to the Maximum Flow Problem. *J. Assoc. Comput. Mach.,* 35 (1988), 921–940. 258

[HN94]   D. S. Hochbaum and J. Naor. Simple and fast algorithms for linear and integer programs with two variables per inequality. *SIAM Journal on Computing,* 23(6) 1179–1192, 1994. 258

[HS90]   D. S. Hochbaum and J. G. Shanthikumar. Convex Separable Optimization is not Much Harder Than Linear Optimization. *Journal of ACM*, 37:4, 843–862, 1990. 266

[Hoc94]   D. S. Hochbaum. Lower and upper bounds for allocation problems. *Mathematics of Operations Research* Vol. 19:2, 390–409, 1994. 257, 258, 263

[Hoc97]   D. S. Hochbaum. The pseudoflow algorithm for the maximum flow problem. Manuscript, UC Berkeley, 1997.

[MM83]   W. L. Maxwell and J. A. Muckstadt, Establishing consistent and realistic reorder intervals in production/distribution systems. Technical report No. 561, School of Industrial Engineering and Operations Research, Cornell University, 1983. 257

[Pic76]   J. C. Picard. Maximal Closure of a Graph and Applications to Combinatorial Problems. *Management Science*, 22 (1976), 1268–1272.

[PQ82]   J.-C. Picard and M. Queyranne. Selected applications of minimum cuts in networks. INFOR 20 (1982) 394-422. 256, 258 257

[PQ85]   J. C. Picard and M. Queyranne. Integer minimization of a separable convex function subject to variable upper bound constraints. Manuscript, 1985. 257, 259

[Rou85]   R. Roundy. A 98%-effective integer-ratio lot-sizing for one-warehouse multi-retailer systems. *Management Science*, 31 1416-1430, 1985.

[SH76]   M. I. Shamos and D. Hoey. Geometric intersection problems. In *Proc. 17th Ann. Symp. on Foundations of Computer Science*, 1976, 208-215.

[SA95]   M. Sharir and P. K. Agarwal. *Davenport-Schinzel Sequences and Their Geometric Applications*. Cambridge University Press, New York, 1995. 265, 266

[Vei81]   A. F. Veinott, Jr. Least d-majorized network flows with inventory and statistical applications. *Management Science*, 17, 1971, 547-567. 257

Preemptive Scheduling with Rejection

(Extended Abstract)

Han Hoogeveen[1], Martin Skutella[2], and Gerhard J. Woeginger[3]

[1] Department of Computer Science, Utrecht University, The Netherlands
[2] Fachbereich Mathematik, Technische Universität Berlin, Germany
[3] Institut für Mathematik, TU Graz, Austria

Abstract. We consider the problem of preemptively scheduling a set of n jobs on m (identical, uniformly related, or unrelated) parallel machines. The scheduler may reject a subset of the jobs and thereby incur job-dependent penalties for each rejected job, and he must construct a schedule for the remaining jobs so as to optimize the preemptive makespan on the m machines plus the sum of the penalties of the jobs rejected.

We provide a complete classification of these scheduling problems with respect to complexity and approximability. Our main results are on the variant with an arbitrary number of unrelated machines. This variant is APX-hard, and we design a 1.58-approximation algorithm for it. All other considered variants are weakly NP-hard, and we provide fully polynomial time approximation schemes for them.

Keywords: Scheduling, preemption, approximation algorithm, worst case ratio, computational complexity, in-approximability.

1 Introduction

Consider a system with $m \geq 2$ (identical, uniformly related, or unrelated) parallel machines $M_1, \ldots, M_m$ and n jobs $J_1, \ldots, J_n$. Job J_j ($j = 1, \ldots, n$) has a *rejection penalty* e_j and a processing time p_{ij} on machine M_i ($i = 1, \ldots, m$). In the case of identical machines, the processing times are machine independent, i.e., $p_{ij} \equiv p_j$. In the case of uniformly related machines, the ith machine M_i runs at speed s_i, and $p_{ij} = p_j/s_i$. In the case of unrelated machines, the processing times p_{ij} are arbitrarily structured. In the standard three-field scheduling notation (see e.g. Lawler, Lenstra, Rinnooy Kan & Shmoys [7]) identical machines are denoted by the letter P, uniformly related machines by Q, and unrelated machines by R.

We consider the following optimization problem in such systems: For each job J_j, we must decide whether to accept that job or whether to reject it. The accepted jobs are to be scheduled on the m machines. Preemption is allowed, i.e., a job may be arbitrarily interrupted and resumed later on. Every machine can process at most one job at a time, and every job may be processed on at most one machine at a time. For the accepted jobs, we pay the makespan of the constructed schedule, i.e., the maximum job completion time in the schedule. For the rejected jobs, we pay the corresponding rejection penalties. In other

M. Paterson (Ed.): ESA 2000, LNCS 1879, pp. 268–277, 2000.
© Springer-Verlag Berlin Heidelberg 2000

	Identical	Uniformly related	Unrelated
m not part of input	weakly NP-hard pseudo-polynomial	weakly NP-hard pseudo-polynomial	weakly NP-hard pseudo-polynomial
m part of input	weakly NP-hard pseudo-polynomial	weakly NP-hard pseudo-polynomial	strongly NP-hard

Table 1. The complexity landscape of preemptive makespan with rejection.

	Identical	Uniformly related	Unrelated
m not part of input	FPTAS	FPTAS	FPTAS
m part of input	FPTAS	FPTAS	1.58-approximation APX-complete

Table 2. The approximability landscape of preemptive makespan with rejection.

words, the objective value is the preemptive makespan of the accepted jobs plus the total penalty of the rejected jobs. We denote this objective function by an entry "Rej + $C_{\max}$" in the third field of the three-field scheduling notation. For example, $P5 \,|\, pmtn \,|\, \mathrm{Rej} + C_{\max}$ denotes this problem on five identical machines; $Qm \,|\, pmtn \,|\, \mathrm{Rej} + C_{\max}$ denotes the problem on uniformly related machines where the number of machines is a fixed constant m that is not part of the input; $R \,|\, pmtn \,|\, \mathrm{Rej} + C_{\max}$ denotes the problem on unrelated machines where the number of machines is part of the input.

Related scheduling problems with rejection have been studied by Bartal, Leonardi, Marchetti-Spaccamela, Sgall & Stougie [2] for non-preemptive makespan on identical machines, by Engels, Karger, Kolliopoulos, Sengupta, Uma & Wein [5] for total weighted job completion time on a single machine, and by Sengupta [8] for lateness and tardiness criteria.

Complexity. Whereas classical preemptive makespan minimization (the problem where all jobs must be accepted) is polynomially solvable even on an arbitrary number of unrelated machines [7], preemptive makespan minimization with rejection is hard even in the case of two identical machines. A complete complexity classification is given in Table 1. In Section 4, we will prove weak NP-hardness of $P2 \,|\, pmtn \,|\, \mathrm{Rej} + C_{\max}$ and strong NP-hardness of $R \,|\, pmtn \,|\, \mathrm{Rej} + C_{\max}$. These two results induce all negative results stated in Table 1. The results in Section 3 on uniformly related machines and the results in Section 2 on unrelated machines yield the existence of pseudo-polynomial time algorithms for $Q \,|\, pmtn \,|\, \mathrm{Rej} + C_{\max}$ and $Rm \,|\, pmtn \,|\, \mathrm{Rej} + C_{\max}$. Perhaps surprisingly, we did not manage to find 'simple' pseudo-polynomial time algorithms for these two problems. Instead, we took a detour and constructed a fully polynomial time approximation scheme (FPTAS); the existence of the FPTAS then implies the existence of a pseudo-polynomial time algorithm. Anyway, these two positive results induce all other positive results stated in Table 1.

Approximability. Our approximability classification is given in Table 2. In Section 3 we will derive an FPTAS for the problem $Q \mid pmtn \mid \text{Rej} + C_{\max}$, and in Section 2 we derive another FPTAS for $Rm \mid pmtn \mid \text{Rej} + C_{\max}$. These two results induce all FPTAS-entries in Table 2. The variant $R \mid pmtn \mid \text{Rej} + C_{\max}$ with an arbitrary number of unrelated machines is APX-complete, even for the case of uniform rejection penalties (cf. Section 4). In Section 2, we construct a polynomial time $e/(e-1)$-approximation algorithm for $R \mid pmtn \mid \text{Rej} + C_{\max}$; note that $e/(e-1) \approx 1.58$.

Organization of the paper. Section 2 contains the positive results on unrelated machines and Section 3 contains the positive results on uniformly related machines. All negative results (NP-hardness and APX-hardness) are proved in Section 4.

2   Unrelated Machines

In this section we derive a polynomial time $e/(e-1)$-approximation algorithm for problem $R \mid pmtn \mid \text{Rej} + C_{\max}$ and an FPTAS for problem $Rm \mid pmtn \mid \text{Rej} + C_{\max}$. Consider the following mixed integer linear programming formulation (1) of $R \mid pmtn \mid \text{Rej} + C_{\max}$. For job J_j, the binary variable y_j decides whether J_j is rejected ($y_j = 0$) or accepted ($y_j = 1$). The variables x_{ij} describe which percentage of job J_j should be processed on machine M_i. The variable T denotes the optimal preemptive makespan for the accepted jobs.

$$\min T + \sum_{j=1}^{n}(1 - y_j)e_j$$

$$
\begin{aligned}
\text{s.t.} \quad & \sum_{j=1}^{n} x_{ij} p_{ij} \leq T && \text{for } i = 1, \ldots, m \\
& \sum_{i=1}^{m} x_{ij} p_{ij} \leq T && \text{for } j = 1, \ldots, n \\
& \sum_{i=1}^{m} x_{ij} = y_j && \text{for } j = 1, \ldots, n \\
& x_{ij} \geq 0 && \text{for } i = 1, \ldots, m \text{ and } j = 1, \ldots, n \\
& y_j \in \{0, 1\} && \text{for } j = 1, \ldots, n
\end{aligned}
\tag{1}
$$

The first set of restrictions states that for every machine the total assigned processing time is at most T. The second set of restrictions states that the total processing time of every job cannot exceed T. The third set of restrictions connects the binary decision variables y_j with the continuous variables x_{ij}. If we want to schedule every job J_j on the m machines according to the values x_{ij}, then we essentially are dealing with a preemptive open shop problem; it is well-known [7] that the smallest number T fulfilling the first two sets of constraints in (1) yields the optimal preemptive makespan. To summarize, every feasible solution of (1) corresponds to a feasible schedule with objective value $T + \sum_{j=1}^{n}(1 - y_j)e_j$.

Now we replace the integrality conditions $y_j \in \{0, 1\}$ in (1) by $0 \leq y_j \leq 1$. This yields the linear programming relaxation LPR which can be solved to optimality in polynomial time. Let x_{ij}^*, y_j^*, and T^* constitute an optimal solution to LPR. From this solution, we compute a *rounded* solution $\tilde{x}_{ij}$, $\tilde{y}_j$, and $\tilde{T}$ for (1) in the following way: We randomly choose a threshold α from the uniform distribution over $[1/e, 1]$. If $y_j^* \leq \alpha$, then we set $\tilde{y}_j := 0$, and otherwise we set $\tilde{y}_j := 1$. Similar *dependent randomized rounding* procedures have already proven useful in other contexts (see e.g. Bertsimas, Teo & Vohra [3]).

For j with $\tilde{y}_j = 0$, we set all variables $\tilde{x}_{ij} = 0$. For j with $\tilde{y}_j = 1$, we set all variables $\tilde{x}_{ij} := x_{ij}^*/y_j^*$. Finally, we set

$$\tilde{T} := \max\{\max_{1 \leq i \leq m} \sum_{j=1}^{n} \tilde{x}_{ij} p_{ij}, \ \max_{1 \leq j \leq n} \sum_{i=1}^{m} \tilde{x}_{ij} p_{ij}\} \ . \tag{2}$$

It can be verified that the values $\tilde{x}_{ij}$, $\tilde{y}_j$, and $\tilde{T}$ constitute a feasible solution of (1): All variables $\tilde{y}_j$ are binary. For j with $\tilde{y}_j = 0$, the variables $\tilde{x}_{ij}$ add up to 0. For j with $\tilde{y}_j = 1$, the variables $\tilde{x}_{ij}$ add up to $\sum_i x_{ij}^*/y_j^* = 1$. Finally, in (2) the value of $\tilde{T}$ is fixed to fulfill the first and the second set of restrictions.

Now let us analyze the quality of the rounded solution. For any fixed value of α, $\tilde{x}_{ij}$ is less than a factor of $1/\alpha$ above x_{ij}^*, and hence by linearity also $\tilde{T}$ is less than a factor of $1/\alpha$ above T^*. Therefore, the expected multiplicative increase in the makespan is at most a factor of

$$\frac{e}{e-1} \int_{1/e}^{1} 1/\alpha \, d\alpha \ = \ \frac{e}{e-1} \ .$$

In the LPR solution, the contribution of job J_j to the total penalty is $(1 - y_j^*)e_j$. The expected contribution of J_j to the penalty in the rounded solution is

$$e_j \cdot \text{Prob}[y_j^* \leq \alpha] \ = \ e_j \int_{\max\{1/e, y_j^*\}}^{1} \frac{e}{e-1} d\alpha$$

$$\leq \ e_j \int_{y_j^*}^{1} \frac{e}{e-1} d\alpha$$

$$= \ \frac{e}{e-1} \cdot (1 - y_j^*)e_j \ .$$

All in all, the expected objective value for the rounded solution is at most a factor of $e/(e-1) \approx 1.58$ above the optimal objective value of LPR. Hence, our procedure yields a *randomized* polynomial time $e/(e-1)$-approximation algorithm. Since the only critical values for the threshold parameter α are the values y_j^* $(j = 1, \ldots, n)$, it is straightforward to derandomize this algorithm in polynomial time.

Theorem 1. *The problem $R \,|\, pmtn \,|\, \text{Rej} + C_{\max}$ possesses a deterministic polynomial time $e/(e-1)$-approximation algorithm.* $\square$

Let us turn to problem $Rm \,|\, pmtn \,|\, \text{Rej} + C_{\max}$. The crucial fact for deriving positive results on this problem is the following discretization lemma.

Lemma 2. *Let δ be a real number with $0 < \delta \le 1/m$, such that $1/\delta$ is integer. Then, the mixed integer linear program (1) possesses a feasible solution, in which the values x_{ij} all are integer multiples of δ^3 and whose objective value is at most a factor of $1 + \delta$ above the optimal objective value of (1).*

Proof. Consider an optimal solution x_{ij}^*, y_j^*, and T^* of the mixed integer linear program (1). Another feasible solution $\tilde{x}_{ij}$ and $\tilde{y}_j$ for (1) is constructed job-wise in the following way. For job J_j, let $\ell(j)$ denote a machine index that maximizes $x_{\ell(j),j}^*$, i.e., an index with $x_{\ell(j),j}^* \ge x_{ij}^*$ for all $1 \le i \le m$. Then for $i \ne \ell$, $\tilde{x}_{ij}$ is the value x_{ij}^* rounded down to the next multiple of δ^3. Moreover, we set $\tilde{y}_j = y_j^*$ and $\tilde{x}_{\ell(j),j} = \tilde{y}_j - \sum_{i \ne \ell(j)} \tilde{x}_{ij}$. Finally, $\tilde{T}$ is computed according to (2). It is straightforward to verify that $\tilde{x}_{ij}$, $\tilde{y}_j$, and $\tilde{T}$ is feasible for (1), and that the values $\tilde{x}_{ij}$ all are integer multiples of δ^3.

We claim that for all $j = 1, \ldots, n$ and $i = 1, \ldots, m$, the inequality $\tilde{x}_{ij} \le (1 + \delta) x_{ij}^*$ is fulfilled. If $y_j^* = 0$, this inequality trivially holds since $\tilde{y}_j = \tilde{x}_{ij} = 0$ for $i = 1, \ldots, m$ then. Otherwise, if $i \ne \ell(j)$, the inequality holds since $x_{ij}^* - \delta^3 < \tilde{x}_{ij} \le x_{ij}^*$. Moreover, for $i = \ell(j)$ we have

$$\tilde{x}_{\ell(j),j} = \tilde{y}_j - \sum_{i \ne \ell(j)} \tilde{x}_{ij} < y_j^* - \sum_{i \ne \ell(j)} (x_{ij}^* - \delta^3) < x_{\ell(j),j}^* + m\delta^3 \le (1 + \delta) x_{\ell(j),j}^* \ .$$

The first inequality follows from the definition of the $\tilde{x}_{ij}$ with $i \ne \ell(j)$. The second inequality is straightforward. The last inequality is equivalent to $m\delta^2 \le x_{\ell(j),j}^*$; this is true since $\delta \le 1/m$ and $x_{\ell(j),j}^* \ge y_j^*/m = 1/m$. Summarizing, the claimed inequalities are indeed fulfilled. Since $\tilde{y}_j \equiv y_j$, the objective value in (1) increases at most by a factor of $1 + \delta$.   □

In the following, we call a feasible solution of (1) where all values x_{ij} are integer multiples of δ^3 as in Lemma 2 a δ-*discrete* feasible solution. Moreover, we assume without loss of generality that all processing times p_{ij} and rejection penalties e_j are integral. Our next goal is to show that the best δ-discrete feasible solution can be computed in pseudo-polynomial time by a dynamic programming approach. A state of the dynamic program encodes a partial schedule for the first k jobs ($1 \le k \le n$). Every state has $m + 2$ components. The first m components store the loads of the m machines in the partial schedule. Component $m + 1$ stores the length of the longest job scheduled so far (i.e., the maximum time that any job needs in the schedule). Component $m + 2$ stores the total penalty of all jobs from $J_1, \ldots, J_k$ that have been rejected so far. The state space S_0 is initialized with the all-zero vector. When job J_k is treated, every state s from the state space S_{k-1} is updated and yields several new states.

- First, job J_k may be rejected. The corresponding new state results from adding the penalty e_k to the last component of s.
- Otherwise, job J_k is accepted. We try all $O(1/\delta^{3m})$ possibilities for the m pieces $x_{1j}, \ldots, x_{mj}$ that are integer multiples of δ^3 and that add up to 1. For each appropriate combination the ith ($i = 1, \ldots, m$) component of s is

increased by $x_{ij}p_{ij}$. The new $(m+1)$th component is the maximum of the old $(m+1)$th component and $\sum_{i=1}^{m} x_{ij}p_{ij}$.

Finally, after treating the last job J_n we compute the objective values for all states in S_n and output the best one; the objective value equals the maximum of the first $m+1$ components plus the last component. The running time of this dynamic program is polynomial in n, $1/\delta$, and in the size of the state spaces. Component i ($i = 1, \ldots, m$) indicates the load of machine i, which is measured in units of δ^3; hence, the number of possible states for component i is $O(\sum_{j=1}^{n} p_{ij}/\delta^3)$. Similarly, the number of possible states for component $(m+1)$ is $O(\sum_{i=1}^{m} p_{ij}/\delta^3)$. Finally, the number of possible states for component $m+2$ is $O(\sum_{j=1}^{n} e_j)$. Clearly, this yields a pseudo-polynomial running time.

Lemma 3. *For any instance of $Rm \mid pmtn \mid \mathrm{Rej} + C_{\max}$ and for any δ with $0 < \delta \leq 1/m$ and $1/\delta$ integer, the best δ-discrete schedule can be computed in pseudo-polynomial time.* □

By applying standard methods, this dynamic programming formulation can be transformed into a fully polynomial time approximation scheme; in fact, the dynamic program belongs to the class of so-called *ex-benevolent* dynamic programs (Woeginger [9]), and therefore automatically leads to an FPTAS for computing the best δ-discrete feasible solution. Finally, let us turn back to the general problem $Rm \mid pmtn \mid \mathrm{Rej} + C_{\max}$. For a given $\varepsilon > 0$, we set $\delta = \min\{1/m, 1/\lceil 3/\varepsilon \rceil\}$ and then compute in fully polynomial time a $(1 + \varepsilon/3)$-approximation for the best δ-discrete feasible solution. It is easily verified that this yields a $(1 + \varepsilon)$-approximation of the optimal objective value; hence there is an FPTAS for $Rm \mid pmtn \mid \mathrm{Rej} + C_{\max}$. Since every sufficiently well-behaved optimization problem with an FPTAS is solvable in pseudo-polynomial time (see e.g. Theorem 6.8 in Garey & Johnson [6]) and since $Rm \mid pmtn \mid \mathrm{Rej} + C_{\max}$ is well-behaved, we may conclude that $Rm \mid pmtn \mid \mathrm{Rej} + C_{\max}$ is solvable in pseudo-polynomial time.

Theorem 4. *The problem $Rm \mid pmtn \mid \mathrm{Rej} + C_{\max}$ has an FPTAS, and it is solvable in pseudo-polynomial time.* □

3 Uniformly Related Machines

In this section we will construct an FPTAS and a pseudo-polynomial time algorithm for $Q \mid pmtn \mid \mathrm{Rej} + C_{\max}$. Our line of approach is quite similar to that for $Rm \mid pmtn \mid \mathrm{Rej} + C_{\max}$ in Section 2 which also gave an FPTAS and a pseudo-polynomial time algorithm.

Now consider an instance of $Q \mid pmtn \mid \mathrm{Rej} + C_{\max}$ with m machines and n jobs. Without loss of generality we assume that $m = n$ holds: If $m > n$, then the $m - n$ slowest machines will not be used in any reasonable schedule and may be removed from the instance. If $m < n$, then we introduce $n - m$ dummy machines of speed 0; these dummy machines will not be used in any reasonable schedule. Let $s_1 \geq s_2 \geq \cdots \geq s_n$ denote the speeds of the machines (so that processing of a job piece of length L on machine M_i takes L/s_i time). For $i \leq n$ let $S_i = \sum_{k=1}^{i} s_k$ denote the total speed of the i fastest machines.

Let $a_1 \geq a_2 \geq \cdots \geq a_q$ denote the lengths of the q *accepted* jobs in some schedule. For $i \leq q$ let $A_i = \sum_{k=1}^{i} a_k$ denote the total length of the i longest accepted jobs. It is well-known [7] that for $m = n$ machines the optimal preemptive makespan for the accepted jobs equals

$$\max_{1 \leq i \leq q} A_i/S_i \ . \tag{3}$$

This leads to the following dynamic programming formulation of $Q \,|\, pmtn \,|\, \text{Rej} + C_{\max}$. Without loss of generality we assume that $p_1 \geq p_2 \geq \cdots \geq p_n$, i.e., that the jobs are ordered by non-increasing processing times. Every state of the dynamic program consists of four values v_1, v_2, v_3, and v_4 and encodes a schedule for a prefix $J_1, \ldots, J_k$ of the job sequence. Value v_1 stores the total penalty of the jobs rejected so far, value v_2 stores the total processing time of the jobs accepted so far, value v_3 stores the number of accepted jobs, and value v_4 stores the maximum value A_i/S_i over $1 \leq i \leq v_3$. How do we update a state $[v_1, v_2, v_3, v_4]$ for $J_1, \ldots, J_k$, if also job J_{k+1} has to be considered?

- If job J_{k+1} is rejected, we replace v_1 by $v_1 + e_{k+1}$ and leave everything else unchanged. This yields the state $[v_1 + e_{k+1}, v_2, v_3, v_4]$.
- If job J_{k+1} is accepted, we define $v_2^{new} := v_2 + p_{k+1}$ and $v_3^{new} := v_3 + 1$. Moreover, v_4^{new} becomes the maximum of the old component v_4 and v_2^{new} divided by $S_{v_3^{new}}$. This yields the state $[v_1, v_2^{new}, v_3^{new}, v_4^{new}]$.

We handle job by job in this way, until we end up with a state space for $J_1, \ldots, J_n$. Then we extract from every state $[v_1, v_2, v_3, v_4]$ its objective value $v_1 + v_4$. The state with the best objective value gives the solution of $Q \,|\, pmtn \,|\, \text{Rej} + C_{\max}$. The time complexity of this dynamic programming formulation mainly depends on the number of states. Since every component in every state is a number whose size is bounded by the input size, the total number of states is pseudo-polynomial. Moreover, we can prove that this dynamic program belongs to the class of *benevolent* dynamic programming formulations [9]. Hence, it can be transformed into an FPTAS by trimming the state space appropriately.

Theorem 5. *The problem* $Q \,|\, pmtn \,|\, \text{Rej} + C_{\max}$ *has an FPTAS, and it is solvable in pseudo-polynomial time.*   □

4   Negative Results

In this section we prove two negative results, the NP-hardness of $P2 \,|\, pmtn \,|\, \text{Rej} + C_{\max}$ and the APX-hardness of $R \,|\, pmtn \,|\, \text{Rej} + C_{\max}$. The strong NP-hardness of $R \,|\, pmtn \,|\, \text{Rej} + C_{\max}$ follows along the same lines: our L-reduction (from the APX-hard maximum bounded 3-dimensional matching problem) at the same time constitutes a Turing-reduction (from the strongly NP-hard 3-dimensional matching problem).

Theorem 6. *The problem $P2 \mid pmtn \mid \mathrm{Rej} + C_{\max}$ is NP-hard in the ordinary sense.*

Proof. The proof will be given in the full version of the paper.   □

Now we turn to problem $R \mid pmtn \mid \mathrm{Rej} + C_{\max}$. The APX-hardness proof is done for the special case of uniform rejection penalties $e_j \equiv 1$ and so-called *restricted assignment*, where the processing times of jobs are not machine-dependent but each job may only be processed on a subset of machines, i.e., $p_{ij} \in \{p_j, \infty\}$. We provide an *L*-reduction from the APX-hard maximum bounded 3-dimensional matching problem.

MAXIMUM BOUNDED 3-DIMENSIONAL MATCHING (MAX-3DM-B)

Input: Three sets $A = \{a_1, a_2, \dots, a_q\}$, $B = \{b_1, b_2, \dots, b_q\}$ and $C = \{c_1, c_2, \dots, c_q\}$. A subset T of $A \times B \times C$ of cardinality s, such that any element of A, B and C occurs in exactly one, two, or three triples in T. Note that this implies that $q \leq s \leq 3q$.
Goal: Find a subset T' of T of maximum cardinality such that no two triples of T' agree in any coordinate.
Measure: The cardinality of T'.

Without loss of generality, we restrict ourselves to instances of MAX-3DM-B where the value q and the value of an optimal solution both are even. Notice that an arbitrary instance can easily be modified to fulfill these requirements by taking two disjoint copies of the instance. The following simple observation will be useful.

Lemma 7. *For any instance I of MAX-3DM-B we have $\mathrm{OPT}(I) \geq \frac{1}{7}s$.*   □

Let $I = (q, T)$ be an instance of MAX-3DM-B. We construct an instance $R(I)$ of the scheduling problem $R \mid pmtn, e_j \equiv 1, p_{ij} \in \{p_j, \infty\} \mid \mathrm{Rej} + C_{\max}$ with $s + 22q$ jobs and $s + 17q$ machines, where all penalties e_j are 1 and the processing time of job J_j on machine i is either p_j or infinite (i.e., a job can only be processed on a subset of machines). The *core* of the instance consists of $s + 7q$ jobs and $s + 2q$ machines. There are further $15q$ *non-core machines* and $15q$ *non-core jobs*. The non-core jobs are matched to the non-core machines. The processing time of each non-core job is $15q$ on its matching non-core machine, and it is infinite on all other (core and non-core) machines. Processing of a core job on a non-core machine also takes infinite time (and thus is impossible).

Now we continue our description of the core of the instance. There are s machines, which correspond to the triples in T, and therefore are called the *triple machines*. Moreover, there are $2q$ so-called *element machines*. As to the jobs, each a_j, b_j, and c_j element corresponds to an *element job* with processing time $5q$. An element job can be processed on any element machine; moreover, each triple machine can process the element jobs of the elements occurring in the corresponding triple. Each triple machine has its own matching dummy job; processing this dummy job takes $15q$ units of time, and no other dummy job can

be processed on the machine. Each element machine has two matching dummy
jobs with processing times $5q$ and $10q$, respectively; again, no other dummy job
can be processed on an element machine.

As we will see later, the sole purpose of adding the $15q$ non-core machines
with corresponding non-core jobs is to enforce that in the optimal schedule
$C_{\max} \geq 15q$. The following lemma gives the basic intuition of how the reduction
works.

Lemma 8. *If the optimal solution to an instance I of* MAX-3DM-B *consists of
k triples, then there is a solution to the instance $R(I)$ of the scheduling problem
with objective value $16q + (q - k)/2$.*

Proof. Without loss of generality, we assume that the first k triples in T constitute an optimal solution of I. We construct the following solution with
makespane $15q$ to instance $R(I)$. The first k triple machines process the element jobs belonging to their triples; the dummy jobs corresponding to the first
k triple machines are rejected. The remaining $3(q - k)$ element jobs are grouped
into $3(q-k)/2$ pairs which are then processed on an arbitrary subset of $3(q-k)/2$
element machines; the corresponding $3(q - k)/2$ dummy jobs of size $10q$ are rejected. This yields a schedule with $C_{\max} = 15q$ and $k + 3(q - k)/2$ rejected jobs.
Hence, the objective value is equal to $16q + (q - k)/2$. $\square$

Lemma 9. *Let I be an instance of* MAX-3DM-B *and $0 \leq k \leq q$. Given a
solution σ to the scheduling instance $R(I)$ with objective value $c(\sigma) < 16q + (q -
k)/2$, one can construct in polynomial time a solution $S(\sigma)$ to I consisting of at
least $k + 1$ triples.*

Proof. Omitted in this version of the paper. $\square$

Lemmas 8 and 9 together yield the following result.

Lemma 10. *If an optimal solution to the instance I of* MAX-3DM-B *consists
of k triples, then the value of an optimum solution to the instance $R(I)$ of the
scheduling problem is equal to $16q + (q - k)/2$.* $\square$

We can now state the main result of this section (Since the notion of preemption
is not used in the proof of Lemmas 8 and 9, we can use the very same L-
reduction to establish APX-hardness of the nonpreemptive problem $R \,|\, e_j \equiv
1, p_{ij} \in \{p_j, \infty\} \,|\, \mathrm{Rej} + C_{\max})$.

Theorem 11. *The problem $R \,|\, pmtn, e_j \equiv 1, p_{ij} \in \{p_j, \infty\} \,|\, \mathrm{Rej} + C_{\max}$ is APX-
hard.*

Proof. Our L-reduction now looks as follows. Given an instance I of MAX-
3DM-B, we construct the instance $R(I)$ of the problem $R \,|\, pmtn, e_j \equiv 1, p_{ij} \in
\{p_j, \infty\} \,|\, \mathrm{Rej} + C_{\max}$ as described above. The transformation S that maps a given
solution for $R(I)$ to a feasible solution of I is given in the proof of Lemma 9.

Clearly, R and S can be implemented to run in polynomial time. Moreover, we have for any instance I of MAX-3DM-B that

$$\text{OPT}(R(I)) \leq 17q \leq 17s \leq 119\,\text{OPT}(I) \ ;$$

the first inequality follows from Lemma 8 and the last inequality from Lemma 7. Finally, for any feasible schedule σ of $R(I)$, the feasible solution $S(\sigma)$ of instance I fulfills the inequality

$$\text{OPT}(I) - |S(\sigma)| \ \leq \ 2\big(c(\sigma) - \text{OPT}(R(I))\big)$$

by Lemmas 9 and 10. □

Acknowledgement. Gerhard Woeginger acknowledges support by the START program Y43-MAT of the Austrian Ministry of Science.

References

1. S. ARORA AND C. LUND [1997]. Hardness of approximation. In: D.S. Hochbaum (ed.) *Approximation algorithms for NP-hard problems*, PWS Publishing Company, Boston, 399–446.
2. Y. BARTAL, S. LEONARDI, A. MARCHETTI-SPACCAMELA, J. SGALL, AND L. STOUGIE [2000]. Multiprocessor scheduling with rejection. *SIAM Journal on Discrete Mathematics 13*, 64–78. 269
3. D. BERTSIMAS, C. TEO, AND R. VOHRA [1996]. On Dependent Randomized Rounding Algorithms. *Proceedings of the 5th International IPCO Conference*, Springer LNCS 1084, 330–344. 271
4. P. CRESCENZI AND V. KANN [1997]. A compendium of NP-optimization problems. http://www.nada.kth.se/nada/theory/problemlist.html.
5. D.W. ENGELS, D.R. KARGER, S.G. KOLLIOPOULOS, S. SENGUPTA, R.N. UMA, AND J. WEIN [1998]. Techniques for scheduling with rejection. *Proceedings of the 6th European Symposium on Algorithms (ESA'98)*, Springer LNCS 1461, 490–501. 269
6. M.R. GAREY AND D.S. JOHNSON [1979]. *Computers and Intractability: A Guide to the Theory of NP-Completeness*. Freeman, San Francisco. 273
7. E.L. LAWLER, J.K. LENSTRA, A.H.G. RINNOOY KAN, AND D.B. SHMOYS [1993]. Sequencing and scheduling: Algorithms and complexity. In: S.C. Graves, A.H.G. Rinnooy Kan, and P.H. Zipkin (eds.) *Logistics of Production and Inventory*, Handbooks in Operations Research and Management Science 4, North-Holland, Amsterdam, 445–522. 268, 269, 270, 274
8. S. SENGUPTA [1999]. Algorithms and approximation schemes for minimum lateness and tardiness scheduling with rejection. Manuscript, Laboratory for Computer Science, MIT. 269
9. G.J. WOEGINGER [1999]. When does a dynamic programming formulation guarantee the existence of an FPTAS? *Proceedings of the 10th Annual ACM-SIAM Symposium on Discrete Algorithms (SODA'99)*, 820–829. 273, 274

Simpler and Faster Vertex-Connectivity Augmentation Algorithms*

(Extended Abstract)

Tsan-sheng Hsu

Institute of Information Science
Academia Sinica, Nankang 11529
Taipei, Taiwan, ROC
tshsu@iis.sinica.edu.tw

Abstract. This paper presents a new technique to solve the problem of adding a minimum number of edges to an undirected graph in order to obtain a k-vertex-connected resulting graph, for $k = 2$ and 3.

1 Introduction

Finding a smallest set of edges whose addition makes an undirected graph k-vertex-connected is a fundamental problem with many important applications (see e.g., the survey in [5,9]) and has been extensively studied for $k = 2$ [4,13,15,16,18], $k = 3$ [12,22] and $k > 3$ [10,11,14]. For $k = 2$, two previous approaches are known to solve this problem in sequential linear time.

Previous Approaches. One approach [4,13,18] first obtains a forest data structure (so called block-cut node forest and is defined in §2) obtained from the input graph representing the relations between the cut vertices and the maximum subsets of vertices that are biconnected (i.e., 2-vertex-connected). Based on the structure of this forest, a lower bound on the number of edges needed to add is derived. This lower bound is zero if and only if the graph is biconnected. Then an induction argument is used to show that we can always find an edge whose addition reduces this lower bound by one. In order to find the desired edge efficiently, it is needed to maintain the current block-cut node forest. A bucket sort routine and dynamic data structures are used in implementation to achieve an overall linear execution time. In [13], an $O(\log^2 n)$-time parallel algorithm using a linear number of processors on an EREW PRAM is presented by showing that simultaneously a constant fraction of currently needed to be added edges, instead of only one edge as in the sequential algorithm, can be found at a time.

The other approach [15] requires no usage of advanced data structures and dynamically maintaining of the block-cut node forest. Instead of using the lower

---

* Research supported in part by NSC grants 88-2213-E-001-026 and 89-2213-E-001-015.

M. Paterson (Ed.): ESA 2000, LNCS 1879, pp. 278–289, 2000.

bound, a new invariant based on a simple tree obtained from the block-cut node forest is used to guide the finding of the desired edges. Let n be the number of vertices of the input graph. His linear-time algorithm uses a linear-time bucket sort routine. Further, the last phase of his algorithm uses a simple dynamic data structure, which maintains two sets with the largest cardinalities among $O(n)$ sets of cardinality $O(n)$ when an element is deleted from each found set, to find an added edge at a time until the resulting graph becomes biconnected. Because of the above, it appears to us that it is non-trivial to parallelize his algorithm.

Our Approach. Using a different approach, in this paper, we first show a very simple linear-time sequential algorithm to solve the smallest biconnectivity augmentation problem. Let G be the input graph and let T be its block-cut node forest. Given T, it is well-known that we only need to focus on the case when T is a tree [4]. Our algorithm first roots T at a vertex such that no rooted subtree has more than half of the total number of degree-1 vertices in T. For the case of the chosen root representing a maximum biconnected subset of vertices, we show we can easily find a smallest set of edges whose addition biconnects G. Otherwise, the chosen root represents a cut node. We then show we can first add a set of edges to G such that the block-cut node tree T' of the resulting graph G' can be rooted at a node representing a maximum biconnected subset of vertices. Further, the resulting rooted tree has no rooted subtree containing more than half of the total number of degree-1 vertices in T'. We then use the algorithm in the previous case to find a smallest set of edges whose addition biconnects G'. We finally show the union of the two added sets of edges is a smallest set of edges whose addition biconnects G.

To decide the set of added edges in all the cases described above, our algorithm only needs to count the number of vertices in the subtrees rooted at all the vertices of a rooted tree, and to number the degree-1 vertices in a rooted subtree consecutively. Our algorithm is simple and finds the set of added edges simultaneously in a single execution, instead of adding one edge at a time and using information in the resulting data structures to decide the next edge to be added as in the previous two approaches. Hence no dynamic data structure is needed. Our algorithm also requires no usage of any sorting routine. Furthermore, our approach can be applied to find a smallest set of edges whose addition triconnects an undirected graph.

Our Results. Let m be the numbers edges in the input graph. Our algorithm for finding a smallest set of edges whose addition makes the resulting graph biconnected not only runs in linear time, but also can be implemented in the same parallel complexities as the ones for constructing the block-cut node forest and for finding connected components, which are either $O(\log n)$ time and $O((n+m)\cdot\alpha(m,n)/\log n)$ processors on a CRCW PRAM [2] where α is the inverse Ackerman function, or $O(\log n \cdot \log \log n)$ time and $O(n+m)$ processors on an EREW PRAM [1].

For finding a smallest set of edges whose addition makes the resulting graph triconnected (i.e., 3-vertex-connected), Hsu and Ramachandran [12] gave a sequential linear-time algorithm following a strategy that is similar to one in [13,18] for finding a smallest biconnectivity augmentation by maintaining a forest data structure (so called 3-block forest [12]) obtained from the input graph. We extend our simple approach for solving the smallest biconnectivity augmentation problem and show the triconnectivity case can also be solved in the same sequential and parallel complexities.

2 Preliminary

In this paper, all graphs are undirected and have neither self-loops nor multiple edges. We use n and m to represent the numbers of vertices and edges in the graph G, respectively. Given a vertex u in G, let $\mathrm{dg}_G(u)$ be the degree of u, i.e., the number of its neighbors, in G.

Two vertices u and v of a graph are *k-vertex-connected*, $k > 1$, if they are in the same connected component with more than k vertices, and remain connected after the removal of any subset of vertices S such that $|S| < k$ and $u, v \notin S$. For convenience, we use the abbreviation *k-connected* to represent *k-vertex-connected*. For $k = 2$ and $k = 3$, we also use the abbreviations *biconnected* and *triconnected*, respectively.

A set of vertices is *k-connected* if every pair of its vertices are *k*-connected; similarly, a graph is *k-connected* if its set of vertices is *k*-connected. Given G with at least $k+1$ vertices, a *smallest k-connectivity augmentation* for G, denoted by $\mathrm{aug}k(G)$, is a set of edges with the minimum cardinality whose addition results in a *k*-connected graph. A *k-block* of a graph is the induced subgraph of a maximal *k*-connected subset of vertices. A *k*-block is *trivial* if it has exactly one vertex of G. A *k*-block is *isolated* if it consists of the nodes in a connected component of G.

A *cut* node of G is one whose removal results in more connected components than originally. A *cut 2-block* is a trivial one consisting of a cut node. The *2-block forest* blk2(G), a variation of the well-known block-cut node forest [7], of G is a forest constructed from G as follows. L_1 denotes the set of nontrivial 2-blocks of G. L_2 is that of trivial ones that are not cut 2-blocks. L_3 is that of cut 2-blocks. C is that of cut nodes. K is that of cut edges. The node set of blk2(G) is $L_1 \cup L_2 \cup C \cup K$, where L_3 is excluded because if $\{u\} \in L_3$, then $u \in C$. The nodes in blk2(G) corresponding to $L_1 \cup L_2$ are called the *b-nodes*; those corresponding to $C \cup K$ are the *c-nodes*. The edge set of blk2(G) is the union of $\{(b_1, c) \mid b_1 \in L_1 \text{ and } c \in C \text{ such that } c \in L_1\}$, $\{(c, e) \mid c \in C \text{ and } e \in K \text{ such that } c \text{ is an endpoint of } e\}$, and $\{(e, b_2) \mid e \in K \text{ and } b_2 \in L_2 \text{ such that an endpoint of } e \text{ is in } b_2\}$.

Given a PRAM model $\mathcal{M}$, let $\mathcal{T}_\mathcal{M}(n + m)$ be the least logarithmic parallel time needed to compute the connected components of G using $\mathcal{P}_\mathcal{M}(n + m) \leq n + m$ processors. If $\mathcal{M} = CRCW$, then $\mathcal{T}_{CRCW}(n + m) = O(\log n)$ and $\mathcal{P}_{CRCW}(n + m) = O((n + m) \cdot \alpha(m, n) / \log n)$ [2]. If $\mathcal{M} = EREW$, then $\mathcal{T}_{EREW}(n + m) = O(\log n \cdot \log \log n)$ and $\mathcal{P}_{EREW}(n + m) = O(n + m)$ [1]. From [17,19,20], blk2(G) can be constructed in sequential linear time and in parallel

$O(\log n + T_{\mathcal{M}}(n+m))$ time using $O((n+m)/\log n + P_{\mathcal{M}}(n+m))$ processors on an $\mathcal{M}$ PRAM.

In Fact 1, let b_1 and b_2 be two distinct degree-1 b-nodes in the same tree in blk2(G). Let e be an edge not in G whose two endpoints are a non-cut-node vertex in b_1 and one in b_2. Let $G' = G \cup \{e\}$. Let P be the path in the blk2(G) between b_1 and b_2.

Fact 1 ([7,18]) *The cut nodes of G corresponding to c-nodes in P and the nodes of G in the b-nodes on P form a new 2-block b_e in G'. The b-nodes of blk2(G') are b_e plus those of blk2(G) not on P. The c-nodes in blk2(G') are those in blk2(G) excluding the ones on P that are of degree 2 in blk2(G). The edge set of blk2(G') is the union of (1) the set of edges in blk2(G) whose two endpoints remain in blk2(G'), (2) $\{(u, b_e) \mid u \in P$ is a cut node of G an is still in blk2(G')$\}$ and (3) $\{(u, b_e) \mid u \notin P$ is a cut node of G incident to P in blk2(G)$\}$.*

Given a rooted tree T and a vertex v of T, let T_v be the subtree of T rooted at v. Let $T - T_v$ denote the subtree obtained from T after removing T_v. Given a vertex u of T, T_v is a *branch* of u if u is the parent of v. A branch T_v of u is a *chain* of T_v contains exactly one degree-1 vertex of T. T is *normalized* if every branch of the root has at most $\lceil \ell/2 \rceil$ degree-1 vertices.

Fact 2 ([13]) *In sequential linear time or in optimal $O(\log n)$ parallel time on an EREW PRAM model, we can find a vertex r in a tree T with at least three degree-1 vertices such that $\mathrm{dg}_T(r) > 1$ and T rooted at r is normalized.*

3   Smallest Biconnectivity Augmentation

In this section, we assume G has at least 3 vertices, $T = $ blk2(G), and T has ℓ degree-1 vertex. We use d to denote the largest degree of all c-nodes in T. Let z_1 be the number of connected components in G and let z_2 be the number of isolated vertices in T. It is well-known [4,16] that $|\mathrm{aug2}(G)| = \max\{d+z_1-2, \lceil \ell/2 \rceil + z_2\}$ if $z_2 + \ell > 1$. Our algorithm first adds a set A_2 of edges to G such that $G \cup A_2$ is connected. Then we design an algorithm to find $\mathrm{aug2}(G)$ for any connected G. We finally show that $|\mathrm{aug2}(G)| = |A_2| + |\mathrm{aug2}(G \cup A_2)|$.

Fact 3 ([13,18]) *We can connect G by adding a set E_1 of $z_1 - 1$ edges, with the conditions that the edges are chosen to be incident on non-cut-node vertices in 2-blocks corresponding to degree-0 or -1 isolated vertices in blk2(G), and a vertex or 2-block is incident by at most one edge of E_1 unless it is isolated. Then $|\mathrm{aug2}(G \cup E_1)| = |\mathrm{aug2}(G)| - (z_1 - 1)$.*

For the rest of this section, we further assume G is connected. In this case, $|\mathrm{aug2}(G)| = \max\{d - 1, \lceil \ell/2 \rceil\}$. It is trivial to find $\mathrm{aug2}(G)$ when $\ell \leq 2$. Hence we also assume $\ell > 2$, the root of T is r and T is normalized for the rest of this section.

We first handle two special cases of the algorithm and then show our complete algorithm. We add edges to T, instead of G, to biconnect G. The two endpoints of each added edge are degree-0 or -1 vertices. Note that degree-0 and -1 vertices are b-nodes. When an edge (u_1, u_2) is added T, corresponding, an edge (u'_1, u'_2) is added to G where u'_i is a non-cut-node vertex in the 2-block representing by u_i. It is well-known that if T is biconnected, then G is biconnected [4].

3.1 T Is Normalized and Rooted at a B-Node

This subsection finds aug2(G) if T is normalized and is rooted at a b-node.

Lemma 1. *Let $v_0, \ldots, v_{\ell-1}$ be the degree-1 vertices of T. Let $E_2 = \{(v_i, v_{(i+\lceil \ell/2 \rceil) \bmod \ell}) \mid 0 \le i \le \lceil \ell/2 \rceil - 1\}$. If r is a b-node, then (1) $|\text{aug2}(G)| = \lceil \ell/2 \rceil = |E_2|$ and (2) $T \cup E_2$ is biconnected.*

Proof. We first prove part 1. Assume the contrary that $|\text{aug2}(G)| = d - 1 > \lceil \ell/2 \rceil$. Then there is a c-node v whose degree is d. Since the root is a b-node, hence v is not the root. Let T_u be the branch of r that contains v. Since T_v has at least $d - 1$ degree-1 vertices of T, T_u has at least $d - 1$ degree-1 vertices of T. Hence T is not normalized, which is a contradiction. Thus part 1 is true.

Let $T' = T \cup E_2$. We now prove part 2 by showing that T' has no cut node. Let v be a vertex of T. It is clear that if v is not a cut node of T, then it is not one of T'. Assume v is a cut node of T. Note that any two vertices in $T' - T_v$ are connected in $T' - \{v\}$. Let $v_1, \ldots, v_{\text{dg}_T(v)-1}$ be the children of v. For any given i such that $1 \le i \le \text{dg}_T(v) - 1$, note also that any two vertices in T_{v_i} are also connected in $T' - \{v\}$. Each T_{v_i} has at least one degree-1 vertex of T. Let b_{n_i} be a degree-1 vertex of T_{v_i}. Hence either the edge $(b_{n_i}, b_{n_i+\lceil \ell/2 \rceil})$ or $(b_{n_i - \lceil \ell/2 \rceil}, b_{n_i})$ is in E_2. Without loss of generality, assume $(b_{n_i}, b_{n_i+\lceil \ell/2 \rceil})$ is in E_2. Since T is normalized, T_{v_i} has at most $\lceil \ell/2 \rceil$ degree-1 vertices of T. Thus $b_{n_i+\lceil \ell/2 \rceil}$ is not in T_v. Hence a vertex in T_{v_i} is connected to a vertex in $T' - T_v$. This implies that $T' - v$ is connected. Hence T' is biconnected.

3.2 T Is Normalized and Rooted at a C-Node

This subsection assume T is normalized and is rooted at a c-node. We find a set E_4 of edges such that (1) there is a b-node in blk2($G \cup E_4$) and blk2($G \cup E_4$) is normalized if it is rooted at this b-node, and (2) $|\text{aug2}(G)| = |\text{aug2}(G \cup E_4)| + |E_4|$. From § 3.1, aug2($G \cup E_4$) can be found. Hence we derive aug2(G) for this case. We first show a key lemma.

> **Key Lemma.** The following states a key lemma used by our algorithms for finding aug2(G) and aug3(G).

Lemma 2. *Let $s_1, \ldots, s_{f_2}$ be integers that are at least 2. Let $s \ge \sum_{i=1}^{f_2} s_i$.*

1. *For any i^*, $1 \le i^* \le f_2$, if $s_{i^*} \le \lceil s/2 \rceil$, then $s_{i^*} - (\lceil s/2 \rceil - f_2 + 2) \le f_2 - 2$.*
2. *If there is an i^* such that $s_{i^*} > \lceil s/2 \rceil - f_2 + 2$, then for each $i \ne i^*$ $s_i \le (\lceil s/2 \rceil - f_2 + 2)$.*
3. *If there are i_1 and i_2 such that $s_{i_1}, s_{i_2} \ge \lceil s/2 \rceil - f_2 + 2$, then $s_{i_1} = s_{i_2} = \lceil s/2 \rceil - f_2 + 2$ and $s_i = 2$ for all $i \notin \{i_1, i_2\}$.*

Proof.

Part 1: Note that $s_{i*} = s_{i*} - (\lceil s/2 \rceil - f_2 + 2) + (\lceil s/2 \rceil - f_2 + 2)$. Assume the contrary, i.e., $s_{i*} - (\lceil s/2 \rceil - f_2 + 2) > f_2 - 2$. Hence $s_{i*} > (f_2 - 2) + \lceil s/2 \rceil - f_2 + 2 = \lceil s/2 \rceil$. This contradicts to the fact that $s_{i*} \le \lceil s/2 \rceil$.

Part 2: Assume that there exists $j \ne 1$ such that $s_j > (\lceil s/2 \rceil - f_2 + 2)$. Note that $s_{i*} > (\lceil s/2 \rceil - f_2 + 2)$ and $s_i \ge 2$ by assumption. Hence $s \ge s_{i*} + s_j + 2 \cdot (f_2 - 2) > \sum_{i=1}^{2}(\lceil s/2 \rceil - f_2 + 2) + 2 \cdot (f_2 - 2) \ge s$. Hence we have reached a contradiction, which implies that j does not exist.

Part 3: Follows from the facts that $s_i \ge 2$ and $s \ge s_{i_1} + s_{i_2} + 2 \cdot (f_2 - 2) \ge 2 \cdot \lceil s/2 \rceil$.

Our Algorithm. Let G be *balanced* if $d - 1 \le \lceil \ell/2 \rceil$.

Fact 4 ([18]) *Let G be connected and let v^* be a c-node of T such that $\deg_T(v^*) - 1 > \lceil \ell/2 \rceil$. We reroot T at v^*. Let $\delta = d - 1 - \lceil \ell/2 \rceil$. Then we can find at least $2 \cdot \delta + 2$ v^*-chains. Let Q be the set of v^*-chain leaves. By adding a set E_3 of $2 \cdot \delta$ edges to connect $2 \cdot \delta + 1$ vertices of Q, we can obtain a balanced graph G' and $|aug2(G')| = |aug2(G)| - 2 \cdot \delta$. Furthermore, v^* remains to be a c-node of $blk2(G')$ and $blk2(G')$ rooted at v^* is normalized.*

For the rest of this section, we assume G is balanced. We use the following notations.

- Let the children of the root r in T be $b_1, \ldots, b_p, g_1, \ldots, g_q$ where T_{b_i}, $1 \le i \le p$, is a chain and T_{g_i}, $1 \le i \le q$, is not a chain.
- Let ℓ_i be the number of degree-1 vertices of T in T_{g_i}.
- Let $\mu_i = \max\{0, \ell_i - (\lceil \ell/2 \rceil - q + 2)\}$ for all $1 \le i \le q$.
- Let $x_{i,j}$, $1 \le i \le q$ and $1 \le j \le \ell_i$, be the degree-1 vertices of T in T_{g_i}.
- Let $g_1, \ldots, g_q$ be ordered in the way such that $\ell_1 > 1 + \lceil \ell/2 \rceil - (q - 1)$ if such a branch of r exists; otherwise, g_i's are ordered arbitrarily.

Let $E_4 = \{(x_{1,i}, x_{i+1,1}) \mid 1 \le i \le \mu_1 + 1\} \cup \{(x_{i,2}, x_{i+1,1}) \mid \mu_1 + 2 \le i \le q - 1\}$.

Lemma 3.

1. *For each edge $(u, v) \in E_4$, there exists no i such that T_{g_i} contains both u and v.*
2. *Let h_i be the number of degree-1 vertices of T_{g_i} that are not endpoints of the edges in E_4. Then $h_i \le \lceil \ell/2 \rceil - (q - 1)$.*
3. *Let $T' = blk2(G \cup E_4)$. Let ℓ' be the number of degree-1 vertices in T'. Then $\ell' = \ell - 2 \cdot (q - 1)$ and there is a b-node r' such that T' rooted at r' is normalized.*

3.3 The Complete Algorithm

Using the algorithms for the above two special cases as subroutines, our complete algorithm for finding $aug2(G)$ is as follows.

---

1. If G is disconnected, then apply Fact 3 to find a set of edges E_1 such that $G \cup E_1$ is connected; otherwise, let $E_1 = \emptyset$.

2. If blk2($G \cup E_1$) has at most two degree-1 vertices, then find aug2($G \cup E_1$) trivially; return $E_1 \cup$ aug2($G \cup E_1$).

3. Find a normalized blk2($G \cup E_1$). If the found normalized blk2($G \cup E_1$) is rooted at a b-node, then apply Lemma 1 on $G \cup E_1$ to find a set of edges E_2; return $E_1 \cup E_2$.

4. If the found normalized blk2($G \cup E_1$) is rooted at a c-node r, then do the followings.

   (1). If $G \cup E_1$ is not balanced, then apply Fact 4 on $G \cup E_1$ to find a set of edges E_3 such that $G \cup E_1 \cup E_3$ is balanced; otherwise, let $E_3 = \emptyset$. Root blk2($G \cup E - 1 \cup E_3$) at r.

   (2). Apply Lemma 3 on $G \cup E_1 \cup E_3$ to find a set of edges E_4.

   (3). Find a normalized blk2($G \cup E_1 \cup E_3 \cup E_4$) rooted at a b-node. Apply Lemma 1 on $G \cup E_1 \cup E_3 \cup E_4$ to find a set of edges E_2 such that $G \cup E_1 \cup E_2 \cup E_3 \cup E_4$ is biconnected.

   (4). Return $E_1 \cup E_2 \cup E_3 \cup E_4$.

Theorem 1. *Assume G is a graph with at least 3 vertices.*

1. *Given* blk2(G), *we can find* aug2(G) *in* $O(\log n)$ *time using* $O((n+m)/\log n)$ *processors on an EREW PRAM.*

2. *We can find* aug2(G) *in sequential linear time or in parallel* $O(\log n + \mathcal{T}_\mathcal{M}(n + m))$ *time using* $O((n + m)/\log n + \mathcal{P}_\mathcal{M}(n + m))$ *processors on an $\mathcal{M}$ PRAM.*

4   Smallest Triconnectivity Augmentation

In this section, we show the approach we used to find aug2(G) can also be used to find aug3(G). We first give an algorithm to find aug3(G) when G is biconnected. Then we find a set of edges A_3 to add to G such that $G \cup A_3$ is biconnected and $|\text{aug3}(G)| = |A_3| + |\text{aug3}(G \cup A_3)|$.

4.1   Three-Block Tree for a Biconnected Graph

Here we further assume G is biconnected. A structure called *3-block tree* and denoted by blk3(G) that is similar to 2-block tree is known when G is biconnected [12]. A pair of vertices is a *separating pair* if either they are both adjacent to the same degree-2 vertex or the removing of them results in a disconnected graph. Given two vertices $\{u_1, u_2\}$ of G, the *separating degree* $\text{sd}_G(\{u_1, u_2\})$ is the number of connected components in $G - \{u_1, u_2\}$. A sequence of vertices $p_0, \ldots, p_{k-1}$ in G is called a *polygon* if $\{p_i, p_j\}$ is a separating pair with the separating degree 2 for all $j \neq (i + 1) \bmod k$, and either $\{p_i, p_{(i+1) \bmod k}\}$ is a separating pair or $(p_i, p_{(i+1) \bmod k})$ is an edge of G. The separating pairs *represented by* the polygon $p_0, \ldots, p_{k-1}$ are all the separating pairs of the form $\{p_i, p_j\}$.

Due to space limitation, the detailed description and the construction of the 3-block tree is omitted and can be found in [12]. Here we describe the properties of the 3-block tree. The 3-block tree is a tree whose vertex set is the union of P, S and D, representing the set of polygons (π-vertices), separating pairs (σ-vertices) , and 3-blocks (β-vertices), respectively. The edge set of blk3(G) is the union of $\{(d, s) \mid d \in D, s \in S$, and the separating pair represented by s are contained in or adjacent to the 3-block $d\}$ and $\{(p, s) \mid p \in P, s \in S$, and the separating pair s is a pair of vertices in the polygon represented by $p\}$.

From [3,8,17,21], we know that blk3(G) is a tree if G is biconnected. We call this tree the *3-block tree* for G. From [6,17], the size of blk3(G) is $O(n)$ and can be constructed in the same sequential and parallel complexities as the one for finding connected components.

4.2   An Algorithm to Find aug3(G) when G is Biconnected

In this section, let $T_3 = $ blk3(G). Let ℓ_3 be the number of degree-1 vertices in T_3. Let d_3 be the maximum degree of all the σ-vertices of T_3. From [12], aug3(G) = max$\{d_3 - 1, \lceil \ell_3/2 \rceil\}$. It is trivial to find aug3(G) if $\ell_3 < 3$. Hence we assume $\ell_3 \geq 3$ for the rest of the discussion.

We find a vertex r of T_3 such that T_3 is normalized if it is rooted at r. By Fact 2, such an r exists if $\ell_3 \geq 3$. Our algorithm finds aug3(G) that are edges whose endpoints are degree-1 vertices of T_3. A found edge (u_1, u_2) corresponds to an edge (u_1', u_2') in G where u_i' is a vertex of G such that it is in the 3-block representing by u_i, but not in any separating pair of G.

Lemma 4. *Let $v_0, \ldots, v_{\ell_3-1}$ be the degree-1 vertices of T_3 numbered according to the preorder visiting sequence. Let $E_5 = \{(v_i, v_{(i+\lceil \ell_3/2 \rceil)} \bmod \ell_3) \mid 0 \leq i \leq \lceil \ell_3/2 \rceil - 1\}$. During the preorder traveling, let the neighbors of the π-vertices be visited according to the clockwise order they are encountered while the corresponding polygon is traversed. If r is a β- or π-vertex, then (1) $|$aug3(G)$| = \lceil \ell_3/2 \rceil = |E_5|$ and (2) $G \cup E_5$ is triconnected.*

Proof. The proof of $|$aug3(G)$| = \lceil \ell_3/2 \rceil = |E_5|$ is the same with the proof of Lemma 1. We now prove the second part. If r is a β-vertex, the proof is the same with the proof of Lemma 1. We now assume r is a π-vertex. It is trivial to know that any separating pair of $G \cup E_5$ must be a separating pair of G. Assume $G \cup E_5$ is not triconnected and let $\{s_1, s_2\}$ be a separating pair of both G and $G \cup E_5$. If $\{s_1, s_2\}$ is represented by a σ-vertex in blk3(G), then the proof of Lemma 1 can also be applied here. Hence we assume the contrary that s_1 and s_2 are two non-adjacent vertices in the polygon represented by r. In this case, removing $\{s_1, s_2\}$ from G results in a graph with exactly two connected components. Let G_1 and G_2 be the two connected components obtained. Let $G_1' = G_1 \cup \{(s_1, s_2)\}$ and let $G_2' = G_2 \cup \{(s_1, s_2)\}$. Then both blk3($G_1'$) and blk3($G_2'$) contain a degree-1 β-vertex blk3(G). Since the branches of r are visited in a clockwise order when we perform a preorder traverse to number the degree-1 β-vertices. The original indexes of degree-1 β-vertices contained in either blk3(G_1') or blk3(G_2') are consecutive. Assume without loss of generality the formal case holds, i.e., $v_i, v_{i+1}, \ldots, v_j$ are the degree-1 β-vertices of T_3 contained in blk3(G_1'). Hence there is an edge $(v_k, v_h) \in E_5$ such that $v_k \in \{v_i, v_{i+1}, \ldots, v_j\}$ and $v_h \notin \{v_i, v_{i+1}, \ldots, v_j\}$. This implies $(G \cup E_5) - \{s_1, s_2\}$ is connected. Thus $G \cup E_5$ is triconnected.

Lemma 5. *If r is a σ-vertex, then we can use the same algorithm as the one used for finding* aug2(G) *by first applying Lemma 3 and then Lemma 1 to find* aug3(G).

Theorem 2. *Assume G is a biconnected graph with at least 4 vertices.*

1. *We can find* aug3(G) *in sequential linear time.*
2. *Given* blk3(G), *we can find* aug3(G) *in $O(\log n)$ time using $O((n+m)/\log n)$ processors on an EREW PRAM.*

4.3   An Algorithm to Find aug3(G) for any G

We now define blk3(G) for an arbitrary G. We first compute blk2(G). For each 2-block B of blk2(G), we first compute blk3(B) as follows. If B is nontrivial, then it is biconnected and we then compute blk3(B). If B is trivial, then let blk3(B) be a one-β-vertex graph whose single vertex represents the trivial 3-block B. The 3-block graph for G is the forest consisted of the 3-block trees of all the 2-blocks of G. The separating pairs of G are the collection of all the separating pairs represented by σ- or π-vertices in blk3(B).

Given two vertices $\{u_1, u_2\}$ of G such that $\{u_1, u_2\}$ is represented by a σ-vertex s in a 3-block graph, then we also use sd$_G(s)$ to denote sd$_G(\{u_1, u_2\})$. Let $d_3(G)$ be the largest separating degree of all the separating pairs in G.

In the following definitions, let B be a 2-block of G.

- A β-vertex in blk3(B) is *legal* if either (1) it is degree-1 in blk3(B) and does not contain any cut node of G that is not in a separating pair of B, or (2) it is degree-0 in blk3(B) and contains at most two cut nodes of G. For convenience, if a β-vertex is legal, then its corresponding 3-block is also called *legal.*
- A *demanding* vertex of G is a vertex in a legal 3-block D such that it is (1) not a cut node or in any separating pair of G if D is non-trivial, or (2) the vertex in D if D is trivial. It is well-known that each legal 3-block has one demanding vertex.
- Given an induced subgraph Z of B in G, let $w_{3,G}(Z)$ denote the number of legal degree-1 β-vertices in blk3(B) whose corresponding 3-blocks are contained in Z.
- Let the *weight* of B in G, $w_G(B)$, be $\max\{3 - \text{dg}_{\text{blk2}(G)}(B), w_{3,G}(B)\}$.
- Given a connected component H of G, the *weight* of H in G is $W_G(H) = \sum_{\forall \text{ 2-block } B \in H} w_G(B)$.
- The *weight* of a graph G, $W(G)$, is $\sum_{\forall \text{ connected component } H \in G} W_G(H)$ if G is not biconnected; is $w_{3,G}(G)$ if G is biconnected.

From [12], $|\text{aug3}(G)| = \max\{d_3(G) - 1, \lceil W(G)/2 \rceil\}$.

$\boxed{G \text{ is Disconnected.}}$ Let z_1 be the number of connected components of G.

Theorem 3. *[12] We can connect G by adding a set E_7 of $z_1 - 1$ edges, where each edge is chosen to be incident on demanding vertices in different trees in the forest $\text{blk3}(G)$ with the addition constraint that a vertex or 2-block is incident by at most one edge of E_7 unless it is isolated. Then $|\text{aug3}(G \cup E_7)| = |\text{aug3}(G)| - (z_1 - 1)$.*

$\boxed{G \text{ is Connected, but not Biconnected.}}$ We assume G is connected for the rest of this section. Hence $\text{blk2}(G)$ is a tree. Let ℓ be the number of degree-1 b-vertices in $\text{blk2}(G)$. Let $\mathcal{B}$ be the set of 2-blocks that either are degree-1 b-vertices in $\text{blk2}(G)$ or with weights more than $\lceil W(G)/2 \rceil - \ell + 1$. It is trivial to see that $\lceil W(G)/2 \rceil - \ell + 1 \geq 2$ if $\ell > 0$.

Corollary 1. *If there are two 2-blocks in $\mathcal{B}$ with weights greater than $\lceil W(G)/2 \rceil - \ell + 1$, then the weights of these two 2-blocks both equal $\lceil W(G)/2 \rceil - \ell + 2$, the weights of the other 2-blocks in $\mathcal{B}$ all equal 2, and $|\mathcal{B}| = \ell$.*

Proof. Follows from Lemma 2(3) by properly setting the parameters.

We assign indexes to 2-blocks in $\mathcal{B}$ and denote them by $B_1, \ldots, B_{|\mathcal{B}|}$ using the following rules: (1) if there is a 2-block in $\mathcal{B}$ with weight more than $\lceil W(G)/2 \rceil - \ell + 1$, then let B_1 be such a 2-block; and (2) if there are two 2-blocks in $\mathcal{B}$ with weights more than $\lceil W(G)/2 \rceil - \ell + 1$, then let B_1 and B_ℓ be these two 2-blocks. Remark that by Corollary 1, $\ell = |\mathcal{B}|$ in the latter case.

Let $y_{i,1}, \ldots, y_{i,w_{3,G}(B_i)}$ be the legal degree-1 β-vertices in $\text{blk3}(B_i)$ if B_i is nontrivial. If B_i is trivial or is triconnected, then let $y_{i,1} = y_{i,2}$ be the 3-block of $\text{blk3}(B_i)$. Note that if $w_G(B_1) > \lceil W(G)/2 \rceil - \ell + 2$, then B_1 is neither trivial nor triconnected. Let $\lambda_1 = \max\{0, w_G(B_1) - (\lceil W(G)/2 \rceil - \ell + 1)\}$.

 - Let $E_8 = \{(y_{1,i}, y_{i+1,1}) \mid 1 \leq i \leq \min\{\ell - 1, \lambda_1\}\} \cup \{(y_{i,2}, y_{i+1,1}) \mid \min\{\ell - 1, \lambda_1\} < i \leq \ell - 1\}$.

Each endpoint of an edge in E_8 is a 3-block. Correspondingly, E_8 represents a set of edges E_8' where each edge of E_8' is chosen to be incident on demanding vertices of the endpoints of edges in E_8 with the addition constraint that a vertex is not incident by two edges of E_8' unless it is in a trivial 2-block. For ease of description, we use $G \cup E_8$ instead of $G \cup E_8'$ in the discussion. For the rest of the discussion, let $G_8^* = G \cup E_8$.

Lemma 6.

1. G_8^* is biconnected.
2. If G is not biconnected, then $W(G_8^*) = W(G) - 2 \cdot (\ell - 1) \geq 2$.
3. Let s be a separating pair of G_8^*. Then $\text{sd}_{G_8^*}(s) \leq \max\{2, \lceil W(G)/2 \rceil - \ell + 2, \text{sd}_G(s) - \ell + 1\}$.

Proof. (Sketch) We first argue G_8^* contains no cut node. We then number the edges in E_8 in a certain order and then prove each edge addition decreases the weight. Each edge addition also decrease the separating degree of separating pairs with "large" separating degrees.

Theorem 4. $\text{aug3}(G_8^*) = \text{aug3}(G) - |E_8|$.

Proof. It is straightforward to see $|E_8| = \ell - 1$. If G is biconnected, then $\text{aug3}(G_8^*) = \text{aug3}(G) - (\ell - 1)$ holds trivially. Hence we assume G is not biconnected. It is also straightforward to see $\text{aug3}(G_8^*) \geq \text{aug3}(G) - (\ell - 1)$. By Lemma 6, $\text{aug3}(G_8^*) = \max\{d_3(G_8^*) - 1, \lceil W(G_8^*)/2 \rceil\} \leq \max\{\max\{2, \lceil W(G)/2 \rceil - \ell + 2, d_3(G) - \ell + 1\} - 1, \lceil W(G)/2 \rceil - (\ell - 1)\}$. Since $W(G_8^*) \geq 2$, $\text{aug3}(G_8^*) \leq \text{aug3}(G) - (\ell - 1)$. Thus $\text{aug3}(G_8^*) = \text{aug3}(G) - (\ell - 1)$.

The Complete Algorithm.

Theorem 5. *Assume G is a graph with at least 4 vertices.*
1. *Given* $\text{blk3}(G)$, *we can find* $\text{aug3}(G)$ *in* $O(\log n)$ *time using* $O((n+m)/\log n)$ *processors on an EREW PRAM.*
2. *We can find* $\text{aug3}(G)$ *in sequential linear time or in parallel* $O(\log n + \mathcal{T}_\mathcal{M}(n + m))$ *time using* $O((n + m)/\log n + \mathcal{P}_\mathcal{M}(n + m))$ *processors on an $\mathcal{M}$ PRAM.*

Proof. Our algorithm is as follows.

---

1. If G is disconnected, then use Theorem 3 on G to find a set of edges E_7 such that $G \cup E_7$ is connected; otherwise, let $E_7 = \emptyset$.
2. If $G \cup E_7$ is not biconnected, then apply Theorem 4 on $G \cup E_7$ to find a set of edges E_8 such that $G \cup E_7 \cup E_8$ is biconnected; otherwise, let $E_8 = \emptyset$.
3. If $\text{blk3}(G \cup E_7 \cup E_8)$ has at most two degree-1 vertices, then find $\text{aug3}(G \cup E_7 \cup E_8)$ trivially; return $E_7 \cup E_8 \cup \text{aug3}(G \cup E_7 \cup E_8)$.
4. Apply Theorem 2 on $G \cup E_7 \cup E_8$ to find a set of edges E_9 such that $G \cup E_7 \cup E_8 \cup E_9$ is triconnected.
5. Return $E_7 \cup E_8 \cup E_9$.

From Theorems 2, 3 and 4, $|E_7 \cup E_8 \cup E_9| = |\text{aug3}(G)|$. It is trivial to implement the algorithms used to find E_7, E_8 and E_9 in the desired sequential and parallel complexities.

5   Concluding Remarks

This paper gives a very simple algorithm for solving the smallest biconnectivity augmentation problem. Our new approach does not use any dynamic data structure and only requires simple arithmetic computations on trees. Our approach can be generalized to also solve the smallest triconnectivity augmentation problem in linear time. The obtained simple sequential algorithms naturally imply faster parallel implementations than the one given in [13]. The only bottleneck that keeps our parallel implementations from being $O(\log n)$-time-optimal is the subroutine for finding connected components.

References

1. K. W. Chong and T. W. Lam. Finding connected components in $O(\log n \log \log n)$ time on the EREW PRAM. In *Proceedings of SODA*, pages 11–20, 1993. 279, 280
2. R. Cole and U. Vishkin. Approximate parallel scheduling. Part II: Applications to logarithmic-time optimal graph algorithms. *Information and Computation*, 92:1–47, 1991. 279, 280
3. G. Di Battista and R. Tamassia. On-line graph algorithms with SPQR-trees. *Algorithmica*, 15:302–318, 1996. 285
4. K. P. Eswaran and R. E. Tarjan. Augmentation problems. *SIAM Journal on Computing*, 5:653–665, 1976. 278, 279, 281, 282
5. A. Frank. Connectivity augmentation problems in network design. In J. R. Birge and K. G. Murty, editors, *Mathematical Programming: State of the Art 1994*, pages 34–63. The University of Michigan, 1994. 278
6. D. Fussel, V. Ramachandran, and R. Thurimella. Finding triconnected components by local replacements. *SIAM Journal on Computing*, 22(3):587–616, 1993. 285
7. F. Harary. *Graph Theory*. Addison-Wesley, Reading, Massachusetts, 1969. 280, 281
8. J. E. Hopcroft and R. E. Tarjan. Dividing a graph into triconnected components. *SIAM Journal on Computing*, 2:135–158, 1973. 285
9. T.-s. Hsu. *Graph Augmentation and Related Problems: Theory and Practice*. PhD thesis, University of Texas at Austin, 1993. 278
10. T.-s. Hsu. Undirected vertex-connectivity structure and smallest four-vertex-connectivity augmentation (extended abstract). In *Springer-Verlag LNCS Vol. 1004: Proceedings of 6th Intl. Symp. on Alg. and Comp.*, pages 274–283, 1995. 278
11. T.-s. Hsu. On four-connecting a triconnected graph. *Journal of Algorithms*, 35:202–234, 2000. 278
12. T.-s. Hsu and V. Ramachandran. A linear time algorithm for triconnectivity augmentation. In *Proceedings of FOCS*, pages 548–559, 1991. 278, 279, 280, 284, 285, 287
13. T.-s. Hsu and V. Ramachandran. On finding a smallest augmentation to biconnect a graph. *SIAM Journal on Computing*, 22:889–912, 1993. 278, 280, 281, 288
14. T. Jordán. On the optimal vertex connectivity augmentation. *Journal of Combinatorial Theory, Series B*, 63:8–20, 1995. 278
15. T.-H. Ma. On the biconnectivity augmentation problem. In *Proceedings of CTS Workshop on Combinatorics and Algorithms*, pages 66–73. National Center for Theoretical Sciences Taiwan, December 21–23 1998. 278
16. J. Plesník. Minimum block containing a given graph. *ARCHIV DER MATHE-MATIK*, XXVII:668–672, 1976. 278, 281
17. V. Ramachandran. Parallel open ear decomposition with applications to graph biconnectivity and triconnectivity. In J. H. Reif, editor, *Synthesis of Parallel Algorithms*, pages 275–340. Morgan-Kaufmann, 1993. 280, 285
18. A. Rosenthal and A. Goldner. Smallest augmentations to biconnect a graph. *SIAM Journal on Computing*, 6:55–66, 1977. 278, 280, 281, 283
19. R. E. Tarjan. Depth-first search and linear graph algorithms. *SIAM Journal on Computing*, 1:146–160, 1972. 280
20. R. E. Tarjan and U. Vishkin. An efficient parallel biconnectivity algorithm. *SIAM J. Comput.*, 14:862–874, 1985. 280
21. W. T. Tutte. *Connectivity in Graphs*. University of Toronto Press, 1966. 285
22. T. Watanabe and A. Nakamura. A minimum 3-connectivity augmentation of a graph. *Journal of Computer and System Science*, 46:91–128, 1993. 278

Scheduling Broadcasts in Wireless Networks

Bala Kalyanasundaram[*1], Kirk Pruhs[**2], and Mahe Velauthapillai[***1]

[1] Dept. of Computer Science
Georgetown University
Washington D.C. 20057 USA
{kalyan, mahe}@cs.georgetown.edu
[2] Dept. of Computer Science
University of Pittsburgh
Pittsburgh, PA. 15260 USA
kirk@cs.pitt.edu
http://www.cs.pitt.edu/~kirk

Abstract. We consider problems involving how to schedule broadcasts in a pulled-based data-dissemination service, such as the DirecPC system, where data requested by the clients is delivered via broadcast. In particular, we consider the case where all the data items are of approximately equal in size and preemption is not allowed. We give an offline $O(1)$-speed $O(1)$-approximation algorithm for the problem of minimizing the average response time. We provide worst-case analysis, under various objective functions, of the online algorithms that have appeared in the literature, namely, *Most Requests First*, *First Come First Served*, and *Longest Wait First*.

1 Introduction

1.1 Motivation, Problem Statement and Background

Recently the WWW has led an explosive demand for pulled-based data-dissemination services, that is services where multiple clients request information from a source of information. Complementing the demand for such services is the greatly increased bandwidth that is available in satellite, wireless, and cable TV based networks that employ broadcasting to disseminate information. Broadcasting is particularly appropriate for data dissemination applications since broadcasting can exploit commonalities among the requests from the clients to improve service. That is, if multiple clients make identical requests at roughly the same time (imagine the 50,000 requests per second to www.yahoo.com), then the server can satisfy all of these requests in one broadcast. Data broadcasting is used in several commercial systems, including the Intel Intercast System [8], the Hughes DirecPC System [6], the Hybrid System [7], and the AirMedia System [3]. For example, in the Hughes DirecPC system (depicted in figure 1) the

* Supported in part by NSF Grant CCR-9734927 and by ASOSR grant F49620010011.
** Supported in part by NSF Grant CCR-9734927 and by ASOSR grant F49620010011.
*** Supported in part by a gift from AT&T

M. Paterson (Ed.): ESA 2000, LNCS 1879, pp. 290–301, 2000.
© Springer-Verlag Berlin Heidelberg 2000

clients request data items over phone lines, and the data is delivered via satellite broadcast. In this paper we consider the problem of how to schedule such broadcasts to optimize various objective functions, including minimizing the average response time. We provide worst-case analysis for various offline and online algorithms.

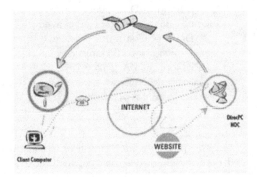

Fig. 1. A depiction of the DirecPC System from their WWW site [6])

The setting for the problems that we consider consists of n possible data items, $P_1, \ldots, P_n$, which we will henceforth call pages. Time is discrete. At each time $t \geq 0$, the server receives $R_i(t)$ new requests for page P_i. At each time t, a k-speed server, can select k pages to broadcast. We say that the $R_i(t)$ requests for page P_i at time t are satisfied at time $C_i(t)$ if $C_i(t)$ is the first time strictly after t when P_i is broadcast. In the online setting, the server is not aware of the requests $R_i(t)$ until time t, and at time t must pick a page to satisfy at time t. Thus in the standard scheduling notation [10] the job environment would be $r_i, x_i = 1$.

For concreteness of explanation, consider for the moment the objective function of minimizing the average response time, which is probably the most commonly used measure of system performance [5]. The *response/flow* time of a request $R_i(t)$ is $(C_i(t) - t)$, which is how long the clients that request page P_i at time t have to wait until P_i is broadcast. The *average response time* is then

$$\frac{\sum_t \sum_i R_i(t) \cdot (C_i(t) - t)}{\sum_t \sum_i R_i(t)}$$

This represents the average time that a user would have to wait for a response. Now consider the example shown in table 1. For this instance, the schedule that minimizes the total response time for a unit speed server is to broadcast P_a at time 1, P_b at time 2, and P_c at time 3. This schedule produces a total response time of 14, and an average response time of $\frac{14}{8}$.

Before stating our results, let us review resource augmentation analysis. Resource augmentation was first proposed in [9] as a method for analyzing schedul-

Input			
$R_i(t)$	$i = a$	$i = b$	$i = c$
$t = 0$	2	3	1
$t = 1$	0	1	1

$R_i(t)(C_i(t) - t)$	$i = a$	$i = b$	$i = c$
$t = 0$	2	6	3
$t = 1$	0	1	2

Table 1. Example Input, and Total Response Time for the schedule P_a, P_b, P_c

ing algorithms. We adopt the terminology proposed in [12]. In the context of broadcast scheduling, an algorithm A is said to be an *s-speed c-approximation* algorithm for a particular minimization objective function if

$$\max_{\text{Inputs } I} \frac{A_s(I)}{\text{OPT}(I)} \leq c$$

where $A_s(I)$ is the value of the objective function of the output of A equipped with a speed s server on input I, and $\text{OPT}(I)$ is value of the objective function in the optimal solution with a unit speed server. If the objective function is a maximization function, A is said to be an *s-speed c-approximation* if

$$\max_{\text{Inputs } I} \frac{\text{OPT}(I)}{A_s(I)} \leq c$$

An algorithm A is a c-approximation, or equivalently a c-competitive, algorithm if A is an 1-speed c-approximation algorithm.

1.2   Related Results

The problem of minimizing the maximum response time was considered in [4] with the job environment $r_i, pmtn$, that is, pages may have variable size (a page of size k would have to be broadcast for k time units to be completed) and preemption is allowed (also it is assumed that clients receive a page in a cyclic order that need not start from the beginning). They give an offline polynomial time algorithm that is $(1 + \epsilon)$-competitive. The key insight is that if the optimal maximum response time OPT was known, then this establishes a deadline for every request, and one can easily write an interval-indexed linear program whose solution gives a schedule with maximum response time OPT. One can then obtain a $(1 + \epsilon)$-competitive algorithm by performing a binary search on the possible values of OPT. They also showed that First Come First Served (FCFS) is 2-competitive, and that no deterministic online algorithm can be better than 2-competitive.

There have also been several empirical, or average case, studies of broadcast scheduling problems (see for example [1] or [2]). In particular, [1] provides a very nice introduction to broadcast scheduling.

1.3   Our Results

To date there have been no worst-case results on how to schedule broadcasts to minimize the average response time, which is perhaps the most common measure of system performance [5].

In section 2 we consider the offline version of the problem of minimizing the average response time. We give an offline $O(1)$-speed $O(1)$-approximation algorithm. Previous techniques fail to provide a good upper bound since the knowledge of OPT does not translate into deadlines for individual jobs. We surmount this difficulty by showing how to express the problem of minimizing the average response time as a (non-time-indexed) integer linear program. The optimal solution LOPT of the relaxed non-integer linear program is then a lower bound for OPT. From LOPT we can determine a deadline for each request that, if met by every request, would essentially guarantee that every request has response time of $O(1)$ times its response time in OPT. We then explain how to meet these deadlines using a variation of Earliest Deadline First scheduling.

In section 3 we consider online broadcast scheduling. We particularly want to analyze the algorithms that have been proposed in the literature (see for example [2]) that are appropriate for our setting, namely:

- Most Requests First (MRF): Broadcast the pages that have the most outstanding requests.
- First Come First Served (FCFS): Broadcast the pages with the earliest unsatisfied requests.
- Longest Wait First (LWF): Broadcast the pages with maximum total waiting time. The waiting time for a page is the total time that all pending requests for that page are waiting. [1]

In subsection 3.1, we show that there is no deterministic online algorithm that is $O(1)$-competitive with respect to average response time. We then turn our attention to $O(1)$-speed $O(1)$-approximation online algorithms. We show that neither MRF nor FCFS are $O(1)$-speed $O(1)$-approximation online algorithms. We conjecture that LWF is in fact an $O(1)$-speed $O(1)$-approximation algorithm with respect to average response time. While we do not provide a proof to this conjecture, we show that this cannot be proved in the standard way. Let $U_A^s(t)$ be the number of requests unsatisfied by algorithm A at time t using a speed s server. To show that an algorithm A is an s-speed c-approximation algorithm, with respect to average response time, it is sufficient to show that $\max_t U_A^s(t)/U_{\text{OPT}}^1(t) \leq c$ [5]. In fact, this is the most commonly used technique to show that an algorithm A is an c-approximation algorithm with respect to average response time (e.g. [5,11,9]). We show, in subsection 3.2, that there is no deterministic online algorithm A that can bound $\max_t U_A^s(t)/U_{\text{OPT}}^1(t)$ by any constant. Thus, there must be times when any $O(1)$-speed online algorithm will have a factor of $\omega(1)$ more unsatisfied requests than the adversary. And thus,

---

[1] The $R \times W$ algorithm introduced in [2] is an attempt to find an algorithm that in some sense approximates LWF, but that is easier to implement than LWF.

one can not prove that an online algorithm is an $O(1)$-speed $O(1)$-approximation algorithm with respect to average response time in this standard way.

While FCFS and MRF may perform badly with respect to average response time, we show in the subsection 3.3 and subsection 3.4 that there are natural objective functions under which MRF is superior to FCFS and LWF, and under which FCFS is superior to MRF and LWF.

We show in subsection 3.3 that while an $O(1)$-speed online algorithm can not be competitive with respect to the number of unsatisfied requests, FCFS is an $O(1)$-speed $O(1)$-approximation algorithm with respect to minimizing the number of unsatisfied pages. More precisely, let $\tilde{U}_A^s(t)$ be the number of pages that have outstanding requests at time t using algorithm A and an s speed server. Consider the objective function of minimizing $\max_t \tilde{U}_A^s(t)/\tilde{U}_{OPT}^1(t)$. We show that FCFS is a 4-speed 4-approximation algorithm for this objective function. That is, FCFS using a 4-speed server guarantees that at every time the number of pages with pending requests is at most 4 times the number of pages with pending requests for the adversary with unit speed server. We show that this is tight by showing the FCFS is not a 3-speed $O(1)$-approximation algorithm for this objective function. It is easy to see that MRF and LWF are not $O(1)$-speed $O(1)$-approximation algorithms for this objective function.

In subsection 3.4 we consider the problem of maximizing the number of satisfied requests. Let $S_A^s(t)$ be the number of requests satisfied by algorithm A by time t using an s speed server. We now consider the objective function of minimizing $\max_t S_{OPT}^1(t)/S_A^1(t)$. We show that MRF is 2-competitive with respect to this objective function, that is, at all times, MRF has satisfied at least half as many requests to date as has the adversary. It is easy to see that neither FCFS nor LWF are $O(1)$-competitive with respect to this objective function.

Thus there are reasonable objective functions under which each of the three algorithms proposed in the literature seems better than the other two.

Due to space limitations, many proofs could not be included in this version of the paper. A fuller copy of the paper may be found on the second authors www site.

2   Offline Approximation

In this section we give a 3-speed 3-approximation algorithm that produces a schedule LEDF. We start by reducing the broadcast problem to an integer program IP. The program IP can be best be thought of as a type of min-cost max-flow problem in a flow graph $G = (V, E)$.

Let T be the latest time that a page was requested. The vertex set V contains one vertex $v_i(t)$ for each page P_i, $1 \leq i \leq n$, and each time t, $0 \leq t \leq T+n$. The set V also contains a source vertex s, and a destination vertex d. For the ease of notation, we use $v_i(-1)$ to denote s and $v_i(T+n+1)$ to denote d, $1 \leq i \leq n$.

There is a directed edge $e_i(-1, 0)$ in E from s to each $v_i(0)$, $1 \leq i \leq n$, with capacity 1 and cost 0. For each $0 \leq t < t' \leq T + n$, and for each $1 \leq i \leq n$, there is a directed edge $e_i(t, t')$ in E from $v_i(t)$ to $v_i(t')$ with capacity 1, and cost $w_i(t, t') = \sum_{j=t}^{t'-1}(t' - j)R_i(j)$; A flow of 1 on this edge corresponds to broadcasting page P_i at time t, and then next broadcasting page P_i at time t', and thus, $w_i(t, t')$ is the contribution to the total flow time for requests to page P_i that arrive between t and $t' - 1$, inclusive. For each $T + 1 \leq t \leq T + n$, and for each $1 \leq i \leq n$, there is a directed edge $e_i(t, T + n + 1)$ in E from $v_i(t)$ to d with capacity 1, and cost 0. The left picture in figure 2 shows the portion of the flow graph G consisting of s, d, the vertices of the form $v_b(*)$ for the input shown in table 1.

Call a schedule k-*feasible* if it broadcasts at most k pages at any one time. Note that the reduction of the broadcast problem to the min-cost max-flow is not complete since a flow may not correspond to a 1-feasible broadcast schedule. But it is easy to add 1-feasibility as a linear constraint. Thus we get the following integer program IP:

$$\min \sum_{t=0}^{T+n} \sum_{t'=t+1}^{T+n} \sum_{i=1}^{n} w_i(t, t')f_i(t, t') \tag{1}$$

$$\sum_{i=1}^{n} f_i(-1, 0) = n \tag{2}$$

$$\sum_{t'<t} f_i(t', t) = \sum_{t'>t} f_i(t, t') \qquad 1 \leq i \leq n \qquad 0 \leq t \leq T + n \tag{3}$$

$$\sum_{t'<t} \sum_{i=1}^{n} f_i(t', t) \leq 1 \qquad 1 \leq t \leq T + n \tag{4}$$

$$f_i(t, t') \in \{0, 1\} \qquad 1 \leq i \leq n \qquad 0 \leq t < t' \leq T + n \tag{5}$$

$$f_i(t, n + T + 1) \in \{0, 1\} \qquad 1 \leq i \leq n \qquad T + 1 \leq t \leq T + n \tag{6}$$

Constraint 2 guarantees that the total flow is n. Constraint 3 guarantees that flow is conserved at each vertex. Constraint 4 guarantees that the schedule is 1-feasible. The right picture in figure 2 shows a (non-optimal) feasible solution corresponding to the MRF schedule b, a, c, b for the example in table 1. The total response time for this feasible solution is 15.

Lemma 1. *The optimal value of the objective function in IP is the minimum total response time for the broadcast problem.*

Proof. A feasible solution to IP can be converted to a broadcast schedule by broadcasting P_i at time t if and only if there exists a positive flow out of $v_i(t)$. One can convert a feasible solution to the broadcast problem to a feasible solution to IP by routing the flow through vertices of the form $v_i(t)$ where P_i is broadcast at time t. If a page P_i is last broadcast at a time $t \leq T$, the last vertex in the flow through the page P_i vertices can be routed through an arbitrary vertex $v_i(k)$,

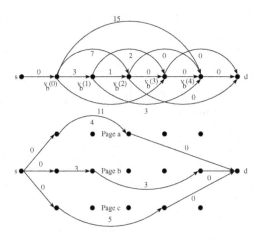

Fig. 2. The flow graph

$T + 1 \leq k \leq T + n$ with the property that no other page is broadcast at time k (such a k must exist since there are n times between $T + 1$ and $T + n$, inclusive). ∎

We relax IP to get a linear program LP in standard way, that is, we replace constraints 5 and 6 by the following constraints:

$$0 \leq f_i(t, t') \leq 1 \qquad 1 \leq i \leq n \quad 0 \leq t < t' \leq T + n \tag{7}$$

$$0 \leq f_i(t, T + n + 1) \leq 1 \qquad 1 \leq i \leq n \quad T + 1 \leq t \leq T + n \tag{8}$$

Let the optimal solution to LP be denoted as LOPT. We also use LOPT to denote the value of the objective function in this optimal solution to LP. It is obvious that LOPT $\leq$ OPT, and hence to approximate OPT well, it is sufficient to approximate LOPT well. One can think of LOPT as a flow, and then decompose this flow into paths in the standard way. Let the paths through the vertices of the form $v_i(*)$ be $W_i(1), \ldots, W_i(g(i))$, and let $F_i(j)$ be the flow on $W_i(j)$. One can also think of LOPT as a probabilistic n-feasible solution to the broadcast problem in the following way: With probability $F_i(j)$, path $W_i(j)$ determines the broadcasts for page P_i, that is, if $W_i(j)$ passes through vertex $v_i(t)$ then page P_i is broadcast at time t. One can easily see that the expected total response time of this n-feasible probabilistic solution is exactly LOPT.

We now consider another way to interpret this probabilistic solution. Let $p_i(t) = \sum_{t' < t} f_i(t', t)$ be the flow into each vertex $v_i(t)$, for $1 \leq i \leq n$, and for $1 \leq t \leq T + n$. The value $p_i(t)$ can be thought of as the probability that page P_i broadcast at time t. We define $n_i(t)$ to be the earliest time t', such that $t' > t$ and such that $\sum_{j=t+1}^{t'} p_i(t) \geq 1$. We will show that $n_i(t)$ is the latest possible time that the requests $R_i(t)$ are satisfied in LOPT. Also, let $\ell_i(t) = 1 - \sum_{j=t}^{n_i(t)-1} p_i(t)$.

Intuitively, $\ell_i(t)$ is the probability that the requests $R_i(t)$ are satisfied at time $p_i(n_i(t))$ in LOPT.

Now consider the following probabilistic n-feasible solution. The requests $R_i(t)$ are satisfied at time $t + 1$ with probability $p_i(t + 1)$, at time $t + 2$ with probability $p_i(t + 2)$, at time $n_i(t) - 1$ with probability $p_i(n_i(t) - 1)$, ..., and at time $n_i(t)$ with probability $\ell_i(t)$. The expected total response time of this probabilistic solution is then:

$$VL = \sum_{i=1}^{n} \sum_{t=1}^{T} R_i(t) \left[\sum_{t'=t+1}^{n_i(t)-1} [p_i(t') \cdot (t' - t)] + \cdot \ell_i(t)(n_i(t) - t) \right] \qquad (9)$$

We now show that this probabilistic solution is in fact the same as the one we directly obtained from LOPT.

Lemma 2. $VL = \text{LOPT}$

Proof. We first show that $VL \leq \text{LOPT}$. Consider an arbitrary request $R_i(t)$. Consider the $g(i)$ paths $W_i(1), \ldots, W_i(g(i))$ in the flow given by LOPT. Assume that these paths are numbered in non-decreasing order of the time that $R_i(t)$ is serviced in the path. It then follows, by induction on b, that the probability that $R_i(t)$ is serviced by time $t + b$ in LOPT, $b \geq 1$, is at least $p_i(t + 1) + p_i(t + 2) + \ldots + p_i(t + b)$. The fact that $VL \leq \text{LOPT}$ then follows.

To see that $VL \geq \text{LOPT}$ it is sufficient to note that VL may be thought of as a feasible solution to LOPT. ∎

The following lemma states the flow crossing each vertex $v_i(t)$, with $R_i(t) > 0$, in LOPT is at least 1. This can easily be seen by observing the structure of G.

Lemma 3. *Let* $R_i(t) > 0$, $1 \leq i \leq n$, *and* $0 \leq t \leq T$. *Then*

$$\sum_{0 \leq j \leq t < k \leq T+n} f_i(j, k) = 1$$

Let $\epsilon < \frac{1}{2}$ be some constant such that $1/\epsilon$ is an integer. Define the jth *break point* of page P_i, denoted $b_i(j)$, to be the earliest time t such that $\sum_{k=1}^{t} p_i(k) \geq j\epsilon$. Let $h(i)$ be the number of break points for page P_i. Let $I_i(j) = [b_i(j), b_i(j+1)]$. We now consider the collection $\mathcal{I}$ of intervals that contain consecutive break-points on the same page, that is $\mathcal{I} = \{I_i(j) \mid 1 \leq i \leq n, 1 \leq j \leq h(i)\}$. If a page P_i is broadcast at time $t \in I_i(j)$, then we say that interval I is *satisfied* at time t. We now explain how to create the schedule LEDF.

Algorithm LEDF: Assume that we have scheduled the broadcasts before time t, and now wish to schedule the broadcasts at time t. Among the intervals that contain time t, select the $1/\epsilon$ intervals with earliest right endpoint to broadcast at time t.

Lemma 4. *Every interval is satisfied in* LEDF.

Proof. The proof is by minimal counter-example. Consider a minimal instance of k intervals for which the claim is not true. Without loss of generality, let the smallest left endpoint of these intervals be 0, and the smallest right endpoint of an unsatisfied interval be z. No intervals have left endpoints to the right of z by the minimality of the instance.

By construction of the intervals, $\sum_{i=1}^{n} \sum_{t=0}^{z} p_i(t) = \sum_{t=0}^{z} \sum_{i=1}^{n} p_i(t) \geq k\epsilon$. Since LOPT is 1-feasible, $z \geq k\epsilon$. Since LEDF employs a $1/\epsilon$ speed server, LEDF could potentially satisfy up to $\frac{z}{\epsilon} \geq k$ intervals during the time interval $[0, z]$. But since not all k intervals were satisfied, it must be the case that LEDF did not use $1/\epsilon$ broadcasts at some time $m \in [0, z)$. Note that by the definition of z, LEDF broadcasted $1/\epsilon$ pages at time z. By the definition of z, it must be the case that every interval with left endpoint in $[0, m]$ must be satisfied by time m. However, then LEDF would fail on the subset of intervals with left endpoints strictly greater than m, which is a strict subset of all of the intervals since at least one interval starts at time 0. We thus reach a contradiction to the minimality of the counter-example.   ∎

Lemma 5. *The total response time of the $\frac{1}{\epsilon}$-feasible schedule* LEDF *is at most* $\frac{\text{LOPT}}{1-2\epsilon}$.

Proof. Consider an arbitrary page P_i, and an arbitrary time t, such that $R_i(t) > 0$. Consider the earliest breakpoint $b_i(j)$ such that $t < b_i(j)$. One can see that such an interval always exists by lemma 3. Therefore, $t \in [b_i(j-1), b_i(j)]$. By lemma 4, there must exist a time $t' \in [b_i(j), b_i(j+1)]$ such that page P_i is broadcasted at time t'.

By lemma 2, the requests $R_i(t)$ must be satisfied in LOPT at time $b_{j+1}(i)$, or later, with probability at least $1 - 2\epsilon$. Hence, the contribution to the total response time made by $R_i(t)$ in LOPT is at least $(1 - 2\epsilon)R_i(t)(b_i(j+1) - t)$.

By lemma 4, $R_i(j)$ is satisfied in LEDF by time $b_{j+1}(i)$. Hence, the contribution to the total response time made by $R_i(t)$ in LEDF is at most $R_i(t)(b_i(j+1) - t)$. The result then follows.   ∎

By setting $\epsilon = \frac{1}{3}$, we get the following Theorem.

Theorem 1. *There is polynomial time 3-speed 3-approximation algorithm for the broadcast problem where the objective function is to minimize the average response time.*

3   Online Approximation

3.1   Minimizing the Average Response Time

Lemma 6. *The competitive ratio of every online deterministic algorithm A, with respect minimizing the average response time, is $\Omega(n)$.*

Proof. At time 0, every page is requested once. Then no pages are requested until time $n/2$. From time 1 until time $n/2$ the adversary services the requests not serviced by A by time $n/2$, that is, at time $n/2$ the requests unsatisfied by A are disjoint from the requests unsatisfied by the adversary. At time $n/2$, the adversary requests all the $n/2$ pages previously serviced by A (note these pages are not already serviced by the adversary). No more pages are requested until time $n+1$.

At time $n+1$, the adversary has serviced all of the requests, while A has $n/2$ unsatisfied requests. At each time t from $n+1$ to k ($k = n^2$), the adversary requests the page that was satisfied by A at time $t-1$. Hence, at each time in $[n+1, k]$, A had $n/2$ unsatisfied requests. At each time $t \in [n+1, k]$, the adversary can service the request at time $t-1$. Hence, the adversary has at most 1 unsatisfied request at each time $t \in [n+1, k]$.

We now compute the total response time for the adversary. To service the first n pages, the adversary incurs a cost of $\sum_{i=1}^{n} i = \Theta(n^2)$. Similarly the cost for servicing the next $n/2$ is requests is $\sum_{i=1}^{n/2} i = \Theta(n^2)$. The cost for servicing the last $k-n$ requests is $k-n$. Hence the total cost incurred by the adversary is $O(n^2 + k)$.

We now compute the total response time for A. Note that for each time $t \in [n+1, k]$, A has $n/2$ unsatisfied requests. Hence, at each such t the response time of $n/2$ requests increases by one. Hence, the total response time for A is $\Omega(nk)$. The claim then follows by our choice of $k = n^2$. ∎

We now show that a faster server does not benefit either MRF or FCFS.

Lemma 7. *The algorithm MRF is not an $O(1)$-speed $O(1)$-approximation algorithm with respect to average response time.*

Lemma 8. *The algorithm FCFS is not an $O(1)$-speed $O(1)$-approximation algorithm with respect to average response time.*

3.2   Minimizing the Number of Unsatisfied Requests

While we conjecture that LWF is an $O(1)$-speed $O(1)$-approximation algorithm, we now show that this can not be proved in the standard way. That is, no deterministic algorithm A, equipped with an $s = O(1)$ speed server, can guarantee that, at every time t, the number of requests unsatisfied by A at time t is $O(1)$ times the number of requests unsatisfied by the adversary at time t.

Lemma 9. *For all online algorithms A, and for all speeds s ($= O(\log n/\log \log n)$), $\max_t U_A^s(t)/U_{Adv}^1(t) = \omega(1)$.*

Proof. Let $n = \sum_{i=0}^{s}(s+1)^i$. The n pages form $s+1$ disjoint groups, with $(s+1)^i$ pages in group G_i. Initially, each page in G_i is requested n^i times. Until time b (which we specify below), if A broadcasts a page from the group G_i at time t, then the adversary requests this page n^i times at time $t+1$.

We now explain how to find the time b. The adversary maintains a schedule B, that is in some sense a delayed version of A's schedule. The adversary maintains a time-dependent set H_i of delayed pages from G_i. Each set H_i is initially empty. If A broadcasts $k \geq 1$ pages from a group G_i at time t, then the adversary arbitrarily picks one of these k pages to broadcast in B at time t; The other $k - 1$ pages are added to H_i. If A does not broadcast a page from a G_i at time t, and H_i is not empty at time t, then the adversary removes an arbitrary page $P_j \in H_i$, and B broadcasts P_j at time t. The adversary terminates the construction when there is a group G_i, and a time $b \geq (s+1)^i$, such that B does not broadcast a page from G_i at time b.

Assume to reach a contradiction that the adversary did not terminate by time $b = 2(s+1)^{2s}$. Then for each G_i, B must broadcast at least one page from G_i at each time from $(s+1)^i$ to b. Hence, each group must broadcast $2(s+1)^{2s} - (s+1)^{s+1}$ pages by time b. Since there are $(s+1)$ groups, B must broadcast $(s+1)[2(s+1)^{2s} - (s+1)^{s+1}]$ pages by time b. Observe that the total number of broadcasts by A, on or before time b, is at least the total number of broadcasts in B, on or before time b. Hence, A must broadcast $(s+1)[2(s+1)^{2s} - (s+1)^{s+1}]$ pages by time b. But since A only has a speed s server, A can broadcast at most $2s(s+1)^{2s}$ pages by time b. Simple algebra reveals a contradiction.

Let G_i be the group responsible for the termination of the construction at time b. We now claim that the adversary can guarantee that it has serviced all requests to pages in group G_i by time b. To see this consider the time interval $I = [b - (s+1)^i + 1, b]$. First consider the pages in G_i that are broadcast in B during I. For each such page $P_j \in G_i$, find the latest time $t \in I$ such that B broadcasted P_j at time t. The adversary then broadcasts P_j at time $t + 1$. Now consider the pages in G_i that are not broadcast in B during I. These pages are broadcast by the adversary at arbitrary times in I when no other broadcasts of pages in G_i are scheduled in B.

So at time b the adversary has no outstanding requests for pages from G_i, while A has outstanding requests for every page in G_i. Starting from time $b+1$, the adversary will sequentially broadcast all of the pages in groups $G_1, \ldots, G_{i-1}$ until time $d = b + \sum_{j=0}^{i-1}(s+1)^j$. Simple algebra shows that $s\sum_{j=0}^{i-1}(s+1)^j < (s+1)^i$. Therefore A still has at least one page left unsatisfied from G_i at time d. Therefore, A has at least n^i requests unsatisfied at time d, while the adversary has $O(n^{i-1})$ requests unsatisfied at time d. ∎

However it is possible that there exists an algorithm A with the property that $\max_t U_A^s(t)/U_{Adv}^1(t) = O(1)$ if we restrict the adversary to be the one that minimizes the average response time.

3.3 Minimizing the Number of Unsatisfied Pages

We now show that FCFS is a 4-speed 4-approximation algorithm with respect to minimizing the number of unsatisfied pages. It should be obvious that MRF and LWF are *not* $O(1)$-speed $O(1)$-approximation algorithms for this objective function.

Lemma 10. $\max_t \tilde{U}^4_{FCFS}(t)/\tilde{U}^1_{OPT}(t) \le 4$.

We now show that the above lemma is essentially tight by showing that FCFS is *not* a 3-speed $O(1)$-approximation algorithm with respect to minimizing the number of unsatisfied pages.

Lemma 11. $\max_t \tilde{U}^3_{FCFS}(t)/\tilde{U}^1_{OPT}(t) = \omega(1)$.

3.4 Maximizing the Number of Satisfied Requests

We now show that MRF is a 1-speed 2-approximation algorithm with respect to maximizing the number of satisfied pages.

Lemma 12. *For all inputs,* $\max_t S^1_{OPT}(t)/S^1_{MRF}(t) \le 2$.

We now show that LWF is not a 1-speed $O(1)$-approximation algorithm with respect to maximizing the number of satisfied pages.

Lemma 13. $\max_t S_{OPT}(t)/S_{LWF}(t) = \omega(1)$.

The proof of the following lemma is obvious.

Lemma 14. $\max_t S_{OPT}(t)/S_{FCFS}(t) = \omega(1)$.

References

1. S. Aacharya, and S. Muthukrishnan, "Scheduling on-demand broadcasts: new metrics and algorithms", MobiCom, 1998.
   also at: `http://www.bell-labs.com/user/acharya/papers/mobicom98.ps.gz/`. 292
2. D. Aksoy, and M. Franklin, "RxW: A scheduling approach for large-scale on-demand data broadcast", INFOCOM, 1998. 292, 293
3. AirMedia website, `http://www.airmedia.com`. 290
4. Y. Bartal, and S. Muthukrishnan, "Minimizing Maximum Response Time in Scheduling Broadcasts", 558-559, SODA, 2000. 292
5. E. Coffman, and P. Denning, *Operating Systems Theory*, Prentice-Hall, 1973. 291, 293
6. DirecPC website, `http://www.direcpc.com`. 290, 291
7. Hybrid website, `http://www.hybrid.com`. 290
8. Intel intercast website, `http://www.intercast.com`. 290
9. B. Kalyanasundaram, and K. Pruhs, "Speed is as powerful as clairvoyance", *IEEE Symposium on Foundations of Computation*, 214 – 221, 1995. 291, 293
10. E. Lawler, J.K. Lenstra, A. Rinnooy Kan, and D. Shmoys, "Sequencing and Scheduling: Algorithms and Complexity", in S. Graves, A. Rinnooy Kan, and P. Zipkin (eds.), *Logistics of Production and Inventory*, Handbooks in OR & MS 4, Elsevier Science, Chapter 9, 445–522, 1993. 291
11. S. Leonardi, and D. Raz, "Approximating total flow time on parallel machines", *ACM Symposium on Theory of Computing*, 110 – 119, 1997. 293
12. C. Phillips, C. Stein, E. Torng, and J. Wein "Optimal time-critical scheduling via resource augmentation", *ACM Symposium on Theory of Computing*, 140 – 149, 1997. 292

Jitter Regulation in an Internet Router with Delay Consideration

Hisashi Koga

Fujitsu Laboratories Ltd., Kawasaki 211-8588, Japan
and
Department of Information Science, University of Tokyo, Tokyo 113-0033, Japan
koga@flab.fujitsu.co.jp

Abstract. To playback multimedia data smoothly via the world-wide Internet, jitter, the variability of delay of individual packets, must be kept low. We examine on-line algorithms in a router to regulate jitter for a given multimedia stream by holding packets in an internal buffer. The previous work solved the problem with focusing only on the buffer size, allowing a packet to stay in the router infinitely long unless the internal buffer is full, which is unrealistic for many real-time applications.
Our main contribution is to introduce a new constraint that a packet can stay in the router at most for a constant time which we name *the permitted delay time* in order to provide the stream communication with real-time property, besides the conventional constraint about the buffer size. We present a nearly optimal on-line algorithm in terms of competitiveness for this new version of the problem. Our analysis yields the result that the competitiveness of on-line algorithms depends on the permitted delay time rather than the buffer size. Finally we make clear quantitatively how much jitter is removed by our on-line algorithm.

1 Introduction

With the rapid expansion of network communication to public represented by the Internet, there is a growing need for networks with guaranteed quality of service (QoS). Whereas the traditional best-effort networks does not give guarantees on communication performance to the clients, QoS networks allow them to transfer information with a certain level of performance guarantees which are evaluated in terms of delay, jitter, throughput and packet loss rate.

Jitter is defined as the maximal gap between the delay of each packet in the given stream. Keeping jitter low is important for many real-time applications which continuously transmit multimedia data [3] for the next reason: Ideally, in a real-time application, it is desirable that packets generated at the source in a periodic fashion should arrive at the destination also periodically, which means the jitter is 0. However, in reality, as packets pass through intermediate switches or routers, the jitter increases due to variable queuing delays and variable propagation delays. As the result, for stable playback of multimedia data, the real-time applications must prepare a buffer in the destination host to absorb the jitter. The large jitter forces the destination host to prepare a huge buffer.

This paper focuses on the jitter regulation in a single router for a given stream in the context that, by holding the incoming packets in an internal buffer, the router attempts to output the packets at the same intervals whose lengths are X (Fig.1). Here, X denotes the length of the intervals between adjacent packets at the source generating packets periodically. As packets go through the

M. Paterson (Ed.): ESA 2000, LNCS 1879, pp. 302–313, 2000.

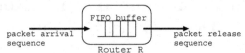

Fig. 1. Jitter regulation at a router

network, the packet intervals become distant from X. We may consider that the router tries to reconstruct the original traffic pattern at the source. The router is assumed to know the value of X a priori. This assumption is natural, because one can inform the router of X during the connection establishment period. Mansour and Patt-Shamir [4] deals with on-line algorithms for this problem under the condition that the number of buffer spaces is restricted, where each buffer space can exactly store one packet. Packets in the buffer must be pushed out obeying a FIFO service discipline, when the buffer becomes full. They presented an on-line algorithm using $2B$ buffer spaces which obtains the same jitter as the optimal off-line algorithm using B spaces. As for lower bounds, they proved that, when on-line algorithms have the same buffer capacity B as the optimal off-line algorithm, for any on-line algorithm, there exists an arrival time sequence for which the on-line algorithm cannot get a jitter less than BX, while the optimal off-line algorithm achieves 0-jitter.

However, they examine the problem allowing a packet to stay in the router infinitely long unless the buffer is full. In fact, their proof for the lower bounds utilizes the technique that the off-line algorithm postpones arbitrarily long sending out a packet, after an on-line algorithm outputs the very packet. This assumption is not tolerable for some real-time applications, because keeping a packet long in the intermediate routers increases the end-to-end delay very much. For example, the bi-directional interactive communication demanded by real-time discussion services will be disabled due to this enormous amount of delay.

To overcome this uselessness, we analyze the jitter regulation problem with introducing the new condition that a packet can stay in the router at most for a constant time L besides the conventional condition about the buffer capacity. That is, if L time units have passed after a packet arrival, the corresponding packet is driven out of the router. This adoption of L is a big contribution of this paper. In a normal situation, $L \gg X$ (at least 20 times larger). Specifically we name this constant L as the *permitted delay time*. This new problem is called *jitter regulation problem with delay consideration (JRDC)* hereafter.

This paper examines the power of on-line algorithms for JRDC using competitive analysis [5] which compares the performance of an on-line algorithm with that of the optimal off-line algorithm. Whereas Mansour and Patt-Shamir [4] computes the size of the additional buffer which must be given to an on-line algorithm in order for it to get the same jitter as the optimal off-line algorithm, we assume on-line algorithms have the same buffer size as the off-line algorithm and evaluate them according to how far their jitter differs from the jitter incurred by the optimal off-line algorithm.

First in Sect.3 we introduce the optimal off-line algorithm. In Sect.4 the lower bound of the competitiveness of deterministic on-line algorithms is investigated. We show, for any on-line algorithm, a packet arrival sequence exists such that the jitter incurred by the on-line algorithm is larger than or equal to that of the optimal off-line algorithm plus $\frac{L}{2} - 2X$. In Sect.5, we present a simple on-line algorithm with the nearly tight upper bound, Interestingly, the competitiveness for JRDC depends only on the permitted delay-time and not on the number of buffer spaces. To be superior to [4], in Sect.6 we also compute how much jitter our on-line algorithm absorbs against individual packet arrival sequences quantitatively. This approach is valuable in practice, because the competitive analysis itself does not give actual quantitative bounds. Our advanced approach finds that, in order for our algorithm to work effectively, a certain amount of buffer spaces should be prepared to the extent that $BX > L$ is satisfied. Especially if the buffer size is sufficiently large, the output stream gets 0-jitter when the jitter in the packet arrival sequence is less than $\frac{L}{2}$. On the contrary, when $BX < L$, our algorithm can work inefficiently.

We conclude the introduction by contrasting this paper with practical QoS processing systems. In the practical QoS systems, routers with jitter control function consist of two components: regulators and a scheduler [1] [2] [6]. The regulators shape the input traffic into the desired traffic pattern by holding packets in a buffer, where one regulator is responsible for one input stream. When a packet is released from a regulator, it is passed to the scheduler which decides the order of transmitting packets from all the streams to the output link. These systems have two primary features.

1. The routers on the distribution path for a given stream cooperate such that a regulator in a router holds packets long enough to absorb the jitter incurred by the scheduler of the previous router whose size grows up to the maximal scheduling delay in the worst case.
2. The total traffic is restricted to suppress the scheduling delay. When there is a request to establish a new stream connection, the entire network judges whether it can afford to accommodate the new stream.

However these two features are difficult to realize in the Internet managed by a lot of different organizations. It is probable that only a part of routers support the QoS mechanism, so the adjacent routers cannot work together. It seems also hard to limit the total traffic. For these reasons, this paper studies regulators only while not considering the schedulers. Especially we concentrate on a single router without considering the entire network, assuming implicitly each router does its best independently rather than in a united fashion.

2   Problem Statement

The jitter-regulation problem is formally defined as follows. We are given a router R and a sequence of packets $p_1, p_2, \ldots p_n$ which arrive at R. Each p_k ($1 \leq k \leq n$) arrives at time $a(k)$. Upon their arrivals, R stores the packets into an internal buffer and releases them later. A jitter-regulation algorithm A decides the release time of packets. The release time of p_k by the algorithm A is denoted as $s_A(k)$. R must release packets obeying the FIFO service discipline. That is, $s_A(k) \leq s_A(k+1)$ for $1 \leq k \leq n-1$. Throughout this paper we refer to the sequence $a(1), a(2), \ldots, a(n)$ as the *arrival time sequence* and the sequence $s_A(1), s_A(2), \ldots s_A(n)$ as the *release time sequence* by the algorithm A.

The purpose of a jitter-regulation algorithm is to make the release time sequence near to the ideal time sequence in which all the packets are spaced X time units apart. The jitter of a time sequence $\sigma: t_1, t_2, \ldots, t_n$ is defined as follows.

$$J^\sigma = \max_{0 \le i, k \le n} \{|t_i - t_k - (i - k)X|\}. \tag{1}$$

This variable measures the distance between σ and the ideal time sequence. Remark that we can define the jitter not only for the release time sequence but also for the arrival time sequence. We express the jitter of the release time sequence by the algorithm A for the arrival time sequence σ as $J_A(\sigma)$. Mansour and Patt-Shamir [4] mentions the next property about the jitter without proof.

Lemma 1 ([4]). *For all m $(1 \le m \le n)$, $J^\sigma = \max_{0 \le i \le n} \{t_i - t_m - (i - m)X\} + \max_{0 \le i \le n} \{t_m - t_i - (m - i)X\}$.*

Proof. First we claim that the value of J^σ is not influenced, even if the absolute value brackets are removed from the definition (1). Suppose that i, k are a pair of indices which maximizes J^σ and that $t_i - t_k - (i - k)X \le 0$. Then, by swapping i and k, we find another pair of indices which achieves the same value of J^σ and satisfies $t_i - t_k - (i - k)X \ge 0$. Hence,

$$\begin{aligned}
J^\sigma &= \max_{0 \le i, k \le n} \{t_i - t_k - (i - k)X\} \\
&= \max_{0 \le i \le n} \{t_i - iX\} - (t_m - mX) + \max_{0 \le k \le n} \{kX - t_k\} + (t_m - mX) \\
&= \max_{0 \le i \le n} \{t_i - t_m - (i - m)X\} + \max_{0 \le i \le n} \{t_m - t_i - (m - i)X\}. \quad \square
\end{aligned}$$

This paper examines the jitter regulation problem under the next two conditions the second of which is newly introduced in this paper.

1. The number of spaces in an internal buffer is exactly B.
2. The permitted delay time is equal to L for every packet.

The first condition says R can store at most B packets simultaneously. Before p_{k+B} arrives, p_k needs to be released. The second condition means each packet cannot stay at R more than L time units. Normally L is far greater than X. In summary, we solve the problem under the condition for the release time that

$$a(k) \le s_A(k) \le \min\{a(k + B), a(k) + L\}, \tag{2}$$

where we define $a(k) = \infty$ for $k > n$.

Because Mansour and Patt-Shamir [4] does not consider the permitted delay time, they solve the problem under the release time condition that

$$a(k) \le s_A(k) \le a(k + B). \tag{3}$$

In terms of interests in algorithmic theory, setting a different condition for the release time yields a different problem.

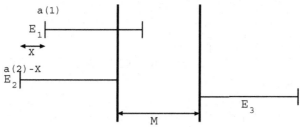

Fig. 2. An example of E_k

3   Optimal Off-Line Algorithm

Our on-line algorithms are compared with the optimal off-line algorithm OFF. OFF is derived by simply extending the optimal off-line algorithm for the case without delay consideration [4].

Algorithm OFF:

1. Let E_k be the time interval such that $E_k = [a(k) - (k - 1)X, \min\{a(k + B), a(k) + L\} - (k - 1)X]$, for $1 \leq k \leq n$.
2. OFF finds a minimal interval M which intersects all the intervals E_k for $1 \leq k \leq n$. Define $P_k = \min(E_k \cap M)$.
3. OFF releases the k-th packet p_k at time $s_{OFF}(k) = P_k + (k - 1)X$.

Figure 2 gives an example of E_k for three consecutive packets p_1, p_2, p_3, where p_1 and p_2 arrive simultaneously and p_3 arrives at R extremely later than the first two packets. Since p_1 and p_2 arrive at the same time, $\min(E_2) = \min(E_1) - X$. The region M is partitioned off by two vertical thick lines.

Theorem 1. OFF is an off-line algorithms which achieves the minimal jitter.

We first need to verify that OFF is a valid algorithm by showing OFF obeys the release time condition (2) and the FIFOness. As the verification procedure is identical with the proof of Theorem 3.1 in [4], it is not described here. From the minimality of M, OFF achieves the minimal jitter.

4   The Lower Bound

This section obtains the lower bound on the competitiveness of deterministic on-line algorithms as Theorem 2. In the proof, an adversary constructs a arrival time sequence σ which annoys on-line algorithms.

Theorem 2. For any deterministic on-line algorithm A, there exists an arrival time sequence σ for which $J_{OFF}(\sigma) = 0$, while $J_A(\sigma) \geq \frac{L}{2} - 2X - \epsilon$ when $B \geq 2\lfloor \frac{L}{X} \rfloor + 1$, where ϵ is a constant which can be arbitrarily close to 0.

The lower bound is proved for particular values of B. However, since, in the next section, we present an on-line algorithm whose jitter is larger at most by $\frac{L}{2}$ than OFF without regard to the value of B, this theorem is sufficient to insist that our on-line algorithm is nearly optimal. As a matter of of fact, the values of B are chosen so that the buffer never becomes full when OFF and A process σ.

Proof. σ is constructed by an adversary in the next way. The adversary passes the first packet p_1 at time 0 to R. Then it passes the subsequent packets at the same interval of X, until A releases p_1. Let T be the time A releases p_1. From the release time condition (2), $T \leq L$. The number of packets passed to R before T is expressed as $\lfloor \frac{T}{X} \rfloor + 1$ from the construction of σ. The adversary alters how to construct the rest of σ according to whether $T \leq \frac{L}{2}$.

(Case I) Suppose that $T \leq \frac{L}{2}$.

The adversary passes the $(\lfloor \frac{T}{X} \rfloor + 2)$-th packet to R at time $L + (\lfloor \frac{T}{X} \rfloor + 1)X$. Since the number of packets composing σ is $\lfloor \frac{T}{X} \rfloor + 2$, the buffer overflow never takes place, because $\lfloor \frac{T}{X} \rfloor + 2 \leq 2\lfloor \frac{L}{X} \rfloor + 1 \leq B$. From the definition of jitter (1),

$$
J_A(\sigma) \geq |s_A(\lfloor \frac{T}{X} \rfloor + 2) - s_A(1) - (\lfloor \frac{T}{X} \rfloor + 1)X|
$$

$$
\geq L + (\lfloor \frac{T}{X} \rfloor + 1)X - T - (\lfloor \frac{T}{X} \rfloor + 1)X = L - T \geq \frac{L}{2}.
$$

On the other hand, OFF gets 0-jitter by releasing p_k at time $L+(k-1)X$. Thus the theorem holds.

(Case II) Suppose that $T > \frac{L}{2}$.

The adversary puts in the next $\lfloor \frac{L}{X} \rfloor$ packets at time T', where T' is an arbitrary time in the interval $(T, (\lfloor \frac{T}{X} \rfloor + 1)X)$. In this case the number of packets included in σ is $\lfloor \frac{T}{X} \rfloor + \lfloor \frac{L}{X} \rfloor + 1$ which is less than $2\lfloor \frac{L}{X} \rfloor + 1$. Thus, the buffer overflow does not occur. OFF gets 0-jitter by releasing p_k at time $(k-1)X$, for $1 \leq k \leq \lfloor \frac{T}{X} \rfloor + \lfloor \frac{L}{X} \rfloor + 1$. On the other hand $J_A(\sigma)$ is at least

$$
J_A(\sigma) = |s_A(\lfloor \frac{T}{X} \rfloor + \lfloor \frac{L}{X} \rfloor + 1) - s_A(1) - (\lfloor \frac{T}{X} \rfloor + \lfloor \frac{L}{X} \rfloor)X|
$$

$$
\geq T + (\lfloor \frac{T}{X} \rfloor + \lfloor \frac{L}{X} \rfloor)X - s_A(\lfloor \frac{T}{X} \rfloor + \lfloor \frac{L}{X} \rfloor + 1)
$$

$$
\geq T + (\frac{T}{X} + \frac{L}{X} - 2)X - (T' + L) > \frac{L}{2} - 2X - (T' - T).
$$

Since $T' - T$ can be arbitrarily close to 0, the theorem holds for (Case II) by replacing $T' - T$ with ϵ. Thus, the entire proof completes. $\square$

5   The Upper Bound

This section describes our competitive on-line algorithm. First we present a technique which narrows the scope of candidates for a competitive algorithm. After that, our on-line algorithm named $HALF$ is introduced among candidates.

The next theorem claims that, in designing an efficient on-line algorithm, we should concentrate on when it releases p_1. The strategy to choose the release times for packets after p_1 can be decided in a rather routine fashion.

Theorem 3. *On condition that an on-line algorithm A releases p_1 at time $s_A(1)$, A gets the best jitter by choosing the release times for packets after p_1 in* *the following way.* Let $s^*(k) = s_A(1) + (k-1)X, (1 \leq k \leq n)$.

$$
s_A(k) = \begin{cases} s^*(k) \quad \text{(TYPE A)} \\ \text{if } a(k) \leq s^*(k) \leq \min\{a(k+B), a(k)+L\}. \\ \\ a(k) \quad \text{(TYPE B)} \\ \text{if } s^*(k) < a(k). \\ \\ \min\{a(k+B), a(k)+L\} \quad \text{(TYPE C)} \\ \text{if } s^*(k) > \min\{a(k+B), a(k)+L\}. \end{cases}
$$

Proof. It is obvious that A behaves as an on-line algorithm. Let σ be an arrival time sequence. From Lemma 1, $J_A(\sigma)$ is equal to

$$
\max_{0 \leq k \leq n} \{s_A(k) - s_A(1) - (k-1)X\} + \max_{0 \leq k \leq n} \{s_A(1) - s_A(k) - (1-k)X\}. \quad (4)
$$

Note that both the first term and the second term of (4) cannot be negative, because they are 0 when we substitute 1 for k. When an index k satisfies $s_A(1) + (k-1)X \in [a(k), \min\{a(k+B), a(k)+L\}]$, both terms of (4) equal 0 by setting $s_A(k)$ to $s_A(1) + (k-1)X$, and, hence, this index k does not contribute to the increase of the jitter. This case corresponds to (TYPE A) in Theorem 3. Otherwise, when an index k satisfies $s_A(1) + (k-1)X < a(k)$, the second term is less than 0 and unrelated to increasing the jitter. To minimize the first term, $s_A(k)$ must be as small as possible. Therefore $s_A(k)$ must be $a(k)$, which corresponds to (TYPE B). Finally, when k satisfies $s_A(1) + (k-1)X > \min\{a(k+B), a(k)+L\}$, the first term is less than 0 and irrelevant to increasing the jitter. To reduce the second term $s_A(k)$ must be as large as possible and chosen to be $\min\{a(k)+L, a(k+B)\}$. This case is associated with (TYPE C). □

Among the algorithms fulfilling the behavior specification in Theorem 3, we present our on-line algorithm named $HALF$ which gets the jitter larger than OFF by at most $\frac{L}{2}$.

Algorithm $HALF$: This algorithm releases p_1 at time $\min\{a(1+B), a(1)+\frac{L}{2}\}$. After that, it constructs the release time sequence obeying the rules mentioned in Theorem 3, where $s^*(k) = \min\{a(1+B), a(1)+\frac{L}{2}\} + (k-1)X$.

Theorem 4. *For any arrival time sequence σ, $J_{HALF}(\sigma) \leq J_{OFF}(\sigma) + \frac{L}{2}$.*

Neglecting the parameter B, OFF can reduce the jitter by L by releasing delayed packets immediately and keeping advanced packets in R during L time units, while on-line algorithm must judge whether p_1 is delayed in an on-line fashion. Intuitively $HALF$ supposes p_1 is neither delayed nor advanced. Though $HALF$ seems to be a naive algorithm because it does not decide $s_{HALF}(1)$ adaptively to the arrival time sequences. But Theorem 2 claims that the efforts for on-line algorithms to be adaptive go to waste. Theorem 2 and Theorem 4 together claim the competitiveness of on-line algorithms for JRDC is dominated by L and is not affected by B.

Proof. As for $HALF$, we classify the values of k into three sets S_A, S_B, S_C by the behavior of $HALF$.

$$S_A = \{k : s_H(k) = s_H(1) + (k-1)X\} \tag{5}$$
$$S_B = \{k : s_H(k) > s_H(1) + (k-1)X\} \tag{6}$$
$$S_C = \{k : s_H(k) < s_H(1) + (k-1)X\} \tag{7}$$

S_A, S_B and S_C correspond to (TYPE A), (TYPE B) and (TYPE C) in Theorem 3 respectively. So we may replace (6) and (7) with (8) and (9).

$$S_B = \{k : a(k) > s_H(1) + (k-1)X\} \tag{8}$$
$$S_C = \{k : \min\{a(k+B), a(k)+L\} < s_H(1) + (k-1)X\} \tag{9}$$

In the subsequent discussion we assume either S_B or S_C is not empty, because, if all indices belong to S_A, $J_H(\sigma) = 0$ and the theorem holds evidently. As is stated in the proof of Theorem 3, regarding (4), we have

- The indices included in S_A make both of the first term and the second term of (4) get the value 0.
- The first term is made positive only by the indices in S_B.
- On the contrary, the second term is made positive only by the indices in S_C.

Therefore (4) may be transformed into the next expression.

$$J_H(\sigma) = \max\{0, \max_{k \in S_B}\{s_H(k) - s_H(1) - (k-1)X\}\}$$
$$+ \max\{0, \max_{k \in S_C}\{s_H(1) - s_H(k) + (k-1)X\}\},$$

where we define $\max_{k \in \emptyset}\{s_H(k) - s_H(1) - (k-1)X\} = \max_{k \in \emptyset}\{s_H(1) - s_H(k) + (k-1)X\} = -1$.

From now on, we start to show $J_H(\sigma) \le J_{OFF}(\sigma) + \frac{L}{2}$. The proof proceeds as divided into two cases depending on whether $J_{OFF}(\sigma) = 0$. Recall the notation in the proof of Theorem 1. To save the space, we simply write H instead of the algorithm name $HALF$, when it appears in mathematical expressions.

(Case I) $J_{OFF}(\sigma) \ne 0$: Suppose i is the value of the index k which maximizes $a(k) - (k-1)X$ and j is the one which minimizes $\min\{a(k+B), a(k)+L\} - (k-1)X$. Since the width of M is not 0, we have $a(i) - (i-1)X > \min\{a(j+B), a(j)+L\} - (j-1)X$. $J_{OFF}(\sigma) = a(i) - (i-j)X - \min\{a(j+B), a(j)+L\}$.

There are three cases depending on the relation between $\min\{a(1)+\frac{L}{2}, a(1+B)\}$ and the open space $(\min\{a(j+B), a(j)+L\} - (j-1)X, a(i) - (i-1)X)$. (i) If $\min\{a(1+B), a(1)+\frac{L}{2}\} \in (\min\{a(j+B), a(j)+L\} - (j-1)X, a(i) - (i-1)X)$, it immediately follows that $i \in S_B$ and $j \in S_C$. Thus neither S_B nor S_C is an empty set. In this case, $J_H(\sigma)$ is computed as follows.

$$J_H(\sigma) = \max_{k \in S_B}\{s_H(k) - s_H(1) - (k-1)X\} + \max_{k \in S_C}\{s_H(1) - s_H(k) + (k-1)X\}$$
$$= \max_{k \in S_B}\{a(k) - (k-1)X\} + \max_{k \in S_C}\{(k-1)X - \min\{a(k+B), a(k)+L\}\}$$
$$= a(i) - (i-j)X - \min\{a(j+B), a(j)+L\}.$$

This indicates $J_{OFF}(\sigma) = J_H(\sigma)$.

(ii) If $\min\{a(1+B), a(1) + \frac{L}{2}\} \geq a(i) - (i-1)X$, we have $S_B = \emptyset$ because for any k, $a(k) - (\min\{a(1+B), a(1) + \frac{L}{2}\} + (k-1)X) \leq a(i) - (i-1)X - \min\{a(1) + \frac{L}{2}, a(1+B)\} \leq 0$. We can calculate $J_H(\sigma)$ in the next way. Note $a(i) - (i-1)X \geq a(1) - (1-1)X = a(1)$.

$$
\begin{aligned}
J_H(\sigma) &= \max_{k \in S_C}\{s_H(1) - s_H(k) + (k-1)X\} \\
&= s_H(1) + (j-1)X - \min\{a(j+B), a(j) + L\} \\
&\leq a(1) + \frac{L}{2} + (j-1)X - \min\{a(j+B), a(j) + L\} \\
&\leq a(i) - (i-j)X - \min\{a(j+B), a(j) + L\} + \frac{L}{2} = J_{OFF}(\sigma) + \frac{L}{2}.
\end{aligned}
$$

The proof for this case completes.

(iii) If $\min\{a(1+B), a(1) + \frac{L}{2}\} \leq \min\{a(j+B), a(j) + L\} - (j-1)X$, $S_C = \emptyset$ because for any k, $\min\{a(1+B), a(1) + \frac{L}{2}\} + (k-1)X - \min\{a(k) + L, a(k+B)\} \leq \min\{a(1+B), a(1) + \frac{L}{2}\} + (j-1)X - \min\{a(j) + L, a(j+B)\} \leq 0$. Thus

$$
\begin{aligned}
J_H(\sigma) &= \max_{k \in S_B}\{a(k) - (k-1)X - s_H(1)\} = a(i) - (i-1)X - s_H(1) \\
&\leq a(i) - (i-1)X - \min\{a(1+B), a(1) + L\} + \frac{L}{2} \\
&\leq a(i) - (i-j)X - \min\{a(j+B), a(j) + L\} + \frac{L}{2} = J_{OFF}(\sigma) + \frac{L}{2}.
\end{aligned}
$$

(Case II) $J_{OFF}(\sigma) = 0$: Our purpose is to show $J_H(\sigma) \leq \frac{L}{2}$. Since the width of M is 0, we have, for any k

- $a(k) - (k-1)X \leq \min\{a(1+B), a(1) + L\}$.
- $a(1) \leq \min\{a(k+B), a(k) + L\} - (k-1)X$.

First we show that either $S_B = \emptyset$ or $S_C = \emptyset$. Suppose that $S_B \neq \emptyset$ and $S_C \neq \emptyset$. Then there exists a pair of indices p, q such that:

1. $a(p) > s_H(1) + (p-1)X$.
2. $\min\{a(q+B), a(q) + L\} < s_H(1) + (q-1)X$.

Therefore $\min\{a(q+B), a(q) + L\} - (q-1)X < a(p) - (p-1)X$. However this inequality indicates the width of M is not 0, contradicting with the fact $J_{OFF}(\sigma) = 0$. Hence at least one of S_B and S_C must be empty. If $S_B = \emptyset$,

$$
\begin{aligned}
J_H(\sigma) &= \max_{k \in S_C}\{s_H(1) + (k-1)X - s_H(k)\} \\
&\leq s_H(1) - a(1) = \min\{a(1) + \frac{L}{2}, a(1+B)\} - a(1) \leq \frac{L}{2}.
\end{aligned}
$$

On the other hand, if $S_C = \emptyset$, $J_H(\sigma)$ is bounded from the above as follows.

$$
\begin{aligned}
J_H(\sigma) &= \max_{k \in S_B}\{s_H(k) - s_H(1) - (k-1)X\} \\
&\leq \min\{a(1+B), a(1) + L\} - \min\{a(1+B), a(1) + \frac{L}{2}\} \leq \frac{L}{2}. \qquad \square
\end{aligned}
$$

6   Quantitative Evaluation of HALF

Although we have evaluated $HALF$ by its competitiveness, these results are not sufficient for router designers, since the competitive analysis does not produce actual quantitative bounds. The router designer will have more interests in how much jitter is removed by $HALF$. We calculate it accurately and then insist that, when $BX < L$, $HALF$ can work inefficiently. This result suggests that a router with jitter control function should prepare a certain amount of buffer spaces to the extent that $BX > L$ is satisfied.

We pick up an arbitrary arrival time sequence σ and obtain the jitter contained in σ_H which is defined as the release time sequence for σ by $HALF$. For $1 \le k \le n$, we define two sequences $b(k) = a(k) - (k-1)X$ and $c(k) = s_H(k) - (k-1)X$. Let $I_r = \max_{1 \le k \le n}\{b(k)\}$ and $I_l = \min_{1 \le k \le n}\{b(k)\}$. It is easy to observe that $I_l \le a(1) \le I_r$ and that $J^\sigma = I_r - I_l$.

Theorem 5. *Consider an arbitrary arrival time sequence σ. If $BX > L$, the jitter contained in σ_H is*

$$J^{\sigma_H} = \max\{I_r, s_H(1)\} - \min\{s_H(1), I_l + L\}, \qquad (10)$$

where $s_H(1) = \min\{a(1+B), a(1) + \frac{L}{2}\}$.

Theorem 5 states that, when $\min\{a(1+B), a(1) + \frac{L}{2}\} \le I_r$, $J^{\sigma_H} = I_r - \min\{s_H(1), I_l + L\}$. This indicates the jitter corresponding to the left side of $s_H(1)$ is removed by up to L.

Proof. Let $O_r = \max_{1 \le k \le n}\{c(k)\}$ and $O_l = \min_{1 \le k \le n}\{c(k)\}$. Then

$$J^{\sigma_H} = O_r - O_l = (O_r - s_H(1)) + (s_H(1) - O_l). \qquad (11)$$

Here we mention an important fact that

- If an index k satisfies $b(k) > s_H(1)$, $c(k) > s_H(1)$.
- If an index k satisfies $b(k) \le s_H(1)$, $c(k) \le s_H(1)$.

Hence the first term of (11) depends on only those indices k satisfying $b(k) \ge s_H(1)$. Similarly, the second term of (11) depends only on those indices k that satisfy $b(k) \le s_H(1)$. Thus, we may calculate the first term and the second term separately.

The first term: There are two cases depending on whether $I_r \le s_H(1)$. If $I_r \le s_H(1)$, O_r is associated with S_A, because $S_B = \phi$. Hence $O_r = s_H(1)$. Otherwise if $I_r > s_H(1)$, O_r is associated with S_B and $O_r = I_r$. Thus,

$$O_r - s_H(1) = \max\{I_r, s_H(1)\} - s_H(1). \qquad (12)$$

The second term: There are two cases depending on whether $I_l + L \ge s_H(1)$.

1. If $I_l + L \ge s_H(1)$, we show O_l is associated with S_A by proving $S_C = \emptyset$. Assume $S_C \ne \emptyset$ and an index k is a member in S_C. As $b(k) \ge I_l$ and $c(k) < s_H(1)$, $c(k) - b(k) < L$. Therefore $c(k)$ is not $a(k) + L - (k-1)X$, but $a(k+B) - (k-1)X$. However, this leads to a following contradiction.

$$b(k+B) = a(k+B) - (k+B-1)X = c(k) - BX < s_H(1) - L \le I_l.$$

Therefore, S_C must be $\emptyset$. Thus, O_l is associated with S_A and $O_l = s_H(1)$.

2. In case $I_l + L < s_H(1)$, O_l depends only on the elements in S_C. We prove O_l is equal to $I_l + L$ by showing either $c(k) = b(k) + L$ or k has influence on neither I_l nor O_l, for each $k \in S_C$. If $s_H(k) = \min\{a(k + B), a(k) + L\}$ is equal to $a(k) + L$, $c(k) = b(k) + L$ trivially. Otherwise, if $s_H(k) = a(k + B)$, it follows that

(I) $b(k + B) = a(k + B) - (k + B - 1)X \le (a(k) + L) - (k + B - 1)X = a(k) - (k - 1)X + (L - BX) < b(k)$.

(II)$c(k + B) \le a(k + B) - (k + B - 1)X + L < a(k + B) - (k - 1)X = c(k)$.

Hence, the index value k is unrelated to both I_l and O_l.

Merging these two cases results in

$$s_H(1) - O_l = s_H(1) - \min\{s_H(1), I_l + L\}. \tag{13}$$

Summing up (12) and (13) yields Theorem 5. □

Specifically, if the buffer size is enough to assure that no buffer overflow takes place, the next corollary holds.

Corollary 1. *If the buffer size is so large that any buffer overflow does not occur and $J^\sigma \le \frac{L}{2}$, $J^{\sigma_H} = 0$ without fail.*

Proof. Since B is sufficiently large, we may replace $\min\{a(1 + B), a(1) + \frac{L}{2}\}$ with $a(1) + \frac{L}{2}$ in the description of $HALF$. Hence, (10) is transformed as follows: $J^{\sigma_H} = \max\{I_r, a(1) + \frac{L}{2}\} - \min\{a(1) + \frac{L}{2}, I_l + L\} = \max\{I_r - \frac{L}{2}, a(1)\} - \min\{a(1), I_l + \frac{L}{2}\}$. Since we have $I_r - \frac{L}{2} \le I_l \le a(1)$ and $I_l + \frac{L}{2} \ge I_r \ge a(1)$ from the assumption $J^\sigma \le \frac{L}{2}$, J^{σ_H} becomes $a(1) - a(1) = 0$. □

The last of this section examines the situation when $BX \le L$.

Theorem 6. *Consider an arbitrary arrival time sequence σ. If $BX \le L$,*

If $S_C = \phi$: $J^{\sigma_H} = \max\{I_r, s_H(1)\} - s_H(1)$. (14)

If $S_C \ne \phi$: $\max\{I_r, s_H(1)\} - (I_l + L) \le J^{\sigma_H} \le \max\{I_r, s_H(1)\} - (I_l + BX)$. (15)

Because Theorem 5 can be interpreted as $J^{\sigma_H} = \max\{I_r, s_H(1)\} - s_H(1)$ if $S_C = \phi$ and $J^{\sigma_H} = \max\{I_r, s_H(1)\} - (I_l + L)$ if $S_C \ne \phi$ when $BX > L$, Theorem 6 certifies the power of $HALF$ is restricted when $BX < L$.

Proof. The outline of the proof resembles to Theorem 5. The equation (11) is brought out again and its first term and its second term are computed separately. About the first term, exactly the same argument as in Theorem 5 yields $O_r - s_H(1) = \max\{I_r, s_H(1)\} - s_H(1)$.

Regarding the second term, $O_l = s_H(1)$ when $S_C = \phi$, because all the indices k satisfying $b(k) \le s_H(1)$ belong to S_A. It remains to show that $I_l + BX \le O_l \le I_l + L$ when $S_C \ne \phi$. The right-hand side of this inequality is directly derived, because $O_l \le c(k') \le I_l + L$ where k' is an index which achieves $b(k') = I_l$. With respect to the left-hand side, it suffices to prove $O_l \ge I_l + BX$ by showing one of the next three conditions always holds for any index $k \in S_C$.

1. $c(k) \ge b(k) + BX$.

2. $\min\{c(k), c(k+B)\} = \min\{b(k), b(k+B)\} + BX$.
3. k has nothing to do with I_l and O_l.

Suppose that $c(k) < b(k) + BX (\leq b(k) + L)$. Then, because $c(k) = a(k+B) - (k-1)X$, it follows that

$$b(k+B) = a(k+B) - (k+B-1)X = c(k) - BX < b(k).$$

Hence, if $c(k+B) \geq c(k)$, $\min\{c(k), c(k+B)\} = \min\{b(k), b(k+B)\} + BX$. Otherwise I_l and O_l never depends on this index k. This concludes the proof. $\square$

In fact, if $BX < \frac{L}{2}$, $HALF$ becomes completely helpless against a certain arrival time sequence.

Theorem 7. *If $BX < \frac{L}{2}$, there exists an arrival time sequence σ for which $J^\sigma = J^{\sigma_H}$.*

The request sequence σ inconvenient to $HALF$ is constructed below. σ stands up the equation mark in the right-hand inequality of (15).

- The first B packets arrives at R such that $a(k) = (k-1)X$.
- At time BX, the next $(\alpha B + 1)$ packets are passed to R all at once, where α is an arbitrary positive integer.

Against σ, the equation $J^\sigma = J^{\sigma_H} = \alpha BX$ holds. The proof is omitted here.

7   Concluding Remarks

In this paper we analyze the jitter regulation problem on the new realistic condition that a packet must be released from the router within the permitted delay time L, in addition to the conventional restriction on the buffer size. This new problem is named JRDC. We presented an on-line algorithm for JRDC whose jitter is larger than OFF by at most $\frac{L}{2}$. We prove that the competitiveness for JRDC depends only on L and not on the buffer size. In addition, we make clear how much jitter is removed by our on-line algorithm. As the result, when $BX < L$, it is proved that our algorithm can perform inefficiently. One interesting open problem is to control jitter for multiple streams while handling the interrelation between them in the scheduler part in the router.

References

1. N. R. Figueira and J. Pasquale. Leave-in-time: A new service discipline for real-time communications in a packet-switching network. In *Proceedings of ACM SIGCOMM'95*, pages 207–218, 1995. 304
2. S. J. Golestani. A stop-and-go queueing framework for congestion management. In *Proceedings of ACM SIGCOMM'90*, pages 8–18, 1990. 304
3. H. Koga and A. Jinzaki. DV stream transmission over world wide internet by video frame recognition. In *Proceedings of 10th International Packet Video Workshop*, 2000. 302
4. Y. Mansour and B. Patt-Shamir. Jitter control in QoS networks. In *Proceedings of 39th IEEE Symposium on Foundations of Computer Science*, pages 50–59, 1998. 303, 304, 305, 306
5. R.E. Tarjan and D.D. Sleator. Amortized efficiency of list update and paging rules. *Communication of the ACM*, 28:202–208, 1985. 303
6. H. Zhang and D. Ferrari. Rate-controlled static-priority queuing. In *Proceedings of IEEE INFOCOM'93*, pages 227–236, 1993. 304

Approximation of Curvature-Constrained Shortest Paths through a Sequence of Points

Jae-Ha Lee[1], Otfried Cheong[2], Woo-Cheol Kwon[3],
Sung Yong Shin[3] and Kyung-Yong Chwa[3]

[1] Max-Planck-Institut für Informatik, Germany. Email:lee@mpi-sb.mpg.de
[2] Dept. of Computer Science, HKUST, Hong Kong. Email: otfried@cs.ust.hk
[3] Dept. of Computer Science, KAIST, Korea
E-mail: {famor,syshin,kychwa}@jupiter.kaist.ac.kr

Abstract. Let B be a point robot moving in the plane, whose path is constrained to forward motions with curvature at most 1, and let $\mathcal{X}$ denote a sequence of n points. Let s be the length of the shortest curvature-constrained path for B that visits the points of $\mathcal{X}$ in the given order. We show that if the points of $\mathcal{X}$ are given *on-line* and the robot has to respond to each point immediately, there is no strategy that guarantees a path whose length is at most $f(n)s$, for any finite function $f(n)$. On the other hand, if all points are given at once, a path with length at most $5.03s$ can be computed in linear time. In the *semi-online* case, where the robot not only knows the next input point but is able to "see" the future input points included in the disk with radius R around the robot, a path of length $(5.03 + O(1/R))s$ can be computed.

1 Introduction

Consider a point robot in the plane whose turning radius is constrained to be at least 1. More formally, given a continuous differentiable path $\mathcal{P} : I \longrightarrow R^2$ parameterized by an arc length $s \in I$, the *average curvature* of $\mathcal{P}$ is defined by $|\mathcal{P}'(s_1)-\mathcal{P}'(s_2)|/|s_1-s_2|$. We require that the robot's path has an average curvature of at most 1 in every interval. This restriction corresponds naturally to constraints imposed by the steering mechanism found in car-like robots, a *non-holonomic constraint*.

In spite of considerable interest and recent work on non-holonomic motion planning problems in the robotics literature, they have received relatively little attention from a theoretical point of view, and in fact appear to be considerably harder than holonomic motion-planning problems.

In this paper, we study the path planning problem for a steering-constrained robot moving through a prespecified sequence of points. We want to minimize the length of the path. Any solution must decide the orientation with which the robot passes each point, and this decision will affect the length of the resulting path.

The problem of finding the shortest path through a sequence of points has not attracted much attention so far, since, in general, the shortest path is simply the union of the shortest paths connecting each pair of consecutive points. In our curvature-constrained case, however, the shortest path through a sequence of points must satisfy the curvature constraint at each "joint," and is not the union of the shortest paths between consecutive points (which are simply line segments).

M. Paterson (Ed.): ESA 2000, LNCS 1879, pp. 314–325, 2000.

Previous results. Most of the previous work considers the curvature-constrained path planning problem between two given configurations (a configuration specifies both a point and a direction of travel). Dubins [7] was perhaps the first to study curvature-constrained shortest paths. He proved that, in the absence of obstacles, a curvature-constrained shortest path from any start configuration to any final configuration consists of at most three segments, each of which is either a straight line or an arc of a unit (radius) circle, assuming that the curvature of the path is upper bounded by 1. Since Dubins' characterization implies a constant-time algorithm to find a shortest path between two configurations, it is easy to obtain a linear-time algorithm for computing a shortest path through a sequence of *configurations*. Reeds and Shepp [11] extended this obstacle-free characterization to robots that are allowed to make reversals, that is, to back up. Using ideas from control theory, Boissonnat, Cerezo, and Leblond [4] gave an alternative proof for both cases.

There has also been work on computing curvature-constrained paths between two configurations in the presence of obstacles [8,9,13]. Agarwal, Raghavan, and Tamaki [2] considered the restricted case of pairwise disjoint *moderate obstacles* (convex obstacles whose boundary has curvature bounded by 1) and gave efficient approximation algorithms. For the same problem, Boissonnat and Lazard [5] gave an $O(n^2 \log n)$-time exact algorithm. Reif and Wang [12] confirmed that the problem of deciding whether there exists a curvature-constrained shortest path amidst general obstacles is NP-hard. Agarwal et al. [1] presented an $O(n^2 \log n)$-time algorithm to compute a curvature-constrained shortest path between two configurations inside a convex polygon. Recently, Ahn et al. [3] characterized the set of all points (without orientations) reachable from a given configuration inside a convex polygon.

Our model and results. Let B be a point robot and $\mathcal{X} = \langle X_1, X_2, \cdots, X_n \rangle$ denote a sequence of points in the plane. We aim to find a curvature-constrained path passing through the points in $\mathcal{X}$ in the given order; there is no restriction on the orientation with which it arrives at (and departs from) each point. The lack of such a restriction, which is the main difference between our problem and the previous work, is a double-bladed sword—on the one hand, the freedom to choose a direction at each point allows for drastically shorter paths, on the other hand, finding such a path is considerably more involved.

A configuration σ for B is a pair $\sigma = (X, \phi)$, where X is a point in the plane representing the location of the robot and ϕ is a unit vector representing its orientation. We emphasize again that a shortest path between two configurations can be computed in constant time, due to Dubins [7]. In order to find a shortest path through $\mathcal{X}$, it suffices to determine the orientation ϕ_i at each point X_i so that

$$\sum_i \|\text{shortest path from } (X_i, \phi_i) \text{ to } (X_{i+1}, \phi_{i+1})\|$$

is minimized, where $\| \cdot \|$ denotes the length of a path.

An algorithm for this problem is called $f(n)$-*competitive* (or $f(n)$-*approximate)* if, given a sequence $\mathcal{X}$ of length n, it computes a path of length at most $f(n)s + c_0$, where s is the length of the shortest path visiting the points of $\mathcal{X}$ and $c_0 \geq 0$ is a constant (independent of $\mathcal{X}$ and n).

We first consider the on-line version of the problem, where the points X_i are given one by one, and the robot has to respond to each new point by moving to it. We show that there is no $f(n)$-competitive algorithm for any finite function $f(n)$, settling an open problem by Agarwal et al. [2].

We then look at the case where the sequence $\mathcal{X}$ is given in advance. An intuitive approach for computing an approximate shortest path would be to choose a discrete set of orientations for each point, and to try all combinations of possible orientations. Somewhat surprisingly, it turns out that this approach cannot guarantee a bounded approximation factor, as we will see in Section 3.

Our main result is a linear time algorithm that, given a sequence $\mathcal{X}$, computes a path whose length is at most 5.03 times the length of the shortest path.

Finally, we mention a $(5.03 + O(1/R))$-competitive algorithm for a *semi-online* version where the robot can "see" the whole prefix of the remaining sequence contained in a disk with radius R around it.

2   Geometric preliminaries

Given a configuration $\sigma = (X, \phi)$, the oriented line passing through X with orientation ϕ is denoted L_σ. A configuration $\sigma = (X, \phi)$ belongs to an oriented path (or curve) $\mathcal{C}$ if $X \in \mathcal{C}$ and L_σ is the oriented tangent line to $\mathcal{C}$ at X. Note that a configuration σ belongs to two oriented unit-radius circles. We denote these two circles by C_σ^+ and C_σ^-, where C_σ^+ is oriented counterclockwise, C_σ^- is oriented clockwise. For a circle C, $closed(C)$ (resp. $open(C)$) denotes the closed (resp. open) disc bounded by C. If X and Y are two points on a simple oriented curve $\mathcal{C}$, then $\mathcal{C}[X, Y]$ denotes the portion of $\mathcal{C}$ from X to Y, including X and Y. For two orientations ϕ and ϕ', let $[\phi, \phi']$ denote the counterclockwise interval of orientations between ϕ and ϕ' (that is, the orientations met when we rotate ϕ counterclockwise to ϕ'). For convenience, let $[0, 2\pi]$ denote the interval of all orientations.

The following fact is a direct by-product of a proof by Pestov and Ionin [10], see also Ahn et al. [3].

Fact 1 *Given a configuration σ and a point Y in the interior of C_σ^+ or C_σ^-. Then the length of any path from σ to Y is at least π.*

Segments and Dubins paths. Let $\mathcal{C}$ be a path. We call a nonempty subpath of $\mathcal{C}$ a C-*segment* (resp. S-*segment*) if it is a circular arc of unit radius (resp. a line segment) and maximal. A *segment* is either a C-segment or an S-segment. Suppose $\mathcal{C}$ consists of a C-segment, S-segment, and a C-segment, in this order; then we will say that $\mathcal{C}$ is of type CSC. Dubins [7] proved the following result.

Fact 2 *Any shortest path between two configurations is of type CCC or CSC, or a substring thereof.*

We will refer to paths of type CCC or CSC or a substring thereof as *Dubins paths*. We emphasize that an optimal path between two given configurations can be computed in constant time.

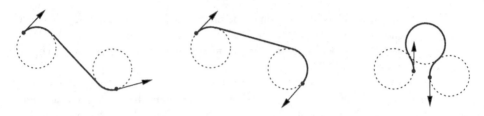

Fig. 1. Different types of Dubins paths.

Direct paths and detour paths. Suppose that X and Y are two consecutive points in $\mathcal{X}$ such that $dist(X, Y) < 2$. We can draw two unit (radius) circles each of which passes both of X and Y, we denote them by C_{XY}^+ and C_{XY}^- (Figure 2). Any path from X to Y that is included in $closed(C_{XY}^+) \cap closed(C_{XY}^-)$ is called a *direct path*. A path that is not direct is called a *detour path*. A detour path will pass some point outside $closed(C_{XY}^+) \cap closed(C_{XY}^-)$ before arriving at Y. Let ϕ_{XY}^U (resp. ϕ_{XY}^L) denote the orientation of the clockwise tangent to C_{XY}^- (counterclockwise tangent to C_{XY}^+) at X. Note that any direct path from X to Y must start with an orientation in $[\phi_{XY}^L, \phi_{XY}^U]$.

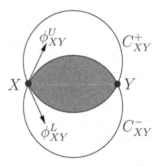

Fig. 2. A path included in $closed(C_{XY}^+) \cap closed(C_{XY}^-)$ is called a direct path.

Lemma 1 *The length of any detour path is at least π.*

Proof: Let $\mathcal{P}$ be any detour path from X to Y. Let $\sigma = (X, \phi)$ be its configuration at X. There are three cases depending on the orientation ϕ. See Figure 3.

-   Case 1. ϕ is directed to the outside of $open(C_{XY}^+) \cup open(C_{XY}^-)$ (Figure 3a).
    Consider $open(C_\sigma^+) \cup open(C_\sigma^-)$. If the path $\mathcal{P}$ passes a point in its interior, the length is longer than π by Fact 1. Otherwise the resulting path must "wind around" C_σ^+ or C_σ^-, so its length is at least π.
-   Case 2. ϕ is directed into $open(C_{XY}^+) \cup open(C_{XY}^-)$ but not $closed(C_{XY}^+) \cap closed(C_{XY}^-)$ (Figure 3b).
    Observe that Y is contained in $open(C_\sigma^+)$ (or $open(C_\sigma^-)$). Therefore the claim follows directly from Fact 1.

– Case 3. ϕ is directed into $closed(C_{XY}^+) \cap closed(C_{XY}^-)$ but later leaves it (Figure 3c).

Suppose that the robot leaves $closed(C_{XY}^+) \cap closed(C_{XY}^-)$ at X'. If we recursively apply this lemma to the path from X' to Y, then the orientation at X' is towards the outside of $closed(C_{X'Y}^+) \cap closed(C_{X'Y}^-)$. Thus a path from X' to Y falls into the above two cases, so its length is at least π.

Any detour path falls into one of the three categories above.   ⊟

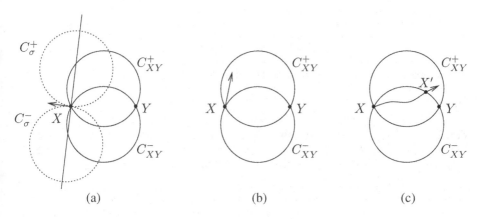

Fig. 3. Proof of Lemma 1

In contrast to the detour path, the length of any direct path from X to Y is bounded by a constant times $dist(X, Y)$. Let us assume that C_{XY}^+ is oriented counterclockwise.

Lemma 2 *The length of any direct path is no greater than* $||C_{XY}^+[X, Y]||$, *which is bounded by* $\frac{\pi}{2} dist(X, Y) (< \pi)$.

Proof: Assume the line XY is parallel to the x-axis. Let $\mathcal{P}$ denote an arbitrary direct path from X to Y. Suppose that $\mathcal{P}$ consists of k segments (each of which is a line segment or a circuclr arc of radius at least 1) such that each segment connects X_i to X_{i+1}, where $X_0(= X), X_1, \cdots X_k(= Y)$ are points on $\mathcal{P}$ occurring in this order. We prove this lemma by induction on k. The base case that $k = 1$ is trivial since $\mathcal{P}$ is one line segment or circular arc in this case. Without loss of generality, assume X_1 is below the line XY. Let X_1' (resp. X_1'') denote the first point on C_{XY}^+ met when we rotate X_1 clockwise centered at X (resp. counterclockwise centered at Y). See Figure 4.

Since $\mathcal{P}[X, X_1]$ is one segment, its length is no greater than $||C_{XY}^+[X, X_1']||$. Moreover, by induction hypothesis, the length of $\mathcal{P}[X_1, Y]$ is no greater than $||C_{XY}^+[X_1'', Y]||$. It is easily seen that x-coordinates of X_1', X_1 and X_1'' are non-decreasing in this order. Therefore, the length of $\mathcal{P}$ is no greater than $||C_{XY}^+[X, Y]||$, which proves this lemma. Since $dist(X, Y)$ is less than 2 from the definition of a direct path, the length of $\mathcal{P}$ is less than π.   ⊟

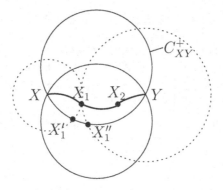

Fig. 4. Proof of Lemma 2.

The proof of the next lemma can be found in the full paper.

Lemma 3 *Given a configuration $\sigma = (X, \phi)$ and a point Y at distance less than 2 from X. The set of orientations with which a direct path from σ to Y can arrive at Y is an interval $[\phi_L, \phi_U]$. The orientation ϕ_L is the direction of the clockwise tangent to C at Y, where C is the circle containing Y that is tangent to C_σ^+ in $closed(C_{XY}^+) \cap closed(C_{XY}^-)$. The orientation ϕ_U is obtained symmetrically. (See Figure 5)*

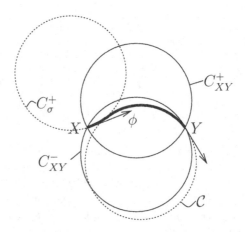

Fig. 5. Lemma 3.

Basic idea. In order to bound the approximation ratio of an algorithm, we need a good lower bound on the value of the optimal solution. The simplest such lower bound is the sum of the Euclidean distances between consecutive points. This bound is good enough when consecutive points are far apart, or if they can be connected with a direct path. The difficulty arises when three consecutive points are close to each other and form a sharp turn—in that case we cannot avoid using a detour path. The length of a detour path is at least π by Lemma 1, which can be arbitrarily longer than the distance of the points. It follows that in the heart of our argument we need a bound on the number of detour paths made by both the optimal path and the path produced by the algorithm. We will first show that if the points are given one-by-one in an on-line fashion, any algorithm can be fooled into producing a large number of detour paths, while the optimal algorithm doesn't need any. Then we will see that if all points are given in advance, we can construct a path using that minimum possible number of detour paths.

3   On-line algorithm

The following on-line version of our problem was posed by Agarwal et al. [2]: Assume the point sequence $\mathcal{X}$ is given one by one in the form of point requests, and the robot has to service each request by moving to the point. There is no constraint on the orientation with which it arrives at the requested point.

We show that even in the absence of obstacles no $f(n)$-competitive on-line algorithm exists, for any finite function $f(n)$.

Theorem 4 *No on-line algorithm can be $f(n)$-competitive, for any finite function $f(n)$.*

Proof: Let $\varepsilon > 0$ and consider some on-line algorithm. We will incrementally construct a sequence $\mathcal{X}$ of $2n + 1$ point requests, making use of the decisions made by the algorithm to serve the requests in constructing the next point. The distance between consecutive points will be ε. The sequence will be such that there is a path visiting all points using only direct paths, but the on-line algorithm will be forced to make at least n detour paths. By Lemma 1 the path returned by the algorithm has length at least $n\pi$. The shortest path, on the other hand, has length bounded by $\frac{\pi}{2}n\varepsilon$ by Lemma 2. Since ε can be chosen arbitrarily small, the theorem follows.

Let $\mathcal{X} = X_1, \cdots, X_{2n+1}$ be the sequence to be constructed, and let X_1 be an arbitrary point in the plane. Suppose that requests $X_1, \ldots, X_{2i-1}$ have already been created, and so the robot is currently at X_{2i-1}.

Let ϕ_{2i-1} be some direction with which a path consisting of direct paths only can arrive at X_{2i-1}. It will follow from our construction that such an orientation exists (in fact, there will be exactly one such orientation).

We create the next point request X_{2i} at distance ε from X_{2i-1} in direction ϕ_{2i-1}, see Figure 6. By Lemma 3, the set of directions with which a direct path starting at the configuration (X_{2i-1}, ϕ_{2i-1}) can arrive at X_{2i} is an interval $[l_{2i}, u_{2i}]$. Recall that the circle $C^+_{(X_{2i}, u_{2i})}$ (resp. $C^-_{(X_{2i}, l_{2i})}$) is tangent to the orientation u_{2i} (resp. l_{2i}).

We now observe the direction ϕ in which the robot controlled by the on-line algorithm arrives at X_{2i}. If ϕ points to the left of the oriented line through X_{2i-1} and X_{2i}, we choose X_{2i+1} to be the point at distance ε from X_{2i} on the circle $C^-_{(X_{2i}, l_{2i})}$ (the lower one in Figure 6). Otherwise, we choose X_{2i+1} symmetrically on the circle $C^+_{(X_{2i}, l_{2i})}$ (the upper one in Figure 6).

The on-line algorithm has no choice but to use at least one detour path while serving the requests X_{2i} and X_{2i+1}. On the other hand, it is clearly possible to visit all points

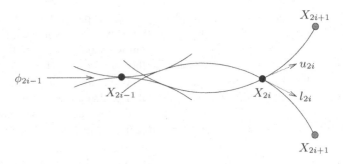

Fig. 6. Proof of Theorem 4. Note that $dist(X_{2i-1}, X_{2i})$ is very small compared to the radius of circles.

using direct paths only.   ⊟

A close look at the construction in this section reveals that any robot arriving at X_{2i-1} with orientation that is different from ϕ_{2i-1} cannot go through X_{2i} and X_{2i+1} using direct paths only. In other words, there is only one possible path from X_{2i-1} to X_{2i+1} that uses direct paths only. This implies that any approximation technique that tries to discretize the set of possible directions at each point before looking at the geometry will have to make at least n detour paths to visit the sequence we constructed, even if it could use a countably infinite set of orientations! It follows that no bounded approximation ratio can be guaranteed by such an algorithm.

4   Approximation algorithm

We now consider the problem where the whole sequence $\mathcal{X}$ is given in advance. We describe a linear-time approximation algorithm that computes a path whose length is within a constant factor of the optimal. The basic idea is simple:

In case two consecutive input points X_i and X_{i+1} are far away from each other, say $dist(X_i, X_{i+1}) \geq 2$, we can arbitrarily choose orientations ϕ_i and ϕ_{i+1} and connect the configurations (X_i, ϕ_i) and (X_{i+1}, ϕ_{i+1}) using the shortest path. The length of the shortest path—a Dubins path—is bounded by a constant times $dist(X_i, X_{i+1})$ (Lemma 7).

The remaining case is that X_i and X_{i+1} are close, that is, $dist(X_i, X_{i+1}) < 2$. There are two apparent ways to connect X_i with X_{i+1}: we can either use a direct path, or a detour path. The length of a direct path is again bounded by a constant times $dist(X_i, X_{i+1})$, but its orientations at X_i and X_{i+1} are restricted. A detour path, on the other hand, can be used for arbitrary orientations, but its length can be far longer than the distance of the two points. In other words, if we connect X_i to X_{i+1} using a detour path, there is no guarantee on the approximation ratio for the path between these two consecutive points. We are nevertheless able to obtain a constant approximation ratio for the total sequence by constructing a path that uses the *minimum possible number of detour paths*.

Consider a *dense* subsequence $X_i, \cdots, X_j$ such that $dist(X_k, X_{k+1}) < 2$ for all $i \leq k < j$. Let Φ_k denote the set of orientations with which the robot can arrive at X_k using direct paths only from X_i. We assume that any orientation can be used in X_i, so $\Phi_i = [0, 2\pi]$. Suppose Φ_k is not empty. In order to go to X_{k+1} using a direct path, the robot must start with an orientation in $\Phi_k \cap [\phi^L_{X_k X_{k+1}}, \phi^U_{X_k X_{k+1}}]$. We can compute Φ_{k+1} using the following lemma. We define the size of an interval $[\phi, \phi']$ of orientations to be the angle between ϕ and ϕ' measured from ϕ counterclockwise.

Lemma 5 *If Φ_k is either $[0, 2\pi]$ or an interval of size less than π, then so is Φ_{k+1}. Furthermore, Φ_{k+1} can be computed from Φ_k in constant time.*

Proof: Under the conditions of the lemma, $\Phi_k \cap [\phi^L_{X_k X_{k+1}}, \phi^U_{X_k X_{k+1}}]$ is an interval; Let $[\phi_L, \phi_U]$ denote it. By Lemma 3, the set of orientations Φ_{k+1} is then also an interval $[\phi'_L, \phi'_U]$, where ϕ'_L is obtained from ϕ_U and ϕ'_U is obtained from ϕ_L.   ☐

Repeated application of Lemma 5 proves that Φ_k is an interval of length less than π, for $k > i$. The following two constants will determine the quality of our approximation.

Definition 1 *Let*

$$\alpha = \sup\{\ell(\sigma, \sigma')/dist(\sigma, \sigma') \mid \sigma, \sigma' \text{ configurations with } dist(\sigma, \sigma') \geq 2\},$$
$$\beta = \sup\{\ell(\sigma, \sigma') \mid \sigma, \sigma' \text{ configurations with } dist(\sigma, \sigma') \leq 2\},$$

where $\ell(\sigma, \sigma')$ denotes the length of a Dubins shortest path between two configurations σ and σ'.

We now describe our main theorem.

Theorem 6 *A $\max(\alpha, \pi/2 + \beta/\pi)$-approximate shortest path subject to the curvature constraint can be computed in linear time.*

Proof: The input sequence $\mathcal{X}$ consists of alternating dense and non-dense subsequences, so that the final point of one subsequence is the first point of the other subsequence. We first determine the orientation at each point in dense subsequences.

Let $X_i, \cdots, X_j$ be a dense subsequence. While scanning it from X_i, we compute Φ_k, for $k = i, \cdots, j$, until it becomes empty. Suppose Φ_{k+1} becomes empty for the first time. Then we construct a path from X_i to X_k using direct paths only. Next, leaving (X_k, X_{k+1}) not connected, we restart from X_{k+1}, as if this is the first point of a dense subsequence, and repeat the same procedure till the point X_j. The remaining dense consecutive points, such as (X_k, X_{k+1}), that have not been connected by a direct path, are then connected by a detour path (the Dubins shortest path respecting the orientations determined by the direct paths). We apply the same procedure to each dense subsequence. Finally, we connect all remaining pairs of consecutive points by Dubins shortest paths, choosing the orientation at each point freely while respecting the predetermined orientations at points in dense subsequences.

We claim that this procedure produces a path using the minimum possible number of detour paths. In fact, consider a maximal sequence of points $X_i, \cdots, X_k$ connected by direct paths by the algorithm. By our construction, any path that uses a direct path to connect X_k with X_{k+1} must use a detour path between X_i and X_k. This implies that no path can use fewer detour paths than the path we constructed. In particular this is also true for the optimal solution.

Our algorithm clearly runs in linear time using Lemma 5. It remains to analyze the approximation ratio.

Let $\mathcal{P}$ denote the oriented path that we generate and let $\mathcal{Q}$ denote an optimal path for the sequence. Let $d_i = 1$ if $dist(X_i, X_{i+1}) < 2$; 0 otherwise. For (X_i, X_{i+1}) with $d_i = 1$, let $D_i = 1$ if $\mathcal{P}[X_i, X_{i+1}]$ is a direct path; 0 otherwise. First, we bound the length of $\mathcal{P}$ restricted to dense subsequences.

$$
\begin{aligned}
\sum_i d_i \cdot \|\mathcal{P}[X_i, X_{i+1}]\| &= \\
&= \sum_i d_i \cdot D_i \cdot \|\mathcal{P}[X_i, X_{i+1}]\| + \sum_i d_i \cdot (1 - D_i) \cdot \|\mathcal{P}[X_i, X_{i+1}]\| \\
&\leq \sum_i d_i \cdot D_i \cdot \pi/2 \cdot dist(X_i, X_{i+1}) + \text{(the number of detour paths in } \mathcal{P}) \cdot \beta \\
&\leq \sum_i d_i \cdot \pi/2 \cdot dist(X_i, X_{i+1}) + \text{(the number of detour paths in } \mathcal{Q}) \cdot \beta \\
&\leq \sum_i d_i \cdot \pi/2 \cdot \|\mathcal{Q}[X_i, X_{i+1}]\| + \beta/\pi \cdot \sum_i d_i \cdot \|\mathcal{Q}[X_i, X_{i+1}]\| \\
&\leq (\pi/2 + \beta/\pi) \cdot \sum_i d_i \cdot \|\mathcal{Q}[X_i, X_{i+1}]\|
\end{aligned}
$$

The first inequality is clear from Lemma 2. The second inequality comes from the fact that our procedure uses the smallest number of detour paths for $\mathcal{X}$. Therefore, we have the following.

$$
\begin{aligned}
\|\mathcal{P}\| &= \sum_i \|\mathcal{P}[X_i, X_{i+1}]\| \\
&= \sum_i (1 - d_i) \cdot \|\mathcal{P}[X_i, X_{i+1}]\| + \sum_i d_i \cdot \|\mathcal{P}[X_i, X_{i+1}]\| \\
&\leq \alpha \cdot \sum_i (1 - d_i) \cdot \|\mathcal{Q}[X_i, X_{i+1}]\| + (\pi/2 + \beta/\pi) \cdot \sum_i d_i \cdot \|\mathcal{Q}[X_i, X_{i+1}]\| \\
&\leq \max(\alpha, \pi/2 + \beta/\pi) \cdot \| \sum_i \mathcal{Q}[X_i, X_{i+1}]\|
\end{aligned}
$$

completing the proof.
$\square$

It remains to prove bounds on the constants α and β.

Lemma 7 $\alpha = 1 + \pi$

Proof: Let $\sigma = (X, \phi)$ and $\sigma' = (Y, \phi')$. Assume that X and Y lie on a horizontal line with X left of Y.

If the distance between X and Y is 2 and $\phi = \phi'$ is the left horizontal direction, there is no path from σ to σ' with length less than $2 + 2\pi$, and so $\alpha \geq 1 + \pi$. This can be seen by considering vertical lines through X and Y and applying Fact 1.

It remains to show $\alpha \leq 1 + \pi$. We prove this by a case analysis according to the orientations ϕ and ϕ'. Details can be found in the full paper. $\Box$

Lemma 8 $\beta \leq (5/2)\pi + 3$

Proof: Let $\sigma = (X, \phi)$ and $\sigma' = (Y, \phi')$. Assume that X and Y lie on a horizontal line with X left of Y. Consider a circle C with radius 2 around Y. We will construct a path from σ to σ' by passing through a configuration $\sigma^* = (X^*, \phi^*)$ on C. Since X^* and Y have distance 2, the shortest path from σ^* to σ' has length at most $2 + 2\pi$ by Lemma 7. It remains to show that σ^* can be chosen such that the length of the path from σ to σ^* is at most $1 + \pi/2$. This can be done by following C_σ^+ or C_σ^- until the orientation is vertical, and then following a segment until C is reached. $\Box$

Plugging the bounds on α and β into Theorem 6, we obtain the following result.

Corollary 9 *A 5.03-approximate shortest path subject to the curvature constraint can be computed in linear time.*

5   Concluding remarks

The on-line and off-line models can be combined into a *semi-online* version. We suppose that the robot not only knows the next request point, but is also able to "see" the following points of the sequence that are included in the disk with radius R around the robot. The semi-online version is a generalization of both the off-line and the on-line problem, since the case $R = 0$ corresponds to the strict on-line problem, while $R = \infty$ is the off-line problem. The following result can be shown using a modification of our off-line approximation algorithm. Details can be found in the full paper.

Theorem 10 *There exists a $(5.03 + O(1/R))$-competitive algorithm for the semi-online version with parameter R.*

One could have defined a different "semi-online" setting, where the robot is told about the following k points whenever it receives a new request. However, our counter-example in Section 3 can be modified to show that no algorithm can achieve a constant competitive ratio even for $k = o(n)$, as in Figure 7.

Our approximation algorithm guarantees that the minimum possible number of detour paths is used. That means that it has to construct the unique optimal path for the construction of Section 3 (the only path that uses only direct paths). Implementing this correctly will probably only be possible if some form of exact arithmetic is used.

Lemma 8 is not tight. A natural conjecture would be $\beta = 2\alpha$, which would imply that our approximation algorithm computes a 4.21-approximation to the shortest path. It would be nice to compute β exactly.

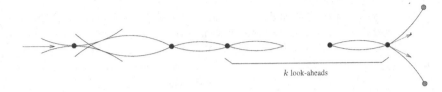

Fig. 7. Extended counter-example

Finally, the shortest path problem through a sequence of points may be of interest in more generalized models that consider the environment with obstacles and/or the restriction on the acceleration [6] instead of the curvature.

Acknowledgement. The part of this research done in KAIST was supported in part by the NRL(National Research Laboratory) of KISTEP (Korea Institute of Science and Technology Evaluation and Planning).

References

1. P. K. Agarwal, T. Biedl, S. Lazard, S. Robbins, S. Suri, and S. Whitesides. Curvature-constrained shortest paths in a convex polygon. In *Proc. 14th SCG*, pages 392–401, 1998. 315
2. P. K. Agarwal, P. Raghavan, and H. Tamaki. Motion planning of a steering-constrained robot through moderate obstacles. In *Proc. 27th STOC*, pages 343–352, 1995. 315, 316, 320
3. H.-K. Ahn, O. Cheong, J. Matoušek, and A. Vigneron. Reachability by paths of bounded curvature in convex polygons. Proc. 16th SCG (to appear), 2000. 315, 316
4. J. D. Boissonnat, A. Cerezo, and J. Leblond. Shortest paths of bounded curvature in the plane. *Interat. J. Intell. Syst.*, 10:1–16, 1994. 315
5. J. D. Boissonnat and S. Lazard. A polynomial-time algorithm for computing a shortest path of bounded curvature admist moderate obstacles. In *Proc. 12th SCG*, pages 242–251, 1996. 315
6. B. Donald, P. Xavier, J. Canny, and J. Reif. Kinodynamic motion planning. *JACM*, 40(5):1048–1066, 1993. 325
7. L. E. Dubins. On curves of minimal length with a constraint on average curvature and with prescribed initial and terminal positions and tangents. *Amer. J. Math.*, 79:497–516, 1957. 315, 315, 316
8. S. Fortune and G. Wilfong. Planning constrained motion. *Ann. Math. Artif. Intell.*, 3:21–82, 1991. 315
9. P. Jacobs and J. Canny. Planning smooth paths for mobile robots. In *Nonholonomic Motion Planning (Z. Li and J Canny, eds.)*, pages 271–342, 1992. 315
10. G. Pestov and V. Ionin. On the largest possible circle imbedded in a given closed curve. *Dok. Akad. Nauk SSSR*, 127:1170–1172, 1959. (In Russian). 316
11. J. A. Reeds and L. A. Shepp. Optimal paths for a car that goes forwards and backwards. *Pacific J. Math.*, 145(2), 1990. 315
12. J. Reif and H. Wang. The complextiy of the two-dimensional curvature-constrained shortest-path problem. In *Proc. of 3rd Workshop on the Algorithmic Foundations of Robotics*, 1998. 315
13. H. Wang and P. K. Agarwal. Approximation algorithm for curvature constrained shortest paths. In *Proc. 7th SODA*, pages 409–418, 1996. 315

Resource Constrained Shortest Paths[*]

Kurt Mehlhorn[1] and Mark Ziegelmann[1][**]

Max-Planck-Institut für Informatik,
Stuhlsatzenhausweg 85, 66123 Saarbrücken, Germany,
{mehlhorn,mark}@mpi-sb.mpg.de
http://www.mpi-sb.mpg.de/~{mehlhorn,mark}

Abstract. The resource constrained shortest path problem (CSP) asks for the computation of a least cost path obeying a set of resource constraints. The problem is NP-complete. We give theoretical and experimental results for CSP. In the theoretical part we present the hull approach, a combinatorial algorithm for solving a linear programming relaxation and prove that it runs in polynomial time in the case of one resource. In the experimental part we compare the hull approach to previous methods for solving the LP relaxation and give an exact algorithm based on the hull approach. We also compare our exact algorithm to previous exact algorithms and approximation algorithms for the problem.

1 Introduction

The resource constrained shortest path problem (CSP) asks for the computation of a least cost path obeying a set of resource constraints. We are given a graph $G = (V, E)$, a source node s and a target t, and k resource limits $\lambda^{(1)}$ to $\lambda^{(k)}$. Each edge e has a cost c_e and uses $r_e^{(i)}$ units of resource i, $1 \leq i \leq k$. Cost and resource usages are assumed to be non-negative. They are additive along paths. The goal is to find a least cost path from s to t satisfying the resource constraints. CSP has applications in mission planning and operations research.

CSP is NP-complete even for the case of one resource [HZ80]. The problem has received considerable attention in the literature [Jok66, Han80, HZ80, AAN83, MR85, Hen85, War87, BC89, Has92, Phi93]. The first step in the exact algorithms of Handler and Zang [HZ80] and Beasley and Christofides [BC89] is the solution of a linear programming relaxation. Hassin [Has92] gives a pseudo-polynomial algorithm for one resource that runs in time $O(nmC)$ for integral edge costs from the range $[0..C]$. He also gives an ε-approximation algorithm that runs in time $O(nm/\varepsilon)$. Other approximation algorithms were given by Phillips [Phi93]. In our experiments the approximation algorithm of Hassin was much slower than our exact algorithm. We discuss the previous work in more detail in later sections of the paper.

In the theoretical part of our paper we study a linear programming relaxation of the problem. It was studied before by Handler/Zang [HZ80] and Beasley/Christofides [BC89] who obtained it by Lagrangean relaxation of the standard ILP formulation. The latter authors use a subgradient procedure to (approximately) solve the LP

---

[*] An extended version of this paper is available at http://www.mpi-sb.mpg.de/~mark/rcsp.ps

[**] Supported by a Graduate Fellowship of the German Research Foundation (DFG).

M. Paterson (Ed.): ESA 2000, LNCS 1879, pp. 326–337, 2000.
© Springer-Verlag Berlin Heidelberg 2000

and the former authors gave a combinatorial algorithm for solving the LP in the case of one resource.

We give a new geometric interpretation of their algorithm and use it to extend the algorithm to the case of many resources. We also show the polynomiality of the algorithm in the case of one resource: the LP can be solved by $O(\log(nRC))$ shortest path computations. This assumes that the c_e and r_e are integral and contained in the ranges $[0..C]$ and $[0..R]$, respectively. Polynomiality of the algorithm in the case of many resources stays open. It follows from the Ellipsoid method that the LP can be solved in polynomial time for any number of resources.

In the experimental part we first compare the hull approach to solving the relaxation with the subgradient method of Beasley/Christofides that approximately solves the Lagrangean relaxation. The hull approach is simultaneously more efficient and more effective in the single resource case; it gives better bounds in less time. For many resources the relation is less clear.

In Section 3 we describe an exact algorithm that follows the general outline of the algorithms of Handler/Zang and Beasley/Christofides and works in three steps: (1) Compute a lower bound LB and an upper bound UB for the objective value by either the hull approach or the subgradient procedure. (2) Use the results of the first step to prune the graph, i.e. to delete nodes and edges which cannot possibly lie on an optimal path. (3) Close the gap between lower and upper bound by enumerating paths.

We use the methods of Beasley/Christofides for the second step. The fact that the hull approach gives better bounds than the subgradient procedure makes pruning more effective. Finally, we discuss three methods for gap closing: (1) Applying the pseudo-polynomial algorithm of Hassin [Has92] to the reduced graph. (2) Enumerating paths in order of increasing reduced cost (the cost function suggested by the optimal solution of the LP) as suggested by Handler and Zang using the k-shortest path algorithm of [JM99]. (3) Enumerating paths in order of increasing reduced cost combined with on-line pruning. This method is new.

Our new method turns out to be most effective in the single resource case when the number of paths to be ranked is large (it otherwise gives similar results as the second method, whereas the first method is only efficient when the pruning reduced the graph drastically). However, the path ranking via k-shortest paths usually outperforms our new method in the multiple resource case.

In our experiments we used random grid graphs, digital elevation models, and road graphs. For all three kinds of graphs the running time of the third step is small compared to that of the first two steps for the single resource case and hence the observed running time of our exact algorithm is $O(m + n \log n) \cdot \log(nCR)$. This compares favorably with the running time of the pseudo-polynomial algorithm ($O(nmC)$) and the approximation algorithm ($O(nm/\varepsilon)$) of Hassin. It is therefore not surprising that the latter algorithms were not competitive in our experiments.

2   Relaxations

We start with a (somewhat unusual) ILP formulation. For any path p from s to t we introduce a 0-1 variable x_p and we use c_p and $r_p^{(i)}$ ($i = 1, \ldots, k$) to denote the cost and resource consumption of p, respectively. Consider

$$\min \sum_p c_p x_p \tag{1}$$

$$\text{s.t.} \sum_p x_p = 1 \tag{2}$$

$$\sum_p r_p^{(i)} x_p \leq \lambda^{(i)} \quad i = 1,\ldots,k \tag{3}$$

$$x_p \in \{0,1\} \tag{4}$$

This ILP models CSP. Observe that integrality together with (2) guarantees that $x_p=1$ for exactly one path, (3) ensures that this path is feasible, i.e. obeys the resource constraints, and (1) ensures that the cheapest feasible path is chosen.

We obtain an LP by dropping the integrality constraint. The objective value LB of the LP relaxation is a lower bound for the objective value of CSP. The linear program has $k+1$ constraints and an exponential number of variables and is non-trivial to solve.

Beasley/Christofides approximate LB from below by relaxing the resource constraint in a Lagrangean fashion and using subgradient optimization to approximately solve the Lagrangean relaxation[1].

The dual of the LP has an unconstrained variable u for equation (2) and k variables $v_i \leq 0$ for the inequalities (3): The dual problem can now be stated as follows:

$$\max \ u + \lambda^{(1)} v_1 + \cdots + \lambda^{(k)} v_k \tag{5}$$

$$\text{s.t.} \ u + r_p^{(1)} v_1 + \cdots r_p^{(k)} v_k \leq c_p \quad \forall p \tag{6}$$

$$v_i \leq 0 \quad i = 1,\ldots,k \tag{7}$$

The dual has only $k+1$ variables albeit an exponential number of constraints.

Lemma 1. *The dual can be solved in polynomial time by the Ellipsoid Method.*

Proof. Let $u^*, v_1^*, \ldots, v_k^*$ be the current values of the dual variables. The separation problem amounts to the question whether there is a path q such that $u^* + v_1^* r_q^{(1)} + \cdots + v_k^* r_q^{(k)} > c_q$ or equivalently $c_q - v_1^* r_q^{(1)} - \cdots - v_k^* r_q^{(k)} < u^*$. This question can be solved with a shortest path computation with the scaled edge costs $\tilde{c}_e = c_e - v_1 r_e^{(1)} - \cdots - v_k r_e^{(k)}$. Hence we can identify violated constraints in polynomial time which completes the proof.

---

[1] For any $v \leq 0$ the linear program $\min \sum_p (c_p + v(\lambda - r_p))$ subject to $x \geq 0$ and $\sum_p x_p = 1$ is easily solved by a shortest path computation (with respect to the edge costs $c_e - v r_e$) and is a lower bound for LB (this is obvious in the presence of the additional constraint $\sum_p r_p x_p \leq \lambda$; dropping the constraint can make the minimum only smaller). Subgradient optimization uses an iterative procedure to approximate the value of v which maximizes the bound. In general, the maximum is not reached, but only approximated. It is known (see, for example, [CCPS98]) that the maximum is equal to the objective value of the LP relaxation.

2.1 The Single Resource Case

The dual now has two variables, u and $v \le 0$ and can be interpreted geometrically.

Standard Interpretation [HZ80]: A pair (u,v) is interpreted as a point in the u-v-plane. Each constraint is viewed as a halfspace in the u-v-plane and we search for the maximal feasible point in direction $(1,\lambda)$ (see Figure 2.1).

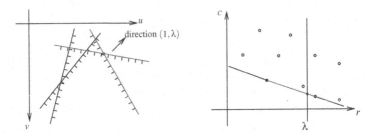

Fig. 1. (left) Find maximal point in direction $(1,\lambda)$ in the intersection of halfspaces. (right) Find line with maximal c-value at $r = \lambda$ which has all points above or on it.

(Geometric) Dual Interpretation: A pair (u,v) is interpreted as a line in r-c-space (the line $c = vr + u$). A constraint $u + r_p v \le c_p$ is interpreted as a point (r_p, c_p) in r-c-space. Hence every path corresponds to a point in r-c-space. We are searching for a line with non-positive slope that maximizes $u + \lambda v$ while obeying the constraints, i.e. which has maximal c-value at $r = \lambda$, and which has all points (r_p, c_p) above or on it (see Figure 2.1), i.e., we are searching for the segment of the lower hull [2] which intersects the line $r = \lambda$.

We now use the new interpretation to derive a combinatorial algorithm for solving the dual. Although equivalent to Handler/Zang's method, our formulation is simpler and more intuitive.

The hull approach: We are searching for the segment of the lower hull which intersects the limit line $r = \lambda$. We may compute points on the lower hull with shortest path computations, e.g. the extreme point in c-direction is the point corresponding to the minimal length path and the extreme point in r-direction is the point corresponding to the minimal resource path.

We start with the computation of the minimum resource path. If this path is unfeasible, i.e. all points lie to the right of the limit line, the dual LP is unbounded and the CSP problem is unfeasible. Otherwise we compute a lower bound, the minimum length path. If this path is feasible, we have found the optimum of CSP, the constraint λ is redundant in that case.

Otherwise we have a first line through the two points that corresponds to a (u,v)-value pair where v is the slope and u the c-abscissae of the line.

---

[2] We use a sloppy definition of lower hull here: there is a horizontal segment to the right incident to the lowest point.

Now we test whether all constraints are fulfilled, i.e. whether all points lie above or on the line. This separation problem is solved with a shortest path computation with the scaled costs $\tilde{c}_e = c_e - v r_e$ and can be viewed as moving the line in its normal direction until we find the extreme point in that direction (see Figure 2). If no point below the

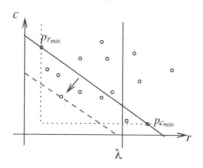

Fig. 2. Finding a new point on the lower hull

line is found, we conclude that no constraint is violated and have found the optimal line. The optimal value of the dual relaxation is its intersection with the limit line $r = \lambda$.

Otherwise we update the line: The new line consists of the new point and one of the two old points such that we again have a line that has all previously seen hull points above it and that intersects the limit line with maximal c-value. We iterate this procedure until no more violating constraint is found.

The Simplex Approach: It is also possible to interpret the hull approach as a dual simplex algorithm with cutting plane generation. We maintain the optimum for a set of constraints. New constraints are generated solving the separation problem with a shortest path computation. The dual simplex algorithm is used to find the optimum for the new set of constraints which is possible in a single pivot step. See the extended version for further details.

2.2   Running Time of the Hull Approach

Now we show that the number of iterations to solve the dual relaxation for integral lengths and resources is logarithmic in the input. This is the first result that gives non-trivial bounds for the number of iterations[3].

We assume that edge costs and resources are integers in the range $[0..C]$ and $[0..R]$, respectively. This implies that the cost and resource of a path lie in $[0..nC]$ and $[0..nR]$, respectively.

---

[3] Anderson and Sobti [AS99] give an algorithm for the table layout problem that works similarly to the hull approach solving parametrized flow problems in each iteration. Using a simple modification of our proof it is also possible to show a logarithmic bound on the number of iterations in their case.

Theorem 1. *The number of iterations of the hull approach is $O(\log(nRC))$. Hence the relaxed problem CSP_{rel} can be solved in $O(\log(nRC)(n\log n + m))$ time which is polynomial in the input.*

Proof. We examine the triangle defining the non-explored region, i.e. the region where we may find hull points. The maximum area of such a triangle is $A_{max} = 1/2n^2RC$, the minimum aera is $A_{min} = 1/2$. We will show in the following Lemma that the area of the triangle defining the unexplored region is at least divided by 4 in each iteration step. Since we have integral resources we may conclude that $O(\log(nRC))$ iterations suffice.

Lemma 2. *Let A_i and A_{i+1} be the area of the unexplored region after step i and $i+1$ of the hull approach, respectively. Then we have $A_{i+1} \le 1/4A_i$.*

Proof. Let A and B be the current feasible and unfeasible hull point in step i and let sl_A and sl_B be the slopes which lead to the discovery of A and B, respectively. The line g_A through A with slope sl_A and the line g_B through B with slope sl_B intersect in the point C. The triangle T_{ABC} defines the unexplored region after step i. Hence $A_i = A(T_{ABC})$.

Wlog consider an update of the current feasible point A[4]. Let A' be the new hull point that was discovered with slope sl_{AB} of the line through A and B.

We assume that A' lies on the segment $\overline{AC}$. The new unexplored region after the update step is defined by A', B, and the intersection point C' of line g_B and the line $g_{A'}$ through A' with slope sl_{AB} (see Figure 4). We have $A_{i+1} = A(T_{A'BC'})$.

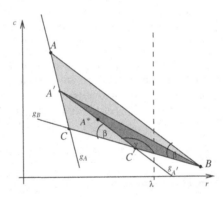

Fig. 3. Area change of unexplored region in an update step.

Now the proportionality principle tells us something about the side lengths of the two triangles. We have $|CA'|/|CA| = |CC'|/|CB|$ and $|CA'|/|CA| = |A'C'|/|AB|$. If we set $|CA'|/|CA| = x < 1$, we get $|C'B| = (1-x)|CB|$ and $|A'C'| = x|AB|$.

The angle β between the segments AB and CB also appears between the segments $A'C'$ and CC', hence the angle γ between the segments $A'C'$ and $C'B$ is $\gamma = \pi - \beta$ (see Figure 4). Thus, $|\sin\gamma| = |\sin\beta|$.

---

[4] The unfeasible update is analogous.

The area of triangle T_{ABC} is $A_i = 1/2|AB||CB|\sin\beta$ and the area of the new triangle $T_{A'BC'}$ is $A_{i+1} = 1/2|A'C'||C'B|\sin\gamma = 1/2x(1-x)|AB||CB|\sin\beta = x(1-x)A_i$.

The product of the two side lengths of the new triangle is maximized for $x = 1/2$, i.e. when choosing A' as midpoint of the segment CA. Hence we have $A_{i+1} \leq 1/4A_i$.

It is easy to see that when we chose the new hull point A^* to lie on the line $g_{A'}$ but in the interior of the triangle T_{ABC} and form the new triangle $T_{A^*BC'}$, its area is smaller than the area of a triangle $T_{A'BC'}$. Hence $A_{i+1} \leq 1/4A_i$ for an arbitray update step.

The same running time for solving the dual relaxation can be obtained by using binary search on the slopes: Starting with a slope s_1 giving a feasible point and a slope s_2 giving an unfeasible point, we do a binary search on the slopes (running Dijkstra with $v = (s_1 + s_2)/2$ and updating accordingly) until the two slopes differ by less than $1/(n^2R^2)$ which means that the line through the two points gives the optimum at $r = \lambda$.

Although binary search and the hull approach have the same worst case bound, the hull approach usually requires much fewer iterations. For example if it has reached the optimal hull segment it detects optimality with a single Dijkstra application, whereas the binary search approach would have to continue the iteration until the slopes differ by less than $1/(n^2R^2)$ to be sure that the current segment is optimal.

2.3   Multiple Resources

We turn back to the general case of k resources. Beasley/Christofides' method also approximates the optimum for the dual relaxation in the multiple resource case whereas Handler/Zang leave this open for future work. We again give two interpretations of the dual relaxation:

Standard interpretation: A tuple $(u, v_1, \ldots, v_k)$ is interpreted as a point in u-v_1-$\cdots$-v_k-space. We view each constraint as a halfspace in u-v_1-$\cdots$-v_k-space and search for the maximal feasible point in direction $(1, \lambda^{(1)}, \ldots, \lambda^{(k)})$. The optimum is defined by $k + 1 - l$ constraints that are satisfied with equality and is parallel to l coordinate axes[5].

(Geometric) Dual interpretation: A tuple $(u, v_1, \ldots, v_k)$ is interpreted as a hyperplane in $r^{(1)}$-$\cdots$-$r^{(k)}$-c-space. A constraint $u + r_p v \leq c_p$ is interpreted as a point $(r_p^{(1)}, \ldots, r_p^{(k)}, c_p)$ in $r^{(1)}$-$\cdots$-$r^{(k)}$-c-space. We are searching for a hyperplane $c = v_1 r^{(1)} + \cdots + v_k r^{(k)} + u$ with nonpositive v_i's, that has all points $(r_p^{(1)}, \ldots, r_p^{(k)}, c_p)$ above or on it, and has maximal c-value at the limit line $(r^{(1)}, \ldots, r^{(k)}) = (\lambda^{(1)}, \ldots, \lambda^{(k)})$.

We again describe two approaches to solve the dual relaxation based on the different interpretations:

Geometric approach: We start off with the artificial feasible point and then compute the minimum cost point. Then in each iteration we have to determine the optimal face of the lower hull[6] seen so far. Since a newly computed hull point q might also see previous hull points not belonging to the old optimal face, we have more candidates for the new optimal solution. This is due to the following fact: The area where new points can be found is a simplex determined by the halfspaces that correspond to the points that define

---

[5] where l is the number of v_i's that are zero

[6] A point is again treated as rays in $r_1, \ldots, r_k$-directions.

the current optimum, and the hyperplane defining the current optimum. In contrast to the single resource case, the current optimal hull facet does not "block" previous points from that region, the new point may also see previous hull points. Hence they might also be part of a new optimal solution.

Thus, we have to determine the q-visible facets of the old hull to update the new lower hull as in an incremental $k + 1$-dimensional convex hull algorithm [CMS93]. The new optimal face is the face incident to q that intersects the limit line. Now we check whether there is a violated constraint. This is done by solving a shortest path problem with scaled costs which again can be seen as moving the facet defining hyperplane in its normal direction until we reach the extreme point in this direction. We stop the iteration when no more violating constraint is found.

Simplex approach: The dual simplex method with cutting plane generation can also be extended to multiple resources. However, the dual simplex algorithm now might take more than one pivot step to update the optimum when adding a new constraint. Details can be found in the extended version.

Bounding the number of iterations remains open for the multiple resource case but Lemma 1 tells us that the dual relaxation can be solved in polyomial time using the Ellipsoid method.

2.4   Comparison Hull Approach – Subgradient Procedure

We have implemented the described methods using LEDA [MN99, MNSU]. We now compare the hull approach and the subgradient method of Beasley and Christofides experimentally[7]. We used three different kinds of graphs: Random grid graphs, Digital elevation models (DEM)[8] and road graphs. Due to lack of space we only report about a small number of experiments using DEMs[9].

We first turn to the single resource case (upper part of Table 9). We observe that the bounds of the hull approach are always better than the bounds of the subgradient procedure. Since the number of iterations of the hull approach scales very moderately and is still below the fixed number of iterations of the subgradient procedure[10], the hull approach is also more efficient. It gives better bounds in less time.

Now we turn to the multiple resource case (lower part of Table 9). Here we cannot compute a trivial feasible upper bound. We implemented two versions of the hull approach, one using CPLEX for the pivot step (solution of a small LP) and one using the incremental d-dimensional convex hull implementation of [CMS93]. The running time is still dominated by the shortest path computations and thus the time of the two versions differs only slightly. The hull approach always computes a feasible upper bound

---

[7] All experiments measuring CPU time in seconds on a Sun Enterprise 333MHz, 6Gb RAM running Solaris compiled with g++ -O2.

[8] Grid graphs where every node has a certain height: costs are randomly chosen from $[10..20]$ and resources are height differences.

[9] The resource limit is chosen to be 110% of the minimal resource path in the single resource case and 120% for 2 resources. Source and target are in the corners.

[10] Beasley/Christofides suggest 11 iterations.

N	UB_{triv}	LB_{triv}	OPT	UB_{MZ}	LB_{MZ}	it	t	UB_{BC}	LB_{BC}	t
50	1524	1206	1358	1387	1354.84	7	0.15	1437	1206	0.23
100	3950	2404	2739	2746	2736.1	7	0.77	2818	2665.85	1.21
150	5374	3584	4277	4304	4275.9	8	2.26	4553	4212.53	3.02
200	7625	4767	5903	5903	5900.47	9	5.07	6019	5884.71	6.07
50	-	1224	1314	1319	1313.27	14	0.27/0.27	-	1224	0.23
100	-	2380	2728	2740	2720.92	14	1.53/1.49	2757	2713.8	1.26
150	-	3591	3984	3992	3981.07	18	4.91/4.73	-	3980.62	3.01
200	-	4642	5321	5326	5318.86	19	10.11/9.95	5358	5307.83	5.54

Table 1. Comparison of the bounds: The first column is the size of the $N \times N$ grid graph, the following three columns contain the trivial upper and lower bound (costs of minimum resource and minimum cost path) and the optimum. Then we have the bounds, number of iterations and time of the hull approach, followed by the bounds and the time of the subgradient procedure. The upper and lower half of the table correspond to the one and two resource case, respectively.

and the gap between the bounds is relatively small albeit generally larger than in the single resource case. The number of iterations roughly doubles. The subgradient procedure with the fixed number of iterations often fails to find a feasible upper bound, the lower bounds are sometimes close but often worse than our lower bounds.

Although it can be shown that our bounds come with no approximation guarantee (see extended version), we observe in our experiments that the gap between upper and lower bound computed by the hull approach is very small.

3   Pruning and Closing the Gap

Solving the relaxation gives us upper and lower bounds for CSP. If we are interested in an exact solution of CSP, we can close the gap between upper and lower bounds with path ranking. To make this efficient we use the bounds for problem reductions to eliminate nodes and edges that cannot be part of an optimal solution. This leads to an exact 3-phase algorithm (bounds computation, pruning and gap closing) for CSP that was first suggested by Handler/Zang.

Pruning: For a theoretical and experimental discussion of pruning consult the extended version of this paper.

Gap Closing: Now we want to show how the gap between upper and lower bound can be closed to obtain the optimal solution of the CSP problem.

Handler/Zang were the first to show how the upper and lower bound of the relaxation can be used to close the gap via path ranking. Beasley and Christofides also used the computed bounds and proposed a depth first tree search procedure. We give more geometric intuition how the bounds are used and propose a new dynamic programming method to close the gap which can be seen as path ranking with online pruning.

Starting from the lower bound: Figure 3 shows the situation after the hull approach. There might be candidates for the CSP optimum in the triangle defined by limit line, upper bound line and the optimal hull segment. To compute the optimum it seems to be

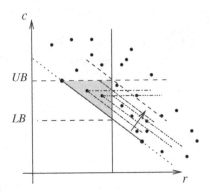

Fig. 4. Closing the gap between upper and lower bound

sensible to rank paths in the normal direction to the hull segment defining the optimum of the relaxation, i.e. we use the lower bound costs as cost matrix. This can be seen as moving the line through the optimal segment in its normal direction. We can stop when the line has reached the intersection between limit line and upperbound line (see Figure 3). When we encounter a feasible path with smaller length than the current upper bound, we update the upper bound and also the point where we can stop the ranking.

The path ranking can be done with a k-shortest path algorithm [Epp99, JM99] as suggested by Handler/Zang [HZ80].

We developed another approach based on dynamic programming. This approach can be seen as a clever way of path ranking, pruning unpromising paths along the way. Details are given in the extended version.

Experiments: We now investigate the practical performance of different gap closing strategies and exact algorithms.

Let us first turn to the single resource case (upper part of Table 3 which is a continuation of Table 9): We observed that the gap closing time using dynamic programming or k-shortest paths is extremely small. The dynamic programming approach performs better when the number of ranked paths is large, whereas the k-shortest path algorithm is slightly faster in "easy" cases. Gap closing with the pseudopolynomial algorithm of Hassin is always much slower. The same happens for the gap closing using Beasley/Christofides relaxation results[11].

We see that the total time for the exact 3-phase algorithm using our bounds is dominated by the bound computation and hence the observed running time is $O(n \log^2 n)$ since we have $N \times N$ grid graphs. It is not surprising that it was more efficient than our implementation of an ε-approximation scheme of Hassin [Has92]. The 3-phase algorithm using Beasley/Christofides bounds is often dominated by the gap closing step and always gives much worse running times.

---

[11] Their suggested tree search procedure for gap closing was not competitive so we used one of the previous approaches giving best performance.

t_{DP}	t_{KSP}	t_{HS}	t_{BC}	t_{BC}	t_{MZ}	t_{DP}	t_{KSP}	t_{ILP}
0.01	0.01	5.65	0.24	0.47	0.23	0.3	-	37.2
0.02	0.08	4.18	1.04	2.25	1.12	1.46	-	-
0.04	0.04	18.9	43.01	46.03	3.29	19.89	-	-
0.04	0.08	1.23	8.13	14.2	7.06	-	-	-
0.05	0.02	-	-	-	0.37	-	-	59.7
55.3	22.4	-	-	-	24.22	-	-	-
22.7	7.02	-	7.48	10.49	12.71	-	-	-
8.23	3.07	-	-	-	14.99	-	-	-

Table 2. Gap closing and total running times: The first three columns contain the time for gap closing with our dynamic programming approach, the k-shortest path implementation of [JM99], the pseudopolynomial algorithm of Hassin [Has92]. Our bounds and the obtained reduced graph is used. The fourth column is the gap closing time starting from the bounds of Beasley and Christofides. The next thwo columns contain the total running time of the exact 3-phase algorithm for different bounds. The last three columns are the times for computing the optimum with our dynamic programming approach and k shortest paths without using the bounds and the times for solving the ILP with CPLEX. A '-' means that the computation was aborted after 60 seconds.

CPLEX solving the ILP takes between 30 and 60 seconds for 50×50 grid graphs and much more for larger instances, hence is not competitive. Our dynamic programming method without using bounds and reduced costs performed surprisingly good for smaller problem instances whereas simple path ranking failed to give solutions even for small problem instances.

Now we turn to the multiple resource case (lower part of Table 3): Here gap closing via k-shortest paths is always more efficient than dynamic programming since with multiple resources there are much less dominating paths that can be pruned. Now the running time of the 3-phase algorithm is dominated by the gap closing in many cases.

The gap closing starting from Beasley/Christofides' bounds is slightly worse when the lower bound is near our lower bound, but drastically slower otherwise.

CPLEX solving the ILP for multiple resources already takes around one minute for 50×50 grid graphs and multiple resource dynamic programming and path ranking without using the bounds is not even efficient for small graphs.

In most cases where gap closing is dominating the total running time in the exact 3-phase algorithm, the optimum is relatively far away from the lower bound hence a large number of paths has to be ranked. Both path ranking approaches can be stopped returning an improved gap if one does not want to wait until the true optimum is reached.

More experiments also using random grid graphs and road graphs will be given in the full version of this paper.

4   Conclusion

We presented an algorithm that computes upper and lower bounds for CSP by exactly solving an LP relaxation. In the single resource case the algorithm was first formulated by Handler/Zang [HZ80]. We extended the algorithm to many resources and proved its polynomiality in the single resource case. For multiple resources we were only able to show that the relaxation can be solved in polynomial time by the Ellipsoid method.

We applied the problem reductions of [AAN83, BC89] using our bounds and proposed an efficient dynamic programming method to reach the optimal solution from the lower bound. In the single resource case our new method is competitive to previously suggested path ranking and even is superior for certain problems when the number of ranked paths is large. Our experiments showed that our 3-phase algorithm is very efficient for the examined graph instances being superior to nonspecialized approaches as ILP solving and path ranking.

References

[AAN83]   Y. Aneja, V. Aggarwal, and K. Nair. Shortest chain subject to side conditions. *Networks*, 13:295–302, 1983. 326, 337

[AS99]   R. Anderson and S. Sobti. The table layout problem. In *Proc. 15th SoCG*, pages 115–123, 1999. 330

[BC89]   J. Beasley and N. Christofides. An Algorithm for the Resource Constrained Shortest Path Problem. *Networks*, 19:379–394, 1989. 326, 337

[CCPS98]   W. Cook, W. Cunningham, W. Pulleyblank, and A. Shrijver. *Combinatorial Optimization*. John Wiley & Sons, Inc, 1998. 328

[CMS93]   K. Clarkson, K. Mehlhorn, and R. Seidel. Four results on randomized incremental construction. *Computational Geometry: Theory and Applications*, 3(4):185–212, 1993. 333

[Epp99]   D. Eppstein. Finding the k shortest paths. *SIAM Journal on Computing*, 28(2):652–673, 1999. 335

[Han80]   P. Hansen. Bicriterion path problems. In G. Fandel and T. Gal, editors, *Multiple Criteria Decision Making: Theory and Application*, pages 109–127. Springer verlag, Berlin, 1980. 326

[Has92]   R. Hassin. Approximation Schemes for the Restricted Shortest Path Problem. *Math. Oper. Res.*, 17(1):36–42, 1992. 326, 327, 335, 336

[Hen85]   M. Henig. The shortest path problem with two objective functions. *European Journal of Operational Research*, 25:281–291, 1985. 326

[HZ80]   G. Handler and I. Zang. A Dual Algorithm for the Constrained Shortest Path Problem. *Networks*, 10:293–310, 1980. 326, 329, 335, 336

[JM99]   V. Jiminez and A. Marzal. Computing the k shortest paths. A new algorithm and an experimental comparison. In *Proc. 3rd Workshop on Algorithm Engineering (WAE99)*, LNCS 1668, pages 25–29, 1999. 327, 335, 336

[Jok66]   H. Joksch. The Shortest Route Problem with Constraints. *Journal of Mathematical Analysis and Application*, 14:191–197, 1966. 326

[MN99]   K. Mehlhorn and S. Näher. *The LEDA platform for combinatorial and geometric computing*. Cambridge University Press, 1999. 333

[MNSU]   K. Mehlhorn, S. Näher, M. Seel, and C. Uhrig. *The LEDA User Manual*. Max-Planck-Institut für Informatik. http://www.mpi-sb.mpg.de/LEDA. 333

[MR85]   M. Minoux and C. Ribero. A heuristic approach to hard constrained shortest path problems. *Discrete Applied Mathematics*, 10:125–137, 1985. 326

[Phi93]   C. Phillips. The Network Inhibition Problem. In *25th ACM STOC*, pages 776–785, 1993. 326

[War87]   A. Warburton. Approximation of pareto-optima in multiple-objective shortest path problems. *Operations Research*, 35(1):70–79, 1987. 326

On the Competitiveness of Linear Search*

J. Ian Munro

Department of Computer Science, University of Waterloo,
Waterloo, Ontario, Canada N2L 3G1
`imunro@uwaterloo.ca`

Abstract. We re-examine offline techniques for linear search. Under a reasonable model of computation, a method is given to perform offline linear search in amortized cost proportional to the entropy of the request sequence; and so, this cost is at most logarithmic. On the other hand, any online technique is subject to linear amortized cost for some sequences. It follows, then, that no online technique can have an amortized cost of that which one could obtain if given the request sequence in advance, i.e., there is no competitive linear search algorithm.

1 Introduction

This is a paper about the competitiveness of algorithms. That is, the extent to which an algorithm can receive and immediately process a sequence of queries almost as well as could be done if all queries were given in advance, and a more global scheme could be developed for their processing. Online algorithms and competitiveness have been a major focus in the theory of query processing over the past 10 or 15 years [1,2,8,9]. Borodin and El Yani [2], in particular, give an excellent treatment of the topic. At the heart of proving anything about "optimal" performance are the details of the costs associated with the model of computation. Most authors, certainly those noted above, make these models very clear. Nevertheless, it is also useful to understand what happens when some constraints are relaxed in a realistic manner. The effects of such relaxation are the focus of this paper.

We look at a very simple, and well studied, problem. Given a linear list of elements, we are to perform searches for elements many times and may reorganize the list as a byproduct of searches. The question is about the extent to which having knowledge of future requests will help in reducing their costs. Our focus is, then, primarily on the problem of the "offline" version of linear search.

The notion of self-organizing data structures has been around since the 1960's. McCabe [5] seems to have been the first to study the behavior of such methods. In particular he introduced the idea of the "move to front" heuristic, by which an element retrieved in a linear list is moved to the front of the list. He, and later Rivest [7], showed that this approach leads to an expected cost of at

---

* This work was supported by the Natural Science and Engineering Research Council of Canada.

M. Paterson (Ed.): ESA 2000, LNCS 1879, pp. 338–345, 2000.

most $2 \sum i p_i$ under the assumption that elements are accessed with independent probabilities and the ith most frequently accessed of n elements is requested with probability p_i. As the optimal (static) arrangement would be to list the elements in decreasing order by probability of access, this means the approach gives an expected performance within a factor of 2 of what we would get if the probabilities were known.

This result was strengthened by Bentley and McGeough [1], who showed that this result holds, not only probabilistically, but in an amortized sense in that for any sufficiently long (length m) sequence in which the ith most frequently accessed element is requested f_i times, the formula above still holds if we substitute f_i/m for p_i.

The theory was further strengthened by Sleator and Tarjan [8], who demonstrated that the approach was 2-competitive under a reasonable, but (as we shall see) still restrictive, model. In particular, they showed that even if we are given the sequence of requests in advance and can make adaptations to the list to make (known) future requests cheaper, the move to front heuristic will process a (suitably long) sequence of requests at a cost within a factor of two of that of any dynamic offline method ... under a particular cost model. The model charges 1 per element in the linear scan for the requested value, and the algorithm can move this value to any earlier location in the list, free of charge. One cannot scan past the requested element to pick up some "soon to be needed" element, although this restriction could probably be removed. The more serious restriction is the cost charged for moving elements around the list. The only way in which this can be done is by swapping consecutive elements at a cost of 1 for each such swap. Therefore, accessing the last element in the list and then reversing the entire list would, for example, cost about $n^2/2$, whereas the "real" cost in an array or a linear linked list would be $O(n)$. Reingold and Westbrook [6] studied the problem of finding the optimal reordering scheme, for the offline problem, under the Sleator and Tarjan model.

In the next section we present an offline algorithm, show that it does much better than the move to front heuristic on a permutation of the keys, and then show that it comes within a constant factor of doing as well as possible (under our model). In Section 3, we demonstrate that the method can easily be implemented with no additional memory, and bound the amortized cost of its runtime on an arbitrary input sequence. In the Conclusion we relate this work to comparable issues on binary search trees, and in particular to the dynamic splay conjecture.

2   Accessing a Permutation of the Elements

We start with the simple case of requesting each element of a linear list once, according to an arbitrary permutation. The rules of the *model* are simple: We must start at the beginning of the list on each search, and charge one for each element accessed. We are free to continue the search past the requested element (at a cost of 1 per element accessed), and may rearrange the accessed elements in any order free of charge. As we shall see, the rearrangements we will actually perform can easily be done on a linear linked list in linear time using a constant number of extra pointers. We employ variations of the following algorithm:

Algorithm Order_by_next_request(x)

- Scan the list to find element x, suppose it is the ith element in the list.
- Continue the scan to the first position at or beyond i that is a power of 2, i.e. position $p \equiv 2^{\lceil \lg i \rceil}$.
- Reorder the elements from position 1 to position p in order of next access.

For purposes of our permutation example, or more to the point understanding its runtime, we adopt two conventions: pre-process by ordering elements in order of next request (so the charge of 16 for the first request in Table 1 is n in general), and, on performing each search, move the requested element to the head of the list. That is view the "current request" as the highest priority by "next request". Table 1 illustrates searching for 16 elements in turn. As the table indicates, and is easily checked in general, every second access costs 2, every fourth costs 4 and every 2^ith costs 2^i, so the amortized cost of an access is roughly $\lg n$, in sharp contrast to the n we could experience with the move to front heuristic. It is easily seen that our method will exhibit the same behavior even without the preprocessing scan, and even if several permutations of accesses are concatenated. The formal proof that this technique has at worst logarithmic amortized cost over any sequence of inputs is given in Section 3, though we will quote here its consequence for the competitiveness of online methods.

In contrast, any online algorithm (i.e., any method that does not know what queries will be asked in the future) could always be asked for the last element in the list as it currently stands. The amortized cost per request would, then, be at n. Therefore we state:

Theorem 1. *Under the previously outlined model, no online sequential list search algorithm is c-competitive with the optimal offline method, for any constant c. In particular any online method must have amortized cost $\Theta(n)$ for some arbitrarily long sequence of queries, while there exists an offline technique with amortized cost $\Theta(\lg n)$.*

Proof. The proof follows from the discussion above and the proof of Theorem 4, where we note r is at most n.   □

We also observe that the reordering need not be constrained to continue to the next position that is a power of 2. We will discuss such tuning later.

While the key point is that we can access an arbitrary permutation of n elements in about $n \lg n$ steps, it is also interesting that this is a lower bound.

Theorem 2. *Under the previously outlined model, $\Theta(n \lg n)$ element inspections are required to separately access each element in a list of length n.*

Proof. Consider an access to position i, or beyond. This can result in the retrieval of the requested element and to the positioning of the $i - 1$ values ahead of

Cost	New Ordering
16	1 2 3 4 5 6 7 8 9 10 11 12 13 14 15 16
2	2 1 ...
4	3 4 1 2 ...
2	4 3 ...
8	5 6 7 8 1 2 3 4...
2	6 5 ...
4	7 8 5 6 ...
2	8 7 ...
16	9 10 11 12 13 14 15 16 1 2 3 4 5 6 7 8
2	10 9 ...
4	11 12 9 10 ...
2	12 11 ...
8	13 14 15 16 9 10 11 12...
2	14 13 ...
4	15 16 13 14 ...
2	16 15 ...

Table 1. Searching for elements in a given permutation.

position i. Hence we must reach position i, or beyond, at least $\lceil n/i \rceil$ times in the processes of retrieving all n elements. Summing this lower bound on the number of times position i must be inspected, we get a lower bound on the total cost as:

$$\sum_{i=1}^{n} \lceil n/i \rceil - 1 \approx n \ln n.$$

□

3   A Bound on the Order by Next Request Heuristic

Our main result is the bound on the behavior of the "order by next request" heuristic presented above. First, however, we make a few other observations about the approach. The list can be viewed as being kept in logical *blocks* of sizes successive powers of some value such as two. Indeed discussion is simpler if we say block 1 contains the first 2 elements and, for $i > 1$, block i contains those from position $2^{i-1}+1$ to position 2^i. Furthermore, we assume that the entire process is preceded by a pass that orders elements by first request. From this point on, each block will remain in this next request order, although elements requested and then not required for a long time will slowly drift back generally making longer and longer stops near the ends of larger and larger blocks. This observation leads to our analysis of the method and also the following observation:

Theorem 3. *The order by next request heuristic can be implemented either in an array or a linear linked list in time linear in the number of elements inspected using a constant number of pointers or indices.*

Proof. The reordering done in any request is simply that of sorting a list or array that is already in sorted blocks of lengths 2^i. If the structure is sorted as a linked list, two pointers suffice to merge two sorted blocks. The blocks are merged into a common list from smallest block to largest (i.e., in the order in which the elements/blocks are stored) in a process involving a total of about $2p$ priority (i.e., next request time) comparisons, if p is the number of elements in the reordered prefix. As each request will be for the first element of a block, if a request is given for the value in in position q, the process will continue to position $p \equiv 2(q - 1)$ for a total of about $4q$ priority comparisons is required.

If the structure is maintained as a simple array, we again have the problem of merging $\lg p$ lists of lengths powers of 2. The process is complicated only by having to do merging in place in linear time by a method such as that of Kronrod [4] or Huang and Langston [3].   □

We return to our main interest, that of proving an amortized bound on the cost of an access request. Consider an element, v, and in particular what happens between two consecutive accesses to v. Suppose r distinct elements (including v itself) are accessed between these two probes. Then our first observation is that v will never be positioned later than position r during this process.

To obtain a good amortized bound, we want to charge something to the elements inspected during the access process. For such a policy to work we would like

- to charge a constant amount to each of the elements that we do charge, hence at least a constant fraction of the elements inspected must be charged, and
- to charge only elements whose "state" changes in some identifiable way, so that the number of times an element is charged can be bounded.

Charging elements near the end of the list (most notably those in the last block which may never be accessed) could result in an unbounded number of charges being made between accesses. On the other hand, there are too few elements near the front to absorb much of the cost. Therefore we adopt a "taxation policy" that can be glibly dubbed "soak the middle class" in proving:

Theorem 4. *The cost of the order by next request heuristic can be amortized as at most $1 + 4\lceil \lg r \rceil$ element inspections for the retrieval of a value, where $r - 1$ other distinct values have been retrieved since the last time the given element was requested. By convention we assume that the list is initialized into order by next request, and that every element is perceived (for purposes of the formula above) as having been first requested at time 0 and is again requested after all requests in the given sequence.*

Proof. If the request is for the first element we charge it 1. Otherwise we charge the entire cost of a p probe request to the $p/4$ elements in the penultimate block inspected, that is to the block immediately prior to the requested element. (Recall a requested element will be the first in its block.) Consider an element,

v, in the charged block, b. We will demonstrate that v can be charged at most once in any block, or equivalently that it cannot be charged again in block b until after v itself has been requested again. There are two cases to consider: that v remains in block b or moves to an earlier block as a consequence of the reordering done during the current request, or that it moves to block $b + 1$.

Consider first the case in which v remains in its block or moves closer to the front. All elements from v's new position to the end of block $b+1$ are ordered by next request. They will remain in the same relative order despite any requests for earlier values, though earlier requested elements may be moved into positions among them after having been accessed. None of these values following v, in any block up to $b + 1$, can be requested before v. Hence v cannot be charged again until after a request is made for an element in a block beyond $b+1$. If this occurs, v will again not be charged, but it could be moved into the block in which the access was made.

Now suppose v is moved "out" to block $b + 1$. It cannot be charged for any access to its block or an earlier one, but is eligible for a charge resulting from an request for an element in block $b + 2$.

It follows, then, that an element is charged at most once in each of the $\lceil \lg r \rceil$ blocks up to position r. As a consequence we have proven an amortized retrieval cost of at most $4\lceil \lg r \rceil$.

$\square$

As noted earlier, there is no particular reason, other than ease of presentation, that successive blocks need double in size. If blocks increase by a factor of a, we perform a scan out to a block of size a^{k+1}, and charge all $(a^{k+2} - 1)/(a - 1)$ probes to the a^k elements in the penultimate block. This results a charge of about $a^2/(a - 1)$ each. As above, we see an element may be charged $\log_a r \equiv \lg r / \lg a$ times between requests. At each step we round the end position of the block to the nearest integer to the value suggested above. Hence a need not be an integer. The cost attributed to an element between two requests is minimized by setting $a = 4.24429..$, leading to an improved bound:

Theorem 5. *The amortized cost of the order by next request heuristic, with block sizes roughly successive powers of 4.24429.. is at most $1 + 2.66241..\lceil \lg r \rceil$ element inspections for the retrieval of a value, where $r - 1$ other distinct values have been retrieved since the last time the given element was requested. Again, we assume that the list is initialized into order by next request, and that every element is perceived (for purposes of the formula above) as having been first requested at time 0 and is again requested after all requests in the given sequence.*

This result can be restated in a couple of other ways:

Corollary 1. *The total cost for a sequence of requests, where the ith request occurs r_i requests after the last request for the same element is at most $n + \sum_{i=1}^{n} 2.66241..\lceil \lg(r_i + 1) \rceil$.*

The following version expresses the result in terms very much like the entropy of a distribution.

Corollary 2. *Let f_i denote the frequency of access to element i. Then the total cost of a sequence of m requests is at most $n + 2.66241.. \sum_{i=1}^{n} \lceil \lg(m/f_i) \rceil$, or if we write $p_i \equiv f_i/m$ then the amortized cost of a search for element i is $O(1 + p_i \lg 1/p_i)$.*

4   Conclusion

We have shown that, under a reasonable model of computation, the move to front heuristic is not close to being competitive. Indeed if all elements occur with the same frequency, move to front takes an average of about $n/2$ probes, while the "offline" order by next request heuristic has an amortized cost of $O(\lg n)$. Furthermore, an extension of Theorem 2 leads to a $\Omega(\lg n)$ lower bound for the offline problem. Of course, we must emphasize that elements must be stored in a linear list to be searched only by starting at the beginning of the list and scanning through.

It is interesting that the behavior of our offline technique is within a constant factor of the entropy of the distribution, if we can interpret relative frequencies from a finite sequence as probabilities. Indeed it would appear that the difference of a constant factor is an artifact of the analysis rather than the method. This leads to questions about offline techniques with binary search trees. If we can achieve the entropy (or come within a constant factor of it) with a linear search, then shouldn't we do much better with a binary search tree? If an offline binary search tree technique could beat the entropy bound by more than a constant factor, then this would lead to a counterexample to the dynamic splay conjecture [9] in the case that requests were always for the deepest element in the splay tree. Perhaps a suitable restriction of this holds, but the most natural extension of our approach to binary search trees is the following:

- Perform a search for the requested element.
- Consider the path from the root to the requested node as a binary search tree, and all branches from this path as external nodes.
- Reorder the path as a binary search tree so as to minimize the path to the next element to be requested element, and subject to this constraint reorder it to minimize the paths to subsequently requested elements. Therefore, if the next node to be requested is on the original path, then that node becomes the root of the new tree. If it is a node off the path, and hence in what we will view as an "external node", then this external node falls between two path nodes. These path nodes will be the root and a child of the root of the new tree. If one of these path nodes is the second value to be requested, then it is moved to the root, otherwise the one whose "other" external node.

We applied this greedy heuristic to a number of binary trees, with given request sequences, and it seemed promising. However, if we apply it to the generalization of the tree in Table 2, no changes are made and the average search cost is about $\lg n$. In the table, we have a balanced binary search tree. Key values are not given; simply assume they are in a valid order. The numbers indicate the order of requests, which we can even take as cyclic. The idea is that on the path to any node, at each node along the way the opposite branch is taken before the current one will be reused. As a consequence, our attempt to reorder the search path to move soon to be accessed elements up the tree, meets with failure ... not just failure, but indeed catatonia.

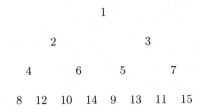

Table 2. Greedy ordering of the search path can lead to no changes

References

1. J. L. Bentley and C. McGeough. Amortized analysis of selforganizing sequestial search heuristics. *Communications of the ACM*, 28(4):404-411, 1985. 338, 339
2. A. Borodin and R. El Yaniv. *Online Computation and Competitive Analysis*, Cambridge University Press, 1998. 338
3. B. C. Huang and M. A. Langston, Practical in-place merging, *Communications of the ACM* **31(3)** (1988) 348-352. 342
4. M. A. Kronrod. Optimal ordering algorithm without operational field. *Soviet Math. Dokl.*, 10:744-746, 1969. 342
5. J. McCabe. On serial files with relocatable records. *Operations Research*, 13:609-618, July 1965. 338
6. N. Reingold and J. Westbrook. Offline algorithms for the list update problem. *Information Processing Letters*, 60:75-80, 1996. 339
7. R. Rivest. On self-organizing sequestial search heuristics. *Communications of the ACM*, 19(2):63-67, February 1976. 338
8. D. D. Sleator and R. E. Tarjan. Amortized efficiency of list update and paging rules. *Communications of the ACM*, 28(2):202-208, 1985. 338, 339
9. D. D. Sleator and R. E. Tarjan. Self-adjusting binary search trees. *Journal of the ACM*, 32:652-686, 1985. 338, 344

Maintaining a Minimum Spanning Tree under Transient Node Failures[*]

Enrico Nardelli[1,2], Guido Proietti[1,2], and Peter Widmayer[3]

[1] Dipartimento di Matematica Pura ed Applicata, Università di L'Aquila, Via Vetoio, 67010 L'Aquila, Italy. E-mail: {nardelli,proietti}@univaq.it
[2] Istituto di Analisi dei Sistemi e Informatica, Consiglio Nazionale delle Ricerche, Viale Manzoni 30, 00185 Roma, Italy.
[3] Institut für Theoretische Informatik, ETH Zentrum, CLW C 2, Clausiusstrasse 49, 8092 Zürich, Switzerland. E-mail: widmayer@inf.ethz.ch

Abstract. Given a 2-node connected, undirected graph $G = (V, E)$, with n nodes and m edges with real weights, and given a minimum spanning tree (MST) $T = (V, E_T)$ of G, we study the problem of finding, for every node $v \in V$, the MST of $G - v = (V \setminus \{v\}, E \setminus E_v)$, where E_v is the set of edges incident to v in G. We show that this problem can be solved in $\mathcal{O}(\min(m \cdot \alpha(n, n), m + n \log n))$ time and $\mathcal{O}(m)$ space. Our solution improves on the previously known $\mathcal{O}(m \log n)$ time bound.

1 Introduction

Let V be a set of n *sites* that must be interconnected, let E be a set of m potential *links* between the sites, and let $w(e)$ be some real *weight* associated with link e. Let $G = (V, E)$ be the corresponding weighted, undirected graph, and assume that G is 2-node connected (i.e., to disconnect G we have to remove at least 2 nodes). Let $\mathcal{N}_G = (V, E'), E' \subseteq E$ be a connected spanning subgraph of G. We call $\mathcal{N}_G$ a *communication network*, or simply a network, in G. A network is generally built with the aim of minimizing some *cost*, which is computed using some criteria defined over $\mathcal{N}_G$ according to the edge weights. For instance, if the network must be built so that the sum of all the edge weights in the network is minimum, then $\mathcal{N}_G$ is a *minimum (weight) spanning tree* (MST) of G. As another example, if one wants the network to be a spanning tree such that the maximum distance between any two nodes is minimum, then $\mathcal{N}_G$ is a *minimum diameter spanning tree* (MDST) of G.

In addition to minimizing the network cost, we want to focus on another important feature of the network that must be taken into account from the very beginning: Its *reliability* or *survivability*, i.e., the ability of the network to remain operational if individual network components (edges or nodes) fail. In the past

---

[*] This work has been partially developed while the second author was visiting the third one, supported by the EU TMR Grant CHOROCHRONOS and by the Swiss National Science Foundation.

M. Paterson (Ed.): ESA 2000, LNCS 1879, pp. 346–355, 2000.

few years, several survivability problems have been studied intensely [7]. Clearly, low cost and high reliability of the network are two conflicting objectives. In the extreme, to maintain costs as low as possible, a network might be designed as a spanning tree of the underlying graph. Such a network, however, will not survive even a single edge failure. Even worse, a single node failure in the network might leave each and every node isolated. Therefore, if the network is operational as long as it is connected, some redundancy in the set of edges must be present. As a result, we design the network on two layers: a first layer, where the primary set of links is activated, in such a way that the cost is minimized, and a secondary layer of inactive *replacement links*, a subset of which will switch to active as soon as a network component fails, so that the network will remain connected. We call the activated set of replacement links the set of *swap edges*, given the failed component, and we call the result of swapping the *replacement (swap) network*.

Transient failures of edges and nodes play an important role in reliability considerations: a component that is down is expected to come up again after a little while. Under this assumption, it may be unlikely that there is more than one failure at any given time. Therefore, we study the problem of dealing with the failure of a single edge or node in the network. Since we want to prepare for the transient failure of an arbitrary component, we precompute all individual component failures. We aim at a set of swap edges that ideally minimizes the cost of the replacement network or at least keeps it low. In the extreme, a minimum cost replacement network might entail the use of many swap edges, at a set-up cost per edge in a real communication network. For instance, if the network is a *single source shortest paths tree* (SPT), then the failure of an edge (node) might completely change the shortest paths to all the nodes beyond the failed edge (node), but a change of the network to the new SPT would induce high set-up costs. After the short duration of the transient failure, the switch back to the old SPT would again have high set-up costs, and so it might be desirable to avoid the expensive switch altogether and cope with a worse network for the short time of the failure. Given that we aim at saving set-up cost, we want to choose a small set of replacement edges which will provide a low but not minimum cost replacement network after a failure. In the extreme, for a tree network one could associate with each edge in the network a single swap edge, so that the resulting swap network is the best possible among all the networks that can be obtained by means of a single swap per edge. Similarly, we associate with each node v having $\delta(v)$ adjacent nodes in the network, a set of $\delta(v) - 1$ swap edges that reconnect the network. This is the problem we study in this paper.

In general, even if we choose the best possible swap edges, this *does not* guarantee that the swap network will be the optimum one in the damaged graph, as the previous example with the SPT shows [10]. A similar example is given by the MDST [8,9]. However, for some specific network topologies (i.e., for some specific cost criteria), the swap network is optimal. A very popular network architecture for which this holds is the MST. In fact, it is easy to see that when an edge e fails in a MST, then the MST of $G - e = (V, E \setminus \{e\})$ can be obtained by means of a single replacement edge. Similarly, when a failure happens to a node v having $\delta(v)$ nodes adjacent in the MST, then the MST of $G - v = (V \setminus \{v\}, E \setminus E_v)$, where E_v is the set of edges incident to v in G, can be obtained by means of $\delta(v) - 1$ replacement edges. Thus, reliability in MSTs can be accomplished by using all the best swap edges for both the set of edges and the set of nodes originally in the network.

Not surprisingly then, the problem of computing efficiently all these best swap edges has been studied in the past, in the two settings in which either edge or node failures are considered, starting from different perspectives, though. Let AER (*all edges replacement*) and ANR (*all nodes replacement*) denote the problem of determining the set of swap edges for all the edges and all the nodes in the network, respectively. The fastest solution for solving the AER problem runs in $\mathcal{O}(m)$ time [4]. Note that this positively compares with the fastest dynamic offline algorithm known up to date for maintaining minimum spanning trees, which requires $\mathcal{O}(\log n)$ time per update, and $\mathcal{O}(m)$ space and preprocessing time [5]. In fact, one can always find all the best swaps by considering all the edges of the MST one after the other (and then the algorithm is offline, since we know in advance which edges need to be considered): when the edge e of weight $w(e)$ is considered, we increase its weight to an arbitrary large value, we compute the new MST and we then set back the weight of e to $w(e)$. This will cost a total of $\mathcal{O}(m + n \log n)$ time and $\mathcal{O}(m)$ space.

On the other hand, as far as the ANR problem is concerned, and despite its intrinsic analogy with the AER problem, the best known algorithms are slower, and run either in $\mathcal{O}(m \log n)$ time for $m = o\left(\frac{n^2}{\log n}\right)$ [5], or in $\mathcal{O}(n^2)$ time otherwise [2]. Das and Loui [3] have shown that if the edge weights are sorted in advance, then the ANR can be solved in $\mathcal{O}(m \cdot \alpha(m, n))$ time and space, where here and in the rest of the paper $\alpha(m, n)$ is the functional inverse of Ackermann's function as in [12]. In this way, however, the logarithmic factor is shifted to the sorting of the edge weights, although there seems to be no evidence that edge weights must be sorted for solving the ANR problem.

In this paper, we move one step towards a (potential) linear solution of the ANR problem, by providing an $\mathcal{O}(\min(m \cdot \alpha(n, n), m + n \log n))$ time and $\mathcal{O}(m)$ space algorithm. The problem of finding a linear time algorithm (or, alternatively, a superlinear lower bound) for the ANR problem remains a challenging open problem.

The paper is organized as follows: in Section 2 we define the problem we are dealing with more precisely and formally, and we give some basic definitions that will be used throughout the paper, while in Section 3 we describe the algorithm for solving the problem, and we provide an analysis of both correctness and complexity of the proposed algorithm. Finally, in Section 4, we present conclusions and list some open problems.

2   Preliminaries

Let $G = (V, E)$ be an undirected graph, where V is a set of n nodes and $E \subseteq V \times V$ is a set of m edges, with a real length $w(e)$ associated with each edge

$e \in E$. If multiple edges between nodes are allowed, then the graph is called a *multigraph*. A graph $H = (V', E')$ is called a *subgraph* of G if $V' \subseteq V$ and $E' \subseteq E$. If $V' \equiv V$ then H is called a *spanning subgraph* of G.

A *simple path* (or a *path* for short) in G is a subgraph $P = (V', E')$ with $V' = \{v_1, v_2, \ldots, v_k | v_i \neq v_j$ for $i \neq j\}$ and $E' = \{(v_i, v_{i+1}) | 1 \leq i < k\}$, also denoted as $P = \langle v_1, v_2, \ldots, v_k \rangle$. Path P is said to go from v_1 to v_k passing through $v_2, v_3, \ldots, v_{k-1}$. If $v_1 \equiv v_k$, then P is a *cycle*. A graph G is *connected* if, given any two distinct nodes u, v of G, there exists a path from u to v. A graph G is *2-node connected* (*biconnected*, for short) if, given any three distinct nodes u, v, w of G, there exists a path from u to w not passing through v. In other words, a graph G is biconnected if at least 2 of its nodes must be removed to disconnect it. A connected, acyclic spanning subgraph of G is called a *spanning tree* of G. A spanning tree $T = (V, E_T)$ of G is said to be a *minimum weight spanning tree* (MST) of G if the sum of all the edge weights $w(e), e \in E_T$ is minimum for all the spanning trees of G.

Let $T(r)$ denote a MST of a biconnected graph G rooted at an arbitrary node r. Let $v_1, \ldots, v_k$ be the children of node v in $T(r)$, and let v_0 be its parent. Let $G - v = (V \setminus \{v\}, E \setminus E_v)$, where E_v is the set of edges incident to v in G. Note that $G - v$ is connected, since G is biconnected. For any two node-disjoint subtrees T_1 and T_2 of $T(r)$, let $E(T_1, T_2)$ be the set of non-tree edges having one endpoint in T_1 and the other endpoint in T_2. Let us denote by $T(v)$ the subtree of $T(r)$ rooted at v, and by $\overline{T}(v)$ the tree $T(r)$ with $T(v)$ and the edge (v_0, v) removed. Let $\mathcal{H}_v = \{f \in E \setminus E_T | f \in E(T(v_i), T(v_j)), i, j = 1, \ldots, k, i \neq j\}$, referred to in the following as the set of *horizontal edges* of v, and let $\mathcal{V}_v = \{f \in E \setminus E_T | f \in E(T(v_i), \overline{T}(v)), i = 1, \ldots, k\}$, referred to in the following as the set of *vertical edges* of v.

It is easy to see that the MST $T_{\not{v}}$ of $G - v$ can be computed by making use of the multigraph obtained by *contracting* to a *vertex* each subtree of $T(r)$ created by the removal of v [5]. More precisely, let $\mathcal{W}_v = \{w_0, w_1, \ldots, w_k\}$ be the set of vertices associated with the contraction of the set of subtrees $\{\overline{T}(v), T(v_1), \ldots, T(v_k)\}$. To compute $T_{\not{v}}$, we first compute the MST $T_v = (\mathcal{W}_v, \mathcal{R}_v)$ of the multigraph $G_v = (\mathcal{W}_v, \mathcal{H}_v \cup \mathcal{V}_v)$. Then, we set $T_{\not{v}} = (V \setminus \{v\}, E_T \setminus \{(v_0, v), (v, v_1), \ldots, (v, v_k)\} \cup \mathcal{R}_v)$. Figure 1 illustrates the used notations.

The set $\mathcal{R}_v$ is called the set of *best swap* (or *replacement*) *edges* for v. The *all nodes replacement* (ANR) problem asks for finding $\mathcal{R}_v$ for every node $v \in V$.

3   Solving the ANR Problem

We first give a high-level description of the algorithm, and we then illustrate it in details.

3.1   High-Level Description of the Algorithm

A high-level description of our algorithm is the following. We consider all the non-leaf nodes in $T(r)$ in any arbitrary postorder (if v is a leaf node, then trivially

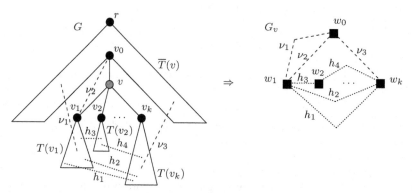

Fig. 1. Node v is removed from G: subtrees $\overline{T}(v), T(v_1), \ldots, T(v_k)$ are contracted to vertices $w_0, w_1, \ldots, w_k$, respectively, joined by vertical edges (dashed) and horizontal edges (dotted), to form G_v.

$\mathcal{R}_v = \emptyset$). Let us now fix such a node v, having parent v_0 and children $v_1, \ldots, v_k$ in $T(r)$, and let $\mathcal{W}_v$, $\mathcal{H}_v$ and $\mathcal{V}_v$ be the corresponding sets of vertices, horizontal and vertical edges, respectively (notice that if $v \equiv r$, then $\mathcal{V}_v = \emptyset$).

Let $f_{v_i} \in E \setminus E_T$ be a *selected vertical edge* for $T(v_i), i = 1, \ldots, k$, defined as an edge for which $w(f_{v_i}) = \min\{w(f)|f \in E(T(v_i), \overline{T}(v))\}$, unless $E(T(v_i), \overline{T}(v))$ is empty, in which case f_{v_i} does not exist. For the sake of avoiding technicalities, assume that f_{v_i} is unique (if it exists). Let $\mathcal{V}'_v \subseteq \mathcal{V}_v$ be the set of existing selected vertical edges associated with v. It is easy to see that $\mathcal{R}_v$ corresponds to the set of edges of the MST of the multigraph $H_v = (\mathcal{W}_v, \mathcal{H}_v \cup \mathcal{V}'_v)$. It will turn out that this selection of edges among the vertical edges is vital to get our claimed time complexity for solving the ANR problem.

It is therefore sufficient to compute the MST of H_v, for all the nodes v of $T(r)$ in postorder.

3.2 Computing Efficiently the Selected Vertical Edges

The efficiency problem is the computation of the selected vertical edges $\mathcal{V}'_v$, for all the nodes $v \in V$. It is clearly prohibitive to simply compute $\mathcal{V}'_v$ from scratch, but it is just as well too expensive to attack the problem in a bottom-up fashion, by using mergeable heaps in which each heap contains all the vertical edges associated with a given subtree of $T(r)$. In fact, it can be proved that $\Omega\left(k \log \frac{n}{k}\right)$ time is needed for deleting k elements from a heap of n elements, assuming that insertions, merges and minimum finds are performed in constant time [11]. Then, it is not hard to see that such an approach would require $\mathcal{O}(m \log n)$ time for solving the ANR problem, since $\Theta(m)$ deletions, spread over $\Theta(n)$ heaps containing $\mathcal{O}(n)$ elements at a time, are needed.

To avoid this problem, we adopt a totally different strategy: We find $\mathcal{V}'_v$ by making use of a transformation of the graph G to a multigraph G' containing less than $2m$ edges. After this transformation, we build a *transmuter* [13], representing the set of *fundamental cycles* of G' with respect to $T(r)$, defined as the cycles of G' containing only a single non-tree edge. This will allow us to compute V'_v in $\mathcal{O}(m \cdot \alpha(m, n))$ time.

We now describe in detail how the transformation works. Let $f = (x, y) \in E \setminus E_T$ denote an arbitrary *non-tree edge* in G. W.l.o.g., we will assume in the following that x precedes y in postorder. Let $nca(x, y)$ denote the *nearest common ancestor* in $T(r)$ of x and y, and let x' and y' denote the children in $T(r)$ of $nca(x, y)$ on the paths (if any) going from $nca(x, y)$ to x and y, respectively (see Figure 2).

Depending on $nca(x, y), x, y, x'$ and y', edge f is either eliminated or is substituted by one or two edges having weight $w(f)$. More precisely, we transform the graph according to the following *substitution rules* (see Figure 2):

(R1): if $nca(x, y) \equiv y$, we substitute f by the edge $f' = (x, x')$;
(R2): if $x' \equiv x$ and $y' \not\equiv y$, we substitute f by the edge $f' = (y, y')$;
(R3): if $y' \equiv y$ and $x' \not\equiv x$, we substitute f by the edge $f' = (x, x')$;
(R4): if $x' \equiv x$ and $y' \equiv y$, f disappears;
(R5): otherwise, f is substituted by the two edges $f' = (x, x')$ and $f'' = (y, y')$.

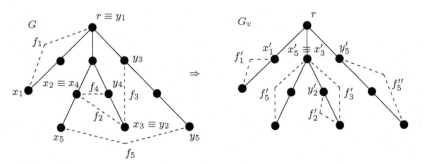

Fig. 2. Substitution of non-tree (dashed) edges $f_i = (x_i, y_i), i = 1, \ldots, 5$, based on $T(r)$ (solid edges), visited in postorder from left to right: edges f_1, f_2, f_3, f_4 and f_5 are substituted according to rules (R1), (R2), (R3), (R4) and (R5), respectively.

The above transformation produces a multigraph $G' = (V, E')$, where E' is the set of original tree edges plus the set of edges substituting the original non-tree edges (*substituting edges*, for short), and therefore $|E'| < 2m$. The decisive property of this transformation is the following:

Lemma 1. *Let $f = (x, y) \in E \setminus E_T$ be a non-tree edge forming a fundamental cycle in G containing a pair of adjacent tree edges $e_v = (v_0, v)$ and $e_{v_i} = (v, v_i)$, where v_0 is the parent of v and v is the parent of v_i in $T(r)$. If x (resp. y) is a descendant of v in $T(r)$, then f is a non-tree edge of minimum weight forming* *a fundamental cycle in G with e_v and e_{v_i}, if and only if $f' = (x, x')$ (resp. $f'' = (y, y')$) is a substituting edge of minimum weight forming a fundamental cycle in G' with e_{v_i}.*

Proof. Let f be a non-tree edge of minimum weight among all the non-tree edges forming a fundamental cycle in G with e_v and e_{v_i}. W.l.o.g., let us assume that x is a descendant of v in $T(r)$ (the case where y is a descendant of v in $T(r)$ can be treated similarly). Since f forms a cycle with e_v and e_{v_i}, it follows that $nca(x, y)$ belongs to the path in $T(r)$ from r to v_0. Hence, the substituting edge f', having weight $w(f) = w(f')$, is such that x' belongs to the path in $T(r)$ from r to v, and therefore f' forms a cycle with e_{v_i}. To show that f' is a substituting edge of minimum weight among all the substituting edges forming a fundamental cycle in G' with e_{v_i}, let us assume that there exists a substituting edge $g' = (u, u')$ forming a cycle with e_{v_i}, and such that $w(g') < w(f')$. Let $g = (u, z)$ be the original non-tree edge substituted by g' (and possibly, by an additional $g'' = (z, z')$). Since g' forms a cycle with e_{v_i}, it follows that u' belongs to the path in $T(r)$ from r to v, and therefore $nca(u, z)$ belongs to the path in $T(r)$ from r to v_0. Hence, g forms a cycle in G with e_v and e_{v_i}, and $w(g) = w(g') < w(f') = w(f)$, a contradiction.

Conversely, w.l.o.g. let f' (the case with f'' can be treated similarly) be a substituting edge of minimum weight among all the substituting edges forming a fundamental cycle in G' with e_{v_i}. The original edge f substituted by f' is such that either $nca(x, y) \equiv v_0$ (according to rules (R1) and (R3)), or $nca(x, y)$ is an ancestor of v_0 in $T(r)$ (according to rule (R5)). In both cases, f forms a fundamental cycle in G with e_v and e_{v_i}, and $w(f) = w(f')$. To show that f is a non-tree edge of minimum weight among all the non-tree edges forming a fundamental cycle in G with e_v and e_{v_i}, let us assume that there exists a non-tree edge $g = (u, z)$ forming a cycle with e_v and e_{v_i}, with u preceding z in postorder and such that $w(g) < w(f)$. Two cases are possible: (1) u is a descendant of v_i in $T(r)$; (2) z is a descendant of v_i in $T(r)$. In the former case, from the fact that $nca(u, z)$ belongs to the path in $T(r)$ from r to v_0, it follows that u' belongs to the path in $T(r)$ from r to v, and therefore g' forms a cycle in G' with e_{v_i}, and $w(g') = w(g) < w(f) = w(f')$, a contradiction. Similarly, in the latter case, it follows that z' belongs to the path in $T(r)$ from r to v, and therefore $g'' = (z, z')$ forms a cycle in G' with e_{v_i}, and $w(g'') = w(g) < w(f) = w(f')$, a contradiction. □

Using the above lemma, we can prove the following result:

Theorem 1. *The selected vertical edges $\mathcal{V}'_v$ for all the non-leaf nodes $v \in V$ can be computed in $\mathcal{O}(m \cdot \alpha(m, n))$ time and $\mathcal{O}(m)$ space.*

Proof. Consider an arbitrary non-leaf node $v \in V$. If $v \equiv r$, then trivially $\mathcal{V}'_v = \emptyset$. Let then v be a non-leaf node other than r, having parent v_0 and children $v_1, \ldots, v_k$ in $T(r)$. The multigraph G' can be computed in $\mathcal{O}(m \cdot \alpha(m, n))$ time and $\mathcal{O}(m)$ space: In fact, $nca(x, y)$ can be found in $\mathcal{O}(\alpha(m, n))$ amortized time for each non-tree edge, using $\mathcal{O}(m)$ space [13]. By slightly modifying this algorithm,

x' and y' can be found in $\mathcal{O}(\alpha(m, n))$ amortized time as well: in fact, assuming that $T(r)$ is visited in a fixed postorder, x' can be computed by postponing the union operations appearing in [13] after all the children and the non-tree edges of a given node have been examined, while y' can be computed by reapplying the modified algorithm, but visiting $T(r)$ in reverse postorder. We omit the details due to lack of space.

Moreover, from Lemma 1, it follows that a selected vertical edge $f_{v_i} \in E \setminus E_T$ for the subtree $T(v_i)$, that is a vertical edge of minimum weight containing $e_v = (v_0, v)$ and $e_{v_i} = (v, v_i)$, corresponds to a substituting edge of minimum weight in G' containing e_{v_i}. Henceforth, we can compute efficiently the set of selected vertical edges by means of a transmuter $D[G', T(r)]$ built on G' with respect to $T(r)$. Basically, $D[G', T(r)]$ contains one source node for each tree edge in G', one sink node for each substituting edge in G', and a number of additional nodes. The fundamental property of a transmuter is that there is a path from a given source node to a given sink node if and only if the associated edges form a fundamental cycle in G'. Recall that $D[G', T(r)]$ can be built in $\mathcal{O}(m \cdot \alpha(m, n))$ time [13]. To find a selected vertical edge for the subtree $T(v_i)$, we label a sink node with the weight of the corresponding substituting edge, and we process the transmuter in reverse topological order, labelling each node with the minimum of the labels of its immediate successors. When the process is complete, the source node representing e_{v_i} is associated with a substituting edge of minimum weight forming a cycle with it, which in its turn is associated with an original non-tree edge forming a cycle of minimum weight with e_v and e_{v_i}. Therefore, the edge e_{v_i} remains associated with a selected vertical edge f_{v_i} for the subtree $T(v_i)$, and $\mathcal{V}'_v = \bigcup_{i=1}^{k} \{f_{v_i}\}$. This can be done in $\mathcal{O}(m \cdot \alpha(m, n))$ time and $\mathcal{O}(m)$ space [13], from which the theorem follows.   □

3.3   Computing All the Best Swap Edges

Once the vertical edges have been selected, to solve the ANR problem we have to compute the MST of $H_v = (\mathcal{W}_v, \mathcal{H}_v \cup \mathcal{V}'_v)$, whose set of edges corresponds to $\mathcal{R}_v$, for every $v \in V$. This leads to the main result.

Theorem 2. *The ANR problem for a minimum spanning tree of a biconnected graph with n nodes and m edges can be solved in $\mathcal{O}(\min(m \cdot \alpha(n, n), m + n \log n))$ time and $\mathcal{O}(m)$ space.*

Proof. Consider an arbitrary node $v \in V$. If v is a leaf node in $T(r)$, then trivially $\mathcal{R}_v = \emptyset$. Let V' be the set of non-leaf nodes. From Theorem 1, computing $\mathcal{V}'_v$ for every $v \in V'$ costs $\mathcal{O}(m \cdot \alpha(m, n))$ time and $\mathcal{O}(m)$ space. On the other hand, since $\mathcal{H}_v = \{f = (x, y) \in E \setminus E_v | nca(x, y) = v\}$, we can associate $\mathcal{H}_v$ with each node v in $\mathcal{O}(m \cdot \alpha(m, n))$ time and $\mathcal{O}(m)$ space [13].

It remains to analyze the total time needed to compute, for every $v \in V'$, the MST of H_v. Let n_v and m_v denote the number of nodes and the number of edges in H_v, respectively. The MST of H_v can be computed in $\mathcal{O}(m_v \cdot \alpha(m_v, n_v))$ time and $\mathcal{O}(m_v)$ space [1]. Notice that each edge in $\mathcal{H}_v$ is associated with only a single node v, and thus it appears just in a single MST computation, while edges in $\mathcal{V}'_v$ can appear in several MST computations. However, $\sum_{v \in V'} |\mathcal{V}'_v| < n - 1$, since we can create at most one selected vertical edge for each tree edge (this also holds for edges leaving the root, with which no selected vertical edges are associated). Notice that this is exactly the reason why vertical edges are selected. In fact, since only $\mathcal{O}(n)$ vertical edges are considered in total, we avoid to consider (in the worst case) $\mathcal{O}(n)$ vertical edges for each of $\mathcal{O}(n)$ failing nodes, i.e., a total of $\mathcal{O}(n^2)$ vertical edges.

It follows that the total time needed to compute the MST of H_v for every $v \in V'$, is

$$\sum_{v \in V'} \mathcal{O}(m_v \cdot \alpha(m_v, n_v)).$$

Since $m_v = \Omega(n_v)$, from the fact that $\alpha(m, n)$ is a monotonically decreasing function in m, it follows that

$$\alpha(m_v, n_v) = \mathcal{O}(\alpha(n_v, n_v)) = \mathcal{O}(\alpha(n, n)).$$

From this and from the fact that $\sum_{v \in V'} m_v = \mathcal{O}(m + n)$, it follows that the total time needed to solve the ANR problem is $\mathcal{O}(m \cdot \alpha(n, n))$, by using $\mathcal{O}(m)$ space.

Notice that the above time bound is worse than $\mathcal{O}(m + n \log n)$ as soon as $m = \omega \left(\frac{n \log n}{\alpha(n,n)} \right)$. Whenever this situation happens, we can use a classic $\mathcal{O}(m_v + n_v \log n_v)$ time and $\mathcal{O}(m_v)$ space algorithm to compute the MST of H_v [6]. This will require $\mathcal{O}(m + n \log n)$ time and $\mathcal{O}(m)$ space to compute $\mathcal{R}_v$ for every $v \in V'$. From this, the thesis follows.   □

4   Conclusions

In this paper we have presented an $\mathcal{O}(\min(m \cdot \alpha(n, n), m + n \log n))$ time and $\mathcal{O}(m)$ space algorithm for solving the ANR problem, improving after several years the previously known $\mathcal{O}(m \log n)$ time bound [5].

A natural open problem is how to reduce the time complexity to $\mathcal{O}(m \cdot \alpha(m, n))$, which might be doable by exploiting the relationships among the various MSTs which are computed in postorder. It appears to be much harder to find a linear time algorithm, at least by using our approach which makes use of an MST computation subroutine. Finally, we plan to extend our approach to different network topologies, analyzing the ANR problem for MDSTs and SPTs.

Acknowledgements – The authors would like to thank Gert Stølting Brodal for helpful discussions on the topic.

References

1. B. Chazelle, A minimum spanning tree algorithm with inverse-Ackermann time complexity, TR NECI 99-099, Princeton University, NJ, 1999. 354
2. F. Chin and D. Houck, Algorithms for updating minimal spanning trees, *J. Comput. System Sci.*, **16**(3) (1978) 333–344. 348
3. B. Das and M.C. Loui, Reconstructing a minimum spanning tree after deletion of any node, TR UILU-ENG-95-2241 (ACT-136), University of Illinois at Urbana-Champaign, IL, 1995. 348
4. B. Dixon, M. Rauch and R.E. Tarjan, Verification and sensitivity analysis of minimum spanning trees in linear time, *SIAM J. Comput.*, **21**(6) (1992) 1184–1192. 348
5. D. Eppstein, Offline algorithms for dynamic minimum spanning tree problem, *2nd Workshop on Algorithms and Data Structures (WADS'91)*, Ottawa, Canada, 1991, Vol. 519 of Lecture Notes in Computer Science, Springer-Verlag, 392–399. A revised version appeared in *J. of Algorithms*, **17**(2) (1994) 237–250. 348, 349, 354
6. M.L. Fredman and R.E. Tarjan, Fibonacci heaps and their uses in improved network optimization algorithms, *J. of the ACM*, **34**(3) (1987) 596–615. 354
7. M. Grötschel, C.L. Monma and M. Stoer, Design of survivable networks, *Handbooks in OR and MS, Vol. 7*, Elsevier (1995) 617–672. 347
8. G.F. Italiano and R. Ramaswami, Maintaining spanning trees of small diameter, *Algorithmica* **22**(3) (1998) 275–304. 347
9. E. Nardelli, G. Proietti and P. Widmayer, Finding all the best swaps of a minimum diameter spanning tree under transient edge failures, *6th European Symp. on Algorithms (ESA'98)*, Venice, Italy, 1998, Vol. 1461 of Lecture Notes in Computer Science, Springer-Verlag, 55–66. 347
10. E. Nardelli, G. Proietti and P. Widmayer, How to swap a failing edge of a single source shortest paths tree, *5th Annual Int. Computing and Combinatorics Conf. (COCOON'99)*, Tokyo, Japan, 1999, Vol. 1627 of Lecture Notes in Computer Science, Springer-Verlag, 144–153. 347
11. D.D. Sleator and R.E. Tarjan, Self-Adjusting Heaps, *SIAM J. Comput.*, **15**(1) (1986) 52–69. 350
12. R.E. Tarjan, Efficiency of a good but not linear set union algorithm, *J. of the ACM*, **22**(2) (1975) 215–225. 348
13. R.E. Tarjan, Applications of path compression on balanced trees, *J. of the ACM*, **26**(4) (1979) 690–715. 351, 352, 353

Minimum Depth Graph Embedding*

Maurizio Pizzonia[1] and Roberto Tamassia[2]

[1] Dipartimento di Informatica e Automazione, Università di Roma Tre, via della
Vasca Navale 79, 00146 Roma, Italy. pizzonia@dia.uniroma3.it
[2] Department of Computer Science, Brown University, 115 Waterman Street,
Providence, RI 02912–1910, USA. rt@cs.brown.edu

Abstract. The depth of a planar embedding is a measure of the topological nesting of the biconnected components of the graph. Minimizing
the depth of planar embeddings has important practical applications to
graph drawing. We give a linear time algorithm for computing a minimum depth embedding of a planar graphs whose biconnected components
have a prescribed embedding.

1 Introduction

Motivated by graph drawing applications, we study the problem of computing
planar embeddings with minimum depth, where the depth of a planar embedding
is a measure of the topological nesting of the blocks (biconnected components) of
the graph. The main result of this paper is a linear time algorithm for computing
minimum-depth embeddings.

In a planar embedding, blocks are inside faces, and faces are inside blocks.
The containment relationships between blocks and faces induces a tree rooted
at the external face. The depth of the planar embedding is the maximum length
of a root-to-leaf path in this tree (see Figure 1).

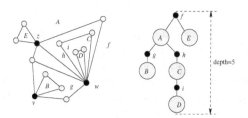

Fig. 1. In this example an embedded graph is shown whose blocks A,B,C,D and
E are connected by means of the cutvertices v, w and z. The embedding has
cutfaces f,g,h and i. The containment relationship between cutfaces and blocks
is represented by a tree of depth 5 rooted at the external face f.

* Research supported in part by the National Science Foundation under grants
CCR–9732327 and CDA–9703080, by the U.S. Army Research Office under grant
DAAH04–96–1–0013, and by "Progetto Algoritmi per Grandi Insiemi di Dati:
Scienza e Ingegneria", MURST Programmi di Ricerca di Rilevante Interesse
Nazionale. Work performed in part while Maurizio Pizzonia was visiting Brown
University.

M. Paterson (Ed.): ESA 2000, LNCS 1879, pp. 356–367, 2000.

The main motivation for studying depth minimization in planar embeddings comes from the field of graph drawing. A widely used technique for constructing orthogonal drawings of general graphs is the *topology-shape-metric* approach [11,26]. This approach has been extensively investigated both theoretically [19,25,16,21,20,9,10,22] and experimentally [13,4]. Also, its practical applicability has been demonstrated by various system prototypes [12,23] and commercial graph drawing tools [1].

The topology-shape-metrics approach consists of three phases. In the first phase, a planar embedding of the input graph is computed (if the graph is not planar, the graph is planarized using dummy vertices to represent crossings). The successive two phases determine the orthogonal shape (angles) and the coordinates of the drawing, respectively, but do not modify the embedding. Hence, the properties of the embedding computed in the first step are crucial for the the quality of the final layout. In particular, the depth of the embedding negatively affects the area and number of bends of the drawing (see Figure 2). Informally speaking, when a block is nested inside a face, the face must "stretch" to accommodate the block inside it. The importance of optimizing the depth of the embedding in the first phase of the topology-shape-metrics approach was already observed in early papers on the GIOTTO algorithm [3,2].

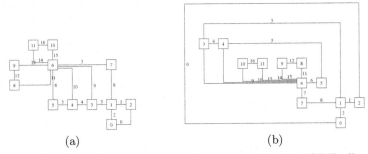

(a)   (b)

Fig. 2. Two drawings of the same graph computed by the GDToolkit system using an algorithm [4] based on the topology-shape-metric approach. Each of the two drawings has bends and area optimized for its embedding. The embedding of drawing (a) has depth 1 while the embedding of drawing (b) has depth 5. Note how the depth of the embedding significantly affects the area, number of bends, and aesthetic quality of the drawing.

Several authors have studied the problem of computing a planar embedding of a graph that is optimal with respect to a certain cost measure. This class of problems is especially challenging because the number of distinct embeddings of a planar graph is exponential in the worst case.

Bienstock and Monma [8] (see also [6,7]) present a general technique to minimize various distance measures over all embeddings of planar graph in polynomial time. They show that minimizing the diameter of the dual of a planar graph, over all planar embeddings, is NP-hard, while it is possible to minimize some distance function to the outer face in polynomial time. Mutzel and Weiskircher [24] present an integer linear programming approach to optimize over all embeddings of biconnected graph a linear cost function on the face cycles.

Computing a planar orthogonal drawing of a planar graph with the minimum number of bends over all possible embeddings is in general NP-hard [17,18]. A polynomial-time algorithm for a restricted class of planar graphs is given in [14]. Heuristic techniques and experimental results are presented in [5,15].

In Section 2 we give basic definitions on embeddings and formally define the concept of depth. In Section 3, we consider a restricted version of the depth minimization problem and present a linear-time algorithm for it. In Section 4 we give a linear time algorithm for the general depth minimization problem, using the algorithm of Section 3 as a building block. Section 5 concludes the paper with open problems.

2   Preliminaries

In this section, we review basic concepts about graphs and embeddings, and give definitions that will be used throughout the paper.

Let G be a connected planar graph. For simplicity, we assume that G has no parallel edges or self-loops. A *cutvertex* of G is a vertex whose deletion disconnects G. Graph G is said to be biconnected if it has no cutvertices. A *block* B of G is a maximal subgraph of G such that B is biconnected. The *block-cutvertex tree* T of G is a tree whose nodes are associated with the blocks and cutvertices of G, and whose edges connect each cutvertex-node to the block-nodes of the blocks containing the cutvertex.

An *embedding* Γ of G is an equivalence class of planar drawings of G with the same circular ordering of edges around each vertex. Two drawings with the same embedding also induce the same circuits of edges bounding corresponding regions in the two drawings. These circuits are called the *faces* of the embedding.

The *dual embedding* Γ' of Γ is the embedded graph induced by the adjacency relations among the faces of Γ through its edges. A *cutface* f of Γ is a face associated with a cutvertex of Γ'. The *block-cutface* tree T^* of Γ is the block-cutvertex tree of Γ'. Since the dual of any biconnected embedding is biconnected T and T^* contains the same set of blocks.

A *planar embedding* is an embedding where a face is chosen as *external face*. We do not consider external faces that are not cutfaces. We consider the block-cutface tree of a planar embedding rooted at the external face.

We now give some definitions about trees. In a tree T the distance between two nodes is the length of the unique path among them. The *eccentricity* of a node v (denoted by $e(v)$) is the maximum among the distances from v to any leaf. The *diameter* of T (diam T) is the maximum among the eccentricity of the leaves. For each node v of T, we have $\frac{\text{diam } T}{2} \le e(v) \le$ diam $T \le 2e(v)$. The *center* of T is the set of nodes with minimum eccentricity.

Assume now that tree T is rooted. The *depth* of T (depth T) is the eccentricity of the root of T. A *depth path* of T is a path from the root to a leaf with maximum distance from the root. The *depth tree* of T is the union of all the depth paths of T. A *diametral path* of T is a path between two leaves with maximum eccentricity. The *diametral tree* of a tree is the union of all the diametral paths of a tree.

Let Γ be an embedding of a planar connected graph G, and let T^* be the block-cutface tree of Γ. We define the *diameter* of Γ (diam Γ) as the diameter of T^*. If Γ is a planar embedding, we define the *depth* of Γ (depth Γ) as the depth of T^*.

We consider the problem of assembling the planar embedding of a connected planar graph G using given embeddings for its blocks, with the goal of minimizing the depth of the embedding.

Let G be a connected planar graph, and assume that we have a prescribed embedding for each block B of G. An embedding Γ of G is said to *preserve the embedding of a block B* if the sub-embedding of B in Γ is the same as the prescribed embedding of B. We say that Γ is *block-preserving for a cutvertex v* if Γ preserves the embedding of each block B containing v, that is, the circular order of the edges of B incident on v is equal to the order in the prescribed embedding of B. Finally, we say that Γ is *block-preserving* if it preserves the embedding of each block. This is equivalent to saying that it is block-preserving for each cutvertex.

We are now ready to define formally the problem studied in this paper:

Problem 1 (Depth Minimization). Given a planar connected graph G and an embedding for each block of G, compute a block-preserving planar embedding Γ of G with minimum depth. □

Solving Problem 1 requires choosing the external face of the planar embedding. We shall find it convenient to study first a restricted version of the problem where the external face must contain a given cutvertex.

Problem 2 (Constrained Depth Minimization). Given a planar connected graph G, an embedding for each block of G, and a cutvertex v of G, compute a block-preserving planar embedding Γ of G, such that v is on the external face and Γ has minimum depth. □

Note that algorithms for Problems 1 and 2 need to access only the circular ordering of the edges around the cutvertices in the prescribed embedding of each block. We now introduce further definitions in order to simplify the description of the solution space of the problems.

Given a cutvertex v of G and block B containing v, we call the pair (B, v) a *cutpair*. The faces of block B containing v are called the *candidate cutfaces* for the cutpair (B, v) since one or more of them will be a cutface in block preserving embedding of the entire graph G.

Lemma 1 *Given a connected planar graph G with a prescribed embedding for its blocks, there exists a block-preserving planar embedding Γ of G with minimum depth such that, for each cutpair (B, v) of G, all the blocks incident to v distinct from B are nested inside the same candidate cutface of (B, v).*

Proof. Let Γ be a minimum-depth block-preserving embedding of graph G. For every cutvertex v of G, if there is more than one cutface incident on v, we choose f among the cutfaces incident on v with minimum distance from the external face. We move all the blocks that are not incident to f in it. This operation can only shorten the distance from the moved blocks to the external face in the block-cutface tree of Γ. Thus, the minimality of the embedding is preserved. □

Motivated by Lemma 1, and observing that the order of the blocks around a cutvertex does not affect the depth of a planar embedding, we represent a block-preserving embedding with the selection of a candidate cutface for each cutpair (see Figure 3):

Definition 2 *Let G be a connected planar graph G with a prescribed embedding for its blocks. A nesting for G is the assignment of candidate cutface f to each each cutpair (B, v). A nesting of G describes a class of planar embeddings of G with the same depth.*

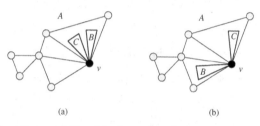

(a)   (b)

Fig. 3. Two embeddings of the blocks around a cutvertex v: (a) embedding described by a nesting with a single cutface; (b) embedding with two cutfaces that is not described by a nesting. Note that the embedding of part (a) has the same depth as the one of part (b).

Using the concept of nesting, Lemma 1 can be restated as follows:

Lemma 3 *Given a connected planar graph G with a prescribed embedding for its blocks, there exists a planar embedding of G with minimum depth that is described by a nesting.*

3   Constrained Depth Minimization

In this section, we present Algorithm 4 (*ConstrainedMinimization*) for solving Problem 2 (Constrained Depth Minimization). It takes as input a connected planar graph G with a prescribed embedding for its blocks, and a cutvertex v. The output is an embedding Γ of G that has v on the external face and minimizes depth Γ. The algorithm considers the block-cutvertex tree T of G rooted at v and builds the planar embedding Γ by means of a post-order traversal of T.

Given a node x of T (it may be a cutvertex or a block), we denote with $G(x)$ the subgraph of G associated with the subtree of T rooted at x. We denote with $\Gamma(x)$ the planar embedding of $G(x)$ computed by method embed(x) of the algorithm. Method embed(x) takes as input graph $G(x)$ and returns a planar embedding $\Gamma(x)$ of $G(x)$ with minimum depth. Let $y_1, y_2, \cdots, y_m$ be the children of x in T. The embedding $\Gamma(x)$ restricted to graph $G(y_i)$ is the embedding $\Gamma(y_i)$ returned by embed(y_i). In other words, we build the embedding $\Gamma(x)$ by assembling the previously computed embeddings $\Gamma(y_i)$ of the children of x.

We describe the embeddings $\Gamma(x)$ and $\Gamma(y_1), \cdots, \Gamma(y_m)$ by means of nestings (see Definition 2). When building $\Gamma(x)$, the information provided by $\Gamma(y_i)$'s is not enough. If x is a cutvertex, we need to select the candidate cutface for cutpairs (y_i, x), $i = 1, \cdots, m$, and if x is a block, we need to select the candidate cutface for cutpairs (x, y_i), $i = 1, \cdots, m$.

We prove the correctness of the algorithm with an inductive argument. The base case is for leaf blocks of T: Given a leaf block B of T, the algorithm correctly returns as $\Gamma(B)$ the prescribed embedding of B, which is trivially minimal.

If B is a non-leaf block of T, we assume, by the inductive hypothesis, the minimality of the planar embeddings of the children of B. We now consider the embedding $\Gamma(B)$ computed by method embed(v) and the block-cutface tree

Algorithm 4 *ConstrainedMinimization*

input A connected planar graph G with block-cutvertex tree T, a prescribed embedding for the blocks of G, and a cutvertex v of G.

output A block-preserving planar embedding Γ of G of minimum depth with v on the external face.

The algorithm computes and returns $\Gamma = $ embed(v).

method embed(v)

input A cutvertex v of T.

output A block-preserving planar embedding $\Gamma(v)$ of minimum depth with v on the external face.

    for all children B of v in T **do**
      if B is a leaf block (i.e., it has no children) **then**
        Let $\Gamma(B)$ be the prescribed embedding of block B, with external face
        equal to one of the candidate cutfaces of (B, v).
      else
        Let $\Gamma(B) = $ embed(B).
      end if
    end for
    for all children B of v in T **do**
      Assign the external face of $\Gamma(B)$ to cutpair (B, v).
    end for
    Construct planar embedding $\Gamma(v)$ by joining the previously computed planar embeddings of the children blocks of v and placing them on the external face.

method embed(B)

input A block B of T.

output A block-preserving planar embedding $\Gamma(B)$ of minimum depth with external face equal to one of the candidate cutfaces of cutpair $(B, \text{parent}(B))$.

    for all children v of B in T **do**
        Let $\Gamma(v) = $ embed(v).
    end for
    for all candidate cutfaces f of $(B, \text{parent}(B))$ **do**
        Let $\delta(f)$ be the maximum depth of the (previously computed) planar embeddings $\Gamma(v)$ of the cutvertices children of B on face f.
    end for
    Let δ_B be the maximum of $\delta(f)$ over all candidate cutfaces of $(B, \text{parent}(B))$. Let f_B be a candidate cutface of $(B, \text{parent}(B))$ such that $\delta(f_B) = \delta_B$. If more than one such cutfaces exist, choose one with the maximum number of cutvertices with deepest embedding (i.e., cutvertices w such that $\Gamma(w)$ has depth δ_B).
    for all children v of B in T **do**
        if v is on face f_B **then**
            Assign cutface f_B to cutpair (B, v).
        else
            Assign an arbitrary cutface to cutpair (B, v).
        end if
    end for
    Construct planar embedding $\Gamma(B)$ by adding to the prescribed embedding of B the previously computed planar embeddings $\Gamma(v)$ of the children of B, where, for a child v of B, embedding $\Gamma(v)$ is placed inside the cutface assigned to (B, v).
    Let f_B be the external face of $\Gamma(B)$.

$T^*(B)$, which is rooted at the external face f_B (which is a cutface of the entire embedding Γ. The children of f_B in $T^*(B)$ are blocks, one of which is B itself. We can write depth $\Gamma(B)$ in the following way:

$$\text{depth } \Gamma(B) = \max\{d_P, d_{NP}\}$$

where

$$d_P = \max_{v \text{ incident to } f_B} \text{depth } \Gamma(v)$$

and

$$d_{NP} = 2 + \max_{v \textbf{ not } \text{incident to } f_B} \text{depth } \Gamma(v)$$

where "incident" refers to the embedding $\Gamma(B)$.

The quantity d_P represents the contribution of the planar embeddings whose external face is f_B in $T^*(B)$, we say that such embeddings are *promoted*. The quantity d_{NP} represents the contribution of the planar embeddings whose external face is a child of B in $T^*(B)$ (see Figure 4).

The minimality of d_P follows from the fact that each $\Gamma(v)$ is minimal. The minimality of d_{NP} can be shown as follows: for any child v of B, the length of a root-to-leaf path in $T^*(B)$ is at most $\Gamma(v) + 2$. The algorithms selects the external face f_B of $\Gamma(B)$ that optimizes the depth by trying to place the deepest sub-embeddings of the children of B on the external face.

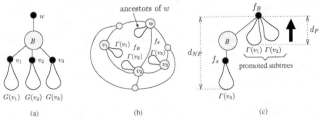

Fig. 4. (a) A block-cutvertex tree. (b) Embedding $\Gamma(B)$ computed by method embed(B). (c) Block cutface-tree of $\Gamma(B)$, where $\Gamma(v_1)$ and $\Gamma(v_2)$ are promoted, while $\Gamma(v_3)$ is not.

Given a cutvertex v, we assume, by the inductive hypothesis, the minimality of the planar embeddings $\Gamma(B)$ of the children of v computed by embed(B). It is easy to see that

$$\text{depth } \Gamma(v) = \max_{B \in \text{children}(v)} \text{depth } \Gamma(B)$$

where children(v) denotes the set of children of v in T. The minimality of $\Gamma(v)$ follows directly from the minimality of depth $\Gamma(B)$ for each child B of v.

The running time of all the executions of method embed(B), excluding the calls to method embed(v), can be written as $\sum_{B \in T} \sum_{f \in B} |f|$ where $|f|$ denotes the number of edges of face f.

The running time of all the executions of method embed(v), excluding the calls to method embed(B), can be written as $\sum_{v \in T} \sum_{B \in \text{children}(v)} |f_B|$ where $|f_B|$ denotes the number of edges of the external face f_B of B.

Each of the above sums can be shown to be equal to the number of edges of a planar graph with $O(n)$ edges, where n is the number of vertices of G. Thus, we conclude that the overall running time of Algorithm 4 (*ConstrainedMinimization*) is $O(n)$.

The main result of this section is summarized in the following theorem:

Theorem 5 *Given a planar connected graph G with n vertices, a prescribed embedding for each block of G, and a cutvertex v of G, Algorithm 4 (ConstrainedMinimization) computes in $O(n)$ time a block-preserving planar embedding Γ of G, such that v is on the external face and Γ has minimum depth.*

4   General Depth Minimization

A brute-force quadratic-time algorithm for Problem 2 (Depth Minimization) consists of repeatedly executing Algorithm 4 (ConstrainedMinimization) for each cutvertex, and then returning the planar embedding with minimum depth.

In this section, we present a linear-time algorithm for depth minimization based on the following approach:

1. compute an embedding with the minimum diameter;
2. select the external face to minimize the depth of the resulting planar embedding.

The reduction used in our approach is summarized in the following theorem:

Theorem 6 *Given a connected planar graph G whose blocks have a prescribed embedding, and a block-preserving embedding Δ of G with minimum diameter, a block-preserving planar embedding of G with minimum depth is obtained from Δ by selecting the external face of Δ as a face with minimum eccentricity in the block-cutface tree of Δ.*

By Theorem 6, we can solve Problem 1 by making an appropriate choice of the external face, provided an efficient algorithm for minimizing the diameter of an embedding is given. Before proving Theorem 6, we state two basic properties of the eccentricity of the nodes of a block-cutvertex tree.

Property 7 *In a block-cutvertex tree, each block-node has even eccentricity while each cutvertex-node has odd eccentricity.*

Property 8 *The center of a block-cutvertex tree contains only one node.*

Since any block-cutface tree is the block-cutvertex tree of the dual embedding of a graph, the above properties hold also for block-cutface trees, where any occurrences of the word cutvertex is replaced by cutface.

In the following, we denote the eccentricity of a face f of an embedding Γ in the block-cutface tree of Γ with $e_\Gamma(f)$. Similarly, we denote with $e_\Gamma(B)$ the eccentricity of a block B.

Lemma 9 *Let Γ be an embedding of a connected planar graph, and let f be a face of Γ with minimum eccentricity. We have:*

$$2 \cdot e_\Gamma(f) - 2 \leq \text{diam } \Gamma \leq 2 \cdot e_\Gamma(f)$$

Proof. If f is the center of the block-cutface tree then it is in the middle of a diametral path of even length, thus diam $\Gamma = 2 \cdot e_\Gamma(f)$.

If f is adjacent to the center B (a block) of the block-cutface tree, we have diam $\Gamma = 2 \cdot e_\Gamma(B)$, and hence diam $\Gamma = 2 \cdot (e_\Gamma(f) - 1)$.   □

We are now ready to prove Theorem 6.

Proof. (of Theorem 6)

We show that, if there exists a planar embedding Γ with external face f_Γ that has smaller depth than the embedding Δ with external face f_Δ, then Δ does not have minimum diameter, which is a contradiction.

Suppose that depth $\Gamma <$ depth Δ. Since the eccentricities of cutfaces are always odd, the difference between the two depths is at least 2, i.e., $e(f_\Gamma) \leq e(f_\Delta) - 2$. Hence, by Lemma 9, we obtain

$$\text{diam } \Gamma \leq 2 \cdot e(f_\Gamma) \leq 2 \cdot e(f_\Delta) - 4 < 2 \cdot e(f_\Delta) - 2 \leq \text{diam } \Delta.$$

□

We say that a planar embedding Γ has *total minimum depth* if $\Gamma(x)$ has minimum depth for each node x of its block-cutface tree. The embedding produced by Algorithm 4 (ConstrainedMinimization) has total minimum depth. We show that this embedding either has optimal diameter, or can be easily transformed into an embedding with optimal diameter.

Intuitively, much of the minimization work performed by Algorithm 4 may be reused to minimize the diameter. Since the algorithm recursively minimizes the depth, each diametral path spans one or two subtrees with minimum depth. If any depth reduction is possible, this should be at the root of such subtrees.

Consider a tree T with root r and its diametral tree T_{diam}. We call *knot* of T the node of T_{diam} closest to r. As a special case, the knot may coincide with r. It is easy to prove that all the diametral paths of T contain the knot.

Theorem 10 *Let G be a connected planar graph with prescribed embeddings for its blocks. Given a planar embedding Γ of G with total minimum depth, there exists a minimum diameter embedding Δ of G such that either*

- *$\Delta = \Gamma$; or*
- *Δ differs from Γ at most for the embeddings of the cutvertices around one block, which is the knot of the block-cutface tree of Γ, and we have diam $\Delta =$ diam $\Gamma - 2$.*

Proof. (sketch) Consider the block-cutface tree T^* of Γ, which is rooted at the external face, and the block-cutvertex tree T of G. We call T_{diam} the diametral tree of T^*, T_{depth} its depth tree, and k its knot. We consider T_{diam} rooted at k, where k may be a block or a cutface. We distinguish two cases:

- *k is a cutface.* In this case, we set $\Delta = \Gamma$, which has minimum diameter. Indeed, any diametral path of T^* may be written as $B_a, \ldots, B_b, k, B_c, \ldots, B_d$, where B_b and B_c are children of k in $T_{diam}(k)$. The subpaths $B_a, \ldots, B_b$ and $B_c, \ldots, B_d$ have minimum length because Γ has total minimum depth and thus $\Gamma(B_b)$ and $\Gamma(B_c)$ are minimum depth embedding of $G(B_b)$ and $G(B_c)$ respectively. Note that selecting different cutfaces for the cutpairs (B_b, k) and (B_c, k) does not change the distance between B_b and B_c.
- *k is a block.* If k is the root of T_{depth}, then $T_{diam} = T_{depth}$ and we set $\Delta = \Gamma$, which has minimum diameter. Else, any diametral path of T^* may be written as $B_a, \ldots, f_a, k, f_b, \ldots, B_b$. The subpaths $B_a, \ldots, f_a$ and $f_b, \ldots, B_b$ have minimum length because Γ has total minimum depth. If we can select the same candidate cutface of the cutpair (k, w) for all the cutvertices w that are children of k in T_{diam}, we obtain an embedding Δ such that diam $\Delta =$ diam $\Gamma - 2$. Else, we set $\Delta = \Gamma$. $\square$

Our algorithm for finding a minimum diameter embedding consists of the following steps.

1. Use Algorithm 4 to construct an embedding Γ with total minimum depth with respect to an arbitrary cutvertex.

2. Find the knot of the block-cutface tree of Γ and, if needed, modify Γ into a minimum diameter embedding as described in the proof of Theorem 10.
3. Apply Theorem 6 to select an external face that minimizes the depth of the resulting planar embedding.

Note that a planar embedding obtained with the above algorithm has minimum depth, but in general does not have total minimum depth. To obtain the latter, we run Algorithm 4 again with a starting cutvertex on the external face. We summarize the main result of this paper in the following theorem:

Theorem 11 *Given a planar connected graph G with n vertices and a prescribed embedding for each block of G, a block-preserving planar embedding of G with minimum depth can be computed in $O(n)$ time.*

5   Conclusions

We have presented an optimal algorithm for the problem of minimizing the depth of a planar embedding whose blocks have a prescribed embedding. Our results have practical applications to graph drawing systems. Future work includes:

- Investigate a weighted version of the minimum-depth embedding problem, where the blocks have positive weights and we want to minimize the weighted depth of the block-cutface tree. This variation of the problem is relevant in graph drawing applications, where the size of the drawings of the blocks should be taken into account.
- Perform an extensive experimental study on the effect of using our depth minimization technique in graph drawing algorithms based on the topology-shape-metric approach.

References

1. D. Alberts, C. Gutwenger, P. Mutzel, and S. Näher. AGD-Library: A library of algorithms for graph drawing. In *Proc. Workshop on Algorithm Engineering*, pages 112–123, 1997. 357
2. C. Batini, E. Nardelli, and R. Tamassia. A layout algorithm for data-flow diagrams. *IEEE Trans. Softw. Eng.*, SE-12(4):538–546, 1986. 357
3. C. Batini, M. Talamo, and R. Tamassia.   Computer aided layout of entity-relationship diagrams. *Journal of Systems and Software*, 4:163–173, 1984. 357
4. G. D. Battista, W. Didimo, M. Patrignani, and M. Pizzonia.   Orthogonal and quasi-upward drawings with vertices of prescribed size. In *Graph Drawing (Proc. GD '99)*, volume 1731 of *Lecture Notes Comput. Sci.* Springer-Verlag, 1999. 357
5. P. Bertolazzi, G. Di Battista, and W. Didimo.   Computing orthogonal drawings with the minimum number of bends. In *Workshop Algorithms Data Struct. (WADS'97)*, volume 1272 of *Lecture Notes Comput. Sci.*, pages 331–344. Springer-Verlag, 1997. 358
6. D. Bienstock and C. L. Monma. On the complexity of covering vertices by faces in a planar graph. *SIAM Journal on Computing*, 17(1):53–76, Feb. 1988. 357

7. D. Bienstock and C. L. Monma. Optimal enclosing regions in planar graphs. *Networks*, 19:79–94, 1989. 357

8. D. Bienstock and C. L. Monma. On the complexity of embedding planar graphs to minimize certain distance measures. *Algorithmica*, 5(1):93–109, 1990. 357

9. U. Brandes and D. Wagner. A Bayesian paradigm for dynamic graph layout. In *Graph Drawing (Proc. GD '97)*, volume 1353 of *Lecture Notes Comput. Sci.*, pages 236–247. Springer-Verlag, 1998. 357

10. U. Brandes and D. Wagner. Dynamic grid embedding with few bends and changes. *Lecture Notes in Computer Science*, 1533, 1998. 357

11. G. Di Battista, P. Eades, R. Tamassia, and I. G. Tollis. *Graph Drawing*. Prentice Hall, Upper Saddle River, NJ, 1999. 357

12. G. Di Battista et al. *Graph Drawing Toolkit*. University of Rome III, Italy. http://www.dia.uniroma3.it/~gdt/. 357

13. G. Di Battista, A. Garg, G. Liotta, R. Tamassia, E. Tassinari, and F. Vargiu. An experimental comparison of four graph drawing algorithms. *Comput. Geom. Theory Appl.*, 7:303–325, 1997. 357

14. G. Di Battista, G. Liotta, and F. Vargiu. Spirality and optimal orthogonal drawings. *SIAM J. Comput.*, 27(6):1764–1811, 1998. 358

15. W. Didimo and G. Liotta. Computing orthogonal drawings in a variable embedding setting. In *Algorithms and Computation (Proc. ISAAC '98)*, volume 1533 of *Lecture Notes Comput. Sci.*, pages 79–88. Springer-Verlag, 1998. 358

16. U. Fößmeier and M. Kaufmann. Drawing high degree graphs with low bend numbers. In *Graph Drawing (Proc. GD '95)*, volume 1027 of *Lecture Notes Comput. Sci.*, pages 254–266. Springer-Verlag, 1996. 357

17. A. Garg and R. Tamassia. On the computational complexity of upward and rectilinear planarity testing. *SIAM J. Computing*. to appear. 358

18. A. Garg and R. Tamassia. On the computational complexity of upward and rectilinear planarity testing. In *Graph Drawing (Proc. GD '94)*, volume 894 of *Lecture Notes Comput. Sci.*, pages 286–297. Springer-Verlag, 1995. 358

19. J. Hopcroft and R. E. Tarjan. Efficient planarity testing. *J. ACM*, 21(4):549–568, 1974. 357

20. M. Jünger, S. Leipert, and P. Mutzel. Pitfalls of using PQ-Trees in automatic graph drawing. In *Graph Drawing (Proc. GD '97)*, volume 1353 of *Lecture Notes Comput. Sci.*, pages 193–204. Springer-Verlag, 1997. 357

21. M. Jünger and P. Mutzel. Maximum planar subgraphs and nice embeddings: Practical layout tools. *Algorithmica*, 16(1):33–59, 1996. 357

22. G. W. Klau and P. Mutzel. Optimal compaction of orthogonal grid drawings. In *IPCO: 7th Integer Programming and Combinatorial Optimization Conference*, volume 1610 of *Lecture Notes Comput. Sci.* Springer-Verlag, 1999. 357

23. H. Lauer, M. Ettrich, and K. Soukup. GraVis - system demonstration. In *Graph Drawing (Proc. GD '97)*, volume 1353 of *Lecture Notes Comput. Sci.*, pages 344–349. Springer-Verlag, 1997. 357

24. P. Mutzel and R. Weiskircher. Optimizing over all combinatorial embeddings of a planar graph. In *IPCO: 7th Integer Programming and Combinatorial Optimization Conference*, volume 1610 of *Lecture Notes Comput. Sci.* Springer-Verlag, 1999. 357

25. R. Tamassia. On embedding a graph in the grid with the minimum number of bends. *SIAM J. Comput.*, 16(3):421–444, 1987. 357

26. R. Tamassia, G. Di Battista, and C. Batini. Automatic graph drawing and readability of diagrams. *IEEE Trans. Syst. Man Cybern.*, SMC-18(1):61–79, 1988. 357

New Algorithms for Two-Label Point Labeling*

Zhongping Qin[1], Alexander Wolff[2], Yinfeng Xu[3], and Binhai Zhu[4]

[1] Dept. of Mathematics, Huazhong University of Science and Technology, Wuhan, China, and Dept. of Computer Science, City University of Hong Kong. zqin@cs.cityu.edu.hk

[2] Institute of Mathematics and Computer Science, Ernst Moritz Arndt University, Greifswald, Germany. awolff@mail.uni-greifswald.de

[3] School of Management, Xi'an Jiaotong University, Xi'an, China. yfxu@xjtu.edu.cn

[4] Dept. of Computer Science, City University of Hong Kong, and Montana State University, Bozeman, MT 59717, USA. bhz@cs.montana.edu

Abstract. Given a label shape L and a set of n points in the plane, the 2-label point-labeling problem consists of placing $2n$ non-intersecting translated copies of L of maximum size such that each point touches two unique copies—its labels. In this paper we give new and simple approximation algorithms for L an axis-parallel square or a circle. For squares we improve the best previously known approximation factor from $\frac{1}{3}$ to $\frac{1}{2}$. For circles the improvement from $\frac{1}{2}$ to ≈ 0.513 is less significant, but the fact that $\frac{1}{2}$ is not best possible is interesting in its own right. For the decision version of the latter problem we have an NP-hardness proof that also shows that it is NP-hard to approximate the label size beyond a factor of ≈ 0.732. As their predecessors, our algorithms take $O(n \log n)$ time and $O(n)$ space.

1 Introduction

Label placement is one of the key tasks in the process of information visualization. In diagrams, maps, technical or graph drawings, features like points, lines, and polygons must be labeled to convey information. The interest in algorithms that automate this task has increased with the advance in type-setting technology and the amount of information to be visualized. Due to the computational complexity of the label-placement problem, cartographers, graph drawers, and computational geometers have suggested numerous approaches, such as expert systems [1,8], zero-one integer programming [22], approximation algorithms [6,11,19,20], simulated annealing [4] and force-driven algorithms [13] to name only a few. The ACM Computational Geometry Impact Task Force report [3] denotes label placement as an important research area. Manually labeling a map is a tedious task that is estimated to take 50 % of total map production time.

---

* This research was conducted during a visit of Z. Qin and Y. Xu to City University of Hong Kong and of A. Wolff to Hong Kong University of Science and Technology. Our work was supported by NSF of China, grant No. 19731001, and by the Hong Kong RGC CERG grants CityU-1103/99E and HKUST-6144/98E.

M. Paterson (Ed.): ESA 2000, LNCS 1879, pp. 368–380, 2000.

In this paper we deal with a relatively new variant of the general label placement problem, namely the 2-label point-labeling problem. It is motivated by maps used for weather forecasts, where each city must be labeled with two labels that contain the city's name and, say, its predicted temperature.

The 2-label point-labeling problem is a variant of the 1-label problem that allows sliding. Sliding labels can be attached to the point they label anywhere on their boundary. They were first considered by Hirsch [13] who gave an iterative algorithm that uses repelling forces between labels in order to eventually find a placement without or only few intersecting labels. Van Kreveld et al. gave a polynomial time approximation scheme and a fast factor-2 approximation algorithm for maximizing the number of points that are labeled by axis-parallel sliding rectangular labels of common height [18]. They also compared several sliding-label models with so-called fixed-position models where only a finite number of label positions (or *label candidates*) per point is considered, usually a small constant like four [4,11,19]. Another generalization was investigated in [6,21], namely arbitrarily oriented sliding labels.

Point labeling with circular labels, though not as relevant for real-world applications as rectangular labels, is a mathematically interesting problem. The 1-label case has already been studied extensively [6,17,7]. For maximizing the label size, the currently best approximation factor is $\frac{1}{3.6}$ [7].

The 2- or rather multi-label labeling problem was first considered by Kakoulis and Tollis who presented two heuristics for labeling the nodes and edges of a graph drawing with several rectangles [14]. Their aim was to maximize the number of labeled features. One of their algorithms is iterative, the other uses a maximum-cardinality bipartite matching algorithm that matches cliques of label candidates with the elements of the graph drawing that are to be labeled. They do not give any runtime bounds or approximation factors.

For the two problems that we will consider in this paper, namely maximizing the size of axis-parallel square and circular labels, two per point, Zhu and Poon gave the first approximation algorithms [20]. They achieved approximation factors of $\frac{1}{4}$ and $\frac{1}{2}$ for square and circle pairs, respectively. Both algorithms rely on the fact that there are disjoint regions around all (pairs of) input points into which the labels can be safely placed. Recently Zhu and Qin improved the result for pairs of square labels to a factor of $\frac{1}{3}$ [21]. They exploit the structure of a graph that has a node for each input point and an edge for each pair of points closer than $\frac{2}{3}$ times an upper bound for the maximum label size.

In this paper we give new and simple approximation algorithms for the 2-square and the 2-circle point-labeling problem. For squares we improve the approximation factor of Zhu and Qin's algorithm from $\frac{1}{3}$ to $\frac{1}{2}$. For circles we present an algorithm with an approximation factor of $\frac{1}{1+\cos 18°} \approx 0.513$. Here the improvement over Zhu and Poon's factor-$\frac{1}{2}$ approximation algorithm is less significant, but the fact that $\frac{1}{2}$ is not best possible is interesting in its own right. For the decision version of the 2-circle labeling problem we have an NP-hardness proof that also shows that it is NP-hard to approximate the label size beyond a factor of ≈ 0.732. However, due to space limitations, we cannot give the proof here. Other than all previous approximation algorithms for 2-label placement, our new algorithms do not necessarily have to compute an upper bound for the maximum label size explicitly. We keep the $O(n \log n)$ time and $O(n)$ space bounds of the previous algorithms.

For the 2-square labeling problem the improved approximation factor is made possible by restricting the search to a subset of the solution space. Within this subset, optimal solutions can be computed easily and their labels are at most by the above mentioned constant factors off the maximum label size. The special case that we solve optimally is the following: we label points with rectangles of height-width ratio 2 of maximum size in one of four positions. Our algorithm is the first *point*-labeling algorithm that solves the size maximization problem for more than two label positions optimally in polynomial time. So far such algorithms have only been know for labeling axis-parallel line segments [16]. Our algorithm also improves the approximation factor of the only known algorithm [12] for Knuth and Raghunathan's Metafont labeling problem [15] from $\frac{1}{3}$ to $\frac{1}{2}$.

Throughout this paper we consider labels being topologically *open*, and we define the *size* of a solution to be the diameter in the case of circular labels and the length of the shorter label edge in the case of rectangular labels. We refer to a label placement as *feasible* if no two labels intersect and as *optimal* if additionally labels have the largest possible size. We will only consider $n > 2$.

2   Two-Square Labeling

Definition 1 (2-square point-labeling problem). *Given a set P of n points in the plane, find a set of $2n$ axis-parallel, uniform, non-intersecting, maximum-size open squares, such that each point touches two unique squares.*

The uniqueness contraint does not forbid a point to touch more than two squares, but ensures there is a function that assigns to each point exactly two squares that touch it.

The first approximation algorithm for the 2-square point-labeling problem was suggested by Zhu and Poon [20]. This algorithm labels pairs of points in order of increasing distance. The labels are placed to the left of the left point and to the right of the right point—except when the two points lie on a vertical, see Figure 1 (a) and (b), respectively. In other words, the algorithm does not really place two square labels at each point but one rectangle of height-width ratio 2 or 1/2. The rectangle is attached to its point in the midpoint of either of its long edges, i.e. the algorithm uses only *four* of the infinitely many possible label positions.

We use this observation as follows. First we device an $O(n \log n)$-time algorithm for the 1-rectangle 4-position point-labeling problem, where all points are labeled with maximum-size rectangles in one of the four discrete positions depicted in Figure 2. Then we show how an optimal solution for the 2-square labeling problem can be transformed into a solution of the 1-rectangle labeling problem by using rectangles of half the square size, see Figure 3. Thus the 1-rectangle labeling algorithm already yields a factor-$\frac{1}{2}$ approximation algorithm for the 2-square labeling problem.

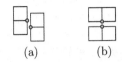
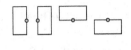
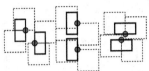

Fig. 1. Label placement of the 2-square labeling algorithm of Zhu and Poon.

Fig. 2. Label candidates for 1-rectangle 4-position labeling.

Fig. 3. Mapping a 2-square labeling to a 1-rectangle 4-position labeling of half the size.

Apart from the improved approximation factor, our approach has the advantage that we do not have to compute an upper bound for the maximum label size explicitly. Both previous algorithms [20,21] first have to determine such an upper bound before they can actually place labels whose size depends on this upper bound. Zhu and Poon use $D_{3,\infty}$, the minimum over the diameters of all 3-subsets of the input points, as an upper bound for the maximum label size. $D_{3,\infty}$ can be computed in $O(n \log n)$ time [5]. Instead of computing $D_{3,\infty}$ we preprocess the input by computing relevant adjacency information, i.e. for each input point we find a constant number of rectilinear nearest neighbors (in the L_∞-metric). For this task a simple $O(n \log n)$ algorithm is known [10]. Of course $D_{3,\infty}$ can be computed in linear time from nearest neighborhood data, but we make better use of this information by solving the following problem optimally.

Definition 2 (1-rectangle 4-position point-labeling problem). *Given n points in the plane, find a set of n congruent, axis-parallel, non-intersecting, maximum-size open rectangles of height-width ratio 2 or 1/2, such that each point touches a unique rectangle.*

The decision version of this problem is the question whether a set of points can be labeled with congruent rectangles of a *given* size. If we encode the four label candidates of each point p by *two* Boolean variables p_1 and p_2 as in Figure 4, we can construct a 2-SAT formula that is equivalent to the decision version of our problem. Thus it can be solved in time and space linear in the number of clauses [9]. Here the number of clauses is at most three times the number of pairs of intersecting label candidates, which must be computed beforehand. A similar strategy has been applied to encode label positions for labeling rectilinear line segments with rectangles of maximum height [16].

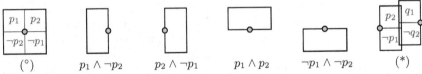

Fig. 4. We encode the four label positions of a point p by the values of two Boolean variables p_1 and p_2 (°). A label intersection (*) can then be read as $\neg((p_2 \wedge q_1) \vee (p_2 \wedge \neg q_2) \vee (\neg p_1 \wedge \neg q_2))$. This equals the 2-SAT formula $(\neg p_2 \vee \neg q_1) \wedge (\neg p_2 \vee q_2) \wedge (p_1 \vee q_2)$.

In order to device an algorithm that maximizes the rectangle size and runs in $O(n \log n)$ time, we use the same strategy as two existing algorithms. The algorithms $AS4a$ [11] and B [19] also try to maximize the label size. They solve the 1-square 4-position point-labeling problem, where labels are restricted to uniform *squares*, and each label must be placed such that one of its *corners* coincides with the point it labels. Like in our problem each point has four label candidates. However, in their problem the candidates only intersect at their borders, while in ours each candidate is completely contained in (the closure of) two others. This is why we can solve our problem optimally, while $AS4a$ and B are factor-$\frac{1}{2}$ approximation algorithms.

Algorithm B proceeds as follows. First it uses the above mentioned simple sweep-line algorithm to detect a constant number of rectilinear nearest neighbors for each point [10]. Then B computes a list of *conflict sizes* from the set N of pairs of neighboring points. A conflict size is a label size for which two label candidates touch but do not contain any input points. Algorithm B does a binary search on the list of conflict sizes. This is sufficient since the intersection graph of the label candidates does not change between two conflict sizes. For each conflict size, B solves the decision problem in three steps. First the algorithm extracts the intersection graph for the current label size from N. Then it uses certain (partly heuristical) rules to simplify the intersection graph until each point has at most *two* label candidates left. Finally B makes a 2-SAT clause for each edge of the intersection graph and tries to find a satisfying truth assignment for the resulting 2-SAT formula. If one exists, it corresponds to a label placement and the search is continued with a greater label size, otherwise with a smaller one.

Asymptotically B does not need more space than the size of N, which is linear in n. The binary search consists of $O(\log n)$ tests. Again due to the linear size of N, each test takes $O(n)$ time. This adds up to $O(n \log n)$ time in total.

Our algorithm for 1-rectangle 4-position point labeling differs from B only in that it is not necessary to reduce the number of candidates of each point from four to two, since we can immediately encode our four label candidates by a 2-SAT formula. All that remains to show is that as for 1-square 4-position point labeling it is sufficient to consider the conflict sizes induced by a constant number of rectilinear nearest neighbors per point.

Lemma 1. *Let r_{opt} be the size of an optimal solution for the 1-rectangle 4-position point-labeling problem. Then for any label size $\rho \leq r_{\text{opt}}$ the label candidates of a point $p \in P$ can intersect only candidates of the 17 points $q \in P \setminus \{p\}$ that are closest to p in the L_∞-metric.*

Since this observation is analogous to [11, Lemma 1], we omit the proof here. The algorithm $AR2$ in that paper, however, needs $O(n \log n)$ time for the *decision* version of a problem similar to our 1-rectangle 4-position labeling problem. (There, only two square labels per point are allowed.) Lemma 1 yields the time and space complexity of our 1-rectangle 4-position labeling algorithm:

Lemma 2. *An optimal solution of the 1-rectangle 4-position point-labeling problem can be computed in $O(n \log n)$ time using linear space.*

Theorem 1. *A feasible solution of the 2-square point-labeling problem of at least half the optimal size can be computed in $O(n \log n)$ time using linear space.*

Proof. Given Lemma 2, we only have to show that an optimal solution of the 1-rectangle 4-position labeling problem always represents a feasible solution of the 2-square labeling problem of at least half the optimal size. In other words, if r_{opt} and s_{opt} are the sizes of optimal solutions of the 1-rectangle and the 2-square labeling problem for the same point set P, then we have to prove $r_{\text{opt}} \geq s_{\text{opt}}/2$. To show this we map a hypothetical optimal solution of the latter problem, i.e. a set S_{opt} of $2n$ squares of size s_{opt} into a feasible solution of the former problem, namely a set $\mathcal{R}$ of n rectangles of size $s_{\text{opt}}/2$. Since the labels in $\mathcal{R}$ cannot be larger than r_{opt}, the size of labels in an optimal solution of the 1-rectangle labeling problem, we have $r_{\text{opt}} \geq s_{\text{opt}}/2$.

The mapping is simple: for each point p we choose from its four label candidates the rectangle R_p such that a rectangle of twice the size of R_p (with p as scaling center) has the largest area of intersection with the two square labels of p in S_{opt}. For our proof it does not matter that R_p is not uniquely defined.

It remains to show why R_p does not intersect the label R_q of some point $q \in P \setminus \{p\}$. Let $s_{\text{opt}} = 1$; the instance can always be scaled such that this is true. Recall that all labels in S_{opt} and $\mathcal{R}$ are topologically open. For $A \subseteq \mathbb{R}^2$ let $\overline{A}$ be the topological closure of A. For $p = (x_p, y_p)$ define the line segments $v(p)$ and $h(p)$ as the intersection of an open unit disk with the vertical and the horizontal through p, respectively. Let T_p and B_p be the top- and bottommost square labels of p in S_{opt}; if their y-coordinates are the same, let T_p be the leftmost. Let $S_p = T_p \cup B_p$. We will use the same notations for q.

Observe that $v(p) \subseteq \overline{S_p}$ or $h(p) \subseteq \overline{S_p}$, otherwise there are points $h \in h(p) \setminus \overline{S_p}$ and $v \in v(p) \setminus \overline{S_p}$ that delimit region E in Figure 5 (b). Since E must contain S_p completely, we would have $T_p \cap B_p \neq \emptyset$.

Now suppose $R_p \cap R_q \neq \emptyset$. Then the distance $d_\infty(p, q)$ of p and q in the L_∞-metric is less than 1 since all points in R_p (R_q) have distance less than $\frac{1}{2}$ from p (q). For this reason $v(p)$ and $h(q)$ as well as $v(q)$ and $h(p)$ intersect, see Figure 5 (c). Since S_p and S_q do not intersect by definition, the observation above yields that $v(p)$ is contained in $\overline{S_p}$ and $v(q)$ in $\overline{S_q}$ or $h(p)$ is contained in $\overline{S_p}$ and $h(q)$ in $\overline{S_q}$. W.l.o.g. we may assume the former, otherwise we rotate the whole instance by 90° around p. Thus T_p lies above p, and B_p below p; the same holds for q. We can also assume that $x_p \leq x_q$, otherwise we mirror our instance.

Let $\Delta x = x_q - x_p$. Then $0 \leq \Delta x < 1$. Let t_p (b_p) be the horizontal distance between the *left* edge of T_p (B_p) and $v(p)$, see Figure 5 (d). Similarly, let t_q (b_q) be the horizontal distance between the *right* edge of T_q (B_q) and $v(q)$.

Now a simple packing argument will yield the contradiction. Since T_p and T_q do not intersect, we have $t_p + \Delta x + t_q \geq 2$. Since B_p and B_q do not intersect, we have $b_p + \Delta x + b_q \geq 2$. These inequalities sum up to $\Sigma := t_p + t_q + b_p + b_q \geq 4 - 2\Delta x$. Since $0 \leq \Delta x < 1$ we only have to consider the following two cases:

A) $\frac{1}{2} \leq \Delta x < 1$. Then $\Sigma > 2$ and $t_p + b_p \leq 1$, otherwise—due to our mapping—R_p lies completely left of $v(p)$ and cannot intersect R_q due to $\Delta x \geq \frac{1}{2}$.

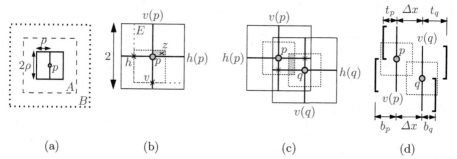

Fig. 5. (a) Only the labels of the 17 nearest neighbors of p can intersect labels of p. (b) If there are points $v \in v(p) \setminus \overline{S_p}$ and $h \in h(p) \setminus \overline{S_p}$ then there is a point $z \in T_p \cap B_p$.   (c) If rectangles R_p and R_q intersect, then $v(p) \cap h(q) \neq \emptyset$ and $v(q) \cap h(p) \neq \emptyset$.   (d) t_p (b_p) is the horizontal distance between the left edge of T_p (B_p) and $v(p)$.

Similarly $t_q + b_q \leq 1$, otherwise R_q lies completely to the right of $v(q)$. Thus $t_p + t_q + b_p + b_q \leq 2$, contradicting $\Sigma > 2$.

B) $0 \leq \Delta x < \frac{1}{2}$. Then $\Sigma > 3$. Since t_p, t_q, b_p, and b_q are all at most 1, we get $t_p + b_p > 1$ and $t_q + b_q > 1$. But then R_p lies completely to the left of $v(p)$ and R_q to the right of $v(q)$—which means they do not intersect since we assumed $x_p \leq x_q$.   ⌑

Obviously the 1-rectangle 4-position algorithm solves an instance P of the 2-square labeling problem optimally if P has an optimal solution that uses only the four label positions depicted in Figure 1, as in the grid $\{(x, 2y) \mid 1 \leq x, y \leq n\}$.

However, our new 1-rectangle 4-position algorithm can also be used to find approximate solutions for another problem, namely the optimization version of the Metafont labeling problem [15]. Knuth developed the program Metafont as a tool for font design. In their paper, Knuth and Raghunathan introduced the name *problem of compatible representatives* for a large class of combinatorial problems that are NP-hard in general. This class of problems has been studied independently in the artificial intelligence community under the name *constraint satisfaction problem* (CSP). Knuth and Raghunathan investigated several special cases of the problem of compatible representatives, among them the Metafont labeling problem, where a set of n grid points is to be labeled with n 2×2 squares such that each square touches exactly one point in the center of one of its edges, see Figure 6. They showed that this 1-square 4-position point labeling problem is NP-complete. Note that the 1-rectangle 4-position problem considered in this section is *not* a more general case of the Metafont labeling problem; each of our rectangular label candidates is the union of two other candidates of the same point, while this is not the case with the square Metafont labels.

Formann and Wagner saw the connection between the Metafont labeling problem and the cartographic map-labeling problem they had attacked earlier [11]. They defined a maximization version of the Metafont labeling problem by dropping the grid constraint that comes from the Metafont application [12] and gave a factor-$\frac{1}{3}$ approximation algorithm similar to the algorithm *AS4a* [11].

Definition 3 (Metafont optimization problem). *Given a set of n points in the plane, find a set of n uniform, axis-parallel, non-intersecting, maximum-size open squares, such that each point touches* exactly *one square in the midpoint of one of its edges.*

Fig. 6. The 4 label positions allowed in Metafont labeling.

Fig. 7. Maping an optimal Metafont solution via rectangles to a Metafont solution of half the optimal size.

Obviously we can approximate the Metafont optimization problem exactly in the same way as the 2-square point-labeling problem, namely by running our 1-rectangle 4-position labeling algorithm. The mapping that transforms an optimal Metafont labeling to a feasible rectangle labeling is the same as for the 2-square problem. Here, however, it's trivial to show the feasibility of the resulting rectangle labeling, since each rectangle (shaded grey) is completely contained in a Metafont label (dashed), and no two Metafont labels intersect, see Figure 7. Finally we map each rectangle back to an inscribed square of half the optimal size (printed bold) in one of the four allowed positions of Figure 6.

This mapping automatically ensures the new uniqueness requirement of Definition 3. In a solution produced via the rectangle-labeling algorithm that is not necessarily the case, but the algorithm can be adjusted to disallow label placements where two labels completely share a long edge. Now Lemma 2 yields:

Theorem 2. *A feasible solution of the Metafont optimization problem of at least half the optimal size can be computed in $O(n \log n)$ time using linear space.*

3   Two-Circle Labeling

Definition 4 (2-circle point-labeling problem). *Given a set P of n points in the plane, find a set of $2n$ uniform, non-intersecting, maximum-size open circles such that each point touches exactly two circles.*

Zhu and Poon [20] have suggested the first approximation algorithm for this problem. Their algorithm always finds a solution of at least half the optimal size. The algorithm is very simple; it relies on the fact that D_2, the minimum Euclidean distance between any two points in P is an upper bound for the optimal label size (i.e. diameter), see Figure 9. On the other hand, given two points p and q in P, open circles $C_{p,D_2/2}$ and $C_{q,D_2/2}$ with radius $\frac{1}{2}D_2$ centered at p and q do not intersect. Thus if each point is labeled within its circle, no two labels will intersect. This allows labels of maximum diameter $\frac{1}{2}D_2$, i.e. half the upper bound for the optimal label size. The difficulty of the problem immediately comes into play when increasing the label diameter d beyond $\frac{1}{2}D_2$, since then the intersection graph of the circles $C_{p,d}$ of all points p in P changes abruptly; the maximum degree jumps from 0 to 6.

Our approach also assigns each point a certain region such that no two regions intersect and each point can be labeled within its region. The regions we use are not circles but the cells of the *Voronoi diagram* of P, a well-known multi-purpose geometrical data structure [2]. We do not compute the Voronoi diagram explicitly—but use its dual, the *Delauney triangulation* [2]. For the description of our algorithm (see Figure 8) we need the following notation.

Definition 5. *Let* $d = \frac{D_2}{1+\cos 18°} \approx 0.513\, D_2$ *and let* $p, q \in P$.

The pair (p, q) *is an edge of the Delaunay triangulation* $DT(P)$ *of* P *if there is a (closed) disk* D *with* $D \cap P = \{p, q\}$. *The edge* (p, q) *is short if* $d(p, q) < 2d$, *long otherwise. Here* $d(p, q)$ *denotes the Euclidean distance of* p *and* q.

$\text{Vor}(p) = \{x \in \mathbb{R}^2 \mid d(x, p) < d(x, q) \ \forall q \in P \setminus \{p\}\}$ *is the Voronoi cell of* p.

Given two lines that intersect at an angle of less than $90°$, *we call the union of the two smaller (open) regions into which the plane is divided a* wedge.

Given two non-parallel short edges incident to a point p *in* P, *we say that the wedge defined by the two lines containing the edges is* free *if it does not contain any short edges incident to* p.

TWO_CIRCLE_POINT_LABELING(P)

Compute $DT(P)$ and D_2 (the length of a shortest edge in $DT(P)$).
Let $d = \frac{D_2}{1+\cos 18°} \approx 0.513\, D_2$ be the label diameter.
Delete all long Delaunay edges (i.e. of length $\geq 2d$).
for all $p \in P$ **do**
    Compute the largest free wedge W of p.
    Place the centers of the labels of p on the bisector of W at distance $\frac{d}{2}$ from p.
end

Fig. 8. Our 2-circle point-labeling algorithm.

In order to show the correctness of our algorithm we first determine the maximum degree in the Delaunay triangulation minus the long edges and then give a lower bound for the size of the largest free wedge of a point p in P.

Fact 1. *The angle between two short edges incident to a point* p *in* P *is at least* $2 \arcsin \frac{D_2}{4d} \approx 58.4°$.

Lemma 3. *Each point* p *in* P *is incident to at most six short edges.*

Proof. Consider an annular ring R with diameters D_2 and $2d$ around p (including the inner and excluding the outer circle). R contains all points in P that share a short edge with p. However, due to Fact 1, R cannot contain more than six points whose pairwise distance is at least D_2. ✐

Lemma 4. *Each point* p *in* P *has a free wedge with an angle of at least* $36°$.

Proof. Due to Lemma 3 we know that p has $k \leq 6$ short edges. There are $k' \leq k$ lines that go through these edges, and there are k' wedges defined by pairs of neighboring lines (if none of them forms an angle $\geq 90°$, but then we would be done). Due to the pigeon-hole principle there must be a wedge with an angle of at least $36°$ if $k' \leq 5$. (Each wedge contributes two angles!) So we only have to consider the case $k = k' = 6$ (i.e. each line contains exactly one short edge) and all of the six wedges have an angle of less than $36°$. Then, however, one of the wedges must be delimited by two short edges on the same side of p, which is a contradiction to Fact 1. ⚭

It remains to show that the labels of a point are placed within its Voronoi cell.

Lemma 5. *Let L be an open circle of diameter d that touches a point p in P. If the center of L lies on the bisector of the largest free wedge of p, then $L \subseteq \mathrm{Vor}(p)$.*

Proof. Lemma 4 guarantees that p has a free wedge W whose supporting lines s_1 and s_2 span an angle α of at least $36°$. The construction in Figure 10 shows for the extremal case $\alpha = 36°$ how the label diameter d is actually chosen, namely such that the center z of the label L of p lies at the intersection of the bisector b of W and the set F of all points with equal distance to p and a line ℓ_1. The line ℓ_1 has distance $\frac{1}{2}D_2$ from p and is perpendicular to s_1. The line ℓ_2 is defined analogously. Let x be the point that lies between p and ℓ_1 on s_1 at a distance of $\frac{D_2-d}{2}$ from p. In the right-angled triangle Δxzp we then have $\cos \frac{\alpha}{2} = \frac{D_2-d}{d}$, and thus $d = \frac{D_2}{1+\cos 18°}$. For a line h let h^+ be the open halfplane that is supported by h and contains p.

Due to our construction L is contained in the union of a disk D and a kite K, see Figure 10. The disk D is centered at p and has radius $\frac{1}{2}D_2$, thus it lies completely within $\mathrm{Vor}(p)$. The kite K is the intersection of W and the two halfplanes ℓ_1^+ and ℓ_2^+. Note that ℓ_1 and ℓ_2 touch D where they intersect the two supporting lines s_1 and s_2 of W at right angles.

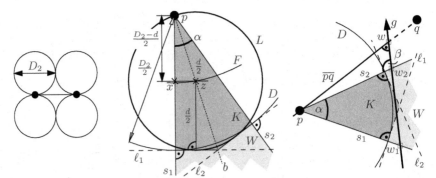

Fig. 9. D_2 is an upper bound for the optimal label size.

Fig. 10. We label symmetrically to a wedge W.

Fig. 11. No Voronoi edge intersects kite K or disk D.

Now suppose L is intersected by a Voronoi edge e of $\mathrm{Vor}(p)$. Let q be the point in P whose Voronoi cell touches that of p in e. If $d(p,q) \geq 2d$ than $d(p,e) = \frac{1}{2}d(p,q) \geq d$ and e does not intersect L. If $d(p,q) < 2d$ then q cannot lie in the wedge W, since W does not contain any short edges. So we can assume $D_2 \leq d(p,q) < 2d$ and $q \notin W$. (p,q) is a Delaunay edge: if the disk with diameter $\overline{pq}$ contained another point r of P, r would be closer than D_2 to p or q.

Let g be the line that contains e. The halfplane g^+ contains $\mathrm{Vor}(p)$ and thus D. Let w_1 and w_2 be the points where g intersects the two supporting lines s_1 and s_2 of W, see Figure 11. Let w be the intersection of g and $\overline{pq}$. w cannot lie between w_1 and w_2, otherwise q would lie in W. So we can assume that w lies outside W and closer to w_2, say. Direct g from w_1 to w_2. By definition g intersects $\overline{pq}$, the Delaunay edge between p and q, in a right angle. Then the angle β that g and s_2 form in the triangle $\triangle pw_2w$ must be less than $90°$. Since ℓ_2 is perpendicular to s_2, this means that g intersects ℓ_2 beyond w_2—and not within W. Thus g^+ contains K, which in turn contains $L \setminus D$. This contradicts our assumption, namely that e and L intersect. $\diamond$

Since the Voronoi cells of a point set are mutually exclusive, Lemma 5 yields the correctness of our 2-circle point-labeling algorithm. Time and space complexity follow from those of $\mathrm{DT}(P)$ and from Lemma 4. Using the same upper bound for an optimal solution as in [20,21], we can summarize as follows.

Theorem 3. *Our algorithm labels a set of n points with $2n$ circles, two per point, of diameter at least $\frac{1}{1+\cos 18°} \approx 0.513$ times the maximum diameter in $O(n \log n)$ time using linear space.*

This algorithm uses the Delaunay triangulation to compute D_2 and to label all points with labels whose size depends on D_2. A different approach would give larger labels in general, although we were not yet able to prove a better approximation factor and keep the runtime of $O(n \log n)$. Instead of computing the Delaunay triangulation and D_2 as an upper bound for the label size, we could directly compute the Voronoi diagram, label each point optimally within its Voronoi cell, and then shrink all labels to the smallest label size we have used. If the Voronoi cell is a regular pentagon, both algorithms actually place labels of the same size.

Due to space limitations we must refer to the full paper for the proof of the following theorem. It does not only provide strong evidence for the necessity to search for approximate solutions of the 2-circle labeling problem, but it also considerably reduces the gap between our approximability result and the polynomial-time non-approximability of 2-circle labeling.

Theorem 4. *It is NP-hard to decide whether a set of points can be labeled with pairs of unit circles, and it is NP-hard to approximate the optimal label size beyond a factor of $\frac{2}{1+\sqrt{3}} \approx 0.732$.*

Acknowledgments. We wish to thank Otfried Cheong, Hong Kong University of Science and Technology, without whose generous support this research would not have been possible.

References

1. J. Ahn and H. Freeman. AUTONAP - an expert system for automatic map name placement. In *Proc. Intl. Symp. on Spatial Data Handling*, pages 544–569, 1984. 368

2. F. Aurenhammer. Voronoi diagrams: A survey of a fundamental geometric data structure. *ACM Comput. Surv.*, 23(3):345–405, Sept. 1991. 376

3. B. Chazelle et al. Application challenges to computational geometry: CG impact task force report. Technical Report TR-521-96, Princeton University, Apr. 1996. 368

4. J. Christensen, J. Marks, and S. Shieber. An empirical study of algorithms for point-feature label placement. *ACM Transactions on Graphics*, 14(3):203–232, 1995. 368, 369

5. A. Datta, H.-P. Lenhof, C. Schwarz, and M. Smid. Static and dynamic algorithms for k-point clustering problems. *J. Algorithms*, 19:474–503, 1995. 371

6. S. Doddi, M. V. Marathe, A. Mirzaian, B. M. Moret, and B. Zhu. Map labeling and its generalizations. In *Proc. of the 8th ACM-SIAM Symp. on Discrete Algorithms (SODA'97)*, pages 148–157, 1997. 368, 369

7. S. Doddi, M. V. Marathe, and B. M. Moret. Point labeling with specified positions. In *Proc. 16th Annu. ACM Sympos. Comput. Geom.*, Hongkong, 2000. to appear. 369

8. J. S. Doerschler and H. Freeman. An expert system for dense-map name placement. In *Proc. Auto-Carto 9*, pages 215–224, 1989. 368

9. S. Even, A. Itai, and A. Shamir. On the complexity of timetable and multicommodity flow problems. *SIAM J. Comput.*, 5:691–703, 1976. 371

10. M. Formann. *Algorithms for Geometric Packing and Scaling Problems*. PhD thesis, Fachbereich Mathematik und Informatik, Freie Universität Berlin, 1992. 371, 372

11. M. Formann and F. Wagner. A packing problem with applications to lettering of maps. In *Proc. 7th Annu. ACM Sympos. Comput. Geom. (SoCG'91)*, pages 281–288, 1991. 368, 369, 372, 374

12. M. Formann and F. Wagner. An efficient solution to Knuth's METAFONT labeling problem. Manuscript available at http://www.math-inf.uni-greifswald.de/map-labeling/papers/fw-eskml-93.ps.gz, 1993. Freie Universität Berlin. 370, 374

13. S. A. Hirsch. An algorithm for automatic name placement around point data. *The American Cartographer*, 9(1):5–17, 1982. 368, 369

14. K. G. Kakoulis and I. G. Tollis. On the multiple label placement problem. In *Proc. 10th Canadian Conf. Comp. Geometry (CCCG'98)*, pages 66–67, 1998. 369

15. D. E. Knuth and A. Raghunathan. The problem of compatible representatives. *SIAM J. Discr. Math.*, 5(3):422–427, 1992. 370, 374

16. C. K. Poon, B. Zhu, and F. Chin. A polynomial time solution for labeling a rectilinear map. *Information Processing Letters*, 65(4):201–207, 1998. 370, 371

17. T. Strijk and A. Wolff. Labeling points with circles. Technical Report B 99-08, Institut für Informatik, Freie Universität Berlin, Apr. 1999. 369

18. M. van Kreveld, T. Strijk, and A. Wolff. Point labeling with sliding labels. *Computational Geometry: Theory and Applications*, 13:21–47, 1999. 369

19. F. Wagner and A. Wolff. A practical map labeling algorithm. *Computational Geometry: Theory and Applications*, 7:387–404, 1997. 368, 369, 372

20. B. Zhu and C. K. Poon. Efficient approximation algorithms for multi-label map labeling. In *Proc. Tenth Annual Intl. Symp. on Algorithms and Computation*

(ISAAC'99), LNCS, pages 143–152, Chennai, India, 1999. Springer-Verlag. 368, 369, 370, 371, 375, 378

21. B. Zhu and Z. Qin. New approximation algorithms for map labeling with sliding labels. Dept. of Computer Science, City University of Hong Kong, 2000. 369, 371, 378

22. S. Zoraster. The solution of large 0-1 integer programming problems encountered in automated cartography. *Operations Research*, 38(5):752–759, 1990. 368

Analysing the Cache Behaviour of Non-uniform Distribution Sorting Algorithms

Naila Rahman and Rajeev Raman*

Algorithm Design Group, Department of Computer Science, King's College London, Strand, London WC2R 2LS, U. K. e-mail: {naila, raman}@dcs.kcl.ac.uk

Abstract. We analyse the average-case cache performance of distribution sorting algorithms in the case when keys are independently but not uniformly distributed. We use this analysis to tune the performance of the integer sorting algorithm MSB radix sort when it is used to sort independent uniform floating-point numbers (floats). Our tuned MSB radix sort algorithm comfortably outperforms a cache-tuned implementation of bucketsort [11] when sorting uniform floats from $[0, 1)$.

1 Introduction

Distribution sorting is a popular alternative to comparison-based sorting which involves placing n input keys into $k \leq n$ *classes* based on their value [7]. The classes are chosen so that all the keys in the ith class are smaller than all the keys in the $(i + 1)$st class, for $i = 1, \ldots, k - 1$, and furthermore, the class to which a key belongs can be computed in $O(1)$ time (e.g. if the keys are floats in the range $[a, b)$, we can calculate the class of a key x as $1 + \lfloor \frac{x-a}{b-a} \cdot k \rfloor$). Thus, the original sorting problem is reduced in linear time to the problem of sorting the keys in each class. A number of distribution sorting algorithms have been developed which run in linear (expected) time under some assumptions about the input keys, such as bucket sort and radix sort. Due to their poor *cache* utilisation, even good implementations—which minimise instruction counts—of these 'linear-time' algorithms fail to outperform general-purpose $O(n \log n)$-time algorithms such as quicksort or mergesort on modern computers [9,11].

Most algorithms are based upon the random-access machine model [1], which assumes that main memory is as fast as the CPU. However, in modern computers, main memory is typically one or two orders of magnitude slower than the CPU [5]. To mitigate this, one or more levels of *cache* are introduced between CPU and memory. A cache is a fast associative memory which holds the values of some main memory locations. If the CPU requests the contents of a memory location, and the value of that location is held in some level of cache (a cache *hit*), the CPU's request is answered by the cache itself in typically 1-3 clock cycles; otherwise (a cache *miss*) it is answered by accessing main memory in typically 30-100 clock cycles. Since typical programs exhibit *locality of reference*

---

* Supported in part by EPSRC grant GR/L92150

M. Paterson (Ed.): ESA 2000, LNCS 1879, pp. 380–392, 2000.

[5], caches are often effective. However, algorithms such as distribution sort have poor locality of reference, and their performance can be greatly improved by optimising their cache behaviour. A number of papers have recently addressed this issue [8,9,11,12,13,14], mostly in the context of sorting and related problems. There is also a large literature on algorithms specifically designed for hierarchical models of memory [15], but there are some important differences between these models and ours (see [9] for a summary).

The cache performance of comparison-based sorting algorithms was studied in [9,13,14] and distribution sorting algorithms were considered in [8,9,11]. One pass of a distribution sort consists of a *count* phase where the number of keys in each class are determined, followed by a *permute* phase where the keys belonging to the same class are moved to consecutive locations in an array. We give an analysis of the cache behaviour of the permute phase, assuming the keys are independently drawn from a non-uniform distribution. We focus on an 'in-place' permute, where the keys are rearranged without placing them first in an auxiliary array[1], but the methods extend easily to an out-of-place permute (omitted here) and also to an average-case analysis of multi-way merge sort. More precisely, we model the above algorithms as probabilistic processes, and analyse the cache behaviour of these processes. For each process we give an exact expression for, as well as matching closed-form upper and lower bounds on, the number of misses.

In previous work on the cache analysis of distribution sorting, [8] have analysed the (somewhat easier) count phase for non-uniform keys, and [11] gave an empirical analysis of the permute phase for uniform keys. The process of *accessing multiple sequences* of memory locations, which arises in multi-way merge sort, was analysed previously by [13,14]. The analysis in [13] assumes that accesses to the sequences are controlled by an adversary; our analysis demonstrates, among other things, that with uniform randomised accesses to the sequences, many more sequences can accessed optimally. In [14] a lower bound on cache misses is given for uniform randomised accesses; our lower bound is somewhat sharper.

In practice there are often cases when keys are not uniform (e.g., they may be normally distributed); our analysis can be used to tune distribution sort in these cases. We consider a different application here: sorting uniform floats using an *integer* sorting algorithm. It is well known that one can sort floats by sorting the bit-strings representing the floats, interpreting them as integers [5]. Since (simple) operations on integers are faster than operations on floats, this can improve performance; indeed, in [11] it was observed that an *ad hoc* implementation of the integer sorting algorithm *most-significant-bit first* radix sort (MSB radix sort) outperformed an optimised version of bucket sort on uniform floats. We observe that a uniform distribution on floating-point numbers induces a non-uniform distribution on the representing integers, and use our new cache analysis to improve the performance of MSB radix sort on our machine. Our tuned 'in-place' MSB radix sort comfortably outperforms optimised implementations of

---

[1] The permute is not strictly in-place, as auxiliary space is used to store the count of keys in each class.

other in-place or 'in-place' algorithms such as Quicksort or FlashSort2P [11], which is a cache-tuned version of bucket sort.

We now introduce some terminology and notation regarding caches. The size of the cache is normally expressed in terms of two parameters, the *block size* (B) and the number of *cache blocks* (C). We consider main memory as being divided into equal-sized *blocks* consisting of B consecutively-numbered memory locations, with blocks starting at locations which are multiples of B. The cache is also divided into blocks of size B; one cache block can hold the value of exactly one memory block. Data is moved to and from main memory only as blocks.

We assume a *direct-mapped* cache, where the value of memory location x can only be stored in cache block $c = (x \text{ div } B) \text{ mod } C$. If the CPU reads location x and cache block c holds the values from x's block (a hit), the CPU reads the value of x from the cache. In case of a write hit, only the copy of x in the cache is updated. In case of a miss on either a read or a write, we assume that the contents of the block containing x are copied into cache block c, *evicting* the current contents of cache block c (the evicted block may be copied back into memory if it contains modified locations, or simply overwritten otherwise). In cache jargon we are assuming a *write-back* cache with *write allocation* [5,4].

We performed our experiments on a Sun UltraSparc-II, whose L2 level cache satisfies the above assumptions and we model only this level of the cache. The L2 cache has a block size of 64 bytes, and its size is 512 kilobytes. Hence, in our machine, $C = (512 \times 1024)/64 = 8192$. It is useful to express B in terms of the number of keys which fit in a block. As this paper deals mainly with single-precision (4 byte) floating-point keys we use $B = 16$ in what follows.

In Section 2 we introduce the process which models the permute phase and analyse it. In Section 3 we explain the relationship between the permute phase and the process. In Section 4 we describe MSB radix sort when applied to uniform floats, explain how to tune the algorithm, and give experimental results.

2   Cache analysis

2.1   Analysing an In-place Permute

In this section we analyse a process which models the permute phase of distribution sorting algorithms. Let k be an integer, $2 \le k \le CB$. We are given k probabilities $p_1, \ldots, p_k$, such that $\sum_{i=1}^{k} p_i = 1$. The process maintains k pointers $D_1, \ldots, D_k$, and there are also k consecutive 'count array' locations, $\mathcal{C} = c_1, \ldots, c_k$. The process (henceforth called *Process A*) executes a sequence of *rounds*, where each round consists in performing steps 1-3 below:

1. Pick an integer x from $\{1, \ldots, k\}$ such that $\Pr[x = i] = p_i$, independently of all previous picks.
2. Access the location c_x.
3. Access the location pointed to by D_x, increment D_x by 1.

We denote the locations accessed by the pointer D_i by $d_{i,1}, d_{i,2}, \ldots$, for $i = 1, \ldots, k$. We assume that:

(a) the start position of each pointer is uniformly and independently distributed over the cache, i.e., for each i, $d_{i,1} \bmod BC$ is uniformly and independently distributed over $\{0, \ldots, BC - 1\}$.

(b) during the process, the pointers traverse sequences of memory locations which are disjoint from each other and from $\mathcal{C}$.

(c) c_1 is located on an aligned block boundary, i.e., $c_1 \bmod B = 0$.

Assuming that the cache is initially empty, the objective is to determine the expected number of cache misses incurred by the above process over n rounds, with the expectation taken over the random choices in Step 1 as well as the starting positions of the pointers. Assumption (b) implies that the pointers must access n distinct locations, and so the process must incur $\Omega(n/B)$ misses. We will say the process is optimal if it incurs $O(n/B)$ expected misses.

We introduce some notation before proceeding with the analysis. We denote the expected value of a function f of a random variable X as $\mathrm{E}[f(X)]$. When we wish to make explicit the distribution D from which the random variable is drawn, we use the notation $\mathrm{E}_{X \sim D}[f(X)]$. All vectors have dimension k (the number of pointers) unless stated otherwise, and we denote the components of a vector $\bar{x}$ by $x_1, x_2, \ldots, x_k$. We now define some probabilities:

(i) For all $i \in \{1, \ldots, k/B\}$, $P_i = \sum_{l=(i-1)B+1}^{iB} p_l$.

(ii) For all $i \in \{1, \ldots, k\}$, we denote by $\bar{a}^i$ the following vector: $a_j^i = 0$ if $i = j$, and $a_j^i = p_j/(1 - p_i)$ otherwise and by $\bar{b}^i$ the following vector: $b_j^i = 0$ if $(i-1)B+1 \le j \le iB$, and $b_j^i = p_j/(1 - P_i)$ otherwise. (Note that $\sum_j a_j^i = \sum_j b_j^i = 1$).

Let $m \ge 0$ be an integer and $\bar{q}$ be a vector of non-negative reals such that $\sum_i q_i = 1$. We denote by $\varphi(m, \bar{q})$ the probability distribution on the number of balls in each of k bins, when m balls are independently put into these bins, and a ball goes in bin i with probability q_i, for $i \in \{1, \ldots, k\}$. Thus, $\varphi(m, \bar{q})$ is a distribution on vectors of non-negative integers. If $\bar{\mu}$ is drawn from $\varphi(m, \bar{q})$, then:

$$\Pr[\mu_1 = m_1, \ldots, \mu_k = m_k] = \left(\prod_{j=1}^{k} q_j^{m_j} \right) m! / \prod_{j=1}^{k} m_j! \tag{1}$$

whenever $\sum_{i=1}^{k} m_i = m$; all other vectors have zero probability[2]. We now define functions $f(x)$ for $x \ge 0$ and $g(\bar{m})$ for a vector $\bar{m}$ of non-negative integers:

$$f(x) = \begin{cases} 1 & \text{if } x = 0 \\ 1 - \frac{x + B - 1}{BC} & \text{if } 0 < x \le BC - B + 1 \\ 0 & \text{otherwise} \end{cases} \tag{2}$$

$$g(\bar{m}) = \frac{1}{C} \sum_{i=1}^{k/B} \min\{1, \sum_{l=(i-1)B+1}^{iB} m_l\} \tag{3}$$

---

[2] We take $0^0 = 1$ in Eq. 1.

Theorem 1. *The expected number X of cache misses in n rounds of process A satisfies $n(p_c + p_d) \leq X \leq n(p_c + p_d) + k(1 + 1/B)$, where:*

$$p_c = \sum_{i=1}^{k/B} P_i \left(1 - \sum_{m=0}^{\infty} P_i(1 - P_i)^m \mathrm{E}_{\bar{\mu} \sim \varphi(m, \bar{b}_i)} \left[\prod_{j=1}^{k} f(\mu_j) \right] \right) \quad \text{and}$$

$$p_d = \frac{1}{B} + \frac{B-1}{B} \sum_{i=1}^{k} p_i \left(1 - \sum_{m=0}^{\infty} p_i(1 - p_i)^m \mathrm{E}_{\bar{\mu} \sim \varphi(m, \bar{a}_i)} \left[(1 - g(\bar{\mu})) \prod_{j=1}^{k} f(\mu_j) \right] \right)$$

Proof. We first analyse the miss rates for accesses to pointers $D_1, \ldots, D_k$. Fix an i, $1 \leq i \leq k$ and a $z \geq 1$. Let μ be the random variable which denotes the number of rounds between accesses to locations $d_{i,z}$ and $d_{i,z+1}$ ($\mu = 0$ if these locations are accessed in consecutive rounds). Clearly, $\Pr[\mu = m] = p_i(1 - p_i)^m$, for $m = 0, 1, \ldots$. Let X_i denote the event that none of the memory accesses in these μ rounds accesses the cache block to which $d_{i,z}$ is mapped. We now fix an integer $m \geq 0$ and calculate $\Pr[X_i | \mu = m]$. Let $\bar{\mu}$ be a vector of random variables such that for $1 \leq j \leq k$, μ_j is the random variable which denotes the number of accesses to D_j in these m rounds. Clearly $\bar{\mu}$ is drawn from $\varphi(m, \bar{a}_i)$ (note that D_i is not accessed in these m rounds by definition).

Fix any vector $\bar{m}$, such that $\Pr[\bar{\mu} = \bar{m}] \neq 0$, and let μ_j be the number of accesses to pointer D_j in these m rounds. Since m_i must be zero, $f(m_i) = 1$, and for $j \neq i$, $f(m_j)$ is the probability that none of the m_j locations accessed by D_j in these m rounds is mapped to the same cache block as location $d_{i,z}$ [13,14]. Similarly $g(\bar{m}) \cdot C$ is the number of count blocks accessed in these rounds, and so $1 - g(\bar{m})$ is the probability that the cache block containing $d_{i,z}$ does not conflict with the blocks from C which were accessed in these m rounds. As the latter probability is determined by the starting location of sequence i and the former probabilities by the starting location of sequences $j, j \neq i$, we conclude that for a given configuration $\bar{m}$ of accesses, the probability that the cache block containing $d_{i,z}$ is not accessed in these m rounds is $(1 - g(\bar{m})) \prod_{j=1}^{k} f(m_j)$. Averaging over all configurations $\bar{m}$, we get that $\Pr[X_i \mid \mu = m] = \mathrm{E}_{\bar{\mu} \sim \varphi(m, \bar{a}_i)}[(1 - g(\bar{\mu})) \prod_{j=1}^{k} f(\mu_j)]$. Finally,

$$\Pr[X_i] = \sum_{m=0}^{\infty} \Pr[\mu = m] \Pr[X_i | \mu = m]$$

$$= \sum_{m=0}^{\infty} p_i(1 - p_i)^m \mathrm{E}_{\bar{\mu} \sim \varphi(m, \bar{a}_i)} \left[(1 - g(\bar{\mu})) \prod_{j=1}^{k} f(\mu_j) \right] \quad (4)$$

If $d_{i,z}$ is at a cache block boundary or if X_i does not occur given that $d_{i,z}$ is not at a cache block boundary ($\Pr[X_i]$ does not change under this condition), then a cache miss will occur. The first access to a pointer is a cache miss. So other than for the first access, the probability p_d of a cache miss for a pointer access is:

$$p_d = \frac{1}{B} + \frac{B-1}{B} \sum_{i=1}^{k} p_i(1 - \Pr[X_i]) \tag{5}$$

We obtain p_c, the probability of a cache miss on an access, other than the first, to a count array block using a similar approach

Including the first access misses, the expected number of cache misses for pointer accesses is at most $np_d + k$, and similarly the expected number of cache misses for count array accesses is at most $np_c + k/B$. Plugging in the values from Eq. 4 into 5 to and the similarly derived p_c we get the upper bound on X, the expected number of cache misses in the processes. The lower bound is obvious.

Remark 1. Although these expressions for the number of misses are not closed-form, they can be estimated fairly rapidly in practice by a simple Monte Carlo method which generates $\bar{\mu}$ according to the distribution $\varphi(m, \bar{a}_i)$. This produces a single estimate of $\Pr[X_i]$ in $O(k + 1/p_i)$ expected time, and a single estimate of p_d in $O(k^2 + \sum_{i=1}^{k} 1/p_i) = O(\sum_{i=1}^{k} 1/p_i)$ expected time.

Theorem 2. *The expected number of cache misses in n rounds of process A is at most $n(p_d + p_c) + k(1 + 1/B)$, where:*

$$p_d \leq \frac{1}{B} + \frac{k}{BC} + \frac{B-1}{BC} \sum_{i=1}^{k} \left(\sum_{j=1}^{k/B} \frac{p_i P_j}{p_i + P_j} + \frac{B-1}{B} \sum_{j=1}^{k} \frac{p_i p_j}{p_i + p_j} \right)$$

$$p_c \leq \frac{k}{B^2 C} + \frac{B-1}{BC} \sum_{i=1}^{k/B} \sum_{j=1}^{k} \frac{P_i p_j}{P_i + p_j}$$

Proof. Again, we consider a fixed i and consider the event X_i defined in the proof of Theorem 1. We now obtain a lower bound on $\Pr[X_i]$. Letting $\Gamma(x) = 1 - f(x)$ and using $\prod_j (1 - x_j) \geq 1 - \sum_j x_j$ if $0 \leq x_j \leq 1$ for all x_j, we can rewrite Eq. 4 as:

$$\Pr[X_i] \geq \sum_{m=0}^{\infty} \Pr[\mu = m] \mathbb{E}_{\bar{\mu} \sim \varphi(m, \bar{a}_i)} \left[1 - g(\bar{\mu}) - \sum_{j=1}^{k} \Gamma(\mu_j) \right] \tag{6}$$

We know that the j-th count block contributes $1/C$ to $g(\bar{\mu})$ if there is an access to that block and $\Pr[j\text{-th count block accessed}|\mu = m] = 1 - (1 - c_j^i)^m$, where $c_j^i = \frac{P_j}{1 - p_i}$. So, $\mathbb{E}_{\bar{\mu} \sim \varphi(m, \bar{a}_i)}[g(\bar{\mu})] = \sum_{j=1}^{k/B} \frac{1}{C}(1 - (1 - c_j^i)^m)$ and we get:

$$\sum_{m=0}^{\infty} \Pr[\mu = m] \mathbb{E}_{\bar{\mu} \sim \varphi(m, \bar{a}_i)}[g(\bar{\mu})] = \frac{1}{C} \sum_{j=1}^{k/B} \frac{P_j}{p_i + P_j} \tag{7}$$

We can show $\mathbb{E}_{\bar{\mu} \sim \varphi(m, \bar{a}_i)}[\Gamma(\mu_j)] = \sum_{l=0}^{m} \Pr[\mu_j = l] \frac{l + B - 1}{BC} - \Pr[\mu_j = 0] \Gamma(0)$
$- \sum_{l=BC-B+1}^{m} \Pr[\mu_j = l] \Gamma(l) \leq \frac{1}{BC} \left(m a_j^i + (B - 1) \left(1 - \left(1 - a_j^i\right)^m \right) \right)$. Using

$\sum_{m=0}^{\infty} mx^m = \frac{x}{(1-x)^2}$ if $|x| < 1$, we get:

$$\sum_{m=0}^{\infty} \Pr[\mu = m] \sum_{j=1}^{k} E_{\bar{\mu} \sim \varphi(m, \bar{a}_i)}[\Gamma(\mu_j)] \leq \frac{1}{BC} \left(\frac{1}{p_i} + (B-1) \sum_{j=1}^{k} \frac{p_j}{p_i + p_j} \right) \quad (8)$$

Substituting Eq. 7 and 8 in Eq. 6 we can write a lower bound for $\Pr[X_i]$ from which we obtain an upper bound on p_d. Using a similar approach we obtain the upper bound on p_c.

Corollary 1. *If $p_1 = \ldots = p_k = 1/k$ then the number of cache misses in n rounds of process A is at most $n \left(\frac{1}{B} + \frac{k(B+3)}{2BC} + \frac{k}{B^2C} + \frac{k}{BC} \right) + k \left(1 + \frac{1}{B} \right)$*

Remark 2. As we will see later, process A models the permute phase of distribution sorting and Corollary 1 shows that one pass of uniform distribution sorting incurs $O(n/B)$ cache misses if and only if $k = O(C/B)$.

Theorem 3. *When $p_i \geq 1/C$ and $k \leq C$ then the expected number of cache misses in process A is at least:*

$$n \left(\frac{1}{B} + \frac{B-1}{B} \frac{k-1}{2k-1} \left(\frac{k}{C} - \frac{k^2}{2C^2} \right) - O \left(e^{-B} \right) \right)$$

Proof. We again consider a fixed i and consider the event X_i defined in the proof of Theorem 1. Let $\bar{\mu}$ be as defined in the proof of Theorem 1. We now obtain an upper bound on $\Pr[X_i]$. In [2] it is shown that the variables μ_j are *negatively associated* [6]. Noting that $f(x)$ is a non-increasing function of x, it follows immediately from the definition of negative association that $E_{\bar{\mu} \sim \varphi(m, \bar{a}_i)}[\prod_{j=1}^{k} f(\mu_j)] \leq \prod_{j=1}^{k} E_{\bar{\mu} \sim \varphi(m, \bar{a}_i)}[f(\mu_j)]$. So we can re-write equation 4 as:

$$\Pr[X_i] \leq \sum_{m=0}^{BC-B} \Pr[\mu = m] \prod_{j=1}^{k} E_{\bar{\mu} \sim \varphi(m, \bar{a}_i)}[f(\mu_j)] + \sum_{m=BC-B+1}^{\infty} \Pr[\mu = m]$$

Assuming $p_i \geq 1/C$ the last term is at most $O(e^{-B})$ and since $\mu_j \leq BC - B$ we get $E_{\bar{\mu} \sim \varphi(m, \bar{a}_i)}[f(\mu_j)] = 1 - \frac{1}{BC}(ma_j^i + (B-1)(1 - (1 - a_j^i)^m)) = 1 - t_j(m)$. Since $0 \leq t_j(m) \leq 1$ we get $e^{-\sum_j t_j(m)} \geq \prod_j (1 - t_j(m))$ so we have:

$$\sum_{m=0}^{BC-B} \Pr[\mu = m] e^{-\sum_j t_j(m)} \leq \sum_{m=0}^{BC-B} \Pr[\mu = m] e^{\frac{-1}{BC}((B-1)k(1-(1-1/k)^m))}$$

$$\leq 1 - \frac{(k-1)}{(2k-1)} \left(\frac{k}{C} - \frac{k^2}{2C^2} \right)$$

These simplifications used $\sum_j (1 - a_j^i)^m \leq k(1 - 1/k)^m$ and $e^{-x} \leq 1 - x + \frac{x^2}{2}$ for $x \geq 0$. Using this in Eq. 5 gives the lower bound on p_d.

Remark 3. For uniforml random keys the lower bound is always within a factor of about 2 of the upper bound for all $k \leq C$, and is much closer if $k \ll C$.

2.2   Accessing k sequences

Accessing k sequences is like to process A except that there is no interaction with a count array, so we delete step 2 and assumption (c). An analogue of Theorem 1 is easily obtained. An easy modification (omitted) to the proof of Theorem 2 gives:

Theorem 4. *The expected number of cache misses in n rounds of sequence accesses is at most* $k + n \left(\frac{1}{B} + \frac{k(B-1)}{B^2 C} + \frac{(B-1)^2}{B^2 C} \sum_{i=1}^{k} \sum_{j=1}^{k} \frac{p_i p_j}{p_i + p_j} \right)$.

Corollary 2. *If $p_1 = \ldots = p_k = 1/k$ then the number of cache misses in n rounds of sequence accesses is at most* $n \left(\frac{1}{B} + \frac{k(B+3)}{2BC} \right) + k$

Remark 4. From Corollary 2, $k = O(C/B)$ random sequences can be accessed incurring an optimal $O(n/B)$ misses. In contrast, [13] shows that in the worst case only $O(\sqrt{C})$ sequences can be accessed optimally. In [14] it is speculated that only $O(\sqrt{C})$ sequences can be accessed optimally on average as well.

Remark 5. Since its derivation ignored the effects of the count array, the lower bound in Theorem 3 applies directly to sequence accesses. The lower bound is always within a factor of about 2 of the upper bound for all $k \leq C$, and is much closer if $k \ll C$. Note that our lower bound is sharper than the lower bound of $0.25(1 - e^{-0.25k/C})$ obtained in [14].

3   Distribution sorting

As noted in the introduction, a distribution pass has two main phases, a *count* phase and a *permute* phase, and our focus here is on the latter. The input to the permute phase is the original array containing the keys (called DATA) and an array called COUNT, initialised as follows: COUNT[1]=0, and for $i > 1$, COUNT[i] equals the number of keys in classes $1, \ldots, i - 1$. After the permute phase, the array DATA should have been permuted so that all elements of class i lie consecutively before all elements of class $i + 1$, for $i = 1, \ldots, k - 1$. We describe an 'in-place' permute phase based on ideas from [7,10], where no auxiliary array is used to store the keys. This moves elements to their final locations along a cycle in the permutation as shown in Fig 1. An invariant is that COUNT[i] always points to the next available location for an element of class i.

The correspondence between Process A of Section 2 and the pseudocode in Fig reffig:distrib is as follows. The array COUNT corresponds to the locations $\mathcal{C}$, and the pointer D_i points to DATA[COUNT[i]]. Each iteration of the inner loop (steps 2.1-2.5) of the pseudocode corresponds to a round of Process A. The variables x in the process and the pseudocode play a similar role. It can easily be verified that in each iteration of the loop in the pseudocode, the value of x is any integer $1, \ldots, k$ with probability $p_1, \ldots, p_k$, independently of its previous values, as in Step 1 of Process A. We count a read of a location immediately followed by a write to the same location as one access. Thus, the read and

```
1    idx := start; key := DATA[idx];
2.1  x := classify(key);
2.2  idx := COUNT[x];
2.3  COUNT[x]++;
2.4  swap key and DATA[idx];
2.5  if idx ≠ start go to 2.1;
3    Find start of next cycle and set
     start to this value. Go to 1.
```

start is set to n initially. The function classify maps a key to a class numbered $\{1, \ldots, k\}$ in $O(1)$ time. To find the start of the next cycle the algorithm makes a copy of COUNT before the start of the permute phase.

Fig. 1. Pseudo-code for the permute phase

increment of COUNT[x] in Steps 2.2 and 2.3 of the pseudocode constitutes one access, equivalent to Step 2 and part of Step 3 of the process. Similarly the "swap" in Step 2.5 of the pseudocode corresponds to the memory access in Step 3 of the process. The process does not model the initial access in Step 1 of the pseudocode, and nor does it model the task of looking for new cycle leaders.

Assumption (b) of the process is clearly satisfied and assumption (c) can normally be made to hold. Assumption (a) of the process, that the starting locations of the pointers D_i are uniformly and independently distributed, is patently false. We may force it to hold by adding random offsets to the starting location of each pointer, at the cost of needing more memory and adding a compaction phase after the permute. This only works if the permute is not in-place, and if k is sufficiently small (e.g. $k \leq n/(CB)$). Assumption (a) was studied empirically in [11] in the context of uniform distribution sorting, where it was noted that for most sufficiently large values of n (a few times greater than CB), calculations based on assumption (a) were quite accurate, and a partial characterisation of 'bad' values of n was given. In Fig 2, we see that predictions made by our analysis are very accurate in the non-uniform case, if n is greater than about $4CB$. The values of n are not known to be bad.

n (thousands)	256	512	1000	2000	4000	8000	16000	32000	64000
Predicted	0.1253	0.1254	0.1259	0.1272	0.1299	0.1354	0.1465	0.1690	0.2065
Simulated	0.1100	0.1255	0.1261	0.1276	0.1301	0.1354	0.1460	0.1676	0.2085

Fig. 2. Predicted and simulated miss rates for permute phase of a distribution sort, where pointer D_i has an access rate $1/(K2^{\lfloor i/K \rfloor})$ and $k = K \lceil 2 \lg \lg n \rceil$, where $K = 2^{\lceil \lg n - \lg 2CB \rceil}$. Predictions made using Monte Carlo method of Remark 2. Cache size is $2^{17} \approx 131{,}000$ keys.

4   MSB Radix Sort

We now consider the problem of sorting n independent and uniformly-distributed floating-point numbers in the range $[0, 1)$ using the integer sorting algorithm

MSB radix sort. As noted earlier, it suffices to sort lexicographically the bit-strings which represent the floats, by viewing them as integers. One pass of MSB radix sort using radix size r groups the keys according to their most significant r bits in $O(2^r + n)$ time. For random integers, a reasonable choice for minimising instruction counts is $r = \lceil \lg n - 3 \rceil$ bits, or classifying into about $n/8$ classes. Since each class has about 8 keys on average, they can be sorted using insertion sort. Using this approach for our problem gives terrible performance even at small values of n (see Fig. 3). As we now show, the problem lies with the distribution of the integers on which MSB radix sort is applied.

Floating-point number representation. A floating-point number is represented as a triple of non-negative integers $\langle i, j, k \rangle$. Here i is called the *sign bit* and is a 0-1 value (0 indicating non-negative numbers, 1 indicating negative numbers), j is called the *exponent* and is represented using e bits and k is called the *mantissa* and represented using m bits. We let $j^* = j - 2^{e-1} + 1$ denote the *unbiased* exponent of $\langle i, j, k \rangle$. Roughly following the IEEE 754 standard, we let the triple $\langle 0, 0, 0 \rangle$ represent the number 0, and let $\langle i, j, k \rangle$, where $j > 0$, represent the number $\pm 2^{j^*}(1 + k2^{-m})$, depending on whether $x = 0$ or 1; no other triple is a floating-point number. Internally each member of the triple is stored in consecutive fields of a word. The IEEE 754 standard specifies $e = 8$ and $m = 23$ for 32-bit floats and $e = 11$ and $m = 52$ for 64-bit floats [5].

We model the generation of a random float in the range $[0, 1)$ as follows: generate an (infinite-precision) random real number, and round it down to the next smaller float. On average, half the numbers generated will lie in the range $[0.5, 1)$ and will have an unbiased exponent of -1. In general, for all non-zero numbers, the unbiased exponent has value i with probability 2^i, for $i = -1, -2, \ldots, -2^{e-1} + 2$, whereas the mantissa is a random m-bit integer. The value 0 has probability $2^{-2^{e-1}+2}$. Clearly, the distribution is not uniform, and it is easy to see that the average size of the largest class after the first pass of MSB radix sort with radix r is $n\left(1 - \frac{1}{2^{2^e - r + 1}}\right)$ if $r < e + 1$, and $n/(2^{r-e})$ if $r \geq e + 1$.

This shows, e.g., that the largest sub-problems in the examples of Figure 3 would be of size $n/2^{\lceil \lg n - 3 \rceil - 11} \approx 2^{14}$, so using insertion sort after one pass is inefficient in this case[3]. To get down to problems of size 8 in one pass requires a radix of about $\lg n + 8$, which is impractical. Also, MSB radix sort applied to random integers has $O(n)$ expected running time independently of the word size, but this is not true for floats. A first pass with $r \ll e$ barely reduces the largest problem size, and the same holds for subsequent passes until bits from the mantissa are reached. As the radix in any pass is limited to $\lg n + O(1)$ bits, we may need $\Omega(e/\lg n)$ passes, introducing a dependence on the word size.

To get around this, we partition the input keys around a value $1/n \leq \theta \leq 1/(\lg n)$, and sort the keys keys smaller than θ in $O(n)$ expected time using Quicksort. We apply MSB radix sort to the remaining keys. We let $e' = \min\{\lceil \lg \lg(1/\theta) \rceil, e\}$ denote the *effective exponent*, since the remaining keys have

---

[3] In fact, the total number of keys in all sub-problems of this size would be $n/2$ on average.

exponents which vary only in the lower order e' bits. This means that keys can be grouped according to a radix $r = e+1+m'$ with $m' \geq 0$ in $O(n+2^{e'+m'})$ time and $O(2^{e'+m'})$ space. Since $e' = O(\lg\lg n)$, we can take up to $\lg n - O(\lg\lg n)$ bits from the mantissa as part of the first radix; as all sub-problems now only deal with bits from the mantissa they can be solved in linear expected time, giving a linear running time overall.

Cache analysis We now calculate an upper bound for the cache misses in the permute phase of the first pass of MSB radix sort using a radix $r = e + 1 + m'$, for some $m' \geq 0$, assuming also that all keys are in the range $[\theta, 1)$, for some $\theta \geq 1/n$. There are $2^{e'+m'}$ pointers in all, which can be divided into $g = 2^{e'}$ groups of $K = 2^{m'}$ pointers each. Group i corresponds to keys with unbiased exponent $-i$, for $i = 1, \ldots, g$. All pointers in group i have an access probability of $1/(K2^i)$. Using Theorem 1 and a slight extension of the methods of Theorem 2 we are able to prove Theorem 5 below, which states that the number of misses is essentially independent of g:

Theorem 5. *Provided $gK \leq CB$ and $K \leq C$ the number of misses in the first pass of the permute phase of MSB radix sort is at most*

$$n\left(\frac{1}{B} + \frac{2K}{BC}\left(2.3B + 2\lg B + \lg C - \lg K + 0.7\right)\right) + gK(1 + 1/B)$$

Tuning We now optimise parameter choices. The smaller the value of θ, the fewer keys are sorted by Quicksort, but reducing θ may may increase e'. A larger value of e' does not mean more misses, by Theorem 5, but it does mean a larger count array. We choose $\theta = 1/(\lg n)^2$ as a compromise, ensuring that Quicksort uses $o(n)$ time. Using our analysis and the results of [8] we are also able to determine an optimal number of classes to use in each sorting sub-problem. We use two criteria of optimality. In the first, we require that each pass incur no more than $(2 + \varepsilon)n/B$ misses for some constant $\varepsilon > 0$, thus seeking essentially to minimise cache misses ($2n/B$ misses is the bare minimum for the count and permute phases). In the second, we trade off reductions in cache misses against extra computation. The latter yields better practical results, and results shown below are for this approach.

In Figure 4 we compare tuned MSB radix sort with memory-tuned Quicksort[9] and Flashsort2P[11]. The algorithms were coded in C and compiled using gcc 2.8.1. The experiments were on a Sun Ultra II with 2×300 Mhz processors and 1GB main memory, and a 512KB L2 direct-mapped cache. We observe that MSB radix sort easily outperforms the other algorithms for the range of values we considered.

Open Problems. Open Problems. It would be interesting to extend the analysis to set-associative caches [5], or to multiple levels of cache, and also to understand why assumption (a) in Section 2 seems to be a good one in practice, even though it is manifestly false.

Acknowledgements. We thank Danny Krizanc and Tomasz Radzik as well as the PC members for some useful comments.

$n =$	1M	2M	4M	8M	16M	32M
Quicksort	0.7400	1.5890	3.3690	7.2430	15.298	32.092
Naive1	7.0620	14.192	28.436	57.082	115.16	233.16

Fig. 3. Running times in seconds of Quicksort and Naive1 MSBRadix sort (single pass MSBRadix sort without partitioning, $r = \lceil \lg n - 3 \rceil$)floating point keys. $M = 10^6$

$n =$	1M	2M	4M	8M	16M	32M	64M
Flash2P	0.8320	1.6860	3.5190	7.2600	14.862	31.839	63.210
Quicksort	0.7400	1.5890	3.3690	7.2430	15.298	32.092	67.861
MSBRadix	0.3865	0.8470	1.9820	5.0300	9.4800	19.436	40.663

Fig. 4. Running times in seconds of Flash2P, Quicksort and MSBRadix sort on a Sun Ultra II using single precision floating point keys. $M = 10^6$

References

1. A. V. Aho, J. E. Hopcroft, and J. D. Ullman. *The Design and Analysis of Computer Algorithms.* Addison-Wesley, Reading, Massachusetts, 1974. 380
2. D. Dubhashi, and D. Ranjan. Balls and Bins: A Study in Negative Dependence. *Random Structures and Algorithms* **13(2)** 1998, pp. 99–124. 386
3. M. Frigo, C. E. Leiserson, H. Prokop and S. Ramachandran. Cache-oblivious algorithms. In *Proc. 40th FOCS*, pp. 285–298, 1999.
4. J. Handy. *The Cache Memory Handbook.* Academic Press, 1998. 382
5. J. L. Hennessy and D. A. Patterson. *Computer Architecture: A Quantitative Approach, 2e,* Morgan Kaufmann, 1996. 380, 381, 381, 382, 389, 390
6. K. Joag-Dev and F. Proschan. Negative association of random variables, with applications. *Ann Statist.* **11** (1983), pp. 286–295. 386
7. D. E. Knuth. *The Art of Computer Programming. Volume 3: Sorting and Searching, 3rd ed.* Addison-Wesley, 1997. 380, 387
8. R. E. Ladner, J. D. Fix and A. LaMarca, The cache performance of traversals and random accesses In *Proc. 10th ACM-SIAM SODA*, pp. 613–622, 1999. 381, 381, 381, 390
9. A. LaMarca and R. E. Ladner, The influence of caches on the performance of sorting. In *Journal of Algorithms* **31** (1999), pp. 66–104. 380, 381, 381, 381, 390
10. K. D. Neubert. The Flashsort1 algorithm. *Dr Dobb's Journal*, pp. 123–125, February 1998. FORTRAN code listing, p. 131, *ibid.* 387
11. N. Rahman and R. Raman. Analysing Cache Effects in Distribution Sorting. In *Proc. 3rd WAE*, LNCS 1668, pp. 184–198, 1999. 380, 380, 381, 381, 382, 388, 390
12. N. Rahman and R. Raman. Adapting radix sort to the memory hierarchy. In *Proc. 2nd ALENEX*, 2000, *www.cs.unm.edu/Conferences/ALENEX00/proceedings.html.* 381
13. P. Sanders. Accessing multiple sequences through set-associative cache. In *Proc. 26th ICALP*, LNCS 1644, pp. 655–664, 1999. 381, 381, 381, 384, 387
14. S. Sen and S. Chatterjee. Towards a theory of cache-efficient algorithms (extended abstract). In *Proc. 11th SODA*, pp. 829–838, 2000. 381, 381, 381, 384, 387, 387
15. J. S. Vitter. External Memory Algorithms and Data Structures: Dealing with Massive Data. To appear in *ACM Computing Surveys.* 381

How Helpers Hasten h-Relations

Peter Sanders[1*] and Roberto Solis-Oba[2**]

[1] Max Planck Insitut für Informatik
Saarbrücken, Germany
sanders@mpi-sb.mpg.de
[2] Department of Computer Science
The University of Western Ontario
London, ON, Canada
solis@csd.uwo.ca

Abstract. We study the problem of exchanging a set of messages among a group of processors, using the model of simplex communication. Messages may consist of different numbers of packets. Let h denote the maximum number of packets that a processor must send and receive. If all the packets need to be delivered directly, at least $\frac{3}{2}h$ communication steps are needed to solve the problem in the worst case. We show that by allowing forwarding, only $\frac{6}{5}h + \mathcal{O}(1)$ time steps are needed to exchange all the messages, and this is optimal. Our work was motivated by the importance of irregular message exchanges in distributed-memory parallel computers, but it can also be viewed as an answer to an open problem on scheduling file transfers posed by Coffmann, Garey, Johnsson, and LaPaugh in 1985.

1 Introduction

Consider a group of P processing elements (PEs) numbered 0 through $P-1$, connected by a complete network. For every pair (i, j) of processing elements there is a message consisting of $m_{ij} \geq 0$ packets that i must sent to j. Let h be the maximum number of packets that a PE must exchange. The *h-relation problem* is to send all messages in the smallest amount of time. Our unit of communication time is the time needed to transmit one packet. We assume synchronized simplex communication, i.e., a PE can exchange information with only one other PE at any given moment. Moreover, we assume that packets can be further subdivided.

This problem has been studied in many variations and under many different names: h-relations [6], file transfer [2], edge coloring [18], and biprocessor task scheduling on dedicated processors [12]. Our original motivation was the study of the function MPI_Alltoallv in the Message Passing Interface (MPI) [19] and its equivalents in other message passing models for parallel computing.

---

* Partially supported by the IST Programme of the EU under contract number IST-1999-14186 (ALCOM-FT).
** Partially supported by Natural Sciences and Engineering Research Council of Canada grant R3050A01.

M. Paterson (Ed.): ESA 2000, LNCS 1879, pp. 392–402, 2000.

The problem can be modeled using an undirected multi-graph $G = (V, E)$ called the *transfer graph*. In this graph $V = \{0, \ldots, P - 1\}$, and for every pair of vertices i, j, there is an edge (i, j) of multiplicity $m_{ij} + m_{ji}$. The maximum degree $h(G)$ of G is a natural measure for the size of the problem and it yields a trivial lower bound for the time needed to exchange all the messages. When there is no confusion, we use h to denote the maximum degree of the transfer graph.

A simple reduction to the chromatic index problem [2] shows that the h-relation problem is NP-hard in the strong sense even when all messages have length 1. In the chromatic index problem, given a graph G and an integer k it is desired to know whether the edges of G can be colored using at most k colors, so that no two edges with the same color share a common endpoint. We note that a coloring of the edges of a transfer graph yields an upper bound on the value of the solution for the h-relation problem, since the color of an edge (i, j) can be interpreted as the time step in which the packet corresponding to this edge should be transmitted. The chromatic index $\chi'(G)$ of G is the minimum number of colors needed to edge-color G as described above. It is known that if G does not have multiple edges then $h(G) \leq \chi'(G) \leq h(G) + 1$ [21]. Gabow et al. [4] give an $\mathcal{O}(m\sqrt{P \log P})$-time algorithm to edge-color a simple graph with at most $h(G) + 1$ colors, where m is the number of edges.

If a message cannot be divided into packets, then the h-relation problem is NP-hard even when the underlying transfer graph is bipartite, or when it is a tree [2]. But the problem can be solved in linear time if the transfer graph is a path.

For the case of multi-graphs, it is known [9] that the chromatic index problem cannot be approximated with a ratio smaller than $\frac{4}{3}$ unless P=NP. There are instances of the problem for which $\lfloor 3h/2 \rfloor$ colors are needed.[1] Nishizeki and Sato [17] present a $\frac{4}{3}$-approximation algorithm for the problem, and Nishizeki and Kashiwagi [18] give an algorithm to edge-color a graph G using at most $1.1\chi'(G) + 0.8$ colors. It is conjectured that there is a polynomial time algorithm that can find a coloring for any graph G with $\chi'(G) + 1$ colors [8,16,18].

In the *regular* h-relation problem every message m_{ij} has length $h/(P-1)$. We show that this problem can be solved using a *1-factorization* [7] of the transfer graph. There are several algorithms that transform an arbitrary instance of the h-relation problem into two "almost" regular h'-relation problems, with $h \approx h'$ [15,20]. In the resulting schedule, almost every packet is forwarded so that the communication time is about $2h$. If $h_{\max}$ is the maximum number of packets that some PE sends *or* receives, and if we assume the duplex model of communication where each PE can send and receive one packet simultaneously, then the problem can be solved optimally using communication time $h_{\max}$ via bipartite edge coloring [14].

All the above results make the assumption that the packets of every message m_{ij} are directly delivered from PE i to PE j. Coffman et al. suggest that

---

[1] For example, consider any transfer graph containing the vertices $\{a, b, c\}$ and $h/2$ copies of the edges (a, b), (b, c), and (c, a).

forwarding messages over different PEs might help speed-up the transmission of the messages, since this gives additional scheduling flexibility. Whitehead [22] shows that when forwarding is *needed* because some of the edges in the transfer graph are not present in the interconnection network, then the h-relation problem is NP-complete even if the transfer graph is a path with edges of arbitrary multiplicity. Goldberg et al. [5] give a randomized algorithm for the h-relation problem with forwarding that with high probability finds a solution of length $\Theta(h + \log \log P)$. This algorithm assumes that each PE can simultaneously send and receive a message, and a PE does not have global information about the h-relation but knows only the messages that is wants to send.

In this paper we study the h-relation problem assuming a complete interconnection network. It might be a little surprising that forwarding can be helpful in this case since it increases the communication volume. We show that the net tradeoff between this disadvantage and the additional flexibility that forwarding provides, is positive.

In Sect. 2 we describe an algorithm that reduces the h-relation problem to the problem of scheduling $\lceil h/2 \rceil$ 2-*relations*. A 2-relation is a set of packets that induce a collection of cycles and paths in the transfer graph. It is easy to schedule a 2-relation in 3 time steps without forwarding, thus this approach yields a solution for the problem of length $3 \lceil h/2 \rceil$. This is optimal for even h if no forwarding is allowed.

In Sect. 3 we explain how to use forwarding to find a solution of length 12/5 for the 2-relation problem when P is even. This yields an algorithm for the h-relation problem with even P, that finds a solution of length $\frac{6}{5}(h + 1)$. The case of odd P is more difficult. By removing some packets from the 2-relations in a "balanced" way, so that many of the removed packets can be concurrently transmitted later, it is possible to find a solution of length $\left(\frac{6}{5} + \mathcal{O}(1/P)\right)(h + 1)$. We also show that there are h-relations where the above bounds cannot be improved much by any algorithm regardless of their forwarding strategy. In Sect. 4 we explain how to modify our algorithms so that they run in strongly polynomial time. In Sect. 5 we outline further improvements and discuss some practical considerations.

2   An Algorithm Based on Bipartite Edge Coloring

We now explain how to translate an h-relation into $\lceil h/2 \rceil$ 2-relations. Besides laying the ground for our main result described in the next section, this also yields a good algorithm without forwarding. Since a 2-relation consists of a collection of disjoint cycles and paths, it is easy to solve it using communication time 3. Therefore, the original h-relation can be solved using communication time at most $3 \lceil h/2 \rceil$. This is an improvements over algorithms which transform a general h-relation into two regular h-relations [15,20], and hence need communication time at least $2h$. Furthermore, since at least h rounds are needed to send the messages, this algorithm achieves a performance ratio no worse than $\frac{3 \lceil h/2 \rceil}{h} \leq \frac{3}{2}(1 + \frac{1}{h})$.

The algorithm for translating an h-relation into a 2-relation first converts an instance of the problem into an edge coloring problem on a bipartite graph with maximum degree $\lceil h/2 \rceil$. Consider the transfer graph $G = (V, E)$. Since the sum of the degrees of the vertices in any graph is even, any graph has an even number of vertices with odd degree. Let us add an edge between every pair of vertices of odd degree, so that every vertex in the graph has even degree. This new graph is Eulerian, and so we can find an Euler cycle for each of its connected components in linear time. This collection of Euler cycles includes all the edges in the graph. By traversing the Euler cycles we can assign orientations to the edges of the graph so that for every vertex its in-degree and out-degree have the same value. Let $G' = (V, E')$ denote the resulting directed graph. In this graph the maximum in-degree and out-degree are $\lceil h/2 \rceil$.

Now build an undirected bipartite graph $\bar{G} = (L, R, \bar{E})$ by making two copies $L = R = V$ of the vertices of G', and adding an edge from vertex $u \in L$ to $v \in R$ whenever $(u, v) \in E'$. In this bipartite graph the two copies of any vertex v have the same degree, and hence the maximum degree is $\lceil h/2 \rceil$. Next, compute an optimum edge coloring for $\bar{G}$. This can be done in time $\mathcal{O}(m \log h) = \mathcal{O}(hP \log h)$ [3]. The coloring of the edges induces a decomposition of $\bar{E}$ into $\lceil h/2 \rceil$ disjoint matchings in which every matching consists of edges of the same color. The edges in each matching induce a 2-relation in G. All these 2-relations together cover all edges in G.

3   Exploiting Forwarding

We now show how the method described in the previous section can be refined to use forwarding to reduce the time needed for exchanging the messages.

Theorem 1. *The h-relation problem can be solved using communication time $\frac{6}{5}(h + 1)$ if P is even, and using time $(\frac{6}{5} + \frac{2}{P})(h + 1)$ if P is odd. This bound is almost tight since for any value h there are problem instances for which at least time $\frac{6}{5}h$ is needed for exchanging the messages when P is even, and at least time $\frac{6}{5}(1 + \frac{3}{5P})h$ is needed when P is odd.*

The remainder of this section is dedicated to proving this result. The first part of the algorithm used for the upper bound is the same as that described in the previous section: we build the bipartite graph $\bar{G}$, color it, and derive $\lceil h/2 \rceil$ 2-relations from the colors. But now we use a more sophisticated algorithm for scheduling the 2-relations. Each 2-relation defines a collection of disjoint paths and cycles in the transfer graph G. To avoid some tedious case distinctions, we add dummy edges closing all paths to cycles. It is easy to schedule the packets in an even length cycle so that they can be transmitted in communication time 2. But it is not so easy to find a good schedule for the messages in an odd length cycle. We use forwarding of some data to solve this problem. Note that from now on we have to reintroduce the direction of a packet.

For example, consider the two 3-cycles shown in Fig. 1, where the packets have been split in five pieces of size 1/5 each. Clearly, no algorithm without

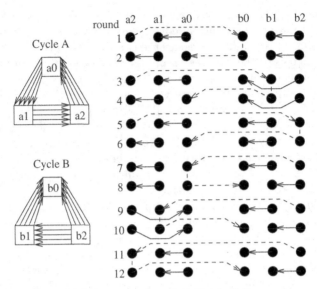

Fig. 1. Example for scheduling two 3-cycles. Dashed lines represent forwarded messages.

forwarding can exchange all the data in less than 3 time steps since at least two PEs will be idle at any point in time. The idea is to exploit the idle times for forwarding data. The schedule in Fig. 1 shows how to transmit all these pieces in 12 rounds using total communication time 12/5. During the first 6 rounds, PE $a2$ sends three pieces of packet $(a2, a0)$ to PE $a0$ with the help of the PEs in cycle B. In the last 6 rounds, PE $b2$ sends 3 pieces of packet $(b2, b0)$ to PE $b0$ with help from the PEs in cycle A.

In the next sections we show how to generalize this idea to reduce the time needed to exchange the messages for any number P of PEs.

3.1   Even Number of PEs

When the number P of PEs is even, there is an even number of odd length cycles (ignoring the directions of the edges). We pair the odd cycles and use the PEs in one cycle to help forward the messages of the other cycle, just like we did in Fig. 1. We now explain how to schedule the packets in a pair of odd length cycles A and B. As in the example of Fig. 1, the packets are split into 5 pieces, and these pieces are exchanged in 12 rounds of length $1/5$ each as described below.

Let us name the PEs of cycle A as $a_0, \ldots, a_{|A|-1}$ in such a way that PE $a_{|A|-1}$ sends a packet to PE a_0 (and not vice versa). Similarly, let $b_0, \ldots, b_{|B|-1}$ denote the PEs in cycle B, and let PE $b_{|B|-1}$ send a packet to b_0. Fig. 2 summarizes the schedule for exchanging the packets of cycles A and B.

In rounds $2i$ and $2i+1$, for $i \in \{0, 1, 2\}$, PE b_i forwards one piece of the packet from PE $a_{|A|-1}$ to PE a_0. Concurrently, three pieces of every other packet in

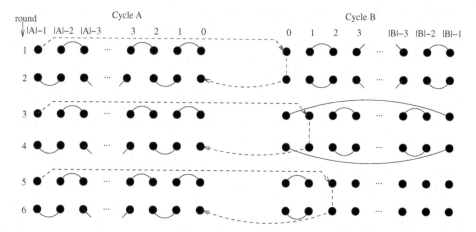

Fig. 2. How odd cycle B helps odd cycle A by forwarding three pieces of a packet. Undirected edges must be given the direction of the corresponding edge in the transfer graph.

cycle A are transmitted directly: in round $2i$ one piece of the packet between a_{2k} and a_{2k+1} is transmitted, and in round $2i+1$, one piece of the packet between a_{2k+1} and a_{2k+2} is transmitted, for every $0 \le k < |A|/2$. Thus, within six rounds, three pieces of every packet in cycle A are transmitted.

As for cycle B, in rounds one and two, two pieces of the packet between PEs b_{2j+1} and b_{2j+2} are transmitted for every $0 \le j < |B|/2$. In rounds three and four, two pieces of the packet between PEs b_{2j} and $b_{2j+1 \bmod |B|}$ are transmitted for every $0 < j < |B|/2$. In rounds five and six, only two pieces of the packet between PEs b_0 and b_1 are transmitted. Thus, within six rounds two pieces of every packet in cycle B are transmitted.

In rounds seven through twelve, cycles A and B switch their roles, so that after round twelve all five pieces of the packets are transmitted. The total time needed to exchange all the messages in the transfer graph G is $\frac{12}{5}\lceil h/2 \rceil \le \frac{6}{5}(h+1)$. This establishes the upper bound of Theorem 1 for even P.

3.2 Odd Number of PEs

If the number of PEs is odd, there will be an odd number of odd cycles in each 2-relation, so it is not possible to pair cycles as before. It is not difficult to see that there are 2-relations with an odd number of PEs that cannot be scheduled in twelve rounds of length $1/5$ each.

We solve this problem by selecting cycles as follows. First we assume that in each 2-relation all PEs are busy. If this is not the case then an idle PE can help to send the packets in an odd cycle in $12/5$ units of time. We choose one of the cycles A and remove one packet from it. Which packet of which cycle is removed is described in the next paragraph. If A is an odd length cycle, the removal of a packet transforms it into a path whose packets can be scheduled in communication time 2 without help from another cycle. Moreover, all the remaining odd length cycles can be paired and their packets exchanged as described in the previous section. If the chosen cycle A has even length, then we can pair it with an odd cycle B. A simple modification of the algorithm described in the previous section can be used to transmit all the remaining packets of A and B in $12/5$ units of time.

What remains to be done is to schedule the packets that have been removed. We maintain the invariant that all the removed packets form a matching M in the transfer graph G. Whenever we select a cycle, we try to choose it so that it has a packet e that maintains this invariant. If this is possible, we just add e to M. Otherwise, if no packet can be added to the matching, then we transmit all the packets in M in one unit of time, emptying the matching M. The process is repeated until all messages are transmitted.

Using this algorithm, we can prove that an additional step for emptying the matching M is only required rarely:

Lemma 1. *Whenever M needs to be emptied, it contains at least $\lceil P/4 \rceil$ edges.*

Proof. In every iteration of the algorithm there are P candidate edges E' to be removed from the cycles. Every edge $e \in M$ can have a common endpoint with at most 4 edges from E'. Hence, if $|M| < P/4$ there must be at least one candidate edge in E' that can be added to M. □

To summarize, the solution produced by the above algorithm needs time $\frac{12}{5} \lceil h/2 \rceil$ for transmitting the packets in the cycles, plus $\lceil h/2 \rceil / \lceil P/4 \rceil$ units of time for for emptying the matchings M. The total length of the solution is then, $(\frac{6}{5} + \frac{2}{P})(h + 1)$.

3.3 Lower Bound

In this section we concentrate on the case of odd P. The case of even P is similar. Consider the following instance of the problem with $P = 3k$, for some $k > 0$. For every $0 \le i < k$, there are messages $m_{3i,3i+1}$, $m_{3i+1,3i+2}$, and $m_{3i+2,3i}$ of length $h/2$, for some h even. All other messages are empty. Consider any algorithm $\mathcal{A}$ for exchanging these messages. Let $D(t) : \mathbb{R}_+ \to \mathbb{N}_+$ denote the number of packets being directly routed by $\mathcal{A}$ to their destinations at time t. We use continuous time here to allow for asynchronous algorithms for which the notion of communication steps make no sense. Note that $D(t) \le P/3$.

There are at most $P - 2D(t)$ other PEs available for forwarding packets. Since $P - 2D(t)$ is odd, these PEs can handle at most $\frac{P-2D(t)-1}{2}$ packets at any time. Since forwarded packets have to be sent at least twice, we define *the progress* made by $\mathcal{A}$ at time t towards delivering the packets to their final destinations to be $D(t) + \frac{P-2D_i-1}{4}$. The integral[2] of the progress over the total communication

---

[2] Note that we get no problems regarding the existence of the integral since the integrand is constant (and hence continuous) almost everywhere.

time T of the solution produced by $\mathcal{A}$ must be equal to the the total volume $hP/2$ of the data. Hence,

$$h\frac{P}{2} = \int_{t=0}^{T}\left(D(t) + \frac{P - 2D(t) - 1}{4}\right)dt = \frac{1}{2}\int_{t=0}^{T}D(t)dt + T\left(\frac{P}{4} - \frac{1}{4}\right)$$
$$\leq \frac{1}{2}\int_{t=0}^{T}\frac{P}{3}dt + T\left(\frac{P}{4} - \frac{1}{4}\right) = T\left(\frac{5P}{12} - \frac{1}{4}\right) \ .$$

Solving this for T yields $T \geq h/(\frac{5}{6} - \frac{1}{2P}) \geq \frac{6}{5}(1 + \frac{3}{5P-3})$. ∎

4 A Strongly Polynomial Time Algorithm

An instance of the h-relation problem can be represented by a weighted transfer digraph $G = (V, E)$ in which the weight w_{ij} of an edge (i,j) is equal to the length of the message m_{ij}. Let m be the number of edges and δ be the maximum degree of G. Define the *load* of a processing element i as the total length of the messages that must be exchanged by i. Let h be the largest PE load. Using this representation, the algorithms described in Sect.s 2 and 3 are not strongly polynomial since the number of edges in the bipartite multi-graph $\bar{G}$ depends on the lengths of the messages that must be exchanged. However it is possible to modify the algorithms so that they run in strongly polynomial time.

We construct $\bar{G}$ by adding edges (i,j) and (j,i) with weight $\bar{w}_{ij} = \lfloor w_{ij}/2 \rfloor$. Next, we build an unweighted multi-graph graph G' containing only the edges of G with odd weight. With G' we proceed as before: add dummy edges to make all degrees even, and find a collection of Euler cycles in $\mathcal{O}(m)$ time to orient the edges in G'. Finally, for every edge $(i,j) \in G'$ we increment the corresponding weight $\bar{w}_{ij}$ in the bipartite graph $\bar{G}$.

Now we find a matching $\bar{M}$ of $\bar{G}$ that covers all nodes corresponding to PEs with maximum load. Such a matching must exist because if we replace every edge (i,j) of $\bar{G}$ by $\bar{w}_{ij}$ copies of weight one each, we obtain a bipartite graph that can be colored with $\lceil h/2 \rceil$ colors. Each color in this coloring induces a matching covering all nodes corresponding to PEs with maximum load. The matching $\bar{M}$ can be found by computing a maximum cardinality matching on the subgraph of $\bar{G}$ formed by those edges incident to nodes from PEs with maximum load. This can be done in $\mathcal{O}(m \log h + \frac{m}{h} \log \frac{m}{h} \log^2 \delta)$ time using the algorithm in [10].

Let $w_{\min}$ be the smallest weight in $\bar{M}$. Let h' be the second largest load among the processing elements. We exchange the messages in $\bar{M}$ as described before, using $\min\{h - h', w_{\min}\}$ packets at once. After doing this we modify the bipartite-graph $\bar{G}$ by decreasing the weight of every edge in $\bar{M}$ by $\min\{h - h', w_{\min}\}$ and discarding all edges of weight zero. Let h be the new largest PE load. The process is repeated until all messages have been sent. Note that each iteration of this process either removes one edge from $\bar{G}$, or it increases by at least one the number of PEs with largest load. Therefore the process requires $\mathcal{O}(m + P)$ iterations.

5   Refinements

Many improvements of our basic forwarding algorithm suggest themselves. In Sect.s 5.1 and 5.2 we outline some of them and in Sect. 5.3 we briefly discuss the usefulness of our results in the context of parallel processing.

5.1   Reducing the Gap between Upper and Lower Bound

The P-dependent term $2(h+1)/P$ in the length of the solution of our algorithm for odd P can be reduced using a more sophisticated algorithm. We can show that instead of removing a entire packet to break a cycle, it suffices to remove four pieces from one packet to achieve the same effect. The most problematic case is if the pieces stem from a 2-cycle. In all other cases it suffices to remove three pieces. Using a more complicated algorithm it can be shown that it is always sufficient to remove at most three pieces. Further special case treatments show that whenever one has to remove pieces from the cycles at all, there are almost only 3-cycles. Using this additional information, one can show that there are at least $P/3$ edges in the matching of removed pieces when $\bar{M}$ needs to be emptied. These measures already move us more than halfway to the lower bound $18h/(25P)$ for the P-dependent part of the communication time. A further reduction might be possible by constructing a new smaller communication problem from the removed edges rather than maintaining matchings. Only at the end, this small communication problem is handled by calling our algorithm recursively.

5.2   Faster Exchange of Easy h-Relations

Some h-relations can be routed in less than $6h/5$ units of time. If the chromatic index of the transfer graph G is smaller than $12h/11$ such a schedule can be found using the edge coloring algorithm of Nishizeki and Kashiwagi [18] without using forwarding. It is an open problem to improve this by combining such general edge coloring algorithms with the idea of forwarding. However, several heuristics suggest themselves. One observation is that two units of time are sufficient for exchanging all packets in those 2-relations that induce only paths and even length cycles. Therefore, one might try to decompose $\bar{G}$ in such a way that many of the resulting 2-relations have this property. If there is only a small number of odd cycles it might even make sense to break them so that again time two suffices to route the rest. Odd cycles of size at least eleven can be scheduled in time $11/5$.

An obvious but useful observation is that instances with $\ell := \min_{i \neq j}\{m_{ij} + m_{ji}\} > 0$ can be decomposed into two instances, one of which consists solely of messages of length ℓ. The packets in this regular instance can be scheduled using total communication time $(P-1)\ell$ (or $P\ell$ for odd P). To our surprise, there seemed to be no optimal algorithm for this purpose in the parallel processing literature, except for the case that P is a power of 2 [13]. Here is a simple optimal algorithm for this problem. For odd P, there are P rounds $0, \ldots, P-1$. In round i, PE j exchanges its data with PE $(i-j) \bmod P$. In each round, the PE with $i-j = j \bmod P$ is (unavoidably) idle. Each PE is idle exactly in one round. For even P, PEs $0, \ldots, P-2$ execute the above algorithm except that the otherwise idle PE exchanges data with PE $P-1$. We do not go into more detail here since this algorithm turns out to be equivalent to a long known graph-theoretical result, namely the factorization of a clique into 1-factors [11,7].

5.3 Is this Practical?

For moderate P and long messages, our algorithm might indeed be useful. For example, consider the main communication phase of sample sort [1] for sorting $n \gg P^2 B$ elements where B is the number of elements fitting into a packet. Sample sort needs an h-relation with $h \approx 2n/(PB)$. This can be a difficult problem since although randomization makes sure that all PEs have to communicate about the same amount of data, the individual message lengths can vary arbitrarily for worst case inputs. In this case, our algorithm can be a large factor faster than a naive direct exchange and still yields a factor of up to $2/\frac{6}{5} = \frac{5}{3}$ speedup over two-phase algorithms [20,15]. The number of PEs should be moderate for several reasons. Firstly, we assume that the network itself is not a bottleneck which is usually only the case for machines up to around 64 PEs.[3] Secondly, the scheduling overhead grows with P. However, in some cases it might be possible to amortize the scheduling overhead over multiple calls for h-relations with identical structure. Many iterative numerical algorithms working on irregular data are of this type.

Message lengths are even more critical. The main requirement is that one fifth of a packet is still so large that the startup overhead for communication is small compared to the transmission overhead itself. For so large packets the simplex-model of communication is also quite realistic.

References

1. G. E. Blelloch, C. E. Leiserson, B. M. Maggs, C. G. Plaxton, S. J. Smith, and M. Zagha. A comparison of sorting algorithms for the connection machine CM-2. In *ACM Symposium on Parallel Architectures and Algorithms*, pages 3–16, 1991. 401
2. E. G. Coffman, M. R. Garey, D. S. Johnson, and A. S. LaPaugh. Scheduling file transfers. *SIAM Journal on Computing*, 14(3):744–780, 1985. 392, 393
3. R. Cole, K. Ost, and S. Schirra. Edge-coloring bipartite multigraphs in $O(E \log D)$ time. *submitted for publication*, 2000. 395
4. H. N. Gabow and O. Kariv. Algorithms for edge coloring bipartite graphs and multigraphs. *SIAM Journal on Computing*, 11(1):117–129, 1982. 393
5. L. A. Goldberg, M. Jerrum, T. Leighton, and S. Rao. Doubly logarithmic communication algorithms for optical-communication parallel computers. *SIAM Journal on Computing*, 26(4):1100–1119, August 1997. 394

---

[3] But note that some machines consist of multiprocessor nodes connected by a crossbar. On such machines our algorithm might be useful for scheduling data exchanges between nodes.

6. M. D. Grammatikakis, D. F. Hsu, M. Kraetzl, and J. Sibeyn. Packet routing in fixed-connection networks: A survey. *Journal of Parallel and Distributed Processing*, 54:77–132, 1998. 392
7. F. Harary. *Graph Theory*. Addison Wesley, 1969. 393, 401
8. D. S. Hochbaum, T. Nishizeki, and D. B. Shmoys. A better than "best possible" algorithm to edge color multigraphs. *Journal of Algorithms*, 7:79–104, 1986. 393
9. Ian Holyer. The NP-completeness of edge-coloring. *SIAM Journal on Computing*, 10(4):718–720, 1981. 393
10. A. Kapoor and R. Rizzi. Edge-coloring bipartite graphs. *Journal of Algorithms*, 34(2):390–396, 2000. 399
11. D. König. *Theorie der endlichen und unendlichen Graphen*. Akademische Verlagsgesellschaft, 1936. 401
12. M. Kubale. Preemptive versus nonpreemptive scheduling of biprocessor tasks on dedicated processors. *European Journal of Operational Research*, 94:242–251, 1996. 392
13. V. Kumar, A. Grama, A. Gupta, and G. Karypis. *Introduction to Parallel Computing. Design and Analysis of Algorithms*. Benjamin/Cummings, 1994. 400
14. G. Lev, N. Pippenger, and L. Valiant. A fast parallel algorithm for routing in permutation networks. *IEEE Trans. on Comp.*, C-30, 2:93–100, 1981. 393
15. W. Liu, C. Wang, and K. Prasanna. Portable and scalable algorithms for irregular all-to-all communication. In *16th ICDCS*, pages 428–435. IEEE, 1996. 393, 394, 401
16. S. Nakano, X. Zhou, and T. Nishizeki. Edge-coloring algorithms. In *Computer Science Today*, number 1000 in LNCS, pages 172–183. Springer, 1996. 393
17. T. Nishizeki and M. Sato. An algorithm for edge-coloring multigraphs. *Trans. Inst. Electronics and Communication Eng.*, J67-D(4):466–471, 1984. (in Japanese). 393
18. Takao Nishizeki and Kenichi Kashiwagi. On the 1.1 edge-coloring of multigraphs. *SIAM Journal on Discrete Mathematics*, 3(3):391–410, August 1990. 392, 393, 400
19. M. Snir, S. W. Otto, S. Huss-Lederman, D. W. Walker, and J. Dongarra. *MPI – the Complete Reference*. MIT Press, 1996. 392
20. L. G. Valiant and G. J. Brebner. Universal schemes for parallel communication. In *Conference Proceedings of the Thirteenth Annual ACM Symposium on Theory of Computation*, pages 263–277, Milwaukee, Wisconsin, 11–13 May 1981. 393, 394, 401
21. V. G. Vizing. On an estimate of the chromatic class of a p-graph (in russian). *Diskret. Analiz*, 3:23–30, 1964. 393
22. J. Whitehead. The complexity of file transfer scheduling with forwarding. *SIAM Journal on Computing*, 19(2):222–245, April 1990. 394

Computing Optimal Linear Layouts of Trees
in Linear Time

Konstantin Skodinis

University of Passau,
94030 Passau, Germany,
skodinis@fmi.uni-passau.de

Abstract. We present a linear time algorithm which, given a tree, computes a linear layout optimal with respect to vertex separation. As a consequence optimal edge search strategies, optimal node search strategies, and optimal interval augmentations can be computed also in $O(n)$ for trees. This improves the running time of former algorithms from $O(n \log n)$ to $O(n)$ and answers two related open questions raised in [7] and [15]. [1]

1 Introduction

The *vertex separation* of graphs has been introduced in [13] in the form of a *vertex separator game*. It is an important concept with theoretical and practical value. Its closer relation to the black-white pebble game of [6] constitutes the theoretical value in complexity theory. Applications in certain layout problems as Weinberger arrays, gate matrix layout, and PLA-folding constitute the practical value in VLSI, see [16]. The *vertex separation* of a graph can be defined using *linear layouts*, see [7]. A linear layout, or simply a layout, L of a graph G is a permutation of the vertices of G. Intuitively L describes how the vertices are to be laid out along a horizontal line. The i-th *vertex cut* of L defines how many of the first i vertices in L are adjacent to a vertex lying right to the i-th vertex in L. The vertex separation of L is the maximum over all vertex cuts of L. The vertex separation of a graph G, $vs(G)$, is the minimum vertex separation of its layouts. L is *optimal* if its vertex separation is $vs(G)$ and therefore as small as possible. Recently the vertex separation of a graph has received more attention due to its closer relation to several well known graph parameters as *edge search number*, *node search number*, *interval thickness*, and *path-width*.

Edge searching has been introduced in [18]. We regard the edges of a graph G as pipes contaminated by a gas. A team of searchers (clearers) has to clear all edges of G. To do so the team has to find an *edge search strategy* which is a sequence of the following activities: (i) place a searcher on a vertex as a guard, (ii) remove a searcher from a vertex, and (iii) slide a searcher along an edge. A contaminated edge becomes cleared if one of its endpoints is guarded and a searcher slides along the edge from the guarded endpoint to the other endpoint of the edge. In [12] it has been shown that recontamination can be disallowed without increasing the search number of any graph. The *edge search number* of G, $es(G)$, is the minimum number of the searchers needed

---

[1] It should be clear that there are some published papers in which similar results are claimed. Later it turned out that the suggested algorithms do not run in $O(n)$ as stated but in $O(n \log n)$.

M. Paterson (Ed.): ESA 2000, LNCS 1879, pp. 403–414, 2000.

to clear G. An edge search strategy of G is *optimal* if it needs the smallest possible number of searchers (this number is $es(G)$).

Node searching is another version of graph searching and has been introduced in [11]. In this slightly different version the third activity of the edge searching disappears. An edge is cleared once both endpoints are simultaneously guarded by searchers. The *node search number* of G is denoted by $ns(G)$.

Interval thickness has been introduced in [10]. Given a set of intervals I of the real line the *interval graph* of I is obtained by representing every interval by a vertex and every intersection of two intervals by an edge between their corresponding vertices. The interval thickness of a graph G, $\theta(G)$, is the smallest maximum clique over all interval supergraphs of G. An interval set I is an *optimal interval augmentation* of G if the interval graph H of I is a supergraph of G and the maximum clique of H has the smallest possible size $\theta(G)$.

Path-width has been introduced in [19,20]. The *path-decomposition* of a graph G is a sequence $D = X_1, X_2, \ldots, X_r$ of vertex subsets of G, such that (i) every edge of G has both ends in some set X_i and (ii) if a vertex of G occurs in some sets X_i and X_j, $i < j$, then the same vertex occurs in all sets X_k with $i < k < j$. The *width* of D is the maximum number of vertices in any X_i minus 1. The *path-width* of G, $pw(G)$, is the minimum width over all path-decompositions of G. A path-decomposition of G is *optimal* if its width is $pw(G)$ which is the smallest possible.

Rather surprisingly all notions mentioned above are closely related to each other. We will explain this in the following discussion: In [7] it has been shown that for every graph G $vs(G) \leq es(G) \leq vs(G) + 2$. The authors have also shown that $es(G) = vs(G')$, where G' is the 2-expansion of G. G' is obtained from G by replacing every edge $\{u, v\}$ by the edges $\{u, x\}$, $\{x, y\}$, and $\{y, v\}$ where x and y are two new vertices. Furthermore an algorithm has been presented which (by using a suitable data structure) transforms a given optimal layout L' of G' into an optimal edge search strategy S of G in linear time. Since the size of L' and G' is $O(size(G))$ we imply that for every graph the computation of its optimal edge search strategy requires no more time than the computation of its optimal layout. In [11] it has been shown that $ns(G) = vs(G) + 1$. The authors have also presented an algorithm which (by using a suitable data structure) transforms a given optimal layout into an optimal node search strategy in linear time. In [10] it has been shown that $\theta(G) = ns(G)$ and that a given optimal node search strategy can be transformed into an optimal interval augmentation in linear time. In [9] it has been shown that $pw(G) = vs(G)$. Additionally an algorithm has been presented which (by using a suitable data structure) transforms a given optimal layout into an optimal path-decomposition in linear time in the output size. Here the transformation time is not necessarily linear in the input size since the size of an optimal path-decomposition may be $pw(G)$ times larger than the size of an optimal layout.

Summarizing we get: For every graph G the above parameters are almost the same. Moreover given an input graph G the computation of an optimal edge search strategy, an optimal node search strategy, and an optimal interval augmentation of G is possible in $O(t(n))$ time where $t(n)$ is the time required by the computation of an optimal layout of G. The computation of an optimal path-decomposition is possible in time $O(t(n) + d)$, where d is the size of the computed path-decomposition.

Unfortunately given a graph G and an integer k the problems whether or not $vs(G) \leq k$ and $es(G) \leq k$ are both NP-complete, see [13] and [15], respectively. By the discussion above we obtain that the problems whether or not $ns(G) \leq k$, $\theta(G) \leq k$, and $pw(G) \leq k$ are also NP-complete. They remain NP-complete even for chordal graphs [8], starlike graphs [8], and for planar graphs with maximum degree three [17,14]. For some special graphs the above problems are solvable in polynomial time. Examples are cographs [4], permutation graphs [3], and graphs of bounded tree-width [2]. Especially if k is fixed then the problems can be solved in $O(n)$ (but exponential in k) [1].

Here we consider trees. It has been shown in [7] that the vertex separation and in [15] that the edge search number can be computed in $O(n)$. Thus the node search number, the interval thickness and the path-width can also be computed in $O(n)$. However the computation of optimal layouts, optimal edge search strategies, optimal node search strategies, optimal interval augmentations, and optimal path-decompositions is more complicated. In [7] an algorithm is given for optimal layouts and in [15] for optimal edge strategies. These algorithms needing $O(n \log n)$ time were the fastest known so far. The authors have raised the question whether this time can be reduced to $O(n)$.

In this paper we establish an algorithm computing an optimal layout of an input tree in $O(n)$. As a consequence of this result and the discussion above the computation of an optimal edge search strategy, an optimal node search strategy, and an optimal interval augmentation of an input tree can be done in $O(n)$. Especially the computation of an optimal path-decomposition can be done in linear time in the output size.

2   Preliminaries

We consider undirected graphs $G = (V, E)$. For convenience loops and multiple edges are disallowed. A *(linear) layout* of G is a one to one mapping $L : V \rightarrow \{1, 2, \ldots, |V|\}$. For an integer $1 \leq i \leq |V|$ the i-th cut of L, denoted by $cut_L(i)$, is the number of vertices which are mapped to integers less than or equal to i and adjacent to a vertex mapped to an integer larger than i. Formally $cut_L(i) = |\{ u \mid \exists(u, v) \in E$ with $L(u) \leq i$ and $L(v) > i \}|$. The *vertex separation* of G with respect to L, denoted by $vs_L(G)$, is the maximum over all cuts of L, i.e., $vs_L(G) = \max_{1 \leq i \leq |V|} \{ cut_L(i) \}$. For convenience $vs_L(G)$ is also called the vertex separation of L. The *vertex separation* of G, denoted by $vs(G)$, is the minimum vertex separation over all layouts of G, i.e., $vs(G) = \min_L \{vs_L(G)\}$. A layout L of G is *optimal* if $vs(G) = vs_L(G)$.

In this paper we deal with rooted trees. Let T be a tree and u a vertex of T. The *branches* of T at u are the tree components obtained by the removal of u from T. $T[u]$ denotes the subtree of T rooted at u and containing all descendants of u in T. The subtrees rooted at the sons of u are called *the subtrees of* u. Let $v_1, \ldots, v_k$ be (arbitrary) vertices of a subtree $T[u]$. $T[u; v_1, \ldots, v_k]$ denotes the tree obtained from $T[u]$ by removing $T[v_1], \ldots, T[v_k]$. Notice that $T[u; v_1, \ldots, v_k]$ is rooted at u.

In [7] the following lemma has been shown.

Lemma 1. *Let T be a tree and $k \geq 1$ an integer. Then $vs(T) \leq k$ iff for all vertices u of T at most two branches of T at u have vertex separation k and all other branches have vertex separation $\leq k - 1$.*

Corollary 1. *Let T be a tree with n vertices. Then $vs(T) \leq \log_3 n + 1 \leq \log n + 1$.*

Next we recall the definition of critical vertices and vertex labelling, see [7].

Definition 1. *Let T be a tree with $vs(T) = k$. A vertex v is k-critical in T if there are two sons v_1 and v_2 of v with $vs(T[v_1]) = vs(T[v_2]) = k$.*

Observe that by Lemma 1 the vertex v in the above definition can not have more than two sons with vertex separation k. By the same lemma we obtain:

Lemma 2. *Let T be a tree with $vs(T) = k$. Then at most one vertex in T is k-critical.*

The following labelling technique has been introduced in [7]. Similar methods were used in [21,5,15].

Definition 2. *Let T be a tree rooted at u with $vs(T) = k_1$. A list of integers $l = (k_1, k_2, \ldots, k_p)$ with $k_1 > k_2 > \ldots > k_p \geq 0$ is a label of u if there is a set of vertices $\{v_1, \ldots, v_{p-1}, v_p = u\}$ in T such that*

(i) For $1 \leq i < p$, $vs(T[u; v_1, \ldots, v_i]) = k_{i+1}$,
(ii) For $1 \leq i < p$, v_i is a k_i-critical vertex in $T[u; v_1, \ldots, v_{i-1}]$,
(iii) either v_p $(= u)$ is the k_p-critical vertex in $T[u; v_1, \ldots, v_{p-1}]$ or there is no k_p-critical vertex in $T[u; v_1, \ldots, v_{p-1}]$.

We call $v_1, \ldots, v_{p-1}, v_p(= u)$ the vertices corresponding to l.

Notice that the vertex separation of T is the largest integer occurring in l. If v_i is k_i-critical (k_i-noncritical) in $T[u; v_1, \ldots, v_{i-1}]$ then we say that k_i is critical (noncritical) in l. By definition all v_i are critical in l except possibly v_p $(= u)$. By Lemma 2 it is easy to see that the label l and the vertices corresponding to l are unique.

Lemma 3. *Let T be a tree rooted at u. The label l of u and the vertices corresponding to l are unique.*

Let T be a tree rooted at u. In [7] a linear algorithm is given computing the label l of u. Since the vertex separation is the largest integer occurring in l this algorithm also computes the vertex separation of T in linear time. The idea of the algorithm is to compute l bottom up. Let $u_1, \ldots, u_d$ be the sons of u and l_i the label of u_i in $T[u_i]$ for $1 \leq i \leq d$. The label l is computed by recursively computing the labels l_i and then by combining them. By definition the label of the leaves is (0). It is interesting to note that by the computation of l some of the old labels $l_1, \ldots, l_d$ must be destroyed. However the computation of an optimal layout of T is more complicated. The best known algorithm needs $O(n \log n)$ time [7] where it was stated as an open question whether or not there exists an $O(n)$ layout algorithm. In the following we give an affirmative answer.

3   Optimal Layouts in Linear Time

We establish an $O(n)$ algorithm computing optimal layouts of trees. The idea is to compute the layout of a tree by first computing the layouts of the subtrees of its root and then by combining them. First we show how to compute an optimal layout of certain trees (we call them simple trees) using the optimal layouts of its subtrees. Then we extend this to arbitrary trees. For convenience we define a layout by a list of vertices and use the concatenation operator & of lists which makes our claims more readable.

3.1 Simple Trees and Their Optimal Layouts

Definition 3. *A tree is k-simple if the label of its root is $l = (k)$. Let T be a k-simple tree. T is called noncritical if k is noncritical in l. Otherwise T is called critical.*

Intuitively, T is noncritical k-simple if $vs(T) = k$ and T contains no k-critical vertex. T is critical k-simple if $vs(T) = k$ and the root of T is the (unique) k-critical vertex. Notice that every noncritical 0-simple tree is a singleton and that there are no critical 0-simple trees. The next two lemmas are more or less consequences of Definition 3.

Lemma 4. *Let T be a noncritical k-simple tree rooted at u. Then at most one of the subtrees of u in T is noncritical k-simple and all other subtrees have vertex separation $\leq k - 1$.*

Lemma 5. *Let T be a critical k-simple tree rooted at u. Then exactly two of the subtrees of u in T are noncritical k-simple and all other subtrees have vertex separation $\leq k-1$.*

We will construct an optimal layout of a tree by efficiently combining the optimal layouts of the subtrees. In order to do so the optimal layouts of the subtrees must have a suitable form. This leads to the introduction of left extendable, right extendable, and sparse layouts. For a tree T let T^w be the tree obtained from T by adding a new vertex w and an edge between w and the root of T.

Definition 4. *Let L be a layout of a tree T. If T is a singleton then L is both left and right extendable. Otherwise L is left extendable if $vs_{(w)\&L}(T^w) = vs_L(T)$ and right extendable if $vs_{L\&(w)}(T^w) = vs_L(T)$.*

Intuitively, L is left (right) extendable if the maximal vertex cut of L does not increase when we add an edge between root u and a new vertex w left (right) to L.

Definition 5. *Let L be a layout of a tree T rooted at u. L is sparse if no vertices v and w in T with $L(v) < L(u) < L(w)$ are adjacent in T.*

Intuitively, L is sparse if no vertex left to u is adjacent to a vertex right to u in L.

Lemma 6. *Every noncritical k-simple tree T has both an optimal left extendable layout lL and an optimal right extendable layout rL.*

Proof. Let u be the root of T and $u_1, \ldots, u_d$ the sons of u. Let L_i be an optimal layout of the subtree $T[u_i]$ for $1 \leq i \leq d$. The proof is by induction on the height h of T.

For $h = 0$ the tree T is a singleton and by Definition 4 the claim is clear.

For the induction step consider T of height $h + 1$. By Lemma 4 at most one subtree of u is noncritical k-simple and all other subtrees have vertex separation at most $k-1$. If no subtree of u is noncritical k-simple then the layout of T in Fig. 1 (a) is both optimal left and optimal right extendable. If a subtree of u, say $T[u_1]$, is noncritical k-simple then let lL_1 be an optimal left and rL_1 an optimal right extendable layout of $T[u_1]$. Both exist by the induction hypothesis. Then the layout lL and rL of T in Fig. 1 (b) and 1 (c) is optimal left extendable and optimal right extendable, respectively.

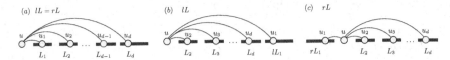

Fig. 1. Optimal extendable layouts of noncritical simple trees. The thick lines represent layouts.

Lemma 7. *Every critical k-simple tree T has an optimal sparse layout sL.*

Proof. Let u be the root of T and $u_1, \ldots, u_d$ the sons of u. By Lemma 5 exactly two subtrees of u, say $T[u_1]$ and $T[u_d]$, are noncritical k-simple and all other subtrees have vertex separation $\leq k - 1$. By Lemma 6, $T[u_1]$ has an optimal right extendable layout rL_1 and $T[u_d]$ has an optimal left extendable layout lL_d. Let L_i be an optimal layout of the subtree $T[u_i]$ for $2 \leq i \leq d - 1$. Then the layout sL in Fig. 3.1 is an optimal sparse layout of T.

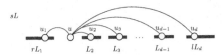

Fig. 2. Optimal sparse layouts of critical simple trees. The thick lines represent layouts.

Remark 1. The proofs of the last two lemmas provide construction schemes for optimal extendable layouts of noncritical simple trees and for optimal sparse layouts of critical simple trees. For the constructions only the optimal extendable layouts of the noncritical simple subtrees and arbitrary optimal layouts of the remaining subtrees are needed.

3.2   Arbitrary Trees and Their Optimal Layouts

The main idea is to partition a tree T into smaller simple trees and to define the coarse layout of T as the list of the optimal (extendable or sparse) layouts of the smaller trees.

Definition 6. *Let T be a tree rooted at u and $l = (k_1, k_2, \ldots, k_p)$ the label of u. Let $v_1, \ldots, v_p$ be the vertices corresponding to l, see Definition 2. The parts $\tilde{T}_{k_i}, 1 \leq i \leq p$, of T are defined as follows, see Fig. 3.*

(i) $\tilde{T}_{k_1}$ is the subtree $T[v_1]$ and
(ii) $\tilde{T}_{k_i}$ is the subtree of $T[u; v_1, \ldots, v_{i-1}]$ rooted at v_i for $2 \leq i \leq p$.

Lemma 8. *Let T be a tree rooted at u and $l = (k_1, k_2, \ldots, k_p)$ the label of u. Then every part $\tilde{T}_{k_i}, 1 \leq i \leq p$, is k_i-simple. Especially, all parts are critical except possibly $\tilde{T}_{k_p}$ which may be noncritical.*

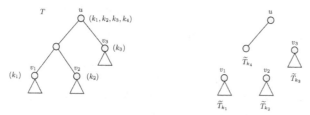

Fig. 3. The parts $\widetilde{T}_{k_i}$ of T

Notation 1 *Let T be a tree and $\widetilde{T}_{k_i}$ a k_i-simple part of T. If $\widetilde{T}_{k_i}$ is noncritical then $l\widetilde{L}_{k_i}$ denotes an optimal left extendable and $r\widetilde{L}_{k_i}$ an optimal right extendable layout of $\widetilde{T}_{k_i}$. If $\widetilde{T}_{k_i}$ is critical then $s\widetilde{L}_{k_i}$ denotes an optimal sparse layout of $\widetilde{T}_{k_i}$. Remember that all these layouts exist since Lemmas 6 and 7.*

Definition 7. *Let T be a tree rooted at u and $l = (k_1, k_2, \ldots, k_p)$ the label of u. The coarse layout of T is defined as*

$$C = \begin{cases} (s\widetilde{L}_{k_1}, s\widetilde{L}_{k_2}, \ldots, s\widetilde{L}_{k_p}) & \text{if } k_p \text{ is critical in } l \\ (s\widetilde{L}_{k_1}, s\widetilde{L}_{k_2}, \ldots, s\widetilde{L}_{k_{p-1}}, (l\widetilde{L}_{k_p}, r\widetilde{L}_{k_p})) & \text{otherwise.} \end{cases}$$

The coarse layout is a flexible representation of an optimal layout. In fact, an optimal layout can be computed from the coarse layout very efficiently. In order to show this we need some new notions: The *focus* of a layout L is a distinguished vertex of L partitioning L into two sublists $pre(L)$ and $suf(L)$. The prefix of L, $pre(L)$, consists of the focus and the vertices of L which are left to the focus. The suffix of L, $suf(L)$, consists of the vertices of L which are right to the focus. Given two layouts L and L' and their focuses, the layout $L \oplus L'$ is defined by $L \oplus L' = pre(L) \& L' \& suf(L)$. The focus of $L \oplus L'$ is the focus of L'. Notice that operation $\oplus$ is associative in the computation of both layouts and their focuses. We will apply $\oplus$ only on layouts of simple trees. The focus of a layout of a simple tree is defined as the root of the simple tree.

Lemma 9. *Let T be a tree and $l = (k_1, k_2, \ldots, k_p)$ the label of its root.*

(i) *If k_p is critical in l then $L = s\widetilde{L}_{k_1} \oplus s\widetilde{L}_{k_2} \oplus \cdots \oplus s\widetilde{L}_{k_p}$ is an optimal layout of T,*

(ii) *If k_p is not critical in l then $L = s\widetilde{L}_{k_1} \oplus \cdots \oplus s\widetilde{L}_{k_{p-1}} \oplus l\widetilde{L}_{k_p}$ is an optimal layout of T.*

Now we are going to show how to compute a coarse layout of a tree using the coarse layouts of its subtrees. The following lemma is important.

Lemma 10. *Let T be a tree rooted at u, $l = (k_1, k_2, \ldots, k_p)$ the label of u, and $u_1, \ldots, u_d$ the sons of u.*

(i) *The simple parts $\widetilde{T}_{k_1}, \widetilde{T}_{k_2}, \ldots, \widetilde{T}_{k_{p-1}}$ of T are exactly the simple parts of $T[u_1]$, $\ldots, T[u_d]$ with vertex separation larger than k_p,*

(ii) No subtree $T[u_i]$, $1 \leq i \leq d$, has a critical k_p-simple part.

From Lemma 10 we directly obtain:

Lemma 11. *Let T be a tree rooted at u and $l = (k_1, k_2, \ldots, k_p)$ the label of u. Let $C_1, \ldots, C_d$ be the coarse layouts of the subtrees of u. The components of $C_1, \ldots, C_d$ with vertex separation larger than k_p can be adopted as the first $p - 1$ components $s\widetilde{L}_{k_1}, s\widetilde{L}_{k_2}, \ldots, s\widetilde{L}_{k_{p-1}}$ of a coarse layout of T.*

Given the layouts $s\widetilde{L}_{k_1}, s\widetilde{L}_{k_2}, \ldots, s\widetilde{L}_{k_{p-1}}$ they must be sorted according to the integers $k_1, \ldots, k_{p-1}$ for the construction of a coarse layout of T. In order to sort as less elements as possible we first find the subtree of u with the largest vertex separation, say $T[u_a]$. Then we find a large prefix of the coarse layout of $T[u_a]$ which is also a prefix of the coarse layout of T. The layouts of this prefix need not be sorted because they are already sorted in the coarse layout of $T[u_a]$. The next lemma shows how to find such a prefix which is large enough for our later complexity considerations.

Lemma 12. *Let T be a tree rooted at u and $l = (k_1, k_2, \ldots, k_p)$ the label of u. Let $u_1, \ldots, u_d$ be the sons of u and l_i be the label of u_i in $T[u_i]$ for $1 \leq i \leq d$. Let $l_a = (m_1, m_2, \ldots, m_r)$ be the largest and $l_b = (n_1, n_2, \ldots, n_s)$ the second largest label over all l_i, where largest means containing the largest element. Then $(k_1, k_2, \ldots, k_j) = (m_1, m_2, \ldots, m_j)$ where j is the largest index $1 \leq j \leq r - 1$ with $m_j > \max\{m_{j+1}, n_1\} + 1$. Hence the first j components of the coarse layout of $T[u_a]$ can be adopted as the first j components of the coarse layout of T.*

Next we show how to compute the last component $s\widetilde{L}_{k_p}$ or $(l\widetilde{L}_{k_p}, r\widetilde{L}_{k_p})$ of the coarse layout of T.

Lemma 13. *Let T be a tree rooted at u and $l = (k_1, k_2, \ldots, k_p)$ the label of u. If the last component k_p of l, the information whether k_p is critical in l, and the coarse layouts of the subtrees of u are given then the last component of the coarse layout of T is computable.*

Proof. Let C be the coarse layout of T, $u_1, \ldots, u_d$ the sons of u, and C_i the coarse layouts of the subtrees $T[u_i]$ for $1 \leq i \leq d$. Let C_i' be the list obtained from C_i by removing all layout components with vertex separation larger than k_p. By Lemma 10 (i) the lists C_i' are the coarse layouts of the subtrees of u in $\widetilde{T}_{k_p}$. There are two cases:

Case 1: $\widetilde{T}_{k_p}$ is noncritical k_p-simple. We have to compute $l\widetilde{L}_{k_p}$ and $r\widetilde{L}_{k_p}$. Assume that u has a noncritical k_p-simple subtree in $\widetilde{T}_{k_p}$, say $\widetilde{T}_{k_p}[u_1]$ (otherwise the proof is similar). By Remark 1 we need the optimal (left and right) extendable layouts of $\widetilde{T}_{k_p}[u_1]$ and the optimal layouts of the other subtrees $\widetilde{T}_{k_p}[u_j]$, $2 \leq j \leq d$. The optimal extendable layouts are in the (only) component of C_1' and the optimal layouts of the other subtrees can be computed from C_j', $1 \leq j \leq d$, using Lemma 9.

Case 2: $\widetilde{T}_{k_p}$ is critical k_p-simple. We have to compute an optimal sparse layout $s\widetilde{L}_{k_p}$ of $\widetilde{T}_{k_p}$. Without loss of generality let $\widetilde{T}_{k_p}[u_1]$ and $\widetilde{T}_{k_p}[u_d]$ be the noncritical k_p-simple subtrees of u in $\widetilde{T}_{k_p}$. By Remark 1 we need the optimal extendable layouts of $\widetilde{T}_{k_p}[u_1]$

and $\widetilde{T}_{k_p}[u_d]$ and the optimal layouts of the other subtrees $\widetilde{T}_{k_p}[u_j]$, $2 \leq j \leq d-1$. The optimal extendable layouts are in the components of C_1' and C_2'. The optimal layouts of the other subtrees can be computed from C_j', $2 \leq j \leq d-1$, using Lemma 9.

3.3 Linear Time Layout Algorithm

We assume that every vertex u of T is marked by the last element k_p of the label $l = (k_1, \ldots, k_p)$ of u in $T[u]$ and by an indicator whether or not k_p is critical in l. Our layout algorithm first recursively computes the coarse layouts of the subtrees of u. Then using the mark of u and the coarse layouts of the subtrees of u it computes the coarse layout of $T[u]$. At the end an optimal layout of T is computed from the coarse layout of T. To achieve linear time we need a suitable representation of the coarse layouts based on the following representation of the labels given in [7]. A label is a list of blocks, where a block contains consecutive integers and is represented by an interval consisting of the endpoints of the block in decreasing order. For example the label $l = (9, 8, 7, 6, 4, 3, 2, 0)$ is represented by $l = ((9, 6), (4, 2), (0, 0))$.

Let $l = ((m_1, n_1), (m_2, n_2), \ldots, (m_{r-1}, n_{r-1}), (m_r, n_r))$ be the representation of the label of u in $T[u]$ where $m_r = n_r$; otherwise we split the last interval (m_r, n_r) into two intervals $(m_r, n_r - 1)$ and (n_r, n_r). The coarse layout of $T[u]$ is represented by a list of r records, one for every interval (m_i, n_i).

The first $r-1$ records consist of four fields. Two fields are of type integer and two are pointers. For $1 \leq i \leq r-1$ the first field of the i-th record is reserved for m_i, the second for n_i, the third for a pointer to the layout $s\widetilde{L}_{m_i, n_i} = s\widetilde{L}_{m_i} \oplus s\widetilde{L}_{m_i-1} \oplus \cdots \oplus s\widetilde{L}_{n_i}$, and the fourth for a pointer to the focus f_i of $s\widetilde{L}_{m_i, n_i}$. For the last r-th record there are two cases. If m_r is critical in l then the r-th record has the same structure as before, but now both integer fields are reserved for m_r since in this case $m_r = n_r$. If m_r is not critical in l then the r-th record has two additional fields which are pointers. Here the first two integer fields are reserved for m_r, the third field is reserved for a pointer to $l\widetilde{L}_{m_r}$, the fourth for a pointer to the focus of $l\widetilde{L}_{m_r}$, the fifth for a pointer to $r\widetilde{L}_{m_r}$ and the sixth for a pointer to the focus of $r\widetilde{L}_{m_r}$. This data structure allows the computation of the prefix and suffix of a layout in constant time.

Now we present our layout algorithm.

function $optimal\_layout(T : tree, u : root) : layout;$
$C := compute\_coarse\_layout(T, u);$
Compute optimal layout L from C according to Lemma 9
return $L;$

function $compute\_coarse\_layout(T : tree, u : root) : coarse\ layout;$
   {Let u be the root of T}
   if (T contains only u) **then** $C := ((0, 0, (u), u, (u), u));$
                 return C
             else {Let $u_1, \ldots, u_d$ be the sons of u in T};
                for all u_i **do** $C_i := compute\_coarse\_layout(T[u_i], u_i);$
                return $combine\_coarse\_layouts(u, C_1, \ldots, C_d);$

function $combine\_coarse\_layouts(u : vertex; C_1, \ldots, C_d : coarse\ layout)$
$: coarse\ layout;$

{Let C_t be the coarse layout of the subtree of u with the largest vertex separation}
1: Excise prefix $\bar{C}_t$ from C_t according to Lemma 12;
{Let k_p be the mark of u}
2: $R :=$ list containing all records (m, n, X, Y) of $C_1, \ldots, C_d$ with $m > k_p$;
3: Sort all records of R according to their first integer fields;
4: Compact R by replacing all consecutive records $(m_k, n_k, X_k, Y_k), (m_{k+1}, n_{k+1}, X_{k+1}, Y_{k+1}), \ldots, (m_r, n_r, X_r, Y_r)$ with $n_i = m_{i+1} + 1$ for $k \leq i < r$ by the
record (m_k, n_r, X, Y), where $X = X_k \oplus X_{k+1} \oplus \cdots \oplus X_r$ and $Y = Y_r$;
5: Compute the last component c of C according to Lemma 13, using the mark of u;
6: $C := \bar{C}_t \& R \& c$;
return C;

Lemma 14. *Given an input tree T compute_coarse_layout correctly computes the representation of a coarse layout of T.*

Proof. Let u be the root of T, $u_1, \ldots, u_d$ the sons of u and $l = (k_1, k_2, \ldots, k_p)$ the label of u. The proof is by induction on the height h of T.

For $h = 0$, T consists only of the root u. In this case *compute_coarse_layout*(T, u) results in $C := ((0, 0, (u), u, (u), u))$ and the claim trivially holds.

Now let T be of height $h + 1$. By the induction hypothesis *compute_coarse_layout* $(T[u_i], u_i)$ correctly computes the representations C_i of the coarse layouts of $T[u_i]$. We show that $C := \bar{C}_t \& R \& c$ computed in step 6 is a correct representation of the coarse layout of T. By Lemma 12, $\bar{C}_t$ computed in step 1 can be adopted as the prefix of C. Now consider the list R computed in step 2. First observe that no C_i has a record (m, n, X, Y) with $m > k_p \geq n$, otherwise subtree $T[u_i]$ would have a critical k_p-simple part which contradicts to Lemma 10 (ii). That means the records of R include exactly those layout components of the coarse layouts of the subtrees which have vertex separation larger than k_p. Now, by Lemma 11 and the associativity of $\oplus$, R after sorting in step 3, and compacting in step 4 can be adopted from C. Finally, by Lemma 13 and the associativity of $\oplus$, the record c computed in step 5 can be adopted as the last record of C. Hence $C := \bar{C}_t \& R \& c$ is a correct representation of the coarse layout of T.

By Lemma 14, Lemma 9 and the associativity of $\oplus$ we conclude:

Theorem 2. *The function optimal_layout correctly computes an optimal layout of an input tree.*

Next we show that the time complexity of our algorithm is linear.

Lemma 15. *Let T be a tree rooted at u and $u_1, \ldots, u_d$ the sons of u. Let C_i, $1 \leq i \leq d$, be the representation of a coarse layout of $T[u_i]$. Let s_i be the ith largest vertex separation of the subtrees of u. The time required by combine_coarse_layouts$(u, C_1, \ldots, C_d)$ to compute the representation of the coarse layout of T is $O(d + 2s_2 + s_3 + \ldots + s_d)$, where s_1 does not appear and s_2 appears twice.*

Proof. First observe that $vs(T[u_i])$ is the integer in the first field of the first record of C_i. The time required in step 1 is $O(s_2 + 1)$. We only need to search C_t from the right to the left until we find the first consecutive records (m_k, n_k, X_k, Y_k) and $(m_{k+1}, n_{k+1}, X_{k+1}, Y_{k+1})$ with $n_k > \max\{m_{k+1}, s_2\} + 1$, see also Lemma 12. Step 2 takes time linear in the number of the records in $C_1, \ldots, C_d$, i.e., time $O(d + 2s_2 + s_3 + \ldots + s_d)$ (note that $\bar{C}_t$ is removed from C_t). In step 3 the sorting of the records of R according to the integers in their first fields can be done in $O(s_2 + 1)$ using bucket sort. To see this observe that these integers are (i) all $\leq s_2$ except possibly one (this would occur in the first field of the first record of C_t) and (ii) all pairwise different since Lemma 10(i). The time required in step 4 is $O(s_2 + 1)$ since R contains at most $s_2 + 1$ records. Step 5 clearly can be done in $O(d + 2s_2 + s_3 + \ldots + s_d)$. Finally step 6 takes constant time. Hence the execution time of *combine_coarse_layouts*$(u, C_1, \ldots, C_d)$ is $O(d + 2s_2 + s_3 + \ldots + s_d)$.

Let T be a tree, u a vertex in T, and j an integer. Let $q_{u,j}$ be the number of the vertices v in $T[u]$ having a brother v' such that $vs(T[v]) = vs(T[v']) \geq j$. In a similar way as in the proof of Lemma 3.1 from [7] we can prove the next lemma.

Lemma 16. *Let T be a tree rooted at u. The time required by compute_coarse_layout* (T, u) *is* $\leq c(n + \sum_{j=0}^{vs(T)} q_{u,j})$ *where c is a constant and n the size of T.*

Proof. Let $u_1, \ldots, u_d$ be the sons of u. The proof is by induction on the height of T.

For $h = 0$, *compute_coarse_layout*(T, u) needs time c_1 where c_1 is the time needed by $C := ((0, 0, (u), u, (u), u))$. Hence the claim holds for every $c \geq c_1$.

Now consider T of height $h+1$. The time needed by *compute_coarse_layout*(T, u) is the time t_1 needed by *combine_coarse_layouts*$(u, C_1, \ldots, C_d)$ plus the time t_2 needed by the calls *compute_coarse_layout*$(T[u_i], u_i)$, $1 \leq i \leq d$. By Lemma 15 and the induction hypothesis we have $t_1 + t_2 \leq c_2(d + 2s_2 + s_3 + \ldots + s_d) + \sum_{i=1}^{d} (cn_i + \sum_{j=0}^{vs(T[u_i])} q_{u_i,j})$, where n_i is the size of $T[u_i]$. Moreover it is easy to verify that

$$2s_2 + s_3 + \ldots + s_d + \sum_{i=1}^{d} \sum_{j=0}^{vs(T[u_i])} q_{u_i,j} = \sum_{j=0}^{vs(T)} q_{u,j}.$$ Defining $c := \max\{c_1, c_2\}$ we get

$$t_1 + t_2 \leq c(d + \sum_{i=1}^{d} n_i) + c(2s_2 + s_3 + \ldots + s_d + \sum_{i=1}^{d} \sum_{j=0}^{vs(T[u_i])} q_{u_i,j}) \leq cn + c\sum_{j=0}^{vs(T)} q_{u,j} \leq$$

$c(n + \sum_{j=0}^{vs(T)} q_{u,j})$ and the proof is complete.

The following has been proved in [7], page 77.

Lemma 17. *Let T be a tree rooted at u. Then* $\sum_{j=0}^{vs(T)} q_{u,j} \leq 4n$ *where n is the size of T.*

By Lemmas 16 and 17 we obtain:

Lemma 18. *The time required by compute_coarse_layout*(T, u) *is linear in the size of T.*

By Corollary 1 a coarse layout of T contains at most $\log n$ records, where n is the size of T. Hence given a coarse layout the computation of an optimal layout according to Lemma 9 takes at most $O(\log n)$ time. This and Lemma 18 imply our last theorem.

Theorem 3. *The time complexity of the function optimal_layout is linear in the size of the input tree.*

References

1. H. Bodlaender. A linear time algorithm for finding tree-decompositions of small treewidth. *SIAM J. Comput.*, 25(6):1305–1317, 1996. 405
2. H. Bodlaender and T. Kloks. Efficient and constructive algorithms for the pathwidth and treewidth of graphs. *J. Algorithms*, 21:358–402, 1996. 405
3. H. Bodlaender, T. Kloks, and D. Kratsch. Treewidth and pathwidth of permutation graphs. *SIAM J. Disc. Math.*, 8:606–616, 1995. 405
4. H. Bodlaender and R. Möhring. The pathwidth and treewidth of cographs. *SIAM J. Disc. Math.*, 6:181–188, 1993. 405
5. M. Chung, F. Makedon, I. Sudborough, and J. Turner. Polynomial algorithms for the min-cut linear arrangement problem on degree restricted trees. *SIAM J. Comput.*, 14(1):158–177, 1985. 406
6. S. Cook and R. Sethi. Storage requirements for deterministic polynomial finite recognizable languages. *J. Comput. System Sci.*, 13:25–37, 1976. 403
7. J. Ellis, I. Sudborough, and J. Turner. The vertex separation and search number of a graph. *Inform. Comput.*, 113:50–79, 1994. 403, 404, 405, 406, 411, 413
8. J. Gustedt. On the pathwidth of chordal graphs. *Disc. Appl. Math.*, 45:233–248, 1993. 405
9. N. Kinnersley. The vertex separation number of a graph equals its path-width. *Inform. Process. Lett.*, 142:345–350, 1992. 404
10. L. Kirousis and C. Papadimitriou. Interval graphs and searching. *Disc. Math.*, 55:181–184, 1985. 404
11. L. Kirousis and C. Papadimitriou. Searching and pebbling. *Theor. Comp. Science*, 47:205–218, 1986. 404
12. A. LaPaugh. Recontamination does not help to search a graph. *J. Assoc. Comput. Mach.*, 40(2):243–245, 1993. 403
13. T. Lengauer. Black-white pebbles and graph separation. *Acta Informat.*, 16:465–475, 1981. 403, 405
14. F. Makedon and I. Sudborough. On minimizing width in linear layouts. *Disc. Appl. Math.*, 23:243–265, 1989. 405
15. N. Megiddo, S. Hakimi, M. Garey, D. Johnson, and C. Papadimitriou. The complexity of searching a graph. *J. Assoc. Comput. Mach.*, 35(1):18–44, 1988. 403, 405, 406
16. R. H. Möhring. Graph problems related to gate matrix layout and PLA folding. In E. Mayr, H. Noltmeier, and M. Syslo, editors, *Computational Graph Theory*, pages 17–51. Springer Verlag, 1990. 403
17. F. Monien and I. Sudborough. Min cut is NP-complete for edge weighted trees. *Theor. Comp. Science*, 58:209–229, 1988. 405
18. T. Parsons. Pursuit-evasion in a graph. In Y. Alavi and D. Lick, editors, *Theory and Applications of Graphs*, pages 426–441. Springer-Verlag, New York/Berlin, 1976. 403
19. N. Robertson and P. Seymour. Graph minors. I. Excluding a forest. *J. Comb. Theory Series B*, 35:39–61, 1983. 404
20. N. Robertson and P. Seymour. Disjoint paths–A servey. *SIAM J. Alg. Disc. Meth.*, 6:300–305, 1985. 404
21. M. Yannakakis. A polynomial algorithm for the min-cut linear arrangement of trees. *J. Assoc. Comput. Mach.*, 32(4):950–988, 1985. 406

Coloring Sparse Random Graphs in Polynomial Average Time[*]

C.R. Subramanian

Max-Planck Institute für Informatik,
Stuhlsatzenhausweg 85, 66123, Saarbrücken, GERMANY.
crs@mpi-sb.mpg.de

Abstract. We present a simple BFS tree approach to solve partitioning problems on random graphs. In this paper, we use this approach to study the k-coloring problem. Consider random k-colorable graphs drawn by choosing each allowed edge independently with probability p after arbitrarily partitioning the vertex set into k color classes of "roughly equal" (i.e. $\Omega(n)$) sizes. Given a graph G and two vertices x and y, compute $n(G, x, y)$ as follows : Grow the BFS tree (in the subgraph induced by $V - y$) from x till the l-th level (for some suitable l) and find the number of neighbors of y in this level. We show that these quantities computed for all pairs are sufficient to separate the largest or smallest color class. Repeating this procedure $k-1$ times, one obtains a k-coloring of G with high probability, if $p \geq n^{-1+\epsilon}$, $\epsilon \geq X/\sqrt{\log n}$ for some large constant X. We also show how to use this approach so that one gets even smaller failure probability at the cost of running time. Based on this, we present polynomial average time ($p.av.t.$) k-coloring algorithms for the stated range of p. This improves significantly previous results on $p.av.t.$ coloring [13] where ϵ is required to be above $1/4$. Previous works on coloring random graphs have been mostly concerned with almost surely succeeding ($a.s.$) algorithms and little work has been done on $p.av.t.$ algorithms. An advantage of the BFS approach is that it is conceptually very simple and combinatorial in nature. This approach is applicable to other partitioning problems also.

1   Introduction

The problem of k-coloring a k-colorable graph is known to be NP-hard even for fixed $k \geq 3$ [6]. The best known approximation algorithms [9] are only guaranteed to use $O(n^{\delta(k)})$ (where $\delta(k) = 1 - 3/(k+1)$) colors on n-vertex graphs. However, it is long known that random k-colorable graphs can be k-colored almost surely in polynomial time [16,3,2,15,1,4]. For some distributions, such a k-coloring can be obtained in polynomial average time ($p.av.t.$) [3,15,13] also. While it is true that random graphs tend to possess nice structural properties, there are several

[*] This research was partially supported by the EU ESPRIT LTR Project No. 20244 (ALCOM-IT), WP 3.3 and also by a fellowship of MPI(Informatik), Saarbrücken, Germany.

M. Paterson (Ed.): ESA 2000, LNCS 1879, pp. 415–426, 2000.

justifications for having an algorithmic study of random graphs (see [5,10]). In addition, algorithmic studies on random instances of NP-hard problems have certain implications to cryptography (See [8,11]).

In this paper, we study the random graph model (denoted by $\mathcal{G}(n,p,k)$) described in the Abstract. There is also a related model (denoted by $\mathcal{GUC}(n,p,k)$) in which each vertex is put into one of the k color classes independently and uniformly randomly. The results of this paper carry over to this model also.

We present both polynomial time $a.s.$ (almost surely succeeding) algorithms and $p.av.t.$ algorithms. While $a.s.$ algorithms provide only one guarantee of failure probability, the $p.av.t.$ algorithms provide a trade-off leading to any desired polynomially low failure probability with a proportional increase in running time (see [15] for more details). While several $a.s.$ k-coloring algorithms have been proposed, only a few $p.av.t.$ algorithms have been proposed so far ([3,15,13]).

In this paper, we present a simple BFS (breadth-first-search) based approach (stated in the abstract) for k-coloring $\mathcal{G}(n,p,k)$ with $p \geq n^{-1+\epsilon}$, $\epsilon \geq X/\sqrt{\log n}$ for some sufficiently large positive constant X. Our approach is based on the following intuition. In such a random graph, consider any vertex x belonging to the largest (or smallest) color class and grow the BFS tree rooted at x. We prove that with high probability, the following holds : The tree grows by a factor of "roughly" np between successive levels till it crosses the threshold of $1/p$. Also, in every even (or odd) numbered level, the contribution of the largest (or smallest) class is larger (or smaller) than other classes by a significant amount. We make use of this phenomenon to separate the largest (or smallest) class and recurse on the remaining graph. Whether we choose the largest or smallest class depends only on p. However, the algorithm does not need to know p.

Our algorithms are conceptually quite simple and combinatorial in nature. Also, because of the way BFS tree grows, one can maintain independence of events thereby simplifying the analysis. We show that our algorithms have very small failure probabilities. For fixed $\epsilon > 0$, it is exponentially low (that is, $\leq e^{-n^\delta}$ for some constant $\delta > 0$) and otherwise at most $e^{-\Omega((\log n)2^{\sqrt{X(\log n)}/2})}$ both of which are asymptotically much smaller than n^{-D} for any fixed D. More importantly, this BFS approach provides a trade-off leading to reduction in failure probabilities at the cost of running time. This helps us in obtaining $p.av.t.$ algorithms for the same range of p.

Having low failure probabilities is not sufficient. Either, it has to be "sufficiently" low, or we need to design techniques which start with $a.s.$ algorithms and successively bring down failure probabilities at the cost of running time, finally yielding $p.av.t.$ algorithms. Four such techniques are presented in [14] (also briefly in [13]). Each of these techniques is characterized by (i) the range of p over which it works and (ii) the failure probability of the polynomial algorithm with which it starts. For example, one technique works as long as $p \geq n^{-1}(\log n)^3$, but requires failure probability to be at most $e^{-(\log n)^3/p}$ which is very small, say, if $p = n^{-1}(\log n)^3$. Another technique works with any exponentially low value but requires $p \geq n^{-1+\epsilon}$, $\epsilon > 1/3$. Using these, [14] presents $p.av.t.$ algorithms for $p \geq n^{-1+\epsilon}$, $\epsilon > 1/4$.

In this paper, we present further progress in this direction by presenting one more technique. It works for $p \geq n^{-1+\epsilon}$, $\epsilon \geq X/\sqrt{\log n}$ and for failure probabilities bounded by $e^{-\Omega((\log n)2^{\sqrt{X(\log n)/2}})}$. Applying this technique, we obtain $p.av.t.$ k-coloring for the same range of p. The new technique is based on an extension of the BFS approach mentioned before. Instead of growing the BFS tree from a single vertex x, one now grows it from an independent set of suitable size m. We can reduce failure probability to $e^{-\Omega(m(\log n)2^{\sqrt{X(\log n)/2}})}$ if we are willing to spend $O(n^{O(m \cdot \log n)})$ time.

An important aspect of our approach is that a simple algorithm based on growing BFS trees works well with very low failure probability and can be made to have any desired low failure probability by spending more time. One should note that as ϵ becomes very small, the amount of information one gets also becomes even little (since there are fewer edges) and hence it becomes more difficult to guarantee very low failure probability.

The paper is organized as follows. In Section 2, we introduce some basic notions, notations and facts. In Sections 3 and 4, we present the $a.s.$ polynomial algorithm. In Section 5, we show how to decrease the failure probabilities. In Section 6, we present and use the new technique to design $p.av.t.$ k-coloring. Finally, in Section 7, we conclude with some remarks.

2   Some Preliminaries

Notations and Notions :

1. Throughout the paper, $G \in \mathcal{G}(n, p, k)$ denotes a random graph. For this paper, we assume that $p \geq n^{-1+\epsilon}$, $\epsilon \geq X/\sqrt{\log n}$. X is a large constant and a tight estimate of X can be worked out from the proof.
2. Let $V_1, \ldots, V_k$, with sizes $n_1, \ldots, n_k$, respectively, be the k color classes used by the model. Assume $n_i \geq n/C$ for all i for some constant $C \geq k$. For all i, $1 \leq i \leq k$, let W_i denote $V_i \cup \ldots \cup V_k$ and G_i denote the subgraph of G induced by W_i. Also for each i, let m_i denote the size of the set W_i, that is, $m_i = n_i + \ldots + n_k$.
3. Often, where it is convenient, we use $e_1 \in (f_l, f_u)e_2$ as a short notation for $f_l e_2 \leq e_1 \leq f_u e_2$. Here, f_l, f_u, e_1 and e_2 are real-valued expressions. We will be using the Chernoff-Hoeffding (CH) bounds and also the following fact.

Fact 2.1 *Let $S(n)$ be a set of vertices of size $M(n)$. For any $u \notin S(n)$, let each edge $\{u, s\}$ ($s \in S(n)$) be selected with probability $p(n)$. If $M = M(n)$ and $p = p(n)$ are such that $Mp = o(1)$ as $n \to \infty$, then*

$$\mathbf{Pr}(u \text{ is adjacent to some vertex in } S(n)) = Mp - M^2 p^2[1 - o(1)]/2 = Mp[1 - o(1)]$$

Without loss of generality, $n_1, \ldots, n_k$ are assumed to be known to the algorithm. This is because there are only polynomially many nonnegative integral solutions to $n'_1 + \ldots + n'_k = n$ and we can enumerate them. Also, assume w.l.o.g. that either $n_1 \leq \ldots \leq n_k$ or $n_1 \geq \ldots \geq n_k$. Consider the following generic algorithm for k-coloring. It uses the standard idea of separating the color classes one by one. Suppose, for each $i \leq k - 1$, there exists some $x_i \in V_i$ and a polynomial time exactly computable quantity $n(x_i, y, G_i)$ such that for all elements y

of W_i, we can correctly infer from $n(x_i, y, G_i)$ whether $y \in V_i$ or not. Then, by enumerating all possible choices of x_i, we can separate the color class V_i in G_i. Now applying this procedure repeatedly for increasing values of i, we can obtain a k-coloring of G. By varying the definition of $n(x_i, y, G_i)$, we get different algorithms for k-coloring. This idea can be formalized to yield an algorithm that runs in polynomial time and it succeeds with probability $1 - f(n)$ if it can be proved that :

(A) For $G \in \mathcal{G}(n, p, k)$, with probability at least $1 - f(n)$, there exist $k - 1$ vertices $x_i \in V_i$ $(i = 1, \ldots, k - 1)$ such that
$$\forall i \ (1 \le i \le k - 1), \forall y \in V_i, y \ne x_i, \forall z \in V_j, j > i, \quad n(x_i, y, G_i) > n(x_i, z, G_i).$$

We can also change the $>$ inequality into $<$. We make use of **(A)** and only show the existence of polynomial time computable quantities $n(x, y, G)$ and suitable vertices of $x_i \in V_i$. The algorithm works by trying to separate the color classes one by one in either decreasing or increasing order (depending on p) of their sizes. In the process, if it finds some k-coloring, it stops.

Throughout the paper, let α denote the value $\sqrt{\log n}/\sqrt{2}^{\sqrt{\log n}}$. For $p \ge n^{-0.1}$, using the results of [14], one can obtain a k-coloring in $p.av.t$. Hence, for the rest of the paper, we assume that $p \le n^{-0.1}$. Thus we have $p = o(\alpha)$. This assumption simplifies the analysis. Given p, one can find a *unique positive* integer l so that we can write $p = n^{-1 + \frac{1}{l+1} + \delta}$ where δ is a positive *real* such that $\tau(l) \le \delta < \frac{1}{l(l+1)} + \tau(l - 1)$ and $\tau(l') \doteq \frac{1}{\sqrt{X}l'(l'+1)}$ for $l' \ge 1$ and $\tau(0) = 0$. Since $p \ge n^{-1 + \epsilon}$ with $\epsilon \ge X/\sqrt{\log n}$, we have $l \le 2\sqrt{\log n}/X$. Because of this, we have $(1 - \alpha)^l = 1 - l\alpha[1 - o(1)]$ and $(1 + \alpha)^l = 1 + l\alpha[1 + o(1)]$. Note that $\tau(l)$ and the upper bound on δ both become smaller as l becomes larger. Initially, we assume that $\tau(l) \le \delta \le (\sqrt{X} - 1)\tau(l)$. Later, we will show how to handle the remaining case.

3   Almost Surely k-Coloring for $\tau(l) \le \delta \le \frac{\sqrt{X} - 1}{\sqrt{X}l(l+1)}$

For an arbitrary graph H and a vertex u of H, define $D_H(u, l')$ (for $l' \ge 0$) as the set of vertices v such that the distance (in H) between u and v is l'.

Given a graph H and two vertices x and y, define $n(x, y, H)$ as the number of y's neighbors in $D_{H-y}(x, l)$ where l is defined before.[1] In other words, $n(x, y, H)$ is the number of y's neighbors in the l-th level of the BFS tree grown from x in the subgraph $H - y$. For the sake of exposition, we assume that l is even and w.l.o.g. that $n_1 \ge \ldots \ge n_k$. For the other case (odd l), the arguments are similar and for this case we assume that $n_1 \le \ldots \le n_k$. Our algorithm is based on the following theorem.

---

[1]   This seems to require that the algorithm know the value of p. However, since $l = O(\sqrt{\log n})$, it is enough to repeat the algorithm for $i = 1, \ldots, \log n$ assuming that $l = i$ during the i-th iteration.

Theorem 3.1. *There exist expressions* $E_l(n_i, \ldots, n_k, j)$, $i = 1, \ldots, k-2$, $j \geq i$, *denoted shortly as* $E_l(i, j)$, *with* $(\geq (n/C)^l)$, *such that the following hold with probability at least* $1 - e^{-\Omega(\alpha^2 \min(np, n^l p^{l+1} C^{-l}))}$ *: for any* i $(1 \leq i \leq k-2)$ *and any* $x_i \in V_i$, *any* $y \in V_i$, $y \neq x_i$, *and any* $z \in V_j (j > i)$,

$$n(x_i, y, G_i) \in \left((1-\alpha)^{l+1}, (1+\alpha)^{l+1}\right) E_l(i, i) p^{l+1} \tag{1}$$

$$n(x_i, z, G_i) \in \left((1-\alpha)^{l+1}, (1+\alpha)^{l+1}\right) E_l(i, j) p^{l+1} \tag{2}$$

$$n(x_i, z, G_i) - n(x_i, y, G_i) = \Omega(n^l p^{l+1} C^{-l} (kC)^{-l/2}) \tag{3}$$

Proof. We will not explicitly give the expressions $E_l(..)$, we will only give an inductive proof showing how to obtain such expressions. It is sufficient to prove for any fixed i, x_i, y and z. Fix these parameters. Let w denote any of y and z and fix w also. For $j \geq i$ and $0 \leq l' \leq l$, define $n_j(x_i, w, l')$ to be the size of $D_{G_i - w}(x_i, l') \cap V_j$, that is, the number of vertices in $(W_i - w) \cap V_j$ which are at a distance of l' from x_i in $G_i - w$. We have the following lemma whose proof is omitted in this abstract due to lack of space.

Lemma 3.1. *There exist expressions* $F_j(n_i, \ldots, n_k, l)$, $i = 1, \ldots, k-2$, $j \geq i$, $\left(\mathbf{1.} \geq (n/C)^l \text{ for } l \geq 2, j \geq i \text{ and } l = 1, j > i\right)$, $\left(\mathbf{2.} \sum_{j \geq i} F_j(n_i, \ldots, n_k, l) \leq n^l, \forall l\right)$, *(***3.*** denoted shortly as $F_j(i, l)$), such that under the assumptions of Theorem 3.1, for x_i and w defined before, with probability at least* $1 - e^{-\Omega(\alpha^2 np)}$, *for all* $j > i$,

$$n_i(x_i, w, l) \in \left((1-\alpha)^l, (1+\alpha)^l\right) F_i(i, l) p^l \tag{4}$$

$$n_j(x_i, w, l) \in \left((1-\alpha)^l, (1+\alpha)^l\right) F_j(i, l) p^l \tag{5}$$

$$F_i(i, l) - F_j(i, l) \geq F_i(i, l)(kC)^{-l/2} \tag{6}$$

Recall our assumption that l is even. Assume Lemma 3.1 and consider $w = y \in V_i - x_i$. We have $|D_{G_i - y}(x_i, l)| = \sum_{j' \geq i} n_{j'}(x_i, y, l)$. But y cannot have neighbors in $D_{G_i - y}(x_i, l) \cap V_i$. Given that Lemma 3.1 holds for $w = y$, we have (by Chernoff's bounds) with probability at least $1 - e^{-\Omega(\alpha^2 n^l p^{l+1} C^{-l})}$,

$$n(x_i, y, G_i) \in ((1-\alpha), (1+\alpha)) \left(\sum_{j' > i} n_{j'}(x_i, y, l)\right) p \tag{7}$$

Now consider $w = z \in V_j$ for some $j > i$. z cannot have neighbors in $D_{G_i - z}(x_i, l) \cap V_j$. As before, given that Lemma 3.1 holds for $w = z$, we have

$$n(x_i, z, G_i) \in ((1-\alpha), (1+\alpha)) \left(\sum_{j' \geq i, j' \neq j} n_{j'}(x_i, z, l)\right) p \tag{8}$$

For each of the cases $w = y$ and $w = z$, the events of Lemma 3.1 and the corresponding inequality appearing before are independent since w is not included in the graph considered in Lemma 3.1. Hence, (7) and (8) hold with probability at least $1 - e^{-\Omega(\alpha^2 \min(np, n^l p^{l+1} C^{-l}))}$. Setting $E_l(i,j) = \sum_{j' \geq i, j' \neq j} F_{j'}(i,l)$ and using (7), (8), we get (1) and (2). For the third inequality, we have

$$n(x_i, z, G_i) - n(x_i, y, G_i) \geq T_1 + T_2 \quad \text{where}$$

$$T_1 = (1-\alpha)^{l+1} F_i(i,l) p^{l+1} - (1+\alpha)^{l+1} F_j(i,l) p^{l+1}$$

$$T_2 = \left(\sum_{j' \geq i, j' \neq i, j} (1-\alpha)^{l+1} F_{j'}(i,l) p^{l+1} - (1+\alpha)^{l+1} F_{j'}(i,l) p^{l+1} \right) \quad (9)$$

The quantity appearing within parentheses in (9) is negative and by choice of α, it is lesser by any desired small factor when compared to the difference of the remaining two terms which is at least $F_i(i,l) p^{l+1} (kC)^{-l/2} [1 - o(1)]$ by (6). Hence (3) follows. This completes the proof of Theorem 3.1.   □

Making use of (3) and **(A)**, we infer that G can be k-colored in polynomial time with probability at least $1 - e^{-\Omega(\alpha^2 \min(np, n^l p^{l+1} C^{-l}))}$. For ϵ bounded below by a constant, this ensures that failure probability is exponentially low and for $\epsilon \geq X/\sqrt{\log n}$, it is at most $e^{-\Omega((\log n) 2^{\sqrt{X(\log n)}/2})}$.

4   Almost Surely k-Coloring for $\delta \geq \frac{\sqrt{X}-1}{\sqrt{X}l(l+1)}$

The problem with this case is the following : When $\delta \geq \frac{1}{l(l+1)}$, we have $n^{l-1} p^{l-1} = \Omega(1/p)$ and hence Fact 2.1 cannot be applied to estimate and bound the sizes of the l-th level after estimating the sizes of the previous $l-1$ levels (this happens in the proof of Lemma 3.1 which is skipped in this abstract). For this case, we compute $n(x_i, w, G_i)$, $x_i \in V_i$, $w \in W_i - x_i$, as follows.

After computing the $(l-1)$-th level $D_{G_i - w}(x_i, l-1)$, we choose a subset D' by randomly and independently including each vertex in level $l-1$ with probability $n^{-\kappa(l)}$, where[2] $\kappa(l) \doteq l(\tau(l) + \tau(l-1))$. For our choice of $\kappa(l)$, with probability at least $1 - e^{-\Omega(\alpha^2 (np/C)^{l-1} n^{-\kappa(l)})}$, for all $j \geq i$,

$$|D' \cap V_j| \in (1-\alpha, 1+\alpha) n_j(x_i, w, l-1) n^{-\kappa(l)}$$

$$|D' \cap V_j| \leq 2p^{-1} n^{-\frac{1}{\sqrt{X}(l+1)}} = o\left(\frac{1}{p}\right) \quad (10)$$

Let D'' be the set of vertices in level l which are adjacent to some vertex of D'. Effectively, we consider l-th level to be grown by considering only edges incident on D'. Now, define $n(x_i, w, G_i)$ as the number of vertices u in D'' such that u and w are adjacent in G_i. Note that now $n(x_i, w, G_i)$ is a random value not a deterministic one.

---

[2]   As explained in the previous section, we repeat the deterministic algorithm for $i = 1, \ldots, \log n$ and if this does not succeed, we repeat the randomized algorithm for $i = 1, \ldots, \log n$.

Applying the arguments used in the proof of Lemma 3.1, one gets, for $j \geq i$,

$$|D'' \cap V_j| \in ((1-\alpha)^3, (1+\alpha)^3) \left(\sum_{r \geq i, r \neq j} \left(\sum_{r' \geq i, r' \neq r} n_{r'}(x_i, w, l) \right) n_r \right) n^{-\kappa(l)} n_j p^2$$

Consider the following changes made to Lemma 3.1. Replace (for $j \geq i$) $n_j(x_i, w, l)$ by $D'' \cap V_j$. Multiply an extra factor of $n^{-\kappa(l)}$ to each $F_j(i, l)$. Also, we multiply an extra factor of $(1\pm\alpha)$ to both sides of inequalities (4) and (5). With these changes to Lemma 3.1 and corresponding changes to Theorem 3.1, one can verify that the random value $n(x_i, w, G_i)$ defined before can be used to obtain a k-coloring in polynomial time with probability (over the input, randomness of the algorithm not taken into account) at least $1 - e^{-\Omega(\alpha^2 \min(np, n^l p^{l+1} n^{-\kappa(l)} C^{-l}))}$. For our choice of $\kappa(l)$, one can verify that failure probability is at most $e^{-\Omega((\log n)2^{\sqrt{X(\log n)}/2})}$. The failure probability of (10) holding is based on the randomness of the algorithm and can be made as small as $e^{-\omega(n)}$ by repeating the experiment (with new and independent random bits for each execution) n times.

Remark 4.1. Note that the algorithm does not need explicitly the values of C and also of the expressions $E_l(i, j)$ and $F_j(i, l)$ for various values of i, j and l. But for a given l, these expressions can be written down explicitly using the recursive definition of $F_j(i, l)$ given in the proofs of Theorem 3.1 and Lemma 3.1. These values are only needed in the analysis.

With these changes, we can arrive at the following conclusions.

1. G can be k-colored in polynomial time with probability at least $1 - e^{-\Omega(\alpha^2 \cdot \min(np, n^l p^{l+1} C^{-l}))}$, if no randomization is employed.
2. G can be k-colored in polynomial time with probability at least $1 - e^{-\Omega(\alpha^2 \cdot \min(np, n^l p^{l+1} n^{-\kappa(l)} C^{-l}))}$, if randomization is employed.
3. This failure probability (with or without randomization), although not exponentially low, is "much lower" than polynomially low values. It is bounded by $e^{-\Omega((\log n)2^{\sqrt{X(\log n)}/2})}$. For ϵ above by a constant, it is exponentially low.
4. Whether or not randomization is needed by the algorithm depends only on p and if it is employed, the failure probability of this randomization not serving its purpose can be made as small as as $e^{-\omega(n)}$ by repeating the algorithm polynomially many times.

5   Intermediate Algorithms for k-Coloring

We k-color in p.av.t. by starting with a polynomial a.s. algorithm and by successively bringing down failure probabilities (at the cost of running time) by applying several intermediate algorithms. Towards this purpose, we introduce an intermediate algorithm *FindColor4* (G, n, k, m) and analyze its performance. For suitable values of m, this algorithm takes $O(n^{O(m \cdot \log n)})$ time and has failure probability at least $1 - e^{-\Omega(mg(n))}$. Here, $g(n) \geq (\log n)2^{\sqrt{X(\log n)}/2}$ for our range of p. In the next section, we show how to use *FindColor4* and the intermediate algorithm *Color2* introduced in [15] to obtain a *p.av.t.* algorithm.

FindColor4 works by finding the color classes one by one in either decreasing or increasing order (depending on p) of their sizes. To separate, say V_1, we do the following for each independent set A of size m : Grow the BFS tree in G (after identifying the vertices of A into a single vertex) till we reach the l-th level for some suitable l. For each remaining vertex y, decide if y belongs to V_1 by computing how many neighbors it has in level l. This gives a reasonably good approximation to V_1 if $A \subset V_1$. The approximation is such that, of the unvisited vertices, at most m of them are wrongly taken in or wrongly left out, with required probability. Further refine this approximation using the partial BFS tree constructed. Now apply the same procedure to the remaining graph. A detailed description is given later.

Additional Notations and Facts : For any set A of size m, for any l $(1 \leq l \leq k)$, let $N_l(A) = \{v \in V_l : v$ is adjacent to some vertex of $A\}$; Also, for any $A \subseteq V_i$, $w \notin V_i$, $\mathbf{Pr}(w$ is adjacent to some vertex of $A) = 1 - (1 - p)^{|A|}$. Let $\alpha_1 \in [0, 1]$ be any value. Using Chernoff bounds, and also Fact 2.1, we derive that the following hold in the random graph G.

P4: For any fixed independent set A of size m such that $mp = o(\alpha_1)$, and for any $l = 1, \ldots, k$ such that $A \cap V_l = \emptyset$, considering only potential edges incident on A, with probability at least $1 - e^{-\Omega(\alpha_1^2 mnp)}$, $|N_l(A)| \in (1 - \alpha_1, 1 + \alpha_1)mn_lp$

For an arbitrary graph $H = (V_H, E_H)$ and a subset $A \subseteq V_H$ define $D_H(A, l')$ (for $l' \geq 0$) as the set of vertices v such that the minimum distance (in H) between v and a vertex of A is l'. That is, the set of vertices at a distance of l' from a new vertex obtained by contracting all vertices of A. Given a graph H, and a subset $A \subset V_H$ and a vertex $y \notin A$, define $n(A, y, H)$ as the number of y's neighbors in $D_H(A, l)$ (here, l depends on $m = |A|$ and is defined below).

Let α be the value defined in Section 2. Consider any integer $m = n^\beta$ such that[3] $mp(\log n) \leq \alpha$. Given m and using $p \geq n^{-1}$, one can find a *unique* positive integer $l(\beta)$ such that $p = n^{-1 + \frac{1-\beta}{l(\beta)+1} + (1-\beta)\delta(\beta)}$ where $\delta(\beta)$ is a positive real such that $\tau(l(\beta)) \leq \delta(\beta) < \frac{1}{l(\beta)(l(\beta)+1)} + \tau(l(\beta) - 1)$ and $\tau()$ is defined before. Since $p \geq n^{-1+\epsilon}$ with $\epsilon \geq X/\sqrt{\log n}$, we have $l(\beta) \leq 2(1 - \beta)\sqrt{\log n}/X$. Initially, we also assume that the chosen m is such that $\delta(\beta) \leq \frac{(\sqrt{X}-1)}{\sqrt{X}l(\beta)(l(\beta)+1)}$.

With these assumptions, it follows that

$$mn^{l(\beta)-1}p^{l(\beta)-1} \leq p^{-1}n^{-\frac{1-\beta}{\sqrt{X}(l(\beta)+1)}} = o\left(\frac{1}{p}\right) \tag{11}$$

$$mC^{-l(\beta)}n^{l(\beta)}p^{l(\beta)} \geq p^{-1}C^{-l(\beta)}n^{(l(\beta)+1)(1-\beta)\delta(\beta)} \geq p^{-1}C^{-l(\beta)}n^{\frac{1-\beta}{\sqrt{X}l(\beta)}} \tag{12}$$

---

[3] We need $mp = o(\alpha)$ to absorb the $[1 - o(1)]$ factor which arises in calculating the size of level 1 in the BFS tree. For higher levels, it is taken care of by (11). Also, as explained at the end of Section 2, we can assume that $p \leq n^{-0.1}$, without loss of generality.

Fix any $i = 1, \ldots, k - 2$ and fix any $A_i \subseteq V_i$, $|A_i| = m$. Using arguments similar to those given in Section 3 and using (11) and (12), we can arrive at the following analogues of Theorem 3.1 and Lemma 3.1. Define

$$F_j(n_i, \ldots, n_k, l(\beta), m) \ (\text{ shortly as } F_j(i, l(\beta), m)) \ \doteq mF_j(i, l(\beta))$$
$$E_{l(\beta)}(n_i, \ldots, n_k, j, m) \ (\text{ shortly as } E_{l(\beta)}(i, j, m)) \ \doteq mE_{l(\beta)}(i, j)$$

As in Section 3, assume $l(\beta)$ is even and also that $n_1 \geq \ldots \geq n_k$. For odd $l(\beta)$, the arguments are similar.

Theorem 5.1. *Let $m, l(\beta)$ be the same defined before with stated assumptions on m. For the expressions $E_{l(\beta)}(i, j, m)$, $j \geq i$, the following hold with probability at least $1 - e^{-\Omega(\alpha^2 \min(mnp, m^2 n^{l(\beta)} p^{l(\beta)+1} C^{-l(\beta)}))}$: For all but at most m vertices $y \in V_i - \cup_{l' < l(\beta)} D_{G_i}(A_i, l')$, and for all but at most m vertices $z \in (W_i - V_i) - \cup_{l' < l(\beta)} D_{G_i}(A_i, l')$, (assuming $z \in V_j, j > i$),*

$$n(A_i, y, G_i) \in \left((1 - \alpha)^{l(\beta)+1}, (1 + \alpha)^{l(\beta)+1} \right) E_{l(\beta)}(i, i, m) p^{l(\beta)+1} \tag{13}$$

$$n(A_i, z, G_i) \in \left((1 - \alpha)^{l(\beta)+1}, (1 + \alpha)^{l(\beta)+1} \right) E_{l(\beta)}(i, j, m) p^{l(\beta)+1} \tag{14}$$

$$n(A_i, z, G_i) - n(A_i, y, G_i) = \Omega(mn^{l(\beta)} p^{l(\beta)+1} C^{-l(\beta)} (kC)^{-l(\beta)/2}) \tag{15}$$

For $j \geq i$ and $0 \leq l' \leq l(\beta)$, define $n_j(A_i, l')$ to be the size of $D_{G_i}(A_i, l') \cap V_j$, that is, the number of vertices in V_j whose minimum distance in G_i from some vertex of A_i is l'. We have the following analogue of Lemma 3.1.

Lemma 5.1. *For the expressions $F_j(i, l(\beta), m)$, $j \geq i$, under the assumptions of Theorem 5.1, for A_i defined before, considering only potential edges incident on levels 0 through $l(\beta) - 1$, with probability at least $1 - e^{-\Omega(\alpha^2 mnp)}$, for all $j > i$,*

$$n_i(A_i, l(\beta)) \in \left((1 - \alpha)^{l(\beta)}, (1 + \alpha)^{l(\beta)} \right) F_i(i, l(\beta), m) p^{l(\beta)} \tag{16}$$

$$n_j(A_i, l(\beta)) \in \left((1 - \alpha)^{l(\beta)}, (1 + \alpha)^{l(\beta)} \right) F_j(i, l(\beta), m) p^{l(\beta)} \tag{17}$$

$$F_i(i, l(\beta), m) - F_j(i, l(\beta), m) \geq F_i(i, l(\beta), m)(kC)^{-l(\beta)/2} \tag{18}$$

Also, with probability at least $1 - e^{-\Omega(\alpha^2 mnp)}$, for each $l' < l(\beta)$, for all but at most m vertices $z \in D_{G_i}(A_i, l') - V_i$, z has some (in fact, $\Omega(np)$) neighbors in $D_{G_i}(A_i, l' + 1) \cap V_i$.

Description of *FindColor4* : We only describe how it separates V_i in G_i for any $i = 1, \ldots, k - 2$. For each independent m-set A in G_i, it computes $D_{G_i}(A, l')$ for $l' = 0, \ldots, l(\beta)$. When $A = A_i$, Theorem 5.1 holds true. After this, it computes $n(A, x, G_i)$ for each $x \in W_i$ which is at a distance of at least $l(\beta)$ from A in G_i and sorts these vertices in increasing or decreasing order (depending on the parity of $l(\beta)$) of their $n(A, x, G_i)$ values. Then, for each position in this sorted order, it assumes that all vertices upto this position in this order are from V_i and the remaining are from $W_i - V_i$. When $A = A_i$, by Theorem 5.1, there exists some position such that the approximation to V_i obtained thus is wrong only for at most m vertices each of V_i and $W_i - V_i$. Hence, by enumerating all subsets of size $\leq m$ each from approximate V_i and approximate $W_i - V_i$, we will correctly separate (during some iteration) all vertices with distance $\geq l(\beta)$ (from A_i) into those belonging to V_i or not. After this, further updating is done by inductively splitting (in the order $l' = l(\beta) - 1, \ldots, 2$) vertices at a distance l' from A_i into those belonging to V_i or not, by looking at the number of neighbors in approximate $D_{G_i}(A_i, l' + 1) \cap V_i$ (Lemma 5.1) and by enumerating all subsets of size $\leq m$. Finally, distance-1 vertices are put into approximate $W_i - V_i$ and A_i is put into approximate V_i. One can verify that this procedure runs in $O(n^{O(m(\log n))})$ time and correctly separates V_i during some iteration, with probability at least $1 - e^{-\Omega(\alpha^2 \min(mnp, m^2 n^{l(\beta)} p^{l(\beta)+1} C^{-l(\beta)}))}$. As in Section 3, the algorithm is repeated for $l' = 1, \ldots, \log n$, each time assuming that $l' = l(\beta)$. The algorithm works because for every enumeration of subsets, there is some iteration during which the approximate partial solution is exactly the actual partial solution the analysis needs. When $\delta(\beta) \geq \frac{(\sqrt{X}-1)}{\sqrt{X} l(\beta)(l(\beta)+1)}$, we can randomize the algorithm as explained in Section 4. We compute a random subset of $(l(\beta)-1)$-th level, grow the $l(\beta)$-th level from this subset and compute $n(A_i, x, G_i)$.

With these, we arrive at the following conclusions.

1. For $m, l(\beta)$ defined before with stated assumptions about them, *FindColor4* (G, n, k, m) k-colors G in $O(n^{O(m(\log n))})$ time.
2. The failure probability is at most $e^{-\Omega(\alpha^2 \min(mnp, m^2 n^{l(\beta)} p^{l(\beta)+1} C^{-l(\beta)}))}$ if no randomization is employed. If randomization is employed, it is at most $e^{-\Omega(\alpha^2 \min(mnp, m^2 n^{l(\beta)} p^{l(\beta)+1} n^{-(1-\beta)\kappa(l(\beta))} C^{-l(\beta)}))}$. Both these values are at most $e^{-\Omega(m(\log n)2^{\sqrt{X(\log n)}/2})}$.
3. For $m = 1$, we can replace each occurrence of "For all but at most m" in Theorem 5.1 and Lemma 5.1 by "For all". Hence, *FindColor4* can be modified to run in polynomial time with the same failure probability stated in Sections 3 and 4. This gives us an alternative *a.s.* polynomial algorithm.

In the next section, we use *FindColor4* to design a new technique for constructing *p.av.t.* algorithms for k-coloring $\mathcal{G}(n, p(n), k)$. One intermediate algorithm *Color2*(G, n, k, m, m) introduced in [15] will also be used in our technique. For $p \geq n^{-1}(\log n)^3$, it runs in $O(n^{3km})$ time and has failure probability at most e^{-nk} for $m \geq (\log n \cdot \log \log n)/p(n)$.

6 k-Coloring in Polynomial Average Time

The broad idea of the *p.av.t.* k-coloring is as follows. We start with a polynomial time algorithm A for k-coloring. If A does not succeed, then we apply a sequence $\langle A_1, \ldots, A_r \rangle$ of intermediate steps with increasing running times and decreasing failure probabilities. If all of them fail, we apply the brute-force coloring method.

Each algorithm A_i is applied when all of the previous intermediate steps $A_j, j < i$, have failed. By properly choosing the intermediate steps, we ensure that this approach yields a *p.av.t.* *k*-coloring algorithm. We choose each intermediate step as a call to *Color2*(G, n, k, m, m) followed by a call to *FindColor4* (G, n, k, m) for increasing values of m. Using arguments similar to those given in [14], one can obtain the following

Theorem 6.1. *Let G, p, ϵ be as before. Let A be a polynomial time (worst-case) algorithm such that A succeeds in *k*-coloring G with probability at least $1 - e^{-\Omega((\log n)2^{\sqrt{X(\log n)}/2})}$. Then there exists another algorithm $B(G, n, k)$ which always *k*-colors G and whose running time is polynomial on the average.*[4]

Corollary 6.1 *If $G \in \mathcal{G}(n, p, k)$ where $p(n) \geq n^{-1+\epsilon}$ with $\epsilon \geq X/\sqrt{\log n}$, then G can be *k*-colored in polynomial average time. The average is with respect to distribution of G.*

In Theorem 6.1, we require that A is a deterministic/randomized polynomial time (worst-case) algorithm which succeeds with probability $1 - o(1)$. In fact, Theorem 6.1 holds true even if A is only a deterministic/randomized polynomial time (average case) (but not necessarily succeeding always) algorithm that *k*-colors G with stated probability. The difference between A and B is that B always *k*-colors, but A need not. Again, the probabilities and average are evaluated with respect to input sample space and does not consider randomness (possibly) employed by A and B.

7   Conclusions and Open Problems

We presented a conceptually very simple BFS tree based approach to solve partitioning problems on random graphs. Our results on *p.av.t.* coloring significantly improve the previous results and hold for a wide range of distributions including those favoring sparse random graphs. We strongly believe that a more careful analysis may enlarge the range of p even further. Currently, the BFS approach is being applied to random instances of several other partitioning problems.

Our results on *k*-coloring can be adapted to the $\mathcal{GUC}(n, p, k)$ model also for the same range of p. The details will appear in the journal version. Some open problems related to these results are : **(i)** Can G be *k*-colored in *p.av.t.* for $\epsilon \geq c/n$, for some suitable positive constant c ? Even though it is known that such graphs can be *k*-colored almost surely in polynomial time, it would be desirable to show that this can be done in *p.av.t.* **(ii)** Can one design *p.av.t.*

---

[4] The probabilities and average are considered with respect to only the input sample space. The randomness which may have been employed by A and B are not taken into account in our statement and its proof. As mentioned before, the failure probability of the randomness (possibly) employed by A and B can be made as small as $e^{-\omega(n)}$ at a cost of polynomial increase in running time, as long as this randomness serves its purpose with at least polynomially low success probability.

algorithms for exact or constant factor approximations to minimum coloring random graphs $\mathcal{G}(n, p)$ where every one of the $\binom{n}{2}$ edges is chosen with equal probability p. For such graphs $\chi(G)$ is concentrated in a few values [7,12],

References

1. N. Alon and N. Kahale, "A spectral technique for coloring random 3-colorable graphs", Proceedings of the *26th Annual Symposium on Theory of Computing*, 1994, 346–355. 415
2. A. Blum and J. Spencer, "Coloring random and semi-random k-colorable graphs", *Journal of Algorithms*, **19**, 204-234, 1995. 415
3. M.E. Dyer and A.M. Frieze, "The solution of some random NP-hard problems in polynomial expected time", *Journal of Algorithms*, 10 (1989), 451–489. 415, 416
4. U. Feige and J. Kilian, "Heuristics for finding large independent sets, with applications to coloring semi-random graphs", *Proceedings of the 39th Annual IEEE Symposium on Foundations of Computer Science (FOCS'98)*, 1998, 674-683. 415
5. A.M. Freize and C. McDiarmid, "Algorithmic Theory of Random Graphs", *Random Structures and Algorithms*, 10:5-42, 1997. 416
6. M.R. Garey and D.S. Johnson, *Computers and Intractability: A Guide to the Theory of NP-Completeness*, W.H. Freemann, San Francisco, 1978. 415
7. G.R. Grimmett and C.J.H. McDiarmid, "On colouring random graphs", *Mathematical Proceedings of Cambridge Philosophical Society*, 77 (1975), 313–324. 426
8. A. Juels and M. Peinado, "Hiding Cliques for Cryptographic Security", Proceedings of the *9th Annual ACM-SIAM Symposium on Discrete Algorithms (SODA'98)*, 1998, 678–684. 416
9. D. Karger, R. Motwani, and M. Sudan, "Approximate Graph Coloring by Semi-Definite Programming", Proceedings of the *35th Annual Symposium on Foundations of Computer Science*, 1994, 2–13. 415
10. R. Karp, "The Probabilistic Analysis of some Combinatorial Search Algorithms", *Algorithms and Complexity*. J.F. Traub, ed., Academic Press, New York, 1976, 1–19. 416
11. L. Kučera, "A generalized encryption scheme based on random graphs", in *Graph Theoretic Concepts in Computer Science, WG'91*, Lecture Notes in Computer Science, LNCS 570, Springer-Verlag, 1991, 180–186. 416
12. E. Shamir and J. Spencer, "Sharp concentration of the chromatic number on random graphs $G_{n,p}$", *Combinatorica*, 7 (1987), 124-129. 426
13. C.R. Subramanian, "Minimum Coloring Random and Semi-Random Graphs in Polynomial Expected Time", Proceedings of the *36th Annual Symposium on Foundations of Computer Science*, 1995, 463-472. 415, 416
14. C.R. Subramanian, "Algorithms for Coloring Random k-colorable Graphs", *Combinatorics, Probability and Computing* (2000) 9, 45-77. 416, 418, 425
15. C.R. Subramanian, M. Furer and C.E. Veni Madhavan, "Algorithms for Coloring Semi-Random Graphs", *Random Structures and Algorithms*, Volume 13, No.2, pp. 125-158, September 1998. 415, 416, 422, 424
16. J.S. Turner, "Almost all k-colorable graphs are easy to color", *Journal of Algorithms*, 9 (1988), 63–82. 415

Restarts Can Help in the On-Line Minimization of the Maximum Delivery Time on a Single Machine
(Extended Abstract)

Marjan van den Akker[1], Han Hoogeveen[2], and Nodari Vakhania[*3]

[1] Department of Mathematical Models and Methods
National Aerospace Laboratory NLR
P.O. Box 90502, 1006 BM Amsterdam, The Netherlands
vdakker@nlr.nl,
[2] Department of Computer Science
Utrecht University
P.O. Box 80089, 3508 TB Utrecht, The Netherlands
slam@cs.uu.nl,
[3] Faculty of Sciences
State University of Morelos
FC Ueam, AV. Universidad 1001, Cuernavaca 62210, Mor. Mexico
nodari@correo.fc.uaem.mx

Abstract. We consider a single-machine on-line scheduling problem where jobs arrive over time. A set of independent jobs has to be scheduled on a single machine. Each job becomes available at its release date, which is not known in advance, and its characteristics, i.e., processing requirement and delivery time, become known at its arrival. The objective is to minimize the time by which all jobs have been delivered. In our model preemption is not allowed, but we are allowed to restart a job, that is, the processing of a job can be broken off to have the machine available to process an urgent job, but the time already spent on processing this interrupted job is considered to be lost. We propose an on-line algorithm and show that its performance bound is equal to 1.5, which matches a known lower bound due to Vestjens. For the same problem without restarts the optimal worst-case bound is known to be equal to $(\sqrt{5}+1)/2 \approx 1.61803$; this is the first example of a situation in which the possibility of applying restarts reduces the worst-case performance bound, even though the processing times are known.

Keywords: on-line algorithm, single-machine scheduling, restart, worst-case analysis.

1 Introduction

Until a few years ago, one of the basic assumptions made in deterministic scheduling was that all of the information needed to define the problem instance was

* Research partially supported by CONACyT grant #473100-5-28937A

M. Paterson (Ed.): ESA 2000, LNCS 1879, pp. 427–436, 2000.
© Springer-Verlag Berlin Heidelberg 2000

known in advance. This assumption is usually not valid in practice, however. Abandoning it has led to the rapidly emerging field of on-line scheduling; for an overview, see Sgall [1998]. Two on-line models have been proposed. The first one assumes that there are no release dates and that the jobs arrive in a list. The on-line algorithm has to schedule the first job in this list before it sees the next job in the list (e.g., see Graham [1966] and Chen, Van Vliet, and Woeginger [1994]). The second model assumes that jobs arrive over time. Besides the presence of release dates, the main difference between the models is that in the second model jobs do not have to be scheduled immediately upon arrival. At each time that the machine is idle, the algorithm decides which one of the available jobs is scheduled, if any. Within the second model we can make a further distinction depending on whether the characteristics of a job, e.g., its processing requirement, become known at its arrival or not. The first variant has been adopted by for example Phillips, Stein, and Wein [1998], Hall, Schulz, Shmoys, and Wein [1997], and Hoogeveen and Vestjens [2000]; see Vestjens [1997] for an overview. Motwani, Phillips, and Torng [1994] and Shmoys, Wein, and Williamson [1995] study the second variant. In this paper we consider a single-machine on-line scheduling problem that falls in the first category of the second model, that is, the jobs arrive over time and the characteristics of each job become known at its arrival time.

Our objective is to minimize the time by which all jobs have been delivered. In this problem, after their processing on the machine, the jobs need to be delivered, which takes a certain *delivery time*. The corresponding off-line problem is strongly NP-hard, but the preemptive version can be solved in polynomial time through the following on-line algorithm, which is known as the Preemptive Largest-delivery-time rule: *always let the machine process the job with the largest delivery time* (e.g., see Lawler, Lenstra, Rinnooy Kan, and Shmoys [1993]). If we adapt this rule to the situation without preemption, then we get the Nonpreemptive Largest-delivery-time rule: from among the available jobs choose the one with the largest delivery time; this algorithm solves the problem if all jobs are released at the same time. For the case with unequal release dates, Kise, Ibaraki, and Mine [1979] prove that it has a performance guarantee of 2, and that this is the best we can hope for if we do not allow the machine to be idle if there is an unprocessed job available. To improve on the ratio of 2, Hoogeveen and Vestjens [2000] developed an algorithm in which a waiting strategy is incorporated; they show that this algorithm has worst-case ratio $(\sqrt{5} + 1)/2 \approx 1.61803$, and that there cannot exist a deterministic on-line algorithm with a better worst-case ratio. Recently, Seiden [1998] has developed a randomized algorithm with worst-case bound 1.55370. This algorithm randomly selects one of two deterministic algorithms and runs it. Furthermore, Seiden has shown that no such algorithm can do better than 1.5.

The weak point of these algorithms from a practical point of view is the idle time strategy: the machine is left idle while there is a job available for processing, just to be prepared for the possible arrival of an urgent job (which may never come). We consider an extension of the standard model by allowing *restarts*: if an urgent job comes in while the machine is busy, then you do not have to wait until the current job is finished, but you can break off the processing of this job and start the urgent one. This is the same as applying preemption, but with one major difference: in case of preemption, you can continue at the place where you were when interrupted, that is, the time that the machine has spent on the job already is subtracted from its processing time, whereas in case of a restart *you have to start all over again with the processing of the job*. The time spent on a job whose processing is broken off is wasted and will be recorded as idle time, but allowing restarts reduces the impact of a wrong decision. Vestjens [1997] has shown that no on-line algorithm can have worst-case ratio smaller than 1.5 in case restarts are allowed. In this paper, we will present an on-line algorithm that achieves this bound.

The concept of allowing restarts is typical of on-line scheduling; it makes no sense if all information is available at time zero. As far as we know, this concept was first mentioned in on-line scheduling in a paper by Davis and Jaffe [1981], who among other problems, study the problem of minimizing the makespan on a set of m uniform parallel machines, where preemption is allowed, but where the processing times are unknown until the job has completed. They provide an instance that shows that no on-line algorithm can have a worst-case ratio less than $\sqrt{m}$, but they remark that this example can easily be countered by allowing restarts. Shmoys, Wein, and Williamson [1995] address the same problem and show that restarts can be of significant help in this case by providing an algorithm with worst-case ratio $O(\log m)$. This algorithm guesses the processing time of a job and runs it until the time at which it should been completed if the guess was right; if the job has not finished by then, then its processing is stopped, the guess is updated, and the job is scheduled anew, possibly on a faster machine. Recently, the concept of restarts has successfully been applied by Hoogeveen, Potts, and Woeginger [1999] for the on-line maximization of the number of on-time jobs on a single machine. They provide an algorithm with worst-case ratio $1/2$ (smaller than one, because it is a maximization problem) and show that this is best possible.

The paper is organized as follows. In Section 2, we repeat the example by Vestjens that shows that the worst-case ratio is bounded from below by 1.5. In Section 3, we present our algorithm and show that its worst-case ratio is equal to 1.5. Finally, in Section 4 we present some conclusions and directions for future research.

2   Why We Cannot Do Better than 3/2

We first introduce some notation that we will use throughout the paper. We use J_j to denote job j, and r_j, p_j, and q_j to denote the release date, processing requirement, and delivery time of J_j, respectively; the delivery time q_j of J_j is also called the *tail* of J_j. We denote by $S_j(\sigma)$, $C_j(\sigma)$, and $L_j(\sigma)$, the starting time, completion time, and the time by which J_j is delivered in schedule σ.

We use σ to denote the schedule produced by the heuristic and π to denote an optimal schedule.

Theorem 1 (Vestjens, 1997). *Any deterministic on-line algorithm for minimizing $L_{\max}$ on a single machine allowing restarts has a performance bound of at least $3/2$.*

Proof. The proof is based on an adversary argument. Given any on-line algorithm, the adversary presents it an instance that is extended over time depending on the choices made by the algorithm.

The initial instance consists of a job J_0 that is released at time 0 with processing time equal to 1 and tail equal to 0. If the algorithm starts processing this job at time $t < 1$, then the adversary releases a job with processing time 0 and tail 1; such a job is called a *jammer*. If the algorithm decides to break off the processing of J_0, execute the jammer, and start J_0 again, then the adversary releases a new jammer until time $t = 1$. The first time that the algorithm decides to start J_0 at a time $t \in [1, 2]$ a job J_1 with processing time $2 - t$ and delivery time 0 is released.

The performance of the algorithm depends on the time S at which it starts J_0 for the last time. If $S < 1$, then $L_{\max}(\sigma) = S + 2$, since the machine completes J_0 at time $S + 1$, after which the last jammer has to be delivered, whereas in π the machine waits for the last jammer to arrive at time $S + \epsilon$, which results in $L_{\max}(\pi) = S + 1 + \epsilon$. Hence, for $S < 1$, the ratio approaches $3/2$, when ϵ approaches zero.

If $S \geq 1$, then $L_{\max}(\sigma) = 3$, since J_1 with $p_1 = 2 - S$ has to wait until time $S + 1$ to get started, whereas in π job J_0 is executed in the interval $[0, 1]$, after which the jammers are processed, and finally J_1 in the interval $[S, 2]$, which results in $L_{\max}(\pi) = 2$. Hence, for $S \geq 1$ the ratio amounts to $3/2$ as well, which implies that no on-line algorithm can choose S such that its performance ratio is smaller than $3/2$.   □

3   An On-Line 3/2-Approximation Algorithm

Throughout the remainder of this paper we will use the following notation:

- For a given set of jobs S, $p(S)$, $q(S)$, and $r(S)$ denote the total processing time, the minimum tail, and the minimum release date of the jobs in S, respectively.
- $Y(t)$ denotes the set of unfinished jobs that are available at time t.
- $p(t)$ denotes the index of the job with the largest processing time in $Y(t)$.
- $q(t)$ denotes the index of the job with the largest tail in $Y(t)$.
- $L(t)$ denotes a lower bound on the maximum delivery time of the the instance known at time t. A job J_0 is called *big* at time t if $p_0 > 1/2 L(t)$.

We first present the algorithm in pseudo-code and after that summarize its working. Note that in our algorithm ties are broken arbitrarily. We will describe how to find $L(t)$ after the summary of the algorithm.

Algorithm

STEP 0. If at time t the machine is not idle and no new job becomes available, then wait until the first such event.

STEP 1. Determine $p(t)$ and $q(t)$. If at time t no new job was released, then go to STEP 3.

STEP 2. Update $L(t)$ and determine whether $J_{p(t)}$ is big. If the currently processed job is big and if $q_{q(t)} > t$, then break off the processing of $J_{p(t)}$, start $J_{q(t)}$, and go to STEP 0.

STEP 3. If $J_{p(t)}$ is not big at time t, then schedule $J_{q(t)}$ and go to STEP 0.

STEP 4. If the processing of $J_{p(t)}$ has been broken off at least once, then go to STEP 8.

STEP 5. If $J_{p(t)}$ is the only unfinished job available, then schedule $J_{p(t)}$ and go to STEP 0.

STEP 6. If $p(t) \neq q(t)$, then schedule $J_{q(t)}$ and go to STEP 0.

STEP 7. If $p(t) = q(t)$, then schedule the job with the second largest tail in $Y(t)$ and go to STEP 0.

STEP 8. Determine t_0 as the last time at which $J_{p(t)}$ got restarted. If $q_{q(t)} > t_0$, then schedule $J_{q(t)}$ and $J_{p(t)}$ otherwise. Go to STEP 0.

We can summarize the working of the algorithm as follows. The basic strategy is to execute the job with the largest tail (STEP 3), with some modifications if there is a big job available. If this big job is processed at some time t and a new job with tail bigger than t gets released, then the algorithm breaks off the processing of the big job and starts the new job (STEP 2). If the machine is idle, then the algorithm checks whether processing has been contiguous since the release date of the big job (STEP 4). If this is the case, then the algorithm processes the available unfinished jobs in order of tail size (STEP 6), except for the big job (STEP 7), which will only be started if there is no other job available (STEP 5), unless somewhere later in time this currently big job is considered to have become an ordinary job instead of a big one. If the processing of the currently big job has been broken off already once, then the algorithm selects the big job, unless there is a job with a tail larger than the last time at which the big job got restarted (STEP 8).

What is left is the derivation of $L(t)$. We compute $L(t)$ as the maximum of two separate lower bounds $L_1(t)$ and $L_2(t)$ on the maximum delivery time of the the instance known at time t. The values $L_i(t)$ $(i = 1, 2)$ are computed as follows.

- Let $I(t)$ denote the set of jobs known at time t. Then $L_1(t) = p(I(t))$.
- Let J_1 be the last released job, other than $J_{p(t)}$, that has a tail that is bigger than its release date. Then $L_2(t) = p_{p(t)} + r_1 + p_1$, where $r_1 = p_1 = 0$ if such a job does not exist.

Lemma 1. $L(t)$ *is a lower bound on the optimum delivery time of the instance known at time* t.

Proof. Let π_t denote an optimum schedule for the instance known at time t. Since the tails are nonnegative, $L_{\max}(\pi)$ is greater than equal to the makespan of π_t for which $L_1(t) = p(I(t))$ forms a lower bound.

Now consider the bound $L_2(t)$. If J_1 does not exist, then $L_2(t) = p_{p(t)}$, which is an obvious lower bound. Otherwise, we know that in π_t $J_{p(t)}$ either precedes or succeeds J_1. In the first case, we have $L_{\max}(\pi_t) \geq p_{p(t)} + p_1 + q_1 > p_{p(t)} + p_1 + r_1 = L_2(t)$, whereas in the second case we find that $L_{\max}(\pi_t) \geq r_1 + p_1 + p_{p(t)} = L_2(t)$. Hence, $L_2(t) \leq L_{\max}(\pi_t)$. □

From now on, we will use the name *jammer* for each small job with a relatively large tail that interrupts the processing of some big job. Before proving the desired worst-case behavior of our algorithm, we first present three lemmas that we will need in the proof of Theorem 2. The first one concerns a well-known lower bound on $L_{\max}(\pi)$ that we will use a lot in our proof.

Lemma 2. $L_{\max}(\pi) \geq r(S) + p(S) + q(S)$ *for any subset* S *of jobset* $I(t)$.

Proof. Since no job in S can start before time $r(S)$, the last job in S will be completed at time $r(S) + p(S)$ at the earliest in any schedule. Since each job in S has a tail of size $q(S)$ at least, the earliest time at which the last job in S can be delivered is time $r(S) + p(S) + q(S)$. □

Lemma 3. *If the release of some jammer* J_1 *at time* r_1 *leads to breaking off the processing of some big job* J_0, *then* $L_{\max}(\pi) \geq 2(r_1 + p_1)$.

Proof. In the proof of Lemma 1 we have shown that $L_{\max}(\pi) \geq p_0 + p_1 + r_1$ in this case. Moreover, we know that at time r_1 job J_0 must have been big, since it was interrupted, which implies that $2p_0 > L(r_1) \geq L_2(r_1) = p_0 + p_1 + r_1$, that is, $p_0 > r_1 + p_1$. Therefore, $L_{\max}(\pi) \geq p_0 + p_1 + r_1 > 2(r_1 + p_1)$. □

Lemma 4. *Suppose that at a given time* t *job* $J_{p(t)}$ *is big. Let* S *denote the set of jobs other than* $J_{p(t)}$ *that were known at time* t. *Then* $p(S) \leq 1/2 L_{\max}(\pi)$.

Proof. Since the tails are nonnegative, $L_{\max}(\pi) \geq p(S) + p_{p(t)}$, which implies that $p(S) \leq L_{\max}(\pi) - p_{p(t)}$. As $J_{p(t)}$ is big at time t, we know that $2p_{p(t)} > L(t) \geq L_1(t) \geq p_{p(t)} + p(S)$, that is, $p(S) < p_{p(t)}$. Adding the two inequalities yields $2p(S) < L_{\max}(\pi)$. □

Theorem 2. *The above algorithm delivers a schedule* σ *with* $L_{\max}(\sigma) \leq 3/2 L_{\max}(\pi)$.

Proof. Let J_c be the critical job, that is, the first job that attains the maximum delivery time in σ. We define the interference job, which we denote by J_b, as the last job in σ before J_c with tail smaller than q_c; *if there is idle time between the completion of* J_b *and* J_c, *then we say that there is no interference job.* We use Q to denote the set of jobs after J_b up to and including J_c in σ; if σ does not contain an interference job, then we let Q contain the last set of contiguously processed jobs in σ.

The structure of the proof will be to assume that $L_{\max}(\sigma) > 3/2L_{\max}(\pi)$ and then find a contradiction. Unfortunately, we need a number of case distinctions. We first show that there must exist an interference job, and that this job must be the last big job in the instance. We put a $\square$ sign to denote the end of the proof of each case and subcase.

Case 1. There is no interference job.
The proof will be given in the full version of the paper. $\square$

Case 2. The interference job J_b is not big at its start time S_b, or at a later time some new big job, which we denote by J_0, gets released.
The proof will be given in the full version of the paper. $\square$

As a consequence, we know that the interference job must be equal to J_0, which is the last big job in the instance, and that it is big at its start time S_0. Hence, $L_{\max}(\sigma) = S_0 + p_0 + p(Q) + q_c$. Again we need to make some case-distinctions. We distinguish between the situation that in π job J_0 precedes at least one of the jobs in Q and the situation that in π job J_0 succeeds all of the jobs from Q.

Subcase A. In schedule π job J_0 precedes one of the jobs in Q.
There are three situations possible, which we will work out below:

(i.) **Schedule σ contains no idle time, or the last idle time period was due to unavailability of jobs.**
(ii.) **J_0 is started immediately after the completion of a jammer.**
(iii.) **Between the completion of the last jammer and S_0 some set A of jobs were scheduled.**

Proof of (i). The proof will be given in the full version of the paper. $\square$

Proof of (ii). We are in the situation that some jammer got completed at time S_0 and $L_{\max}(\sigma) = S_0 + p_0 + p(Q) + q_c$. Because of the jammer that got completed at time S_0, Lemma 3 implies that $L_{\max}(\pi) \geq 2S_0$. Moreover, $L_{\max}(\pi) \geq p_0 + p(Q) + q_c$, since J_0 precedes a job in Q. Hence, we find that $L_{\max}(\sigma) - L_{\max}(\pi) \leq S_0 \leq 1/2L_{\max}(\pi)$.
$\square$

Proof of (iii). Denote the last jammer by J_1; it got started at time S_1 and completed at time C_1. Define A as the set of jobs that were executed in σ in the period $[C_1, S_0]$.

If we take a careful look at our algorithm, then we see that there are two possible reasons for scheduling the jobs from A instead of J_0:

♣ The last jammer J_1 did not break off the processing of J_0 but the processing of some other big job. Hence, the processing of J_0 has never been broken off, which implies that the algorithm prefers any other available job to J_0 as long as J_0 is big.

♠ Job J_1 did break off the processing of J_0, which implies that all jobs from A have a tail is bigger than S_1, which is the last time that J_0 got restarted.

Proof of ♣. Since the last jammer J_1 did not break off the processing of J_0, job J_0 must have been released after the start of J_1, as a jammer only interrupts a big job; hence, $r_0 > S_1$. As J_0 is big at time S_0 and got started though, it must have been the only available unfinished job at that time, which implies that none of the jobs from Q were available at time $S_0 \geq r_0$. Combining this, we find that $L_{\max}(\pi) \geq r_0 + p_0 + p(Q) + q_c$, since J_0 precedes at least one job from Q. But $L_{\max}(\sigma) = S_1 + p_1 + p(A) + p_0 + p(Q) + q_c$, from which we find that $L_{\max}(\sigma) - L_{\max}(\pi) \leq S_1 - r_0 + p_1 + p(A) < p_1 + p(A) \leq 1/2 L_{\max}(\pi)$ because of Lemma 4.   □

Proof of ♠. The last time that J_0 got restarted was at time S_1 when J_1 came in; since J_1 interrupted J_0, we know that $q_1 > S_1$. We further have that $q_j > S_1$ for each job $J_j \in A$, as the jobs in A were preferred to J_0. To prove the bound of $3/2$, we distinguish between the cases $q_c \leq S_1$ and $q_c > S_1$.

First, suppose that $q_c < S_1 \leq q(A)$. Since J_0 precedes a job from Q in π, we find that $L_{\max}(\pi) \geq p_0 + p(A) + p(Q) + q_c$. But this implies that $L_{\max}(\sigma) - L_{\max}(\pi) \leq S_0 - p(A) = C_1 \leq 1/2 L_{\max}(\pi)$ because of the jammer that got completed at time C_1 (Lemma 3).

In the second case, we know that the algorithm would have preferred any job from Q to J_0 if a job from Q would have been around at time S_0 since $q_c > S_1$, which implies that $r(Q) > S_0$. Hence, $L_{\max}(\pi) \geq r(Q) + p(Q) + q_c > S_0 + p(Q) + q_c$, from which we obtain that $L_{\max}(\sigma) - L_{\max}(\pi) \leq p_0$. We compute a second bound on $L_{\max}(\pi)$ based on the jobs in A and Q and the jammer J_1, which all have a tail of size at least equal to S_1. Since J_0 is not the last job in π, we find that $L_{\max}(\pi) \geq p_0 + p(Q) + p(A) + p_1 + S_1 = S_0 + p_0 + p(Q)$, from which we derive that $L_{\max}(\sigma) - L_{\max}(\pi) \leq q_c$. Combining this with the previous expression, we find that $2(L_{\max}(\sigma) - L_{\max}(\pi)) \leq p_0 + q_c \leq L_{\max}(\pi)$, since J_0 precedes a job from Q, which gives us the desired bound.   □

Subcase B. Job J_0 succeeds all jobs from Q in the optimal schedule.
We consider the following two possibilities:

(1.) $r(Q) > S_0$, that is, none of the jobs from Q is available at time S_0.
(2.) Some job J_q from Q is available at time S_0.

Proof of (1). As $r(Q) > S_0$, we find that $L_{\max}(\pi) \geq S_0 + p(Q) + q_c$ and $L_{\max}(\pi) \geq S_0 + p(Q) + p_0$ (since J_0 succeeds all jobs from Q in π). Hence, we find that $L_{\max}(\sigma) - L_{\max}(\pi) \leq p_0$ and that $L_{\max}(\sigma) - L_{\max}(\pi) \leq q_c$, respectively. But as the processing of J_0 was not broken off because of the release of the jobs in Q, we know that $q_c \leq r(Q)$. Hence, $L_{\max}(\pi) \geq r(Q) + p(Q) + p_0 \geq q_c + p_0 \geq 2(L_{\max}(\sigma) - L_{\max}(\pi))$, from which the $3/2$ bound follows immediately.

Proof of (2). As J_0 was started by the algorithm although another job was available, we know that J_0 must have been restarted at least once. Let J_1 be the last jammer in σ; J_1 starts at time S_1. Let A denote the set of jobs that are processed in the interval $[C_1, S_0]$; this set can be empty. Hence, $L_{\max}(\sigma) = S_1 + p_1 + p(A) + p_0 + p(Q) + q_c$.

We will now derive a bound on the makespan of π; this is a lower bound on $L_{\max}(\pi)$. Since the algorithm preferred the jobs in A to J_0, we know that each job $J_j \in A$ has a tail $q_j > S_1$. Hence, we know that each job in A got released at or after time S_1, which also holds for J_1. As J_0 succeeds the jobs in Q, we know that $L_{\max}(\pi) \geq C_{\max}(\pi) \geq \min\{r(Q), S_1\} + p_1 + p(A) + p_0 + p(Q)$. If $r(Q) \geq S_1$, then $L_{\max}(\sigma) - L_{\max}(\pi) \leq q_c \leq S_1$, since J_0 was started at time S_0 although J_q was available, and we are done then because of Lemma 3. Hence, we assume that $r_q < S_1$. We will show now that $r_q \geq q_q$. To that end, we take a closer look at the situation at time r_q. Since the algorithm prefers any job to a big job whose processing has never been broken off, we know that the algorithm started J_0 for the first time before time r_q. If J_0 was being processed at time J_0, then $q_q \leq r_q$, since J_q would have interrupted J_0 (and be completed a long time ago) otherwise. If on the other hand some other job occupied the machine at time r_q, then $q_q \leq t_0 \leq r_q$, where we use t_0 to denote the last time at which J_0 got interrupted before time r_q. Hence, we may conclude that $r(Q) = r_q \geq q_q \geq q_c$. But this implies that $L_{\max}(\pi) \geq r_q + p_1 + p(A) + p_0 + p(Q) \geq p_1 + p(A) + p_0 + p(Q) + q_c$, and we find that $L_{\max}(\sigma) - L_{\max}(\pi) \leq S_1 \leq 1/2L_{\max}(\pi)$ because of Lemma 3.

This last contradiction settles the last case in our proof. $\square$

4 Conclusion and Directions for Further Research

In this paper we have presented an on-line algorithm with worst-case ratio $3/2$. It is easily verified that our algorithm can be made to run in $O(n \log n)$ time, since the bound $L(t)$ can be updated in constant time at each of the at most n times that a new job comes in. As the best possible deterministic on-line algorithm for the case in which restarts are not allowed has worst-case performance ratio $(\sqrt{5}+1)/2 \approx 1.61803$, we have shown that it may be beneficial to allow restarts in case of on-line scheduling. An interesting question is of course to determine the effect of a partial restart, in which a fraction of the work spent on a job whose processing is broken off is subtracted from its processing time. It would particularly interesting to find out the relation between the worst-case ratio and this 'reduction percentage' p; it might be a straight line between 1.5 (complete restart) and 1 (preemption allowed). Woeginger (1999) has recently shown a lower bound of $1 + (1-p)/2$ by adapting the example shown in Lemma 2.

References

1. B. CHEN, A. VAN VLIET, G.J. WOEGINGER (1994). New lower and upper bounds for on-line scheduling. *Operations Research Letters 16*, 221-230.
2. E. DAVIS AND J.M. JAFFE (1981). Algorithms for scheduling tasks on unrelated processors. *Journal of the Association of Computing Machinery 28*, 721-736.
3. R.L. GRAHAM (1966). Bounds for certain multiprocessing anomalies. *Bell System Technical Journal 45*, 1563-1581.
4. L.A. HALL, A.S. SCHULZ, D.B. SHMOYS, J. WEIN (1997). Scheduling to minimize average completion time: off-line and on-line approximation algorithms. *Mathematics of Operations Research 22*, 513-544.
5. J.A. HOOGEVEEN, C.N. POTTS, AND G.J. WOEGINGER (1999). *On-line scheduling on a single machine: Maximizing the number of early jobs*. Report Woe-48, Technical University Graz.
6. J.A. HOOGEVEEN AND A.P.A. VESTJENS (2000). A best possible deterministic on-line algorithm for minimizing maximum delivery time on a single machine. *SIAM Journal on Discrete Mathematics 13*, 56-63.

7. H. KISE, T. IBARAKI, H. MINE (1979). Performance analysis of six approximation algorithms for the one-machine maximum lateness scheduling problem with ready times. *Journal of the Operations Research Society Japan 22*, 205-224.

8. E.L. LAWLER, J.K. LENSTRA, A.H.G. RINNOOY KAN, D.B. SHMOYS (1993). Sequencing and scheduling: algorithms and complexity. S.C. Graves, P.H. Zipkin, A.H.G. Rinnooy Kan (eds.). *Logistics of Production and Inventory; Handbooks in Operations Research and Management Science, Vol. 4*, North-Holland, Amsterdam, 445-522.

9. R. MOTWANI, S. PHILLIPS, E. TORNG (1994). Non-clairvoyant scheduling. *Theoretical Computer Science 130*, 17-47.

10. C. PHILLIPS, C. STEIN, J. WEIN (1998). Minimizing average completion time in the presence of release dates. *Mathematical Programming (B) 28*, 199-223.

11. S.S. SEIDEN (1998). *Randomized Online Scheduling with Delivery Times*, Report Woe-24, Technical University Graz.

12. J. SGALL (1998). On-line scheduling. A. Fiat, G.J. Woeginger (eds.). *On-Line Algorithms: the State of the Art*, Lecture Notes in Computer Science 1442, Springer, Berlin, 196-231.

13. D.B. SHMOYS, J. WEIN, D.P. WILLIAMSON (1995). Scheduling parallel machines on-line. *SIAM Journal on Computing 24*, 1313-1331.

14. A.P.A. VESTJENS (1997). *On-line machine scheduling*, PhD Thesis, Eindhoven University of Technology.

15. G.J. WOEGINGER (1999). Personal communication.

Collision Detection Using Bounding Boxes: Convexity Helps

Yunhong Zhou and Subhash Suri

Department of Computer Science, Washington University, St. Louis, MO 63130
{yzhou,suri}@cs.wustl.edu

Abstract. We consider the use of bounding boxes to detect collisions among a set of *convex* objects in R^d. We derive tight bounds on the ratio between the number of box intersections and the number of object intersections. Confirming intuition, we show that the performance of bounding boxes improves significantly when the underlying objects are all convex. In particular, the ratio is $\Theta(\alpha^{1-1/d}\,\sigma_{\text{box}}^{1/2})$ if each object has aspect ratio at most α and the set has scale factor σ_{box}. More significantly, the bounding box performance ratio is

$$\Theta\left(\alpha_{\text{avg}}^{\frac{2(1-1/d)}{3-1/d}}\ \sigma_{\text{box}}^{\frac{1}{3-1/d}}\ n^{\frac{1-1/d}{3-1/d}}\right)$$

if only the *average* aspect ratio α_{avg} of the n objects is known. These bounds are the best possible as we show matching lower bound constructions. The case of convex objects is interesting for several reasons: first, in many applications, the objects are either naturally convex or are approximated by their convex hulls for convenience; second, in some applications, the penetration of convex hulls is interpreted as collision; and finally, the question is interesting from a theoretical standpoint.

1   Introduction

Bounding boxes are commonly used in geometric algorithms as inexpensive filters to avoid processing object pairs that are trivially non-intersecting. Due to their simpler shape, intersecting bounding boxes is almost always more efficient than intersecting complex objects. Thus, bounding boxes allow an algorithm to quickly perform a "trivial reject" test that prevents more costly processing in unnecessary cases. For instance, many rendering algorithms use bounding boxes, such as the visible-surface determination [3], the view-frustum culling [5] and the reprojected pixel-imaging [10]. Many modeling algorithms also use bounding boxes for defining complex shapes as boolean combinations of simpler shapes [7], or for verifying the clearance of parts in an assembly [4]. Animation algorithms use bounding boxes for collision detection in path planning [9] or simulation of physically-based motion [1,8,11].

The use of bounding boxes in collision detection is representative of its use in other algorithms, so we will focus on collision detection. Many collision detection algorithms break their work into two phases, which we call *broad phase*

M. Paterson (Ed.): ESA 2000, LNCS 1879, pp. 437–448, 2000.
© Springer-Verlag Berlin Heidelberg 2000

and *narrow phase*. In the broad phase, the algorithm determines all pairs whose bounding boxes intersect. The narrow phase then performs a detailed intersection test on each pair found by the broad phase. The two phases have distinct characteristics, and have often been treated as independent problems for research. In particular, suppose we have n objects, each of size at most m. An efficient broad phase algorithm must avoid checking all $\Omega(n^2)$ pairs, while an efficient narrow phase algorithm must avoid the $\Omega(m^2)$ complexity.

Consider a set $\mathcal{S}$ of n geometric objects in d dimensions. While several different definitions of a bounding box are possible, we will use the smallest enclosing L_∞-norm ball, namely, the cube. (The choice of this box is purely for technical convenience, and our result hold for all the standard forms of bounding boxes, including the commonly used axis-aligned rectangular box.) Given a geometric object P, let $b(P)$ denote the smallest L_∞ ball containing P. Let $K_o(\mathcal{S})$ denote the number of intersecting object-pairs in $\mathcal{S}$, and let $K_b(\mathcal{S})$ denote the number of pairs whose bounding-boxes intersect. Then, the *effectiveness* of bounding boxes as a broad phase heuristic can be measured by the following ratio:

$$\rho(\mathcal{S}) \; = \; \frac{K_b(\mathcal{S})}{n + K_o(\mathcal{S})} \; .$$

The denominator represents the *best-case* work done by an ideal object intersection algorithm, so the ratio can be seen as the relative performance measure of the heuristic.

Empirical evidence suggests that bounding boxes are effective in practice, meaning that ratio ρ is quite small. On the other hand, simple examples show that $\rho(\mathcal{S})$ can be $\Omega(n)$ in the worst-case. (See Figure 1.) Such examples are normally written off as being "pathological" and practitioners would insist that they do not occur in practice. However, it does beg the question: what is a pathological input, and whether there is a quantifiable parameter that measures the degree to which an input scene is pathological. The recent work by Suri, Hubbard and Hughes [12] and Zhou and Suri [13] investigates the phenomenon and shows that the effectiveness of bounding boxes in collision detection can be precisely calibrated by two natural shape parameters, *aspect ratio* and *scale factor*. An early result by Edelsbrunner [2] also considers this problem, in a different context, and proves a weaker version of the theorem in [12]. In [2], both the aspect ratio and the scale factor are assumed to be constant. By contrast, the result in [12] derives the explicit dependence on the ratio ρ in terms of the two parameters.

In this paper, we extend the work of [13] in one important direction: analyze the performance of bounding boxes when all objects in the collection are *convex*. At first glance, our result is perhaps not all that surprising. We show that the ratio of box-box intersections to object-object intersection shrinks, implying that convexity helps. However, the result is technically quite difficult, requiring a rather delicate analysis as well as new geometric insights. We are able to establish tight asymptotic bounds on the performance of bounding boxes, both when all objects have a bounded aspect ratio, as well as when *only the average aspect ratio* is bounded. It is worth noting that even for convex objects in R^d, there is no good algorithm for determining all intersecting pairs in an *output sensitive fashion*. Even in two dimensions, the best algorithm for finding all intersecting pairs in a set of n convex polygons takes $O(n^{4/3} + K_o)$ time [6].

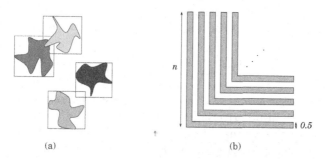

(a)   (b)

Fig. 1. (a) An example of four objects, with $K_b = 2$ and $K_o = 1$. (b) $\Omega(n^2)$ bounding-box intersections without any object intersection.

The case of convex objects is interesting for several reasons. First, in many applications, the objects are either naturally convex or are approximated by their convex hulls for convenience. Second, in some applications, the penetration of convex hulls is interpreted as collision; for instance, interlocking human figures might be legitimately considered colliding. Finally, the question is interesting from a theoretical standpoint—convexity is one of the most fundamental geometric properties. While convexity does indeed improve the performance of bounding boxes, *proving* that fact is not at all trivial. In particular, the earlier analyses have relied on the fact that small aspect ratio implies a large core, and used that to conclude that many box-box intersections must lead to many core-core intersections. In order to show a better bound on the ratio $\rho(\mathcal{S})$ for convex objects, we need to go beyond core-core intersections, and that requires non-trivial exploitation of convexity. Before stating our results precisely, let us introduce some necessary definitions.

1.1   Aspect Ratio and Scale Factor

Given a solid object P in d-space, let $b(P)$ denote the smallest L_∞ ball containing P, and let $c(P)$ denote the largest L_∞ ball contained in P. The *aspect ratio* of P is defined as

$$\alpha(P) = \frac{\text{vol}(b(P))}{\text{vol}(c(P))} \; ,$$

where vol(P) denotes the d-dimensional volume of P. We will call $b(P)$ the *enclosing box*, and $c(P)$ the *core* of P. Thus, the aspect ratio measures the volume of the enclosing box relative to the core. For a set $\mathcal{S} = \{P_1, P_2, ..., P_n\}$

of objects in d-space, its *average* aspect ratio is defined as:

$$\alpha_{\text{avg}}(\mathcal{S}) = \frac{1}{n} \sum_{i=1}^{n} \alpha(P_i) .$$

We also need a bound on the *scale factor*, which measures the *disparity* between the largest and the smallest objects. We take the ratio between the largest box and the smallest box, and call it the *box scale factor* of $\mathcal{S}$.

$$\sigma_{\text{box}}(\mathcal{S}) = \max_{i,j} \frac{\text{vol}(b(P_i))}{\text{vol}(b(P_j))} .$$

(One can also define a *core scale factor* as the ratio between the largest to the smallest core. We also derive our results in terms of σ_{cor}, and mention them in the conclusions.) Our main result on the performance of broad phase with convex objects are summarized in the following two theorems.

Theorem 1. *Let $\mathcal{S}$ be a set of n convex objects in R^d, where d is a constant. If each object of $\mathcal{S}$ has aspect ratio at most α and $\mathcal{S}$ has box scale factor σ_{box}, then* $\rho(\mathcal{S}) = \Theta(\alpha^{1-1/d}\, \sigma_{\text{box}}^{1/2})$.

Theorem 2. *Let $\mathcal{S}$ be a set of n convex objects in R^d, where d is a constant. If the objects of $\mathcal{S}$ have average aspect ratio α_{avg} and box scale factor σ_{box}, then* $\rho(\mathcal{S}) = \Theta(\alpha_{\text{avg}}^{\frac{2(1-1/d)}{3-1/d}}\, \sigma_{\text{box}}^{\frac{1}{3-1/d}}\, n^{\frac{1-1/d}{3-1/d}})$.

Thus, for instance, if a set of n convex objects has average aspect ratio α_{avg}, then $\rho(\mathcal{S})$ is $\Theta(\alpha_{\text{avg}}^{2/5}\, \sigma_{\text{box}}^{2/5}\, n^{1/5})$ in two dimensions, and $\Theta(\alpha_{\text{avg}}^{1/2}\, \sigma_{\text{box}}^{3/8}\, n^{1/4})$ in three dimension. By contrast, *without the convexity assumption*, the best bound possible is $\Theta(\alpha_{\text{avg}}^{2/3}\, \sigma_{\text{box}}^{1/3} n^{1/3})$ [13], showing that the performance guarantee significantly improves if the objects are known to be convex.

1.2   Organization

Section 2 proves several key packing and intersection lemmas, which crucially exploit the convexity of objects. In Section 3, we analyze the performance of bounding boxes when the aspect ratio of each object is bounded. In Section 4, we extend this analysis to the case when only the average aspect ratio is bounded. In Section 5, we mention several extensions and related results.

2   Packing and Intersection Lemmas

We establish several key geometric facts in this section, which rely heavily on the convexity of objects. These lemmas are crucial in our analysis of bounding boxes. Let $\mathcal{S} = \{P_1, P_2, \ldots, P_n\}$ be a set of n convex objects in R^d, where the dimension d is assumed to be a constant. Recall that $b(P)$ and $c(P)$, respectively, denote the smallest enclosing and largest enclosed L_∞-boxes for P. Our first lemma gives a lower bound on the volume of a convex object P as a function of the volumes of $b(P)$ and $c(P)$. Due to lack of space, we omit most of the proofs from this extended abstract.

Lemma 1. *Given a convex object P in R^d with its bounding box volume $\alpha = \mathrm{vol}(b(P))$ and core volume $\beta = \mathrm{vol}(c(P))$. Then, $\mathrm{vol}(P) = \Omega\left(\beta(\frac{\alpha}{\beta})^{1/d}\right)$.*

Our next lemma formulates a type of packing result: lower bounding the number of intersections among convex objects packed in a box.

Lemma 2. *Consider a size α box B in R^d containing m objects, each with volume at least v. Then, the number of pairwise intersections among these objects is at least $\frac{1}{2}\frac{m^2}{\alpha/v} - \frac{m}{2}$.*

Proof. Let $\mathcal{S} = \{P_1, P_2, \ldots, P_m\}$ be the set of objects. Without loss of generality, assume that all objects have the same volume, namely, $\mathrm{vol}(P_i) = v$ for all i; otherwise, we can shrink each object appropriately, without increasing the number of intersections. For each point x in B, let $f(x)$ denote the number of objects in $\mathcal{S}$ that contain x. The function f is nonnegative, bounded by m. Furthermore, those m objects partition B into a finite number of small regions where f is constant among each small region. Thus both f and f^2 are integrable. $\int_B f(x)\,dx = \sum_{1 \le i \le m} \mathrm{vol}(P_i) = mv$. Also observe that $\int_B 1\,dx = \alpha$. Using Cauchy's Inequality, $\left(\int_B f(x)^2\,dx\right)\left(\int_B 1\,dx\right) \ge \left(\int_B f(x)\,dx\right)^2$, we get $\int_B f(x)^2\,dx \ge \frac{m^2v^2}{\alpha}$. Thus,

$$\int_B \binom{f(x)}{2}\,dx = \frac{1}{2}\int_B f(x)^2\,dx - \frac{1}{2}\int_B f(x)\,dx \ge \frac{m^2v^2}{2\alpha} - \frac{mv}{2} \qquad (1)$$

The integral in (1) sums over each point of B the number of intersecting pairs. This counts a pair of intersecting objects many times, once for each point in their common intersection. Given two objects P_i and P_j, their intersection is counted for each point $x \in P_i \cap P_j$. Since each object has volume v, the intersection of two objects cannot have volume more than v. Thus, if we let K denote the number of distinct pairs of intersecting objects, then $\int_B \binom{f(x)}{2}\,dx \le vK$.

By combining this inequality with (1), we get the result:

$$K \ge \frac{1}{v}\int_B \binom{f(x)}{2}\,dx \ge \frac{m^2v}{2\alpha} - \frac{m}{2} = \frac{1}{2}\frac{m^2}{\alpha/v} - \frac{m}{2}$$

$\square$

In our analysis of the bounding boxes, we will need an upper bound on the number of intersecting boxes, and a lower bound on the number of intersecting objects. Towards this end, we will need to group the objects depending on their positions. We will use a tiling of the plane to achieve this grouping.

Pick a volume g, and consider a tiling of the space by size g boxes. We are only interested in the finite portion of the tiling that covers the space occupied by the bounding boxes of the objects, namely, $\bigcup b(P_i)$. Let $B_1, B_2, \ldots, B_p$ denote the boxes in this finite portion of the tiling. See Fig. 2. We assume that each box is semi-open, so that the boundary shared by two boxes belongs to the one on the left, or above. Thus, each point of the plane belongs to at most one tiling box. *We assign each object $P_i \in \mathcal{S}$ to the unique tiling box that contains the center of its bounding box $b(P_i)$.*

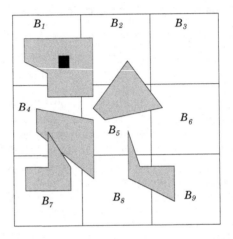

Fig. 2. A tiling. The dark square in the middle of the object in B_1 depicts its core.

Lemma 3. *Consider a tiling of the space with $g = k\alpha$ volume boxes. Suppose a box B of this tiling is assigned m convex objects, each of which has box size α and core size at least β. Then, there are at least $c\frac{m^2}{k(\alpha/\beta)^{1-1/d}} - \frac{m}{2}$ pairwise intersections among the m objects assigned to B, where c is a constant dependent only on the dimension d.*

3   Analysis for Worst-Case Aspect Ratio

We now prove our first main result: a tight bound on the ratio ρ for a set of convex objects $\mathcal{S} = \{P_1, P_2, \ldots, P_n\}$, where each object has aspect ratio at most α, and the scale factor of $\mathcal{S}$ is σ_{box}. Our analysis has three steps: we first deal with the case when all objects have roughly the same size enclosing box; next we deal with the case when objects belong to one of the two extreme classes, a bounding box of size either α or $\alpha\sigma_{\text{box}}$; finally, we handle the general case. The next two lemmas deal with the first two cases. Recall that $K_b(\mathcal{S})$ denotes the number of intersecting bounding box pairs, and $K_o(\mathcal{S})$ denotes the number of intersecting object pairs.

Lemma 4. *Let S be a set of n convex objects in R^d, where the bounding box of each object has size α and the aspect ratio of each object is at most α. Then, $\rho(S) = O(\alpha^{1-1/d})$.*

Lemma 5. *Let S be a set of n convex objects in R^d where each object has aspect ratio at most α, and the bounding box of each object has size either α (small) or $\alpha\sigma_{\mathrm{box}}$ (large). Then, $\rho(S) = O(\alpha^{1-1/d}\sigma_{\mathrm{box}}^{1/2})$.*

We are now ready to prove our first theorem.

Theorem 3. *Let S be a set of n convex objects in R^d, where d is a constant. If each object of S has aspect ratio at most α and S has scale factor σ_{box}, then $\rho(S) = \Theta(\alpha^{1-1/d}\,\sigma_{\mathrm{box}}^{1/2})$.*

Proof. Clearly we can first scale all of these objects such that the largest bounding boxes have size σ_{box} and the smallest ones have size at least 1. Then we partition the set S into $O(\log\sigma_{\mathrm{box}})$ classes such that P belongs to the ith class if $\mathrm{vol}(b(P)) \in (\alpha 2^{i-1}, \alpha 2^i]$. Let K_b^{ij}, for $0 \le i \le j \le \log\sigma_{\mathrm{box}}$, denote the number of intersecting box-pairs where one object is in class i and the other one is in class j. Then $K_b = \sum_{i \le j} K_b^{ij}$. Applying Lemma 5 to the union of classes i and j, we get

$$\frac{K_b^{ij}}{n + K_o(S)} \le c_1\alpha^{1-1/d}(2^{j-i})^{1/2} \ .$$

Thus,

$$\rho(S) = \frac{\sum_{i \le j} K_b^{ij}}{n + K_o(S)} \le \sum_{0 \le i \le j \le \log\sigma_{\mathrm{box}}} c_1\alpha^{1-1/d}\, 2^{(j-i)/2} = O\left(\alpha^{1-1/d}\sigma_{\mathrm{box}}^{1/2}\right) \ ,$$

while the last inequality uses the following fact:

$$\sum_{0 \le i \le j \le \log\sigma_{\mathrm{box}}} 2^{(j-i)/2} = \sum_{0 \le j \le \log\sigma_{\mathrm{box}}} 2^{j/2} \sum_{0 \le i \le j} 2^{-i/2} = O\left(\sum_{0 \le j \le \log\sigma_{\mathrm{box}}} 2^{j/2}\right) = O\left(\sigma_{\mathrm{box}}^{1/2}\right).$$

This completes the proof of the theorem. □

A Matching Lower Bound

We show that the bound in Theorem 1 is the best possible by giving a matching lower bound. Figure 3 shows the construction. The construction is placed inside a cubic box B of volume $\alpha\sigma_{\mathrm{box}}$. We define a *large* object to be a rectangular box, whose core size is σ_{box} and whose smallest enclosing cube has volume $\alpha\sigma_{\mathrm{box}}$. The long side of each rectangle is parallel to the dth dimension, and has length $\sigma_{\mathrm{box}}^{1/d}\alpha^{1/d}$. Each of the remaining sides of the rectangle has length $\sigma_{\mathrm{box}}^{1/d}\alpha^{1/d}$. We partition the left half of B into $\Theta\left(\alpha^{1-1/d}\right)$ cells, each of which is a homothet of our

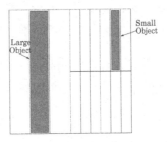

Fig. 3. The lower bound construction showing $\rho(\mathcal{S}) = \Omega\left(\alpha^{1-1/d}\,\sigma_{\text{box}}^{1/2}\right)$.

large object; observe that each large rectangle has volume $\sigma_{\text{box}}\alpha^{1/d}$. We evenly distribute $X = \alpha^{(1-1/d)/2}n^{1/2}$ large rectangles among these cells. Therefore, we have $K_o^\ell = \Theta\left(\frac{X^2}{\alpha^{1-1/d}}\right) = \Theta(n)$.

Similarly, we define a *small* object to be a rectangular box with core size 1 and smallest enclosing cube volume α. The side length of the small rectangle along the dth axis is $\alpha^{1/d}$, while the other sides have length 1. We partition the right half of B into $\Theta\left(\alpha^{1-1/d}\sigma\right)$ cells, each of which is a homothet of the small object, and has volume $\alpha^{1/d}$. We evenly distribute $Y = \alpha^{(1-1/d)/2}\sigma_{\text{box}}^{1/2}n^{1/2}$ small objects in these cells. We get $K_o^s = \Theta\left(\frac{Y^2}{\alpha^{1-1/d}\sigma_{\text{box}}}\right) = \Theta(n)$. There are no object intersections among large and small objects, and thus $K_o^{s\ell} = 0$. Finally, every small-large pair of objects contributes a bounding box intersecting pair, and so

$$K_b \geq XY = \Theta\left(\alpha^{1-1/d}\sigma_{\text{box}}^{1/2}n\right).$$

Thus,

$$\rho(\mathcal{S}) = \frac{K_b(\mathcal{S})}{n + K_o(\mathcal{S})} = \Omega(\alpha^{1-1/d}\sigma_{\text{box}}^{1/2}) \ .$$

Remark

In the preceding construction, the smallest enclosing cube of each object is significantly larger than the smallest enclosing *rectangular* bounding box. This is purely for the convenience of presentation. We can simply rotate each object so that the main diagonal of each rectangle is oriented along the vector $(1, 1, \cdots, 1)$, which ensures that the smallest rectangular bounding box is also the smallest cubic box, implying that our construction is fully general.

4   Analysis for the Average Aspect Ratio

We prove our most significant result in this section: a tight bound on the ratio ρ when only the *average aspect ratio* is bounded. In practice, it is far more reasonable to assume that the overall aspect ratio of a scene is small than to assume that *each individual* object has small aspect ratio. Thus, we believe that the result of this section has much broader applicability than Theorem 1. Our proof for the average case follows the same general outline as the one in the previous section, but the technical details are significantly more complicated and the proof is far more delicate: a bound on the average aspect ratio does not constrain individual objects. In addition to the convexity-based lemmas of Section 2, we also need the following technical lemma for the average case.

Lemma 6. *Let $\{a_{ki}\}$ be a finite sequence of nonnegative numbers, where $1 \le k \le p$, and $1 \le i \le q$, such that $\sum_{k=1}^{p} a_{ki} \le n/2^i$, for a positive integer n and all i's. Let $a_k = \sum_i a_{ki}$ for all k's. Then the following bound holds, for any t, $1 \le t \le n$:*

$$\frac{\sum_k a_k^2}{n + \sum_k \sum_i \frac{a_{ki}^2}{t \, 2^{i(1-1/d)}}} = O\left(t^{\frac{2}{3-1/d}} n^{\frac{1-1/d}{3-1/d}}\right) .$$

Lemma 7. *Let S be a set of n convex objects in R^d with average aspect ratio α_{avg}. Let $S' \subseteq S$ be a subset in which each object has the same size bounding box. Then,*

$$\frac{K_b(S')}{n + K_o(S)} = O(\alpha_{\text{avg}}^{\frac{2(1-1/d)}{3-1/d}} \, n^{\frac{1-1/d}{3-1/d}}) .$$

Lemma 8. *Let S be a set of n convex objects in R^d with average aspect ratio α_{avg}. Let $S' \subseteq S$ be a subset where each object's bounding box has size either α_{avg} or $\alpha_{\text{avg}}\sigma_{\text{box}}$. Then,*

$$\frac{K_b(S')}{n + K_o(S)} = O\left(\alpha_{\text{avg}}^{\frac{2(1-1/d)}{3-1/d}} \, \sigma_{\text{box}}^{\frac{1}{3-1/d}} \, n^{\frac{1-1/d}{3-1/d}}\right) .$$

We can now prove our second theorem.

Theorem 4. *Let S be a set of n convex objects in R^d, where d is a constant. If the objects of S have average aspect ratio α_{avg} and box scale factor σ_{box}, then*

$$\rho(S) = \Theta(\alpha_{\text{avg}}^{\frac{2(1-1/d)}{3-1/d}} \, \sigma_{\text{box}}^{\frac{1}{3-1/d}} \, n^{\frac{1-1/d}{3-1/d}}) .$$

Proof. We partition the set S into $O(\log \sigma_{\text{box}})$ classes, such that P belongs to the ith class if $vol(b(P)) \in (\alpha_{\text{avg}}2^{i-1}, \alpha_{\text{avg}}2^i]$. Let K_b^{ij}, for $0 \le i \le j \le \log \sigma_{\text{box}}$, denote the number of intersection pairs where one object is in class i and the other one is in class j. Then $K_b = \sum_{i \le j} K_b^{ij}$. If we apply Lemma 5 to the union of classes i and j, then

$$\frac{K_b^{ij}}{n + K_o(S)} \le c_1 \alpha_{\text{avg}}^{\frac{2(1-1/d)}{3-1/d}} (2^{j-i})^{\frac{1}{3-1/d}} n^{\frac{1-1/d}{3-1/d}}$$

Thus,

$$\rho(\mathcal{S}) = \frac{\sum_{i\leq j} K_b^{ij}}{n + K_o(\mathcal{S})} \leq c_1 \sum_{0\leq i\leq j\leq \log \sigma_{\mathrm{box}}} \alpha^{1-1/d}\, 2^{\frac{j-i}{3-1/d}} n^{\frac{1-1/d}{3-1/d}} = O\!\left(\alpha_{\mathrm{avg}}^{\frac{2(1-1/d)}{3-1/d}}\, \sigma_{\mathrm{box}}^{\frac{1}{3-1/d}}\, n^{\frac{1-1/d}{3-1/d}}\right),$$

while the last inequality uses the following fact:

$$\sum_{0\leq i\leq j\leq \log \sigma_{\mathrm{box}}} 2^{\frac{j-i}{3-1/d}} = \sum_{0\leq j\leq \log \sigma_{\mathrm{box}}} 2^{\frac{j}{3-1/d}} \sum_{0\leq i\leq j} 2^{\frac{-i}{3-1/d}} = O\!\left(\sum_{0\leq j\leq \log \sigma_{\mathrm{box}}} 2^{\frac{j}{3-1/d}}\right) = O\!\left(\sigma_{\mathrm{box}}^{\frac{1}{3-1/d}}\right).$$

This completes the proof.

A Matching Lower Bound

We show that the bound in Theorem 2 is the best possible by giving a matching lower bound. The construction follows the general outline of our earlier construction, but the details are more complicated. We omit the construction from this abstract.

5   Extensions and Concluding Remarks

We studied the effectiveness of bounding boxes as a filtering mechanism in collision detection. Many other applications also use bounding boxes in much the same way as collision detection, and our results should apply to them as well. The main focus of our paper was to study collision detection among convex objects. We gave tight bounds on the ratio $\rho(\mathcal{S}) = \frac{K_b(\mathcal{S})}{n+K_o(\mathcal{S})}$ both when the aspect ratio of each object is bounded as well as when only the average aspect ratio of the whole set is bounded. In fact, our results precisely calibrate the worst-case effectiveness of bounding boxes in terms of the shape parameters α and σ_{box}.

An important feature of our shape parameters is that they constrain only the individual objects and their relative sizes in the set, but *put no constraints on the distribution or placement* of objects. In other words, we make no assumptions about the motion, trajectories, or relative spacing among the objects. The advantage of our minimalist assumption is that our results should apply to a broad spectrum of situations. On the negative side, it is possible that in many situations, the additional constraints may lead to a better performance than our results would indicate. For instance, it may be possible that objects tend to be sparsely distributed in space, and have very few intersections. On the other hand, this can be a highly artificial and unnatural assumption in other situations. For instance, many group dynamics exhibit a *flocking tendency*, such as the motion of a school of fish, a flock of birds, visitors in a museum, or commuters in a subway station.

In defining the scale factor, we used the ratio between the largest and the smallest bounding boxes. An alternative definition might be the ratio between the largest and the smallest cores. Somewhat surprisingly, the choice does matter, and the precise bounds on the ratio ρ turn out to be different. While the proofs are similar in format, they are different in many details. We just state them here for the sake of completeness. Define the *core scale factor* of $\mathcal{S}$ as

$$\sigma_{\text{cor}}(\mathcal{S}) = \max_{i,j} \frac{\text{vol}(c(P_i))}{\text{vol}(c(P_j))} \ .$$

Then, we prove the following tight bounds.

Theorem 5. *Let $\mathcal{S}$ be a set of n convex objects in R^d, where d is a constant. If each object of $\mathcal{S}$ has aspect ratio at most α and $\mathcal{S}$ has core scale factor σ_{cor}, then $\rho(\mathcal{S}) = \Theta(\alpha^{1-1/2d}\sigma_{\text{cor}}^{1/2})$.*

Theorem 6. *Let $\mathcal{S}$ be a set of n convex objects in R^d, where d is a constant. If the objects of $\mathcal{S}$ have average aspect ratio α_{avg} and core scale factor σ_{box}, then*
$$\rho(\mathcal{S}) = \Theta(\alpha_{\text{avg}}^{\frac{2(1-1/2d)}{3-1/d}} \ \sigma_{\text{cor}}^{1/2} \ n^{\frac{1-1/2d}{3-1/d}}).$$

Finally, we also have some minor results concerning the narrow phase computation. The narrow phase of collision detection deals with determining the detailed intersection between pairs of objects whose bounding boxes are found to intersect. In some applications, it may be sufficient to simply *detect* whether the intersection is non-empty, while in others a complete feature-by-feature intersection may be needed. In particular, the complete intersection is required for computing the collision *response*. In this paper, by the narrow phase computation, we mean the problem of determining the complete intersection.

Consider two (possibly non-convex) polyhedral objects P and Q in three dimensions. We put a bounding box around each face of P and Q; determine the intersecting box pairs; and then perform the detailed intersection for those face-pairs whose boxes intersect. Suppose there are a total of m faces in P and Q, the number of bounding box intersections is K_b, and the number of face-pair intersections between the faces of P and Q is K_o. Then, what is the worst-case bound on the ratio $\rho = \frac{K_b}{m+K_o}$? Clearly, without any assumptions on P and Q, the ratio can be $\Omega(m)$. We consider below a natural shape parameter, which helps capture the pathological cases.

Given a face f of a polyhedron P, let $b(f)$ denote the smallest cubic box containing f. Let size(f) denote the side length of the box $b(f)$. Given two faces f and g, let dist(f, g) denote the smallest distance between a point of f and a point of g. We define the *face packing density* of f, denoted $\mu(f)$, as follows:

$$\mu(f) = |\{g \in F(P) \ : \ \text{size}(g) \geq \text{size}(f) \text{ and } \text{dist}(f, g) \leq \text{size}(f)\}| \ , \qquad (\dagger)$$

where $F(P)$ is the set of faces of P. Here we are using $|A|$ to denote the cardinality for a set A. We say that a *face packing density* of P, denoted $\mu(P)$, is the smallest μ such that $\mu(f) \leq \mu$, for all $f \in F(P)$. Formally, $\mu(P) = \max_{f\in F(P)} \mu(f)$. A large face packing density implies that faces are packed relatively tightly (like a closed accordion), a situation not likely to arise in practice.

A key observation is that face packing density is a property of the *individual* polyhedron, and it does not depend on the other polyhedra with which it may intersect. Our main result is the following theorem, which implies that the performance ratio ρ is bounded by face packing density μ.

Theorem 7. *Let P and Q be two 3-dimensional polyhedra, with a total of m faces. If P and Q both have face packing density at most μ, then $\rho = O(\mu)$. In fact, if K_b is the total number of intersections between the bounding boxes of the faces of P and Q and K_o is the total number of face intersections between P and Q, then $K_o \leq K_b = O(\mu\, m)$.*

References

1. J. D. Cohen, M. C. Lin, D. Manocha and M. K. Ponamgi. I-COLLIDE: An interactive and exact collision detection system for large-scale environments. *Proc. of ACM Interactive 3D Graphics Conference*, pp. 189–196, 1995. 437

2. H. Edelsbrunner. Reporting intersections of geometric objects by means of covering rectangles. *Bulletin of the EATCS*, 13, pp. 7–11, 1981. 438

3. J. D. Foley, A. van Dam, S. K. Feiner, J. F. Hughes *Computer Graphics: Principles and Practice (2nd Edition in C)*. Addison Wesley, 1996. 437

4. A. Garcia-Alonso, N. Serrano and J. Flaquer. Solving the Collision Detection Problem. *IEEE Computer Graphics and Applications*, 14, pp. 36–43, 1995. 437

5. N. Greene. Detecting Intersection of a Rectangular Solid and a Convex Polyhedron. *Graphics Gems IV*, pp. 83–110, 1994. 437

6. P. Gupta, R. Janardan and M. Smid. Efficient algorithms for counting and reporting pairwise intersection between convex polygons. Technical report, Computer Science, King's College, UK, 1996. 439

7. C. Hoffmann. *Geometric and Solid Modeling*. Morgan Kaufmann, 1989. 437

8. J. T. Klosowski, M. Held, J. S. B. Mitchell, H. Sowizral, and K. Zikan. Real-time collision detection for motion simulation within complex environments. *ACM SIGGRAPH'96 Visual Proceedings*, pp. 151, 1996. 437

9. J.-C. Latombe. *Robot Motion Planning*. Kluwer Academic Publishers, 1991. 437

10. L. McMillan. *An Image-Based Approach to Three-Dimensional Computer Graphics*. Ph.D. Thesis, University of North Carolina at Chapel Hill, 1997. 437

11. M. P. Moore and J. Wilhelms. Collision Detection and Response for Computer Animation. *Computer Graphics*, 22, pp. 289–298, 1988. 437

12. S. Suri, P. M. Hubbard and J. F. Hughes. Collision Detection in Aspect and Scale Bounded Polyhedra. *Proc. of 9th Annual Symposium on Discrete Algorithms*, 1998. *ACM Transactions on Graphics*. In Press. 438

13. Y. Zhou and S. Suri. Analysis of a Bounding Box Heuristic for Object Intersection. *Journal of the ACM*, 46(6): 833-857, 1999. 438, 440

Author Index

Agarwal, Pankaj K., 20
Ageev, Alexander A., 32
van den Akker, Marjan, 427
Ambühl, Christoph, 42, 52
Andrews, Matthew, 64

Barrière, Lali, 76
Benczúr, András A., 88
de Berg, Mark, 167
Bespamyatnikh, S., 100
Bhattacharya, B., 100
Brass, Peter, 112
Buchsbaum, Adam L., 120

Carr, Robert, 132
Chakraborty, Samarjit, 52
Cheong, Otfried, 314
Chwa, Kyung-Yong, 314
Clementi, A.E.F., 143
Czumaj, Artur, 155

Dickerson, Matthew, 179
Duncan, Christian A., 179

Ferreira, A., 143
Flammini, Michele, 191
Flato, Eyal, 20
Fleischer, Rudolf, 202
Fraigniaud, Pierre, 76
Fujito, Toshihiro, 132
Fülöp, Ottilia, 88

Gärtner, Bernd, 52
Gaur, Daya Ram, 211
Gavoille, Cyril, 76
Goodrich, Michael T., 120, 179
Govindarajan, Sathish, 220
Gudmundsson, Joachim, 167, 232
Guruswami, Venkatesan, 244

Haas, Walter A., 256
Halperin, Dan, 20
Hammar, Mikael, 167, 232
Henzinger, Monika, 1
Hochbaum, Dorit S., 256
Hoogeveen, Han, 268, 427

Hsu, Tsan-sheng, 278

Ibaraki, Toshihide, 211

Kalyanasundaram, Bala, 290
Keil, J. Mark, 100
Kirkpatrick, D., 100
Koga, Hisashi, 302
Konjevod, Goran, 132
van Kreveld, Marc, 232
Krishnamurti, Ramesh, 211
Kwon, Woo-Cheol, 314

Lee, Jae-Ha, 314
Lengauer, Thomas, 9
Lukovszki, Tamás, 220

Maheshwari, Anil, 220
Mans, Bernard, 76
Mehlhorn, Kurt, 326
Munagala, Kamesh, 64
Munro, J. Ian, 338

Nardelli, Enrico, 346
Nicosia, Gaia, 191

Overmars, Mark, 167

Parekh, Ojas, 132
Penna, P., 143
Perennes, S., 143
Pizzonia, Maurizio, 356
Proietti, Guido, 346
Pruhs, Kirk, 290

Qin, Zhongping, 368
Queyranne, Maurice, 256

Rahman, Naila, 380
Raman, Rajeev, 380
Robson, John M., 76

Sanders, Peter, 392
Segal, M., 100
Shin, Sung Yong, 314
Silvestri, R., 143
Skodinis, Konstantin, 403

Skutella, Martin, 268
Sohler, Christian, 155
Solis-Oba, Roberto, 392
Subramanian, C.R., 415
Sudan, Madhu, 244
Suri, Subhash, 437
Sviridenko, Maxim I., 32

Tamassia, Roberto, 356

Vakhania, Nodari, 427
Velauthapillai, Mahe, 290

Wahl, Michaela, 202
Westbrook, Jeffery R., 120
Widmayer, Peter, 346
Woeginger, Gerhard J., 268
Wolff, Alexander, 368

Xu, Yinfeng, 368

Zeh, Norbert, 220
Zhou, Yunhong, 437
Zhu, Binhai, 368
Ziegelmann, Mark, 326
Ziegler, Martin, 155

Lecture Notes in Computer Science

For information about Vols. 1–1804
please contact your bookseller or Springer-Verlag

Vol. 1805: T. Terano, H. Liu, A.L.P. Chen (Eds.), Knowledge Discovery and Data Mining. Proceedings, 2000. XIV, 460 pages. 2000. (Subseries LNAI).

Vol. 1806: W. van der Aalst, J. Desel, A. Oberweis (Eds.), Business Process Management. VIII, 391 pages. 2000.

Vol. 1807: B. Preneel (Ed.), Advances in Cryptology – EUROCRYPT 2000. Proceedings, 2000. XVIII, 608 pages. 2000.

Vol. 1809: S. Biundo, M. Fox (Eds.), Recent Advances in AI Planning. Proceedings, 1999. VIII, 373 pages. 2000. (Subseries LNAI).

Vol. 1810: R.López de Mántaras, E. Plaza (Eds.), Machine Learning: ECML 2000. Proceedings, 2000. XII, 460 pages. 2000. (Subseries LNAI).

Vol. 1811: S.W. Lee, H.. Bülthoff, T. Poggio (Eds.), Biologically Motivated Computer Vision. Proceedings, 2000. XIV, 656 pages. 2000.

Vol. 1813: P.L. Lanzi, W. Stolzmann, S.W. Wilson (Eds.), Learning Classifier Systems. X, 349 pages. 2000. (Subseries LNAI).

Vol. 1815: G. Pujolle, H. Perros, S. Fdida, U. Körner, I. Stavrakakis (Eds.), Networking 2000 – Broadband Communications, High Performance Networking, and Performance of Communication Networks. Proceedings, 2000. XX, 981 pages. 2000.

Vol. 1816: T. Rus (Ed.), Algebraic Methodology and Software Technology. Proceedings, 2000. XI, 545 pages. 2000.

Vol. 1817: A. Bossi (Ed.), Logic-Based Program Synthesis and Transformation. Proceedings, 1999. VIII, 313 pages. 2000.

Vol. 1818: C.G. Omidyar (Ed.), Mobile and Wireless Communications Networks. Proceedings, 2000. VIII, 187 pages. 2000.

Vol. 1819: W. Jonker (Ed.), Databases in Telecommunications. Proceedings, 1999. X, 208 pages. 2000.

Vol. 1820: J.-J. Quisquater, B. Schneier (Eds.), Smart Card Research and Applications. Proceedings, 1998. XI, 381 pages. 2000.

Vol. 1821: R. Loganantharaj, G. Palm, M. Ali (Eds.), Intelligent Problem Solving. Proceedings, 2000. XVII, 751 pages. 2000. (Subseries LNAI).

Vol. 1822: H.H. Hamilton, Advances in Artificial Intelligence. Proceedings, 2000. XII, 450 pages. 2000. (Subseries LNAI).

Vol. 1823: M. Bubak, H. Afsarmanesh, R. Williams, B. Hertzberger (Eds.), High Performance Computing and Networking. Proceedings, 2000. XVIII, 719 pages. 2000.

Vol. 1824: J. Palsberg (Ed.), Static Analysis. Proceedings, 2000. VIII, 433 pages. 2000.

Vol. 1825: M. Nielsen, D. Simpson (Eds.), Application and Theory of Petri Nets 2000. Proceedings, 2000. XI, 485 pages. 2000.

Vol. 1826: W. Cazzola, R.J. Stroud, F. Tisato (Eds.), Reflection and Software Engineering. X, 229 pages. 2000.

Vol. 1827: D. Bert, C. Choppy, P. Mosses (Eds.), Recent Trends in Algebraic Development Techniques. Proceedings, 1999. X, 477 pages. 2000.

Vol. 1829: C. Fonlupt, J.-K. Hao, E. Lutton, E. Ronald, M. Schoenauer (Eds.), Artificial Evolution. Proceedings, 1999. X, 293 pages. 2000.

Vol. 1830: P. Kropf, G. Babin, J. Plaice, H. Unger (Eds.), Distributed Communities on the Web. Proceedings, 2000. X, 203 pages. 2000.

Vol. 1831: D. McAllester (Ed.), Automated Deduction – CADE-17. Proceedings, 2000. XIII, 519 pages. 2000. (Subseries LNAI).

Vol. 1832: B. Lings, K. Jeffery (Eds.), Advances in Databases. Proceedings, 2000. X, 227 pages. 2000.

Vol. 1833: L. Bachmair (Ed.), Rewriting Techniques and Applications. Proceedings, 2000. X, 275 pages. 2000.

Vol. 1834: J.-C. Heudin (Ed.), Virtual Worlds. Proceedings, 2000. XI, 314 pages. 2000. (Subseries LNAI).

Vol. 1835: D. N. Christodoulakis (Ed.), Natural Language Processing – NLP 2000. Proceedings, 2000. XII, 438 pages. 2000. (Subseries LNAI).

Vol. 1836: B. Masand, M. Spiliopoulou (Eds.), Web Usage Analysis and User Profiling. Proceedings, 2000, V, 183 pages. 2000. (Subseries LNAI).

Vol. 1837: R. Backhouse, J. Nuno Oliveira (Eds.), Mathematics of Program Construction. Proceedings, 2000. IX, 257 pages. 2000.

Vol. 1838: W. Bosma (Ed.), Algorithmic Number Theory. Proceedings, 2000. IX, 615 pages. 2000.

Vol. 1839: G. Gauthier, C. Frasson, K. VanLehn (Eds.), Intelligent Tutoring Systems. Proceedings, 2000. XIX, 675 pages. 2000.

Vol. 1840: F. Bomarius, M. Oivo (Eds.), Product Focused Software Process Improvement. Proceedings, 2000. XI, 426 pages. 2000.

Vol. 1841: E. Dawson, A. Clark, C. Boyd (Eds.), Information Security and Privacy. Proceedings, 2000. XII, 488 pages. 2000.

Vol. 1842: D. Vernon (Ed.), Computer Vision – ECCV 2000. Part I. Proceedings, 2000. XVIII, 953 pages. 2000.

Vol. 1843: D. Vernon (Ed.), Computer Vision – ECCV 2000. Part II. Proceedings, 2000. XVIII, 881 pages. 2000.

Vol. 1844: W.B. Frakes (Ed.), Software Reuse: Advances in Software Reusability. Proceedings, 2000. XI, 450 pages. 2000.

Vol. 1845: H.B. Keller, E. Plöderer (Eds.), Reliable Software Technologies Ada-Europe 2000. Proceedings, 2000. XIII, 304 pages. 2000.

Vol. 1846: H. Lu, A. Zhou (Eds.), Web-Age Information Management. Proceedings, 2000. XIII, 462 pages. 2000.

Vol. 1847: R. Dyckhoff (Ed.), Automated Reasoning with Analytic Tableaux and Related Methods. Proceedings, 2000. X, 441 pages. 2000. (Subseries LNAI).

Vol. 1848: R. Giancarlo, D. Sankoff (Eds.), Combinatorial Pattern Matching. Proceedings, 2000. XI, 423 pages. 2000.

Vol. 1849: C. Freksa, W. Brauer, C. Habel, K.F. Wender (Eds.), Spatial Cognition II. XI, 420 pages. 2000. (Subseries LNAI).

Vol. 1850: E. Bertino (Ed.), ECOOP 2000 – Object-Oriented Programming. Proceedings, 2000. XIII, 493 pages. 2000.

Vol. 1851: M.M. Halldórsson (Ed.), Algorithm Theory – SWAT 2000. Proceedings, 2000. XI, 564 pages. 2000.

Vol. 1852: T. Thierauf, The Computational Complexity of Equivalence and Isomorphism Problems. VIII, 135 pages. 2000.

Vol. 1853: U. Montanari, J.D.P. Rolim, E. Welzl (Eds.), Automata, Languages and Programming. Proceedings, 2000. XVI, 941 pages. 2000.

Vol. 1854: G. Lacoste, B. Pfitzmann, M. Steiner, M. Waidner (Eds.), SEMPER — Secure Electronic Marketplace for Europe. XVIII, 350 pages. 2000.

Vol. 1855: E.A. Emerson, A.P. Sistla (Eds.), Computer Aided Verification. Proceedings, 2000. X, 582 pages. 2000.

Vol. 1857: J. Kittler, F. Roli (Eds.), Multiple Classifier Systems. Proceedings, 2000. XII, 404 pages. 2000.

Vol. 1858: D.-Z. Du, P. Eades, V. Estivill-Castro, X. Lin, A. Sharma (Eds.), Computing and Combinatorics. Proceedings, 2000. XII, 478 pages. 2000.

Vol. 1860: M. Klusch, L. Kerschberg (Eds.), Cooperative Information Agents IV. Proceedings, 2000. XI, 285 pages. 2000. (Subseries LNAI).

Vol. 1861: J. Lloyd, V. Dahl, U. Furbach, M. Kerber, K.-K. Lau, C. Palamidessi, L. Moniz Pereira, Y. Sagiv, P.J. Stuckey (Eds.), Computational Logic – CL 2000. Proceedings, 2000. XIX, 1379 pages. (Subseries LNAI).

Vol. 1862: P.G. Clote, H. Schwichtenberg (Eds.), Computer Science Logic. Proceedings, 2000. XIII, 543 pages. 2000.

Vol. 1863: L. Carter, J. Ferrante (Eds.), Languages and Compilers for Parallel Computing. Proceedings, 1999. XII, 500 pages. 2000.

Vol. 1864: B. Y. Choueiry, T. Walsh (Eds.), Abstraction, Reformulation, and Approximation. Proceedings, 2000. XI, 333 pages. 2000. (Subseries LNAI).

Vol. 1865: K.R. Apt, A.C. Kakas, E. Monfroy, F. Rossi (Eds.), New Trends Constraints. Proceedings, 1999. X, 339 pages. 2000. (Subseries LNAI).

Vol. 1866: J. Cussens, A. Frisch (Eds.), Inductive Logic Programming. Proceedings, 2000. X, 265 pages. 2000. (Subseries LNAI).

Vol. 1867: B. Ganter, G.W. Mineau (Eds.), Conceptual Structures: Logical, Linguistic, and Computational Issues. Proceedings, 2000. XI, 569 pages. 2000. (Subseries LNAI).

Vol. 1868: P. Koopman, C. Clack (Eds.), Implementation of Functional Languages. Proceedings, 1999. IX, 199 pages. 2000.

Vol. 1869: M. Aagaard, J. Harrison (Eds.), Theorem Proving in Higher Order Logics. Proceedings, 2000. IX, 535 pages. 2000.

Vol. 1872: J. van Leeuwen, O. Watanabe, M. Hagiya, P.D. Mosses, T. Ito (Eds.), Theoretical Computer Science. Proceedings, 2000. XV, 630 pages. 2000.

Vol. 1876: F. J. Ferri, J.M. Iñesta, A. Amin, P. Pudil (Eds.), Advances in Pattern Recognition. Proceedings, 2000. XVIII, 901 pages. 2000.

Vol. 1877: C. Palamidessi (Ed.), CONCUR 2000 – Concurrency Theory. Proceedings, 2000. XI, 612 pages. 2000.

Vol. 1878: J.P. Bowen, S. Dunne, A. Galloway, S. King (Eds.), ZB 2000: Formal Specification and Development in Z and B. Proceedings, 2000. XIV, 511 pages. 2000.

Vol. 1879: M. Paterson (Ed.), Algorithms – ESA 2000. Proceedings, 2000. IX, 450 pages. 2000.

Vol. 1880: M. Bellare (Ed.), Advances in Cryptology – CRYPTO 2000. Proceedings, 2000. XI, 545 pages. 2000.

Vol. 1881: C. Zhang, V.-W. Soo (Eds.), Design and Applications of Intelligent Agents. Proceedings, 2000. X, 183 pages. 2000. (Subseries LNAI).

Vol. 1883: B. Triggs, A. Zisserman, R. Szeliski (Eds.), Vision Algorithms: Theory and Practice. Proceedings, 1999. X, 383 pages. 2000.

Vol. 1886: R. Mizoguchi, J. Slaney /Eds.), PRICAI 2000: Topics in Artificial Intelligence. Proceedings, 2000. XX, 835 pages. 2000. (Subseries LNAI).

Vol. 1889: M. Anderson, P. Cheng, V. Haarslev (Eds.), Theory and Application of Diagrams. Proceedings, 2000. XII, 504 pages. 2000. (Subseries LNAI).

Vol. 1892: P. Brusilovsky, O. Stock, C. Strapparava (Eds.), Adaptive Hypermedia and Adaptive Web-Based Systems. Proceedings, 2000. XIII, 422 pages. 2000.

Vol. 1893: M. Nielsen, B. Rovan (Eds.), Mathematical Foundations of Computer Science 2000. Proceedings, 2000. XIII, 710 pages. 2000.

Vol. 1896: R. W. Hartenstein, H. Grünbacher (Eds.), Field-Programmable Logic and Applications. Proceedings, 2000. XVII, 856 pages. 2000.

Vol. 1897: J. Gutknecht, W. Weck (Eds.), Modular Programming Languages. Proceedings, 2000. XII, 299 pages. 2000.

Vol. 1899: H.-H. Nagel, F.J. Perales López (Eds.), Articulated Motion and Deformable Objects. Proceedings, 2000. X, 183 pages. 2000.

Vol. 1900: A. Bode, T. Ludwig, W. Karl, R. Wismüller (Eds.), Euro-Par 2000 Parallel Processing. Proceedings, 2000. XXXV, 1368 pages. 2000.

Vol. 1912: Y. Gurevich, P.W. Kutter, M. Odersky, L. Thiele (Eds.), Abstract State Machines. Proceedings, 2000. X, 381 pages. 2000.

Vol. 1913: K. Jansen, S. Khuller (Eds.), Approximation Algorithms for Combinatorial Optimization. Proceedings, 2000. IX, 275 pages. 2000.